S0-AHA-585

ADP-Ribose Transfer Reactions

Myron K. Jacobson
Elaine L. Jacobson
Editors

ADP-Ribose
Transfer Reactions

Mechanisms and
Biological Significance

With 283 Illustrations

Springer-Verlag
New York Berlin Heidelberg
London Paris Tokyo Hong Kong

Myron K. Jacobson
Department of Biochemistry
Texas College of Osteopathic Medicine
Fort Worth, Texas 76107-2690
U.S.A.

Elaine L. Jacobson
Department of Medicine
Texas College of Osteopathic Medicine
Fort Worth, Texas 76107-2690
U.S.A.

QP
625
,A29
A34
1989

Library of Congress Cataloging-in-Publication Data
ADP-ribose transfer reactions : mechanisms and biological significance
/ Myron K. Jacobson, Elaine L. Jacobson, editors.
    p.   cm.
    ISBN 0–387–97087-8
    1. Adenosine diphosphate ribose—Metabolism.   2. NAD (Coenzyme)—
Metabolism.   I. Jacobson, Myron K.   II. Jacobson, Elaine L.
QP625.A29A34 1989
574.19′25—dc20

© 1989 Springer-Verlag New York Inc.
Copyright not claimed for employees of the United States Government.

All rights reserved. This work may not be translated or copied in whole or in part without the written permission of the publisher (Springer-Verlag New York, Inc., 175 Fifth Avenue, New York, NY 10010, U.S.A.), except for brief excerpts in connection with reviews or scholarly analysis. Use in connection with any form of information storage and retrieval, electronic adaptation, computer software, or by similar or dissimilar methodology now known or hereafter developed is forbidden.
The use of general descriptive names, trade names, trademarks, etc., in this publication, even if the former are not especially identified, is not to be taken as a sign that such names, as understood by the Trade Marks and Merchandise Act, may accordingly be used freely by anyone.

Text prepared by the editors in camera-ready form.
Printed and bound by Edwards Brothers, Ann Arbor, Michigan.
Printed in the U.S.A.

9 8 7 6 5 4 3 2 1

ISBN 0-387-97087-8 Springer-Verlag New York Berlin Heidelberg
ISBN 3-540-97087-8 Springer-Verlag Berlin Heidelberg New York

This monograph is dedicated
to Professor Helmuth Hilz
of the University of Hamburg.

We would like to take this opportunity to acknowledge
his many important contributions to our understanding of
ADP-ribose transfer reactions and to express our appreciation
for his advice, counsel and friendship.

Elaine L. Jacobson
Myron K. Jacobson

Preface

The history of our understanding of the function of niacin and niacin-derived molecules in biological processes has evolved in three distinct phases. The first phase occurred between 1900 and 1940 and led to the discovery of nicotinic acid and nicotinamide as anti-pellagra factors. The second phase followed during the 1940's and 1950's and led to our understanding of the fundamental role of nicotinamide adenine dinucleotide, NAD, in hydride transfer reactions central to the energy metabolism of all living organisms. The topic of this monograph is the third and current phase: the role of NAD in ADP-ribose transfer reactions. While these reactions are still poorly understood, our current knowledge indicates that ADP-ribose transfer reactions are fundamentally involved in the regulation of many physiological processes. The contributions in this monograph are representative of current research in this rapidly expanding phase of our understanding of niacin metabolism.

The editors of this monograph are deeply indebted to the able and tireless work of Kay Hartman and Erica Castle, without whom this effort would not have been possible. We also thank Robin Chambers, Patrick Rankin and Ronaye Hubbard for their generous assistance.

Contents

Part One: Enzymology of ADP-Ribosylation

Part Two: ADP-Ribosylation and Chromatin Function

Part Three: Carcinogenesis and Differentiation

Part Four: NAD Metabolism and Chemotherapy

Part Five: ADP-Ribosylation and Signal Transduction

xiv

Part Six: Molecular Genetic Approaches to ADP-Ribosylation

Amino Acid-Specific ADP-Ribosylation: Purification and Properties of an Erythrocyte ADP-Ribosylarginine Hydrolase

Joel Moss, Su-Chen Tsai, Ronald Adamik, Hao-Chia Chen, and Sally J. Stanley

Laboratory of Cellular Metabolism, NHLBI and Endocrinology Reproduction Research Branch, NICHD, National Institutes of Health, Bethesda, Maryland 20892 USA

Introduction

There is a class of mono-ADP-ribosyltransferases that are distinguished by their ability to utilize as ADP-ribose acceptors the free amino acid arginine, other simple guanidino compounds, and proteins. Several transferases of this type were identified in and purified from turkey erythrocytes (1-3). The enzymes displayed different physical, kinetic and regulatory properties and were localized to the soluble, membrane and nuclear compartments (1-3). Similar NAD:arginine ADP-ribosyltransferases have been observed in other tissues and organ systems from a variety of species (1-5). The enzymes are found in viruses, bacteria and animal cells (1-7).

It appears that the ADP-ribosylation of arginine residues is a reversible modification of protein (8-10). Enzymatic activities have been found that catalyze the degradation of ADP-ribosylarginine or ADP-ribosyl (arginine) protein with release of the free amino acid or protein (8-10). Such activities have been found in avian, bovine, porcine and human cells or tissues (8-11, data not shown). To understand the mechanism of action of these enzymes, currently termed ADP-ribosylarginine hydrolases, they have been extensively purified from turkey erythrocytes and characterized with regard to their kinetic and physical properties.

Results and Discussion

Purification and characterization of ADP-ribosylarginine hydrolases. ADP-ribosylarginine hydrolases were extensively purified from turkey erythrocytes by chromatography on DE-52, phenyl-Sepharose, hydroxylapatite, Ultrogel AcA 54 and mono Q (Purification Procedure A). An alternative purification procedure (B) employed an organomercurial-agarose chromatography step with thiol elution of hydrolase immediately prior to Ultrogel AcA 54 chromatography. The overall purification was approximately 50,000-fold, and yielded an enzyme with a specific activity of approximately 2 μmol ADP-ribosylarginine hydrolyzed per min per mg at 30°C. In the mono Q eluate, hydrolase activity comigrated with a protein peak. This purified hydrolase exhibited one major band on sodium dodecyl

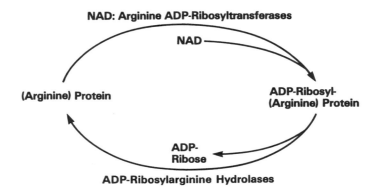

NAD: Arginine ADP-Ribosyltransferases

NAD

(Arginine) Protein

ADP-Ribosyl-
(Arginine) Protein

ADP-
Ribose

ADP-Ribosylarginine Hydrolases

Fig. 1. ADP-ribosylation, a reversible modification of (arginine) proteins. NAD:arginine ADP-ribosyltransferases catalyze the stereospecific ADP-ribosylation and phospho-ADP-ribosylation of arginine residues in proteins. ADP-ribosylarginine hydrolases cleave the ADP-ribosylarginine linkage, releasing ADP-ribose and regenerating arginine.

sulfate-polyacrylamide gels which had a mobility compatible with a molecular weight of 39,000. The k_{av} for the hydrolase by gel permeation chromatography corresponded to a native protein of similar size. Thus, it appears that the hydrolase exists as a monomeric species.

The purified hydrolases were stimulated greater than 10-fold by Mg^{2+} and dithiothreitol (9). Optimal Mg^{2+} concentration was approximately 5 to 10 mM (9). The dithiothreitol sensitivity of the hydrolase was dependent on the purification procedure. Following hydroxylapatite chromatography, the impure enzyme was relatively insensitive to dithiothreitol; activation under the standard assay procedure required approximately 10 mM thiol (9). The enzyme isolated by procedure A exhibited a similar sensitivity (9). In contrast, the hydrolase purified by procedure B was dramatically more sensitive to dithiothreitol; approximately 50 μM produced considerable activation. Both hydrolases were partially resistant to sulfhydryl reagents such as N-ethylmaleimide (NEM) unless subjected to prior incubation with thiol; as expected, higher concentrations of thiol were required to convert hydrolase A to an NEM-sensitive state than was the case with hydrolase B.

It appears that the organomercurial-agarose chromatography step may be responsible for the conversion of the thiol-resistant hydrolase to a thiol-sensitive species. A partially purified, thiol-resistant hydrolase was incubated in the presence of $HgCl_2$. The reaction was terminated by addition of thiol; the hydrolase was then purified by gel permeation chromatography on Ultrogel AcA 54. The $HgCl_2$-treated enzyme exhibited a thiol-sensitive state similar to the hydrolase purified by the procedure employing organomercurial agarose.

2

Substrate specificity of ADP-ribosylarginine hydrolase. The hydrolase was identified based on its cleavage of the (ADP-ribose)-arginine linkage, releasing ADP-ribose and arginine. The enzyme exhibited a relatively broad specificity, degrading ADP-ribosylguanidine and a number of ADP-ribosylproteins (10). At the ADP-ribose acceptor site, therefore, the hydrolase did not appear to have a structural requirement, beyond the presence of a guanidino group (10). In agreement with this finding was the observation that guanidino analogues, such as arginine and agmatine were relatively poor inhibitors of catalytic activity. The hydrolytic reaction proceeded with generation of an intact ADP-ribose acceptor site; the product of hydrolase cleavage was ADP-ribosylated in the presence of NAD and either the turkey erythrocyte NAD:arginine ADP-ribosyltransferase or the bacterial toxin, choleragen (9, data not shown).

Although the ADP-ribose acceptor moiety of the substrate did not appear to be critical for recognition by the hydrolase, the ADP-ribose group was a major determinant of enzymatic activity (10). Degradation of the ADP-ribose moiety of the substrate by phosphodiesterase action at the pyrophosphate linkage to yield phosphoribosylarginine or by combined action of phosphodiesterase and phosphatase to yield ribosylarginine yielded products that were poor substrates for the hydrolase. Similarly, ADP-ribose was a potent competitive inhibitor of hydrolysis; AMP, ADP and ATP were far less effective (10). It would thus appear that an intact ADP-ribose group is necessary for substrate recognition.

In contrast, the guanidino portion of the substrate was not as critical for hydrolytic activity. Both ADP-ribosylarginine and ADP-ribosylguanidine served as substrates (10); these two compounds did, however, exhibit different kinetic constants and pH optima (10). These results are in agreement with the prior finding that primary recognition by the hydrolase for these model substrates is of the ADP-ribose moiety. It is not clear when ADP-ribosylated proteins serve as substrates whether the protein, as well as the ADP-ribose, is a determinant of reactivity.

ADP-ribosylation: a reversible modification of proteins. ADP-ribosylation of arginine residues appears to be a reversible modification of proteins. NAD:arginine ADP-ribosyltransferases catalyze the formation of the ADP-ribosyl (arginine) protein, while ADP-ribosylarginine hydrolases cleave the ADP-ribosylarginine linkage, leading to the formation of ADP-ribose and the regenerated (arginine) protein. If the NAD:arginine ADP-ribosyltransferases and the ADP-ribosylarginine hydrolases are linked in an ADP-ribosylation cycle, these enzymes should exhibit a compatible product-substrate relationship (Fig. 1).

In fact, the family of ADP-ribosyltransferases in turkey erythrocytes utilized both NAD and NADP as donors of ADP-ribose and phospho-ADP-ribose respectively, leading to the formation of ADP-ribosylarginine and phospho-ADP-ribosylarginine (12). As noted, some of the transferases

3

exhibited a more stringent substrate specificity, clearly preferring NAD over NADP, whereas another transferase readily used NADP as well as NAD (12, data not shown). The transferase-catalyzed reaction is stereospecific; in the presence of ß-NAD and arginine, the α-anomeric ADP-ribosylarginine is synthesized (12). In solution, this product readily anomerized to yield an α, ß mixture. α- and ß-ADP-ribosylarginine were distinguished and quantified by monitoring the resonance of the ribosyl anomeric proton. In a racemic α, ß mixture the turkey erythrocyte ADP-ribosylarginine hydrolase preferentially degraded the α–anomer with appearance of new resonances corresponding to the α- and ß-anomeric protons of ADP-ribose (10). Consistent with this stereospecificity was the observation that hydrolysis of α-ADP-ribosylarginine was inhibited by α-NAD>>ß-NAD (data not shown). The avian erythrocyte hydrolase also catalyzed the degradation of phospho-ADP-ribosylarginine but with a significantly lower V_{max} (10). The degradation of phospho-ADP-ribosylarginine appeared not to reflect prior de-phosphorylation of the substrate by contaminating phosphatases; purification of the products of the hydrolase-catalyzed reaction by HPLC did not reveal the presence of ADP-ribosylarginine.

These studies are consistent with the hypothesis that hydrolase activity in erythrocytes possesses stereospecificity and substrate specificity compatible with those of the NAD:arginine ADP-ribosyltransferases and thus these enzymatic activities could function in opposing arms of an ADP-ribosylation cycle (7).

Effect of NAD:arginine ADP-ribosyltransferases on ADP-ribosylarginine hydrolase activity. NAD:arginine ADP-ribosyltransferases catalyze the modification of a number of purified proteins; in some cases ADP-ribosylation results in a loss of enzymatic activity. Incubation of either the thiol-resistant or thiol-sensitive hydrolase with NAD and erythrocyte NAD:arginine ADP-ribosyltransferase A decreased its catalytic activity when assayed in the presence of Mg^{2+} and DTT. When the incubation with NAD and transferase was supplemented with Mg^{2+} and DTT, no loss of hydrolase activity was detected in the subsequent assay. When the reaction mix contained [^{32}P]NAD, and the products were analyzed by SDS-PAGE, it was observed that the amount of [^{32}P]ADP-ribose incorporated into a protein of 39 kDa was considerably reduced by the presence of Mg^{2+} and DTT. These studies are compatible with the hypothesis that the hydrolase is resistant to transferase-catalyzed ADP-ribosylation in the Mg^{2+}- and DTT-activated state. It is equally possible that activation of the hydrolase does not alter its ability to serve as a substrate for the transferase but, in the presence of Mg^{2+} and DTT, the active hydrolase can remove the ADP-ribose moiety and reverse inactivation. Based on these data, it would appear that active forms of transferase and hydrolase can coexist.

Possible role for ADP-ribosylarginine hydrolase in recovery from cholera. As noted elsewhere in this monograph (13), the diarrheal disease cholera results in large part from the secretion by *Vibrio cholerae* of a bacterial enterotoxin, choleragen or cholera toxin. Cholera toxin causes fluid and electrolyte abnormalities by activating adenylate cyclase in intestinal cells and thereby increasing intracellular cyclic AMP (7). Activation of adenylate cyclase results from the ADP-ribosylation by the A_1 protein of toxin of a regulatory guanine nucleotide-binding component of the cyclase system, termed $G_{s\alpha}$ (7). ADP-ribosyl-$G_{s\alpha}$ is clearly more active than is the unmodified protein (7). Based on the facts that choleragen catalyzes the ADP-ribosylation of free arginine (14), that the toxin modified an arginine residue in $G_{t\alpha}$ of transducin (15), a guanine nucleotide-binding protein present in retinal rod outer segments, and that the deduced amino acid sequence of $G_{s\alpha}$ has an arginine residue in a similar position to the one modified in $G_{t\alpha}$ (16), it has been proposed that the ADP-ribosylation site on $G_{s\alpha}$ is arginine. To determine whether ADP-ribosylarginine hydrolysis might have a role in the recovery from cholera, the ability of hydrolase to cleave the ADP-ribose-$G_{s\alpha}$ linkage was examined. Purified $G_{s\alpha}$ was ADP-ribosylated in the presence of [^{32}P]NAD, choleragen, GTP and ADP-ribosylation factor. [^{32}P]ADP-ribosyl-$G_{s\alpha}$ was then incubated with purified erythrocyte ADP-ribosylarginine hydrolase, Mg^{2+} and DTT; release of the ADP-ribose was monitored by SDS-PAGE followed by autoradiography. The hydrolase did, indeed, cleave the ADP-ribose-arginine linkage (17). It is not clear whether these *in vitro* effects occur *in vivo*. First, the soluble hydrolase may not have access to the ADP-ribose moiety in membrane-associated $G_{s\alpha}$; in the experiment described here, detergent-solubilized and partially purified $G_{s\alpha}$ was utilized. Second, the rates of ADP-ribosylation and de-ADP-ribosylation may depend on local levels of toxin and hydrolase activities; in the presence of excess toxin, the ADP-ribosylation reaction may be favored. These studies are compatible, however, with a possible role for the enzyme in reversing the toxin-catalyzed reaction responsible for intoxication and thus promoting recovery from cholera.

References

1. Moss, J., Stanley, S.J., Watkins, P.A. (1980) J Biol Chem 255: 5838-5840
2. Yost, D.A., Moss, J. (1983) J Biol Chem 258: 4926-4929
3. West, R.E., Jr., Moss, J. (1986) Biochemistry 25: 8057-8062
4. Soman, G., Mickelson, J.R., Louis, C.F., Graves, D.J. (1984) Biochem Biophys Res Commun 120: 973-980
5. Tanigawa, Y., Tsuchiya, M., Imai, Y., Shimoyama, M. (1984) J Biol Chem 259: 2022-2029
6. Goff, C.G. (1974) J Biol Chem 249: 6181-6190
7. Moss, J., Vaughan, M. (1988) Adv Enzymology 61: 303-379.
8. Smith, K.P., Benjamin, R.C., Moss, J., Jacobson, M.K. (1985) Biochem Biophys Res Commun 126: 136-142
9. Moss, J., Jacobson, M.K., Stanley, S.J. (1985) Proc Natl Acad Sci USA 82: 5603-5607
10. Moss, J., Oppenheimer, N.J., West, R.E., Jr., Stanley, S.J. (1986) Biochemistry 25: 5408-5414
11. Chang, Y-C., Soman, G., Graves, D.J. (1986) Biochem Biophys Res Commun 139: 932-939.
12. Moss, J., Stanley, S.J., Oppenheimer, N.J. (1979) J Biol Chem 254: 8891-8894

13. Tsai, S.-C., Noda, M., Adamik, R., Moss, J., Vaughan, M. (1987) This volume
14. Moss, J., Vaughan, M. (1977) J Biol Chem
    252: 2455-2457
15. Van Dop, C., Tsubokawa, M., Bourne, H.R., Ramachandran, J. (1984) J Biol Chem
    259: 696-698
16. Robishaw, J.D., Russell, D.W., Harris, B.A., Smigel, M.D., Gilman, A.G. (1986)
    Proc Natl Acad Sci USA 83: 1251-1255
17. Moss, J., Jacobson, M.K., Stanley, S.J. (1985) Clin Res 33: 564A

Abbreviations:

Gsα -the α subunit of a guanine nucleotide-binding protein that stimulates the adenylate cyclase catalytic unit
Gtα -the α subunit of a guanine nucleotide-binding protein from rod outer segments
DTT - dithiothreitol
SDS-PAGE - sodium dodecyl sulfate-polyacrylamide gel electrophoresis

ADP-Ribosyltransferase and Endogenous Acceptor Proteins in Animal Muscle Tissues: ADP-Ribosylation of Ca^{2+}-Dependent ATPase in Rabbit Skeletal Muscle Sarcoplasmic Reticulum and the Effect of Basic Peptides

Makoto Shimoyama, Nobumasa Hara[1], Mikako Tsuchiya, Koichi Mishima, and Yoshinori Tanigawa

Department of Biochemistry, Shimane Medical University, Izumo 693, Japan

Introduction

Guanidino compound-specific ADP-ribosyltransferase activity is present in certain eukaryote tissues and cells, and the enzymes from turkey erythrocytes and hen liver nuclei have been purified and characterized (1, 2). These experiments were done using exogenous acceptors and much less is known of the endogenous acceptor proteins for ADP-ribosylation.

In 1984, Graves and associates obtained evidence for ADP-ribosyltransferase activity in the sarcoplasmic reticulum (SR) from rabbit skeletal muscle (3). The SR contains predominantly Ca^{2+}-dependent ATPase, accounting for 70 to 80% of the membrane (4). Therefore, we examined whether or not the ADP-ribosylation of Ca^{2+}-dependent ATPase occurs in the rabbit skeletal muscle SR, and obtained evidence for the ADP-ribosylation of the ATPase and enhancement of this modification by basic peptides such as poly L-lysine and poly L-ornithine (5). We also noted, *in vitro*, the ADP-ribosylation of Ca^{2+}-dependent ATPase blocks the enzyme activity.

Results and Discussion

ADP-ribosyltransferase activities in various muscles. Graves and associates detected ADP-ribosyltransferase activity in different skeletal muscles (3). We confirmed this and observed that different cardiac muscles and some smooth muscles, rabbit esophagus and hen gizzard, contained significant ADP-ribosyltransferase activities as did the skeletal muscles. These experiments were carried out with the exogenous acceptor, L-arginine, and the product, ADP-ribosyl L-arginine, was determined by HPLC (6). The activities were not detected in the rabbit and rat uterus. We next investigated the ADP-ribosylation of endogenous acceptor protein in rabbit skeletal muscle SR.

---

[1]Central Research Laboratories, Shimane Medical University, Izumo 693, Japan

Table 1. ADP-ribosyltransferase activity in various muscles

Cell Type		Activity
		μmol/g wet weight/hr.
Skeletal muscle	Rat	2.12
	Rabbit	1.42
	Hen	0.76
Cardiac muscle	Hen	3.35
	Rabbit	1.74
	Rat	0.47
Smooth muscle	Rabbit esophagus	3.52
	Hen gizzard	1.30
	Rabbit uterus	Nil
	Rat uterus	Nil

One g of each freshly removed muscle tissue was minced and homogenized in a Polytron with 3 volumes of 0.1 M KCl containing 5 mM Tris-Cl buffer (pH 8.0) and 2 mM mercaptoethanol. The suspension was used as the enzyme source for the ADP-ribosyltransferase assay. The reaction mixture contained 50 mM Tris-Cl buffer (pH 7.5), 5 mM NAD, 25 mM L-arginine and appropriate amounts of the enzyme, in a total volume of 50 μl. The mixtures were incubated at 25°C for 1 hr. Following the incubation, a sample was diluted 10-fold with 0.1% trifluoroacetic acid and passed through a Millipore filter (pore size: 0.22 μm). The filtrate was used for HPLC analysis and the amount of ADP-ribosyl arginine was determined as described (6). Values are the mean of two experiments.

Effect of poly L-lysine, L-arginine or both on ^{32}P incorporation from [adenylate-^{32}P] NAD into proteins of SR vesicles. Recently, the basic peptide-induced activation of some key enzymes present in the membrane of *Xenopus laevis* oocytes has been reported (7). To determine whether poly L-lysine affects the ADP-ribosyltransferase of SR vesicles, we added poly L-lysine to the incubation mixture (Fig. 1). Poly L-lysine (100 μg/ml) dramatically enhanced the incorporation of ^{32}P from labeled NAD into endogenous acceptor proteins, and poly L-arginine reduced it in the absence and presence of poly L-lysine at the same rate. When poly L-lysine, L-arginine or both were added after termination of the incubation, the effects on ^{32}P incorporation were nil (data not shown). These results indicate that during the incubation, poly L-lysine enhanced ^{32}P incorporation into the SR proteins and that arginine residues of the endogenous proteins serve as the acceptor for ADP-ribosylation. The effect of poly L-lysine on the modification of exogenous acceptor L-arginine was not observed (Fig. 1). Therefore, we assume that poly L-lysine probably affects the intrinsic acceptor proteins, but not the enzyme *per se*.

Effect of increasing concentrations of basic peptides, divalent metals and spermine on the ^{32}P incorporation from labeled NAD into SR proteins. The effects of increasing the concentration of several compounds

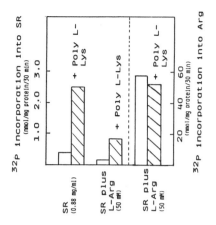 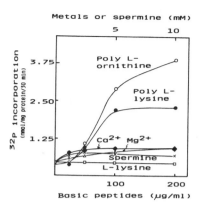

Fig. 1. (left) Effect of poly L-lysine, L-arginine or both on ^{32}P incorporation from [adenylate-^{32}P] NAD into proteins of SR vesicles. The reaction mixture containing 0.88 mg/ml of SR and 0.5 mM labeled NAD (2 ci/mol) in the absence and presence of 50 mM L-arginine was incubated, with or without 100 μg/ml of poly L-lysine, at 25°C for 30 min, and radioactivity of the acid-insoluble fraction was measured by filter assay (2). The minute amount of [^{32}P]NAD which bound to poly L-lysine during the incubation was detected. Thus, this radioactivity was subtracted from the value obtained in the presence of SR plus poly L-lysine. The ADP-ribosylation of L-arginine was determined by HPLC analysis (6).

Fig. 2. (right) Effect of increasing concentrations of basic peptides, divalent metals and spermine on the ^{32}P incorporation from [^{32}P] NAD into SR vesicles. Indicated amounts of compounds were added to the reaction mixture as described in the legend to Fig. 1, except that L-arginine was not included. Other experimental conditions were as described in the legend to Fig. 1.

including basic peptides, divalent metals and spermine on labeling of proteins of SR were investigated. As shown in Fig. 2, poly L-lysine exhibited dose dependent effects on the modification of SR vesicles and the maximal stimulation of ^{32}P incorporation by this peptide occurred in the presence of 100 μg/ml poly L-lysine. Poly L-ornithine further enhanced ^{32}P incorporation into the SR proteins. L-lysine did not affect the modification of SR vesicles at concentrations of 1 to 10 mM, but 5 to 10 mM Mg[2+], Ca[2+] or spermine did enhance it by a factor of 2.

To confirm that the incorporated radioactive moiety was mono(ADP-ribose), the acid-insoluble fractions of SR vesicles incubated with labeled NAD in the presence and absence of poly L-lysine were treated with alkali and the respective supernatant was analyzed by reverse phase HPLC. In both cases, radioactive peaks co-eluted with authentic ADP-ribose and 5'AMP (data not shown).

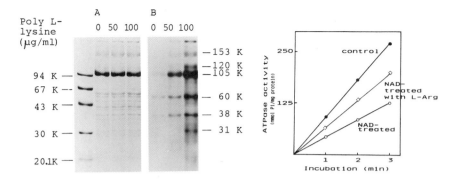

Fig. 3. (left) SDS-PAGE of [^{32}P]NAD treated SR vesicles in the absence and presence of poly L-lysine. 0.88 mg/ml of SR was incubated at 25°C for 60 min with 20 μM [^{32}P]NAD (2.8 ci/mmol) and 0, 50 or 100 μg/ml of poly L-lysine. Radiolabeling of the acid insoluble fraction (25 μg protein) from each sample was analyzed by SDS-polyacrylamide gel electrophoresis (8). The Coomassie brilliant blue-staining pattern (A) and an autoradiogram of the same gel (B) are shown. Molecular weight markers: phosphorylase b (94 K), bovine serum albumin (67 K), ovalbumin (43 K), carbonic anhydrase (30 K) and soybean trypsin inhibitor (20.1 K).

Fig. 4. (right) NAD-induced suppression of Ca^{2+}-dependent ATPase activity and protection by L-arginine in the reconstitution system for ADP-ribosylation of the enzyme. ADP-ribosyltransferase was partially purified from rabbit skeletal muscle SR. Details of the purification and properties of the enzyme will be described elsewhere. In brief, purified SR (50 mg protein) were suspended in 450 ml of 10 mM potassium phosphate buffer (pH 6.8) containing 7% Trition X-100 and 50 mM NaCl, homogenized, and centrifuged at 105,000 x g for 60 min to obtain the extract. The enzyme was then partially purified ca 300-fold by subsequent chromatography, successively on hydroxyapatite, Con A-Sepharose (Pharmacia) and Red-Toyopearl (Toyo Soda). The specific activity of the transferase in the final preparation, determined in the presence of 50 mM L-arginine as substrate, under the conditions described (6) was 50.0 μmol/mg protein/hr. Ca^{2+}-dependent ATPase was prepared from rabbit skeletal muscle SR by the method of MacLennan (4). The reconstitution system for ADP-ribosylation of Ca^{2+}-dependent ATPase contained 5 mM dithiothreitol, 1 mM NAD, 100 μg/ml of poly L-lysine, 7.1 μg/ml of partially purified ADP-ribosyltransferase and 600 μg/ml of ATPase and in some cases 20 mM L-arginine. For the control, both NAD and L-arginine were omitted. The mixture was preincubated at 25°C for 60 min, following which aliquots of the mixture containing 1.2 μg ATPase preparation were directly added to the Ca^{2+}-dependent ATPase assay system containing 50 mM Tris-Cl$^-$ buffer (pH 7.5), 0.1 M KCl, 5 mM MgCl$_2$, 50 μM CaCl$_2$ and 10 μM [α-^{32}P]ATP (20-40 Ci/mol) in a total volume of 0.1 ml. The incubation was carried out at 25°C for 0, 1, 2 and 3 min, respectively and the reaction was quenched by adding 2 ml of 10% trichloroacetic acid containing 0.5 mM ATP and 0.2 mM P$_i$ (14). The denatured sample was centrifuged, and [^{32}P]P$_i$ in the supernatant was extracted as the phosphomolybdate complex with an isobutyl alcohol-benzene mixture, as described by Martin and Doty (15). One mM NAD, 20 mM L-arginine or both, had no effect on the ATPase activity.

10

Evidence for ADP-ribosylation of Ca²⁺-dependent ATPase in SR vesicles and the effect of poly L-lysine. To further observe the effect of poly L-lysine on the ADP-ribosylation of SR proteins, SR vesicles were incubated with labeled NAD, with different concentrations of poly L-lysine and the acid-insoluble fractions were subjected to SDS-polyacrylamide gel electrophoresis. As shown in Fig. 3, the staining pattern of proteins was characteristic of the purified SR; such bands include the major one corresponding to the molecular weight of 105 kDa Ca²⁺-dependent ATPase and several minor bands. These profiles did not vary when poly L-lysine was present during the incubation. An autoradiogram of the labeled NAD-treated SR vesicles in the absence of poly L-lysine showed the major labeled band corresponding to the Ca²⁺-dependent ATPase and proteins of 60 and 38 kDa and other minor bands. The labeling of these proteins was greatly enhanced with increases in poly L-lysine. The extent of these changes determined by densitometric scan showed increases in the ADP-ribosylation of Ca²⁺-dependent ATPase by a factor of 20, while those of both 60 and 38 kDa proteins were enhanced by a factor of 10 (data not shown).

NAD-induced suppression of Ca²⁺-dependent ATPase activity in the reconstitution system for ADP-ribosylation of the enzyme and protection by L-arginine. The active transport of Ca²⁺ from the cytoplasm into the SR lumen occurs at the expense of ATP hydrolysis through Ca²⁺-dependent ATPase (9, 10). The hydrolysis of ATP occurs through a phosphoenzyme closely linked to the Ca²⁺ transport process (11). However, it is still unknown how the concentration of Ca²⁺ in the SR is regulated.

We reported that when phosphorylase kinase or histone H1 was ADP-ribosylated by hen liver nuclear ADP-ribosyltransferase, the modified proteins served as less effective acceptors for the phosphorylation (12, 13). Therefore, we investigated the effect of ADP-ribosylation on Ca²⁺-dependent ATPase activity. We prepared a reconstitution system for ADP-ribosylation of Ca²⁺-dependent ATPase containing a partially purified Ca²⁺-dependent ATPase (4) and ADP-ribosyltransferase from rabbit skeletal muscle and poly L-lysine. When the reconstitution mixture was incubated with 1 mM NAD, there was a significant decrease in the ATPase activity (Fig. 4). The same concentration of nicotinamide or ADP-ribose did not influence the ATPase activity (data not shown). If, however, the NAD treatment involved the use of 20 mM L-arginine, the NAD-dependent decrease in the Ca²⁺-dependent ATPase activity was largely overcome. In this case, we confirmed that high ADP-ribosylation of the Ca²⁺-dependent ATPase preparation by the ADP-ribosyltransferase was reduced in the presence of L-arginine. Thus, the ADP-ribosylation of SR vesicles may participate in the regulation of Ca²⁺ transport in skeletal muscle, through the ADP-ribosylation-induced decrease in Ca²⁺-dependent ATPase activity. It is also possible to speculate that ADP-ribosylation of Ca²⁺-dependent ATPase

11

in SR may regulate the rate of Ca^{2+} release from these vesicles, because the Ca^{2+} release channel is activated by Ca^{2+} and ATP or nonhydrolyzable ATP analogues in skeletal and cardiac muscle SR (16).

From these results, we conclude that the ADP-ribosylation of Ca^{2+}-dependent ATPase by endogenous ADP-ribosyltransferase occurs in rabbit skeletal muscle SR vesicles and that this reaction seems to be important for regulating Ca^{2+} transport in skeletal muscle. The basic peptides poly L-ornithine and poly L-lysine are proving to be useful tools for identifying acceptor proteins in the sarcoplasmic reticulum.

Acknowledgements. This work was supported in part by grants-in-aid for Scientific and Cancer Research from the Ministry of Education, Science and Culture, Japan.

References

1. Moss, J., Stanley, S.J., Watkins, R.A. (1980) J Biol Chem 255: 5838-5840
2. Tanigawa, Y., Tsuchiya, M., Imai, Y., Shimoyama, M. (1984) J Biol Chem 259: 2022-2029
3. Soman, G., Michelson, J.R., Luis, C.F., Graves, D.J. (1984) Biochem Biophys Res Commun 120: 973-980
4. MacLennan, D.H., Seeman, P., Iles, G.H., Yip, C.C. (1971) J Biol Chem 246: 2702-2710
5. Hara, N., Mishima, H., Tsuchiya, M., Tanigawa, Y., Shimoyama, M. (1987) Biochem Biophys Res Commun 144: 856-862
6. Tsuchiya, M., Tanigawa, Y., Mishima, K., Shimoyama, M. (1986) Anal Biochem 157: 381-384
7. Gatica, M., Allede, C.C., Antonelli, M., Allede, J.E. (1987) Proc Natl Acad Sci USA 84: 324-328
8. Weber, K., Osborn, M. (1969) J Biol Chem 244: 4406-4412
9. Ebashi, S., Lipman, E. (1962) J Cell Biol 14: 389-400
10. Weber, A. (1966) Curr Top Bioenerg 1: 203-2548
11. Tsuchiya, M., Tanigawa, Y., Ushiroyama, T., Matsuura, R., Shiomyama, M. (1985) Eur J Biochem 147: 33-40
12. DeMeis, L. and Vianna, A.L. (1979) Annu. Rev. Biochem. 48: 257-292
13. Ushiroyama, T., Tanigawa, Y., Tsuchiya, M., Matsuura, R., Ueki, M., Sugimoto, O., Shimoyama, M. (1985) Eur J Biochem 151: 173-177
14. Takakuwa, Y., Kanazawa, T. (1979) Biochem Biophys Res Commun 88: 1209-1216
15. Martin, J.B., Doty, D.M. (1949) Anal Chem 21: 965-967
16. Meissner, G., Henderson, J.S. (1987) J Biol Chem 262: 3065-3073

Endogenous ADP-Ribosylation of Proteins at Cysteine Residues

Elaine L. Jacobson, Mingkwan Mingmuang, Nasreen Aboul-Ela, and Myron K. Jacobson

Departments of Biochemistry and Medicine, Texas College of Osteopathic Medicine, University of North Texas, Fort Worth, Texas 76107 USA

Introduction

NAD$^+$ is a substrate for many dehydrogenases that catalyze hydride transfer reactions central to energy metabolism. It is also the substrate for enzymes that catalyze the cleavage of the linkage between nicotinamide and ribose and the transfer of ADP-ribose to a nucleophilic acceptor. Such ADP-ribose transfer reactions represent a versatile mechanism for the posttranslational modification of proteins. For example, poly(ADP-ribose) polymerase catalyzes transfer of ADP-ribose to protein carboxylate groups and to the ribosyl hydroxyls of ADP-ribose resulting in the modification of proteins with ADP-ribose polymers (1). While all other ADP-ribose transfer enzymes catalyze the transfer of only single ADP-ribose groups to an acceptor, they also show a wide range of specificity for acceptors. The best understood mono-ADP-ribosyl transferases are the bacterial toxins which transfer ADP-ribose to specific amino acid residues of specific target proteins including arginine, modifed histidine and cysteine residues (2-10). However, several endogenous mono-ADP-ribosyl transferases also have been detected that have the same range of amino acid acceptor specificity as those exhibited by the bacterial toxins (11-14).

Evidence that the endogenous arginine specific mono-ADP-ribosyl transferases actually modify proteins *in vivo* has come from the detection of proteins that are covalently modified with ADP-ribose by linkages indistinguishable from ADP-ribosyl-arginine (15). Endogenous protein linkages with properties indistinguishable from ADP-ribosyl carboxylate ester have also been detected (15). Reported here is the discovery of a third class of linkages which appear to be characteristic of an ADP-ribose modification at cysteine. These results provide further evidence that the modification of proteins by ADP-ribose is a versatile mechanism to posttranslationally modify proteins.

Results and Discussion

Conditions for the selective release of ADP-ribose from carboxylate ester linkages to glutamate or aspartate and for release from glycosylic linkages to the guanidinium group of arginine were reported by this laboratory earlier (15). Under conditions whereby ADP-ribosyl-carboxylate

13

ester and/or ADP-ribosyl-arginine bonds are hydrolyzed, cysteine and diphthamide linkages to ADP-ribose are stable. It has been reported previously that thioglycosides can be cleaved in the presence of mercuric ion (16). In order to search for endogenous proteins modified by ADP-ribose on cysteine, experiments were conducted to determine whether this reagent could be used to cleave ADP-ribosyl-cysteine linkages in crude extracts containing total liver proteins and, if so, whether ADP-ribose was released intact. For this purpose, transducin modified with radiolabeled ADP-ribose by the action of pertussis toxin was used as a model conjugate. ADP-ribosyl transducin was added to crude extracts of liver proteins and treated with mercuric ion. Fig. 1, Panel A shows that the radiolabel was released as a function of time in the presence of 10 mM mercuric ion. Approximately 90% of the label was released following a 10 min incubation at 37°C. Panel B shows the effect of increasing concentrations of mercuric ion during a 10 min incubation. Approximately 90% of the linkages were released in the presence of 5 mM mercuric ion and no further release was observed with increasing concentrations. Treatment with mercuric ion released intact ADP-ribose from transducin as judged by strong anion exchange HPLC analysis of the released material.

The selectivity of the release of ADP-ribose by mercuric ion for different ADP-ribosyl linkages was also examined. The presence of mercuric ion had no effect on ADP-ribosyl-arginine or ADP-ribosyl-diphthamide. Thus, these conditions provide a selective release of cysteine residues in the presence of ADP-ribosyl-arginine and ADP-ribosyl-diphthamide. The effect of mercuric ion on the cleavage of ADP-ribosyl-carboxylates was examined by analyzing cell extracts for carboxylate-ester like linkages both prior to and following incubation in the presence of mercuric ion. The presence of mercuric ion did not cause the release of carboxylate ester linkages to ADP-ribose. Thus, for the known protein ADP-ribose linkages, mercuric ion catalyzes the selective release of ADP-ribose from cysteine.

Next, total rat liver proteins were examined for the presence of linkages characteristic of ADP-ribosyl-cysteine. Trichloroacetic acid insoluble extracts of liver tissue were treated to remove non-covalently bound ADP-ribose and subsequently treated with mercuric ion and analyzed for ADP-ribose. Fig. 2 shows such an analysis along with control experiments. Panel A shows that analysis of an extract treated with mercuric ion exhibited a fluorescent peak that migrated at the expected elution position of etheno-ADP-ribose, the fluorescent derivative used for quantitative determination of ADP-ribose. Panel B shows the analysis in which mercuric ion was omitted from a parallel sample of liver extract. Panel C shows the result obtained when chloroacetaldehyde, which is required for the formation of the fluorescent derivative of ADP-ribose, was omitted. This control rules out the possibility of endogenously fluorescent compounds present in the cell extract that were released by mercuric ion. Panel D shows that a small amount of authentic etheno-ADP-ribose added to extracts prepared as in

Panel A resulted in an enhancement of the peak. Taken together, the results of Fig. 2 demonstrate that endogenous proteins are modified with ADP-ribose by linkages indistinguishable from ADP-ribosyl-cysteine. Table 1 shows quantitative data with regard to the total amount of the different types of linkages of ADP-ribose to protein in rat liver.

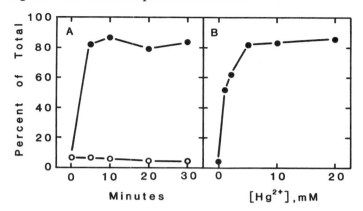

Fig. 1. Release of cysteine-linked ADP-ribose by mercuric ion. The acid insoluble fraction of rat liver (15) was dissolved in 98% ice cold formic acid and radiolabeled mono-ADP-ribosylated protein was added. The solution was diluted with five volumes of ice cold H_2O and precipitated by addition of 100% (w/v) ice cold trichloroacetic acid to a final concentration of 20% (w/v). The sample was held on ice for 10 min and subjected to centrifugation. The precipitate was resuspended in ice cold, 98% formic acid and stored at -20°C for subsequent use. To release cysteine-linked ADP-ribose, the sample in ice cold 98% formic acid was diluted with an equal volume of ice cold H_2O or a freshly prepared solution of 20 mM mercuric acetate and the resulting solution was incubated at 37°C 10 min. The samples were then placed on ice, 5 volumes of ice cold H_2O were added followed by 100% (w/v) trichloroacetic acid to a final concentration of 20%. After 10 min on ice, the samples were collected by centrifugation and the supernatant was removed. A sample was taken to determine released radiolabeled mono-ADP-ribose. The pellet containing the remaining protein-bound mono-ADP-ribose was dissolved in 250 mM ammonium acetate, 10 mM EDTA and 6 M guanidine before sampling for radioactivity. (●) presence, (○) absence of mercuric ion. Panel A shows a time course at 10 mM mercuric ion and Panel B shows a 10 min incubation at the indicated concentrations of mercuric ion.

Table 1. Protein-bound ADP-ribosyl linkages in rat liver

Linkage	ADP-Ribose pmol/mg Protein	
Arginine	19.2 ± 2.6	(12)
Cysteine	14.2 ± 1.6	(6)
Carboxylate	16.2 ± 0.3	(6)

The acid-insoluble fraction of rat liver was analyzed for linkages characteristic of ADP-ribosyl-cysteine, ADP-ribosyl-arginine and ADP-ribosyl-carboxylate. Standard deviations are shown and the number of replicate analyses is shown in parenthesis.

15

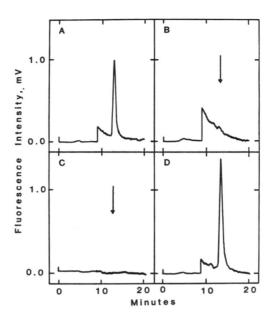

Fig. 2. Detection of endogenous cysteine-like ADP-ribose in rat liver proteins. Panel A, complete treatment. Panel B, omission of mercuric ion. Panel C, omission of chloroacetaldehyde. Panel D, complete treatment with addition of authentic etheno-ADP-ribose. The arrows in panel B and C are for reference and indicate the elution position of etheno(ADP-ribose).

These results are of interest with respect to the recent detection of a mono-ADP-ribosyl transferase in human erythrocytes that catalyzes the ADP-ribosylation of free cysteine and of the GTP binding protein $G_{i\alpha}$ (14). The results presented here provide evidence that endogenous enzyme(s) specific for transfer to cysteine can also use protein molecules as substrates *in vivo*. The present study allows an estimate of overall protein modification by ADP-ribose on arginine and cysteine residues. Assuming an average molecular weight for protein of 50 kDa and that an endogenous protein molecule carries only a single ADP-ribose, the estimated frequency of protein molecules modified by ADP-ribose is approximately 1 in 1,000 protein molecules modified at arginine and 1 in 1,400 protein molecules carrying an ADP-ribose at cysteine.

The covalent posttranslational modification of specific amino acid side chains in protein can be subdivided into those that are metabolically stable and those that are metabolically reversible (17). In general, metabolically reversible modifications are involved in metabolic signaling systems. A major unanswered question with regard to the modification of proteins by ADP-ribose is whether these modifications are metabolically reversible. With regard to modification of arginine residues, enzymatic activities have been detected in cultured mouse cells which catalyze the release of ADP-ribose from protein arginine residues (18). This supports the possibility that this modification is, at least in part, metabolically reversible. The studies of Ludden and co-workers (19-21) have shown that a metabolically reversible modification cycle occurs in procaryotic cells. The enzyme nitrogenase in *R. rubrum* is inactivated and reactivated via cycles of ADP-ribose addition to and removal from a specific arginine residue of the enzyme. The detection of proteins modified at cysteine reported here raises the question of whether these modifications are reversible or not *in vivo*.

Acknowledgements. This work was supported by Grant No. CA43894 from the National Institutes of Health and by the Texas VFW Ladies Auxiliary Cancer Research Fund. We are grateful to Professor Helmuth Hilz for helpful discussions and to Dr. Joel Moss for the (ADP-ribosyl) transducin and mono-ADP-ribosyl transferase A.

References

1. Althaus, F.R., Richter, C.W. (1987) ADP-Ribosylation of Proteins. Springer-Verlag
2. Collier, J. (1982) in ADP-Ribosylation Reactions: Biology and Medicine. O. Hayaishi, K. Ueda eds. pp 575-592. Academic Press, N.Y
3. Cassel, D., Pfeuffer, T. (1978) Proc. Natl. Acad. Sci. USA 75, 2669-2673
4. Moss, J., Vaughan, M. (1977) J. Biol. Chem. 252, 2455-2457
5. Katada, T., Ui, M. (1982) J. Biol. Chem. 257, 7210-7216
6. West, R.E., Moss, J., Vaughan, M., Liu, T., Liu, T.-Y. (1985) J. Biol Chem. 260, 14428-14430
7. Bokoch, G.M., Katada, T., Northup, J.K., Hewlett, E.L., Gilman, A.G. (1983) J. Biol Chem. 258, 2072-2075
8. Aktories, K., Bärmann, M., Ohishi, I., Tsuyama, S., Jakobs, K.H., Habermann, E. (1986) Nature 322, 390-392
9. Ohishi, I., Tsuyama, S. (1986) Biochem. Biophys. Res. Commun. 136, 802-806
10. Aktories, K., Frevert, J. (1987) Biochem. J. 247, 363-368
11. Moss, J. (1987) Clin. Res. 35, 451-458
12. Sayham, O., Özdemirli, M., Nusten, R., Bermek, E. (1986) Biochem. Biophys. Res. Commun. 139, 1210-1214
13. Iglewski, W.J., Lee, H., Muleer, P. (1984) Febs Lett. 173, 113-118
14. Tanuma, S., Kawashima, K., Endo, H. (1988) J. Biol. Chem. 263, 5485-5489
15. Payne, D.M., Jacobson, E.L., Moss, J., Jacobson, M.K. (1985) Biochemistry 24, 7540-7549
16. Krantz, M.J., Lee, Y.C. (1976) Anal. Biochem. 71, 318-321
17. Rucker, B.R., Wold, F. (1988) FASEB J. 2, 2252-2261
18. Smith, K.P., Benjamin, R.C., Moss, J., Jacobson, M.K. (1985) Biochem. Biophys. Res. Commun. 126, 136-142
19. Saari, L, Triplett, E.W., Ludden, P.W. (1984) J. Biol. Chem. 259, 15502-15508
20. Kanemoto, R.H., Ludden, P.W. (1984) J. Bact. 158, 713-720
21. Pope, M.R., Murrell, S.A., Ludden, P.W. (1985) Proc. Natl. Acad. Sci., U.S.A. 82, 3173-3177

Reversible ADP-Ribosylation of Dinitrogenase Reductase from *Rhodospirillum rubrum* Regulates the Activity of the Enzyme *In Vivo* and *In Vitro*

Paul W. Ludden, Scott A. Murrell, Robert G. Lowery,
Wayne P. Fitzmaurice, Mark R. Pope, Leonard R. Saari, Roy H. Kanemoto,
and Gary P. Roberts

Departments of Biochemistry and Bacteriology, University of Wisconsin - Madison
Madison, Wisconsin 53706 USA

Introduction

The nitrogenase activity of *Rhodospirillum rubrum* is regulated by reversible mono(ADP-ribosyl)ation of the dinitrogenase reductase component of the enzyme (1, 2). *R. rubrum* is a gram-negative, non-sulfur photosynthetic bacterium that is capable of growth on N_2 as the N-source in pure culture. The organism is capable of diverse modes of growth including anaerobic fermentation, aerobic respiration and anoxygenic photosynthesis.

Nitrogen fixation by *R. rubrum* was discovered by Gest and Kamen (3) who also noted that fixed sources of nitrogen inhibited nitrogen fixation; the molecular basis for this observation is now known to be the reversible ADP-ribosylation of one of the nitrogenase proteins. The assimilation of N in *R. rubrum* is via nitrogenase, glutamine synthetase and glutamate synthase as shown in Fig. 1. It is interesting to note that the first two enzymes of this metabolically extensive pathway are regulated by two different covalent modification cascades. A long-term goal of our research program will be to understand how these two cascades recognize each other's operation and how they are integrated into metabolism as a whole.

The enzyme nitrogenase consists of two proteins: dinitrogenase reductase (alias, the iron protein and abbreviated Rr2) and dinitrogenase (alias, the molybdenum-iron protein, Rr1). Dinitrogenase is the site of substrate reduction and is not a target for ADP-ribosylation. Rr2 is an α_2 dimer of 60,000 kDa that has a single 4Fe-4S cluster. It binds two molecules of MgATP and hydrolyzes them concomitant with transfer of a single electron to dinitrogenase (4,13). In addition to the reduction of N_2 to ammonia, the nitrogenase complex (Rr1 + Rr2) carries out the reduction of protons to H_2 and acetylene to ethylene; the latter non-physiological reaction provides a fast, easy assay for the enzyme *in vitro* as well as *in vivo*. Rr2 is the target for ADP-ribosylation. The present model for reversible ADP-ribosylation of Rr2 is shown in Fig. 2. ADP-ribose is transferred to Rr2 by inactivating enzyme (IE) in a M^{2+} ADP-dependent reaction. Rr2-ADP-ribose is activated by dinitrogenase reductase activating glycohydrolase (DRAG; formerly activating enzyme or AE) in an M^{2+} and MgATP dependent reaction.

18

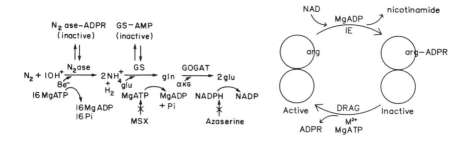

Fig. 1. (left) The path of ammonia assimilation in *Rhodospirillum ruburm.*

Fig. 2. (right) Model for reversible ADP-ribosylation of dinitrogenase reductase from *Rhodospirillum ruburm.*

Results and Discussion

Properties of Rr2-ADP-ribose. ADP-ribose is attached to Rr2 by an N-α-glycosidic bond to the guanidinium group of arginine-100 of the protein primary sequence. This site is 3 amino acid residues removed from cysteine-97 that is a ligand to the iron-sulfur cluster of Rr2 (5). The linkage is thermolabile (6) and mutarotates with a $t_{1/2}$ of approximately 6 hrs. This mutarotation may be physiologically significant.

Purification and properties of DRAG. DRAG is a 32 kDa monomer that is present in extremely low levels in the cell (7). It was first observed by its ability to activate inactive nitrogenase in crude extracts (8, 9). DRAG is associated with membranes in crude extracts and can be solubilized by treating the membrane fractions with 0.5 M NaCl. The enzyme has been purified 12,000 fold by a combination of ion exchange, hydroxylapatite, dye matrix affinity and gel filtration chromatography (Fig. 3, lane 4). The enzyme is extremely oxygen-labile with a $t_{1/2}$ in air of 90 sec; this observation leads to the suggestion that the enzyme may be controlled by redox in the cell. The basis for the oxygen lability is not known, but dithiothreitol does not protect the enzyme from oxygen.

Activities of DRAG. DRAG removes ADP-ribose from Rr2-ADP-ribose. It is specific for the α configuration of the N-glycosidic bond (15). DRAG will also cleave ADP-ribose from the hexapeptide gly-arg-(ADP-ribose)-gly-val-ile-thr obtained by proteolysis of Rr2-ADP-ribose. The

19

products are ADP-ribose and the regenerated guanidinium group of arginine. Although MgATP is required for removal of ADP-ribose from native Rr2, it is not required for glycohydrolysis of the hexapeptide-ADP-ribose or other low molecular weight substrates. MgATP is not hydrolyzed during activiation of Rr2, thus it is concluded that MgATP acts as an allosteric effector that binds to Rr2 and allows DRAG access to the arg-ADP-ribose site.

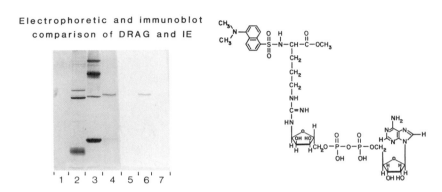

Electrophoretic and immunoblot comparison of DRAG and IE

Fig. 3. (left) Electrophoretic and immunoblot comparison of DRAG and IE. Lanes 1 - 4; Silver stained SDS polyacrylamide gel of protein fractions. Lanes 5 - 7; Immunoblot of fractions developed with anti-DRAG antibody and peroxidase conjugate anti-rabbit IgG. Lane 1, blank; Lane 2, Inactivating Enzyme; Lane 3, Molecular Weight Standards (BSA, ovalbumin, carbonic anhydrase, myoglobin); Lane 4, Purified DRAG; Lane 5, IE; Lane 6, Purified DRAG; Lane 7, blank.

Fig. 4. (right) Dansylarginine methylester adenosine diphosphoribose.

DRAG will hydrolyze dansylarginine methylester-ADP-ribose (Fig. 4). It is not necessary that the arginine carboxyl group be esterified for DRAG to recognize the compound as a substrate but the arginine amino group must be blocked or the K_m increases 1000-fold. The enzyme is non-discriminating at the adenine end of the molecule; guanine and ethenoadenine may be substituted with no effect on K_m or V_{max}. Hydrolysis of dansylarginine methylester etheno-ADP-ribose by DRAG (or by phosphodiesterase) results in a greatly enhanced fluorescence and provides a sensitive continuous assay for the enzyme (14).

Purification and properties of IE. The inactivating enzyme (IE) is detected by its ability to transfer ADP-ribose to Rr2, inactivating it in the process (2). The enzyme has been purified greater than 12,000-fold but is not yet pure (Fig. 3), although the protein has been identified on SDS gels as a 29 kDa band. Purification involves ion exchange, dye matrix affinity, gel filtration and HPLC ion exchange chromatography. The protein is in the soluble fraction of extracts and is stable to oxygen. Of the purified proteins tested, only dinitrogenase reductases function as substrates. Interestingly, the heterologous dinitrogenase reductase from *Klebsiella pneumoniae* is the best substrate (Table 1). Other proteins and small molecules tested to date do not function as ADP-ribose acceptors (Table 1).

Table 1. Nucleotide and receptor specificity for IE

FE Protein	2 mM Nucleotide	% IE Activity*
Rr2	None	1
Rr2	ADP	100
Rr2	ATP	7
Rr2	AMP	3
Rr2	GDP	4
Rr2	ADP-BS	60
Rr2	8 BR-ADP	3
Rr2	ETHENO-ADP	2
Rr2	DEOXY-ADP	120
Rr2	ADP, UDP, TDP, CDP, ADPR	0
Av2	None	33
Av2	ADP	100
Av2	ADP-BS	116
Av2	Deoxy-ADP	102

FE Proteins	nmol ADPR/min/mg
Rr2 (native)	2.82
Rr2 (O_2-treated)	0.0
Av2	2.79
Kp2	6.90

OTHER PROTEINS AND SMALL MOLECULES

Lysozyme, BSA, Ovalbumin, Histone, Polyarginine	0.0
Water	0.0
Arginine, Agmatine, Arginine Methyl Ester, DNS Arginine, Benzyl Arginine Ethyl Ester, Tosyl Arginine Methyl Ester	0.0

*Determined by ^{32}P-NAD/Filter Assay

Nucleotide requirements for IE. Like DRAG, IE requires a nucleotide to carry out its activity on Rr2 (Table 1). MgADP is much preferred over MgATP. Our hypothesis is that MgADP is required as an effector that binds to Rr2. Justification for this hypothesis is based on the observation that ADP is not hydrolyzed in the reaction, that analogues that inhibit the electron transfer activity of Rr2 also allow its ADP-ribosylation by IE (for

21

example, ADP-β-S and deoxy-ADP) whereas those that do not inhibit Rr2 enzyme activity do not allow its ADP-ribosylation, and finally, that dinitrogenase reductase from other sources (*Klebsiella pneumoniae,* [Kp2] and *Azotobacter vinelandi,* [Av2]) show decreased requirements for ADP. Proof of this point awaits the development of small molecule substrates for IE as has been done for DRAG. IE appears not to require free divalent metal. NAD is the donor for ADP-ribose and incorporation of [^{32}P]ADP-ribose into Rr2 is inhibited by a number of NAD analogs including etheno-NAD (98% inhibition at 2 mM) and nicotinamide guanidine dinucleotide (99% inhibition at 2 mM). Desamido-NAD showed little inhibition of activity.

DRAG and IE are distinct gene products. DRAG and IE activities are found in different fractions of cell extract, show differing sensitivity to oxygen and do not co-purify. These observations all suggest that the enzymes are distinct gene products. Further evidence for this point is shown in Fig. 3. DRAG and IE do not co-electrophorese (lanes 2 and 4) and polyclonal antibody against DRAG does not recognize IE (lane 5). Final proof of this point comes from the fact that mutants which lack DRAG activity exhibit IE activity (see below).

The effect of ADP-ribosylation on enzymatic activity of dinitrogenase reductase. The dinitrogenase reductase from *Clostridium pasteurianum* (Cp2) is known to form a tight complex with the dinitrogenase from *A. vinelandii* (Av1) (10, 15). We asked if Cp2 was a substrate for IE and if ADP-ribosylation of Cp2 would disrupt the inhibitory complex. Table 2 shows that Cp2-ADPR no longer inhibits the activity of Av1 + Av2. This result suggests that ADP-ribosylation of Rr2 inhibits its ability to form functional complexes with its electron acceptor, Rr1.

Table 2. Effect of ADP-ribosylation of dinitrogenase reductase from *C. pasteurianum* on *A. vinelandii* nitrogenase activity

Av1	Av2	Cp2	Cp2-ADPR	Activity (nmoles C_2H_2 reduced/20 min)
100 µg	50 µg	0	0	551
100 µg	50 µg	50 µg	0	214
100 µg	0	50 µg	0	0
100 µg	50 µg	0	50 µg	517
100 µg	0	0	50 µg	0

Evidence for reversible operation of DRAG/IE *in vivo*. Nitrogenase activity can be monitored *in vivo* by acetylene reduction assay and the modification state of Rr2 can be assessed on SDS gels. This allows the demonstration of reversible regulation of Rr2 *in vivo*. In an experiment in which acetylene reduction activity and modification of Rr2 are followed, there are three cycles of dark and light modification and loss of activity observed in all three cycles. Because there are two potential sites for modification on Rr2 (an α_2 dimer), the modification during the third cycle indicates that the site for modification is being regenerated *in vivo* (11). This observation is consistent with the demonstration that DRAG removes ADP-ribose and regenerates the arginine guanidinium group *in vitro* (12).

Table 3. Isolation of the gene for drag from *R. rubrum*

Amino terminus sequence for DRAG:
      Gly Pro Ser Val His Asp Arg Ala Leu Gly Ala Phe Leu
      Gly Leu Ala Val Gly Asp Ala Leu Gly Ala Thr <u>Val Glu</u>
      <u>Phe Met Thr Lys</u> Gly Glu Ile *** Gln Gln

Probe: _____
5' GTN GAA TTC ATG ACN AA 3'
     G   T

Sequence and translation of RI-Ea fragment of insert in pWPF102:
GAA TTC ATG ACC AAG GGC GAG ATC GTC CAG TCG ATA CGG CAT
<u>GLU PHE MET THR LYS</u> GLY GLU ILE VAL GLN SER ILE ARG HIS

7.4 kb insert in pWPF102:

A		'D'	B	'E'	C	'

                                      ' _____ '
                                        DRAG

Isolation and mutagenesis of the gene encoding DRAG. The amino terminal sequence of DRAG was determined by Dr. Ron Niece of the University of Wisconsin Biotechnology Center (Table 3). The sequence val-glu-phe-met-thr-lys was selected to design the 64-fold degenerate probe shown in Table 3. This probe was used to identify lambda clones potentially containing the gene for DRAG, designated *draG*. A 7.4 kb *Bam*HI fragment was subcloned into pBR322 to yield pWPF102. A fragment of this clone was identified as the region hybridizing to the probe and this fragment was partially sequenced. As shown in Table 3 the sequence agrees with the amino terminal sequence of DRAG. The plasmid pWPF104 was mutagenized in *E. coli* by Tn5 and the mutagenized gene was reisolated, subcloned into pRK290 and mobilized into *R. rubrum*. This conjugation required the helper plasmid PRK2013. After selection for the appropriate antibody resistances and demonstration that *draG* gene was present only in the Tn5-containing form, the mutant strain (UR135 was analyzed for its ability to synthesize and ADP-ribosylate nitrogenase. The cells showed a long lag before growth and Fig. 5 shows the result of the experiment. Both wild-type and UR135 showed vigorous nitrogenase activity and both exhibited modification of the dinitrogenase reductase protein in the dark and upon ammonia treatment (upper subunit formation, labeled U in Fig. 5). However, the mutant strain showed only slow, incomplete recovery, as expected for a strain lacking DRAG. Measurements of *in vivo* nitrogenase activity confirmed this lack of recovery.

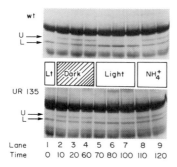

Fig. 5. Time course of dark and ammonia switch-off of nitrogenase in *R. rubrum* wild type and *draG⁻* strain UR135.

In the experiment described above, strain UR135 showed a long lag before growth initiated under selective (kanamycin) conditions. Strains containing *draG*::Tn5 do not exhibit photosynthetic growth and our hypothesis is that the strain analyzed was a pseudorevertant variant of UR135 that allowed photosynthetic growth but not the normal regulation of nitrogenase.

The inability of *draG*::Tn5 lines to grow suggests that DRAG/IE ADP-ribosylation system is involved in regulation of a variety of metabolic pathways.

Acknowledgements. This work was supported by the College of Agricultural and Life Sciences at the University of Wisconsin-Madison and by grants from the USDA-competitive grants program (85-CRCR-1-1668) and NSF (DMB-87-40296).

References

1. Pope, M.R., Murrell, S.A.., Ludden, P.W. (1985) Proc Natl Acad Sci USA 82: 3173-3177
2. Lowery, R.G, Saari, L.L., Ludden, P.W. (1986) J Bacteriol 16: 513-518
3. Gest, H., Kaman, M.D. (1949) Science 109: 558
4. Hageman, R.V.., Burris, R.H. (1978) Proc Natl Acad Sci USA 75: 2699-2702
5. Hausinger, R.P., Howard, J. (1983) J Biol Chem 258: 13486-13492
6. Dowling, T.G., Preston, G.G., Ludden, P.W. (1982) J Biol Chem 257: 13987-1399
7. Saari, L.L., Triplett, E., Ludden, P.W. (1984) J Biol Chem 259: 15502-15508
8. Ludden, P.W., Burris, R.H. (1976) Science 194: 424-426
9. Nordlund, S., Erickson, U., Baltscheffsky, H. (1977) Biochim Biophys Acta 462: 187-195

10. Emerich, D.W.., Burris, R.H. (1978) J Bacteriol 134: 936-943
11. Kanemoto, R.H., Ludden, P.W. (1984) J Bacteriol 158: 713-720
12. Saari, L.L., Pope, M.R., Murrell, S.A., Ludden, P.W. (1986) J Biol Chem 261: 4973-4977
13. Ljones, T., Burris, R.H. (1972) Biochim Biophys Acta 275: 98-101
14. Pope, M.R., Ludden, P.W. (1987) Anal Biochem 160: 68-77
15. Pope, M.R., Saari, L.L., Ludden, P.W. (1986) J Biol Chem 261: 10104-10111
16. Emerich, D.W., Ljones, T., Burris, R.H. (1978) Biochim Biophys Acta 527: 359-369

Phospho ADP-Ribosylation and Phospho Adenylation of Proteins

Helmuth Hilz

Institut für Physiologische Chemie, Universität Hamburg, 2000 Hamburg 20, West Germany

Introduction

In 1941 the late Fritz Lipmann formulated his now famous concept of energy-rich bonds (1). Although he developed his thesis for phosphate compounds, it has proven extremely fruitful in other areas as well. When looking through his eyes at the structure of pyridine nucleotides, we see two activated groups (Fig. 1): Active ADP-ribose and active AMP in NAD, and the corresponding groups in NADP as well, but extended by a 2'-phosphate residue. Recognition of NAD as a group transferring coenzyme dates back to the mid 1960's when poly(ADP-ribose) was detected by Chambon *et al.* (3), Nishizuka *et al.* (4) and Sugimura *et al.* (5), and when Olivera and Lehman reported the use of NAD by bacterial DNA ligase to reseal DNA breaks (2). Although it seems reasonable to assume that analogous reactions exist for NADP as well, no group transfer from NADP to acceptor proteins has been described so far.

Results and Discussion

Phospho ADP-ribosylation of proteins. Non-enzymatic incorporation of free phospho ADP-ribose into polypeptides occurs to a significant degree in mitochondrial preparations of bovine and rat liver (6). This reaction still proceeded when the proteins were denatured by heating. The first enzymic transfer of phospho ADP-ribose to acceptor proteins was shown when we analyzed, in cooperation with J. Moss, the arginine-specific ADP-ribosyl transferase from turkey erythrocytes (7). With histone H1 as a substrate, incorporation of adenine equivalents from NADP was markedly higher than that from NAD. Use of NADP labeled in the nicotinamide-proximal ribose showed equivalent transfer of label into histone H1 thus demonstrating incorporation of the entire phospho ADP-ribose group. To our knowledge, no other example of phospho ADP-ribosylation has been found so far and the biological significance of this reaction remains unknown.

Protein modification by phospho adenylylation. When we compared incorporation of adenine equivalents from labeled NAD and labeled NADP into the acid-insoluble fraction of rat liver microsomes, NADP proved to be a much better precursor than NAD. SDS gel electrophoresis revealed a

26

highly specific modification of a polypeptide with an apparent molecular weight of about 40 kDa (p40) when NADP was used as a substrate. However, no incorporation took place when NADP labeled in the nicotinamide-proximal ribose was used as a substrate (Fig. 1). Synthesis of several NADP species labeled at different parts of the molecule and their application to the analysis of the reaction with p40 demonstrated that radioactivity was only incorporated when it was present in the 2'-phospho AMP moiety of the molecule (8). This demonstrated a novel type of covalent modification: phospho adenylylation.

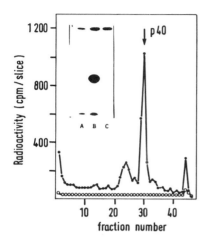

Fig. 1. Incorporation of radioactivity into p40 from differently labeled NADP precursors. -- · [5'-phosphate-^{32}P]NADP; o--o [NMN-ribose-^{14}C]NADP. Incubation and processing was performed under standard conditions as described in (8). Inset: Autoradiogram of standard incubations, A: [5'-phosphate-^{32}P]NAD; B: [5'-phosphate-^{32}P]NADP; C: [5'-phosphate-^{32}P] ADP-ribose.

The enzyme leading to the modification of p40 does not use NADP directly. The substrate is free phospho ADP-ribose produced by a glycohydrolase present in the microsomal preparation. Addition of isonicotinate hydrazide, an inhibitor of NADP glycohydrolase, prevented incorporation from NADP while incorporation from phospho ADP-ribose was hardly affected (Table 1). Also, reduction to NADPH, which is not a substrate for the glycohydrolase, eliminated incorporation into the acceptor polypeptide. Non-enzymatic attachment of labeled phospho ADP-ribose (6) was excluded by the inefficiency of free ADP-ribose to serve as a precursor (Fig. 1, inset), and by denaturation experiments (Table 1): Thermic treatment as well as short exposure to alkali or acid at 0° completely eliminated incorporation (8).

In rat liver, most of p40 and the modifying transferase are firmly associated with the microsomal fraction, where they resisted extraction of proteins with Triton X114. In contrast, a similar system in EAT cells was localized in the cytosolic compartment. Reversibility of p40 modification can be demonstrated in both compartments: In EAT cytosol, modification proceeds to a maximal value between 2 and 3.5 hr, and then gradually declines. At 24 hr, no modifying group was left. In Triton-extracted microsomes, the highest value was reached later, and label persisted much longer. Nucleotides with a phosphate in the 2'-position act as inhibitors. The most potent representatives are 2'-phospho AMP and coenzyme A (Fig. 2). Furthermore, divalent cations like Ca^{2+} or Mn^{2+} as well as phosphate or pyrophosphate ions can block the reaction. Partial purification of the enzyme was achieved by affinity chromatography on immobilized 2'-phospho AMP.

Table 1. Different sensitivities towards various treatments of phospho adenylyl transfer from NADP versus P-ADP-ribose.

Treatment	P-AMP Transfer, cpm Incorporation Relative to Controls	
	[^{3}H] NADP	[^{3}H]P-ADP-ribose
None	100	100
Isonicotinic acid hydrazide, 20 mM	11	80
56°C, 10 min	21	70
95°C, 5 min	0.1	0.1
NaOH at 100 mM, 0°C, 5 min	5	1

Triton X114-extracted microsomes (600 µg of protein) were treated as indicated followed by incubation (2 hr, 37°; ± isonicotinic acid hydrozide) in the presence of 100 µM labeled precursor (about 2 x 10^6 cpm). Incorporation into p40 was analyzed. Taken from (8).

The molecular weight of p40 is similar to that of the α-subunit of G$_i$ protein. We therefore considered a possible role of p40 in connection with G proteins. Such a function would fit Martin Rodbell's concept of programmable messengers (9), in which the α-subunits become subject to regulation by covalent modification. If one adds the notion that the two toxins, pertussis toxin and cholera toxin misuse a normal pathway when they ADP-ribosylate Gα subunits, it might well be that they do so by using the wrong substrate. However, it could be shown that plasma membranes from rat liver or rat brain in the presence of NADP did not modify Giα or one of the other known Gα subunits. Instead, we observed under these special conditions (inclusion of nucleotides and use of 10 to 100 fold higher specific radioactivities) that the liver membranes alone catalyzed the modification of multiple polypeptides, independent of the addition of pertussis or cholera toxin. This reaction again appeared to be specific for NADP since with NAD only traces of labeled polypeptides were found.

The modification of multiple acceptors in the plasma membrane fraction of rat liver with NADP as a substrate prompted a study of other compartments. The analysis was run with [^{32}P]NADP and [^{32}P]phospho ADP-ribose in parallel in order to differentiate between transfer of phospho

ADP-ribose from NADP and transfer of phospho AMP from free phospho ADP-ribose. In homogenates, modification of multiple acceptors proceeded equally well from phospho ADP-ribose and from NADP (Fig. 3). To exclude interference by NAD dependent reactions, the label in both substrates was at the 2'-phosphate group. Unfortunately, the amount and specific radioactivity of phospho ADP-ribose was only two-thirds of that of NADP. These differences must be taken into consideration when the labeling intensities of modified polypeptides are compared. As shown in Fig. 3, modification of polypeptides proceeded from both substrates to a similar extent. Most of the polypeptides labeled in the homogenate were also found in a crude nuclear fraction containing the plasma membranes. Purified nuclei on the other hand were practically devoid of these transfer reactions. Other compartments like mitochondria and microsomes were also able to modify endogenous polypeptides with NADP or phospho ADP-ribose as a substrate. In liver cytosol, the most prominent acceptor was a 90 to 95 kDa polypeptide. In this compartment, however, phospho ADP-ribose was a significantly better precursor than NADP, indicating that conversion of NADP to phospho ADP-ribose was the rate-limiting step in the NADP-dependent reaction. The different compartment patterns of acceptors are not the consequence of proteolysis. Peptide mapping of four major acceptors indicated unrelated structures.

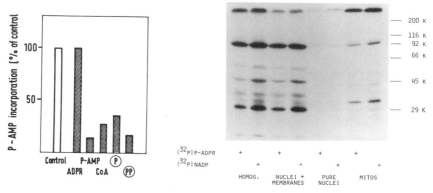

Fig. 2. (left) Influence of various compounds on phospho adenylylation of p40. Triton X114-extracted rat liver microsomes were incubated under standard conditions (8); [³H]P-ADP-ribose at 100 μM concentration in the presence of the compounds listed (ADP-ribose: 1 mM; P-AMP: 3 mM; CoA: 1 mM; Na-phosphate: 10 mM; Na-pyrophosphate: 10 mM). Data from (8).

Fig. 3. (right) Modification of polypeptides in subcellular compartments of rat liver by NADP or phospho ADP-ribose. Fractions (8-20 μg protein) were incubated with 10 μM [2'-phosphate-³²P] NADP; 3550 cpm/pmol; or 6.5 μM phospho [2'-phosphate-³²P]ADP-ribose; 3550 cpm/pmol) for 60 min at 30°C and processed by a similar procedure as that described in (8).

The fact that free phospho ADP-ribose can serve as a substrate for the modification of these polypeptides clearly excludes phospho ADP-ribosylation as the mode of reaction. Phospho ADP-ribose with its single energy-rich bond only allows the transfer of the activated phospho AMP residue. Additional evidence for phospho adenylylation of multiple acceptors came from experiments with the inhibitor of NAD(P) glycohydrolase, isonicotinic acid hydrazide. Similar to the p40 system, isonicotinic acid hydrazide prevented incorporation from NADP, but not from phospho ADP-ribose. Sensitivity to heat was higher than that of p40 modification: Incubation at 50°C for 10 min eliminated practically all transfer activity, except a small residual activity modifying high molecular weight material.

Phospho adenylylation appears to be a prevailing type of pyridine nucleotide dependent modification present in all major compartments and involving the modification of specific acceptors. It also surpasses ADP-ribosylation, by far, at least under the conditions applied.

For phospho adenylylation to proceed *in vivo*, an adequate supply of the substrate phospho ADP-ribose must be guaranteed. No exact data as to the intracellular concentration of phospho ADP-ribose are available. Preliminary data indicate rather low levels (< 50 µM) in rat liver. Since the K_m value for the transferase modifying p40 is about 200 µM, it appears that the activity of NADP glycohydrolase may be a rate-limiting factor, as indeed seen in rat liver cytosol when NADP serves as a precursor. On the other hand, abundant glycohydrolase activity was present in all other compartments. Whether phospho adenylyl transferase forms a complex with NADP glycohydrolase, to allow direct delivery of phospho ADP-ribose from glycohydrolase to transferase, remains to be determined. Steady state concentrations of phospho ADP-ribose may also depend on other enzymes of NADP metabolism. NAD kinase as well as NADP-dependent oxidoreductases govern the actual concentration of intracellular NADP, and these enzymes could contribute as well to the regulation of phospho ADP-ribose transfer.

Several factors have had a retarding influence on the development of enzymology in this area. This includes separation of incubated samples on SDS gels and determination of radioactivity in many gel slices as well as considerable expense required for the synthesis of substrates of sufficiently high specific radioactivity. It is the enthusiasm of my coworkers, Werner Fanick, Karin Klapproth and Helga Ehmcke, which prevented us from dropping the subject long before we finally found solid ground and a novel type of covalent modification.

In conclusion, it can be said that NAD as a group transferring coenzyme is primarily used as an ADP-ribose leading to mono- and poly(ADP-ribosyl) proteins. Only in one case (DNA ligase) does it serves as an active AMP. In contrast, NADP appears to be hardly used *per se* as a group transferring coenzyme. However, our data suggest that the second energy-rich bond

30

present in the pyrophosphate group of NADP serves to modify multiple acceptors after the coenzyme has been converted to free phospho ADP-ribose.

Acknowledgements. This work was supported by the Deutsche Forschungsgemeinschaft.

References

1. Lipmann, F. (1941) Advances in Enzymology 1: 99-162
2. Olivera, B.M., Lehman, I.R. (1967) Proc Natl Acad Sci USA 57: 1700-1709
3. Chambon, P., Weill, J.D., Doly, J., Strosser, M.T., Mandel, P. (1966) Biochem Biophys Res Commun 25: 638-643
4. Nishizuka, Y., Ueda, K., Nakazawa, K., Hayaishi, O. (1967) J Biol Chem 242, 3164-3171
5. Sugimura, T., Fugimura, S., Hasegawa, S., Kawamura, Y. (1967) Biochim Biophys Acta 138: 438-441
6. Hilz, H., Koch, R., Fanick, W., Klapproth, K., Ademietz, P. (1984) Proc Natl Acad Sci USA 81: 3929-3933
7. Moss, J., Stanley, S.J., Watkins, P.A. (1980) J Biol Chem 255: 5838-5840
8. Hilz, H., Fanick, W., Klapproth, K. (1986) Proc Natl Acad Sci USA 83: 6267-6271
9. Rodbell, M. (1985) TIBS 10: 461-464

ADP-Ribosyl Transferase of Mitochondria and of Cytoplasmic Ribonucleoprotein Particles Containing Silent Messenger RNA

P. Mandel, A. Masmoudi, C. Chypre, and H. Thomassin

Centre de Neurochimie du C.N.R.S., 67084 Strasbourg Cedex, France

Introduction

Poly(ADP-ribose) polymerase transferase was initially investigated in the nuclei of eucaryotic cells (1-4). Considering the developing but still limited understanding regarding the presence and the function of ADP-ribosylation reactions in extra nuclear compartments (5-7), we became interested in ADP-ribosylation of two sub-cellular entities: mitochondria, and free ribonucleoprotein particles: mRNP particles; the latter carry silent messenger RNAs (mRNAs).

The mitochondria are self-contained biochemical entities. The discovery of their own genetic system reinforced the notion of mitochondrial autonomy. Growth and division of pre-existing mitochondria is a likely mechanism of formation of these particles. We know, however, very little about the regulatory systems controlling DNA duplication, mRNA processing and translation within the mitochondria (8). Does ADP-ribosylation play a role in these processes in mitochondria, as this enzyme does in the nucleus? Characterization of such a control might be a step to a better understanding of mitochondrial biogenesis and functions. Transfer of ADP-ribose from NAD to an acceptor protein in mitochondria was first reported by Kun *et al.* (9) and by Burzio *et al.* (10). Apparently the crude mitochondrial ADP-ribosyl transferase synthesizes only a short chain, producing mono and oligo ADP-ribose protein adducts. The reaction appeared to be insensitive to inhibition by thymidine or stimulation due to added DNA (10, 11). Molecular mass of the ADP-ribosylated protein was between 90 and 110 kDa. A significant portion of ADP-ribosylated proteins appeared intimately associated to the inner mitochondrial membrane. Considering the high activity of a specific inner membrane associated NAD-glycohydrolase acting as a transglycosydase, Kun and Kirsten concluded that this enzyme might be responsible for ADP-ribosylation in the hydrophobic membrane environment of the mitochondria (11).

Recent reports question the assumption that mitochondrial ADP-ribosylation is a true ADP-ribosyl transferase catalyzed enzymic reaction. This is mainly because mono-ADP-ribose produced by NAD-glycohydrolase can be bound to mitochondrial protein acceptors nonenzymatically (12).

According to Richter *et al.* in intact mitochondria at least three classes of proteins are ADP-ribosylated *in vivo*. In isolated inner mitochondrial membrane mono(ADP-ribosyl)ation of a protein apparently occurs through NAD glycohydrolase and subsequent binding of free ADP-ribose (13). Richter *et al.* also suggested that NAD(P) stimulated mono(ADP-ribosyl)ation is involved in calcium release from liver mitochondria (14).

Cytoplasmic poly(ADP-ribosyl)ation has been found in several cell types. DNA dependent poly(ADP-ribose) polymerase activity associated with free ribosomes and polysomes in HeLa cells has been described by Roberts *et al.* (18). Burzio *et al.* (19) reported that poly(ADP-ribose) activity was stimulated by DNA 3 to 4 times and was associated with the microsome, ribosome fraction of rat, mouse, carp and bull testis. It was suggested that these ADP-ribosyl transferase activities may correspond to newly synthesized DNA dependent nuclear enzyme later transferred to the nucleus (18). We have focused our investigations on ADP-ribosyl transferase of free ribonucleoprotein particles (mRNP) which contain silent messenger RNA (mRNA). Spirin *et al.* (20, 21) provided the first demonstration that mRNA is capable of existing in cytoplasm apart from polyribosomes, in the form of nucleoprotein particles (mRNP). These mRNP storage units of mRNA, not immediately translated in the usual translation systems, was termed by Spirin as informosomes (20). These particles sediment more slowly as compared to sedimentation of ribosomes. Their buoyant density in CsCl gradient after fixation with formaldehyde, was lower than that of ribosomal subunits, 1.42 to 1.45 versus 1.52 g/ml, and became an important criterion for their isolation. Numerous observations of mRNPs in a wide variety of organisms have been reported (22, 23). In order to avoid protein denaturation by formaldehyde used in CsCl density gradient we developed a method for isolation of mRNP in a sucrose D_2O gradient based on the principle of density equilibrium sedimentation. By this method we could separate native cytoplasmic informosomes from mouse plasmacytoma cells (24).

Results and Discussion

ADP-ribosylation in mitochondria. We investigated ADP-ribosyl transferase and NAD-glycohydrolase activities in rat liver mitochondria and mitoplasts (15) as well as in rat brain synaptic and non-synaptic mitochondria (16). The time course of ADP-ribosyl transferase activity in synaptic and non-synaptic rat brain mitochondria was similar to that reported in rat liver mitochondria. A linear relationship between the concentration of mitochondrial proteins used and enzyme activity could be demonstrated. The velocity of the reaction reached a plateau at approximately 400 μg of protein. On protein basis the total ADP-ribosyl transferase activity in rat liver and rat brain synaptic and non-synaptic mitochondria was quite similar: 23.7 ± 1.8; 22.7 ± 2.2; 19.7 ± 1.5 pmol/mg protein at 37°C, 10 min, respectively.

A striking difference between the activities of ADP-ribosyl transferase and NAD-glycohydrolase (NADase) was observed when ADP-ribosyl transferase inhibitors nicotinamide and 3-aminobenzamide were used. In rat liver mitochondria NAD-glycohydrolase activity was no more detectable when 10 mM nicotinamide or 1 mM 3-aminobenzamide were used as inhibitor. Nevertheless, the major part of ADP-ribosyl transferase activity was maintained. Moreover, when lithium dodecyl sulfate polyacrylamide gel electrophoresis followed by autoradiography was performed despite the inhibition of NAD-glycohydrolase the transfer of ADP-ribose to proteins ranging in 50 to 55 kDa molecular weight still occurred irrespective of the presence of inhibitors (15).

Further support for the presence of ADP-ribosyl transferase activity in rat liver mitochondria came from experiments in which NAD-glycohydrolase was separated from solubilized submitochondrial proteins by hydroxylapatite and DEAE-cellulose chromatography (17). As could be shown after lithium dodecyl sulfate polyacrylamide gel electrophoresis and autoradiography the fractions enriched in ADP-ribosyl transferase activity catalyze the transfer of ADP-ribose from NAD to, at least five acceptor proteins. In contrast, the fractions enriched in NAD glycohydrolase activity could only produce a non-enzymatic ADP-ribosylation of one single protein of molecular weight of ~50 kDa (17).

Nicotinamide (10 mM) and 3-aminobenzamide (5 mM) were also much more efficient inhibitors of NAD-glycohydrolase than of ADP-ribosyl transferase activity in brain mitochondria although the effect of 3 aminobenzamide on NAD-glycohydrolase was less potent (Fig. 1). Nevertheless, while using nicotinamide at 10 mM concentration, NAD-glycohydrolase activity was no more detectable in synaptic mitochondria and was inhibited by 90% in non-synaptic mitochondria, while the inhibition of ADP-ribosyl transferase activity was only of about 30-40%.

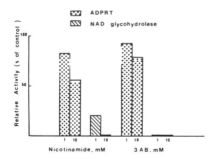

Fig. 1. Inhibition of rat liver mitochondria ADP-ribosyl transferase and NAD glyco-hydrolase activities by nicotinamide (1 and 10 mM) and 3-aminobenzamide (3AB) (1 and 10 mM). For methods see (15).

Incubation of synaptic and non synaptic mitochondria with [^{32}P]NAD, resulted in a labeling of several polypeptides while NAD glycohydrolase was strongly inhibited or no more detectable. After lithium dodecyl sulfate gel electrophoresis and autoradiography four major bands of molecular weight of 30 to 60 kDa were observed. The most intense radio-labeling was seen in one band with apparent molecular weight of 50 kDa. The pattern of radio-labeling seems not to differ in synaptic and non-synaptic mitochondria. The increase in the NAD concentration (from 10 μM to 100 μM) affected similarly ADP-ribosylation of synaptic or non-synaptic mitochondria; only the intensity of labeling increased, as expected. All of these data support the existence, in highly purified rat brain mitochondria, of endogenous enzymatic ADP-ribosyl transferase analogous to that of rat liver mitochondria.

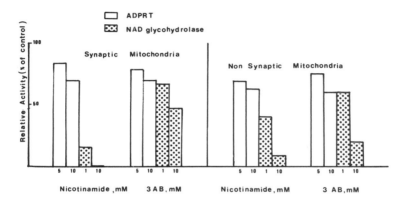

Fig. 2. Inhibition of rat brain mitochondria (synaptic and non synaptic) ADP-ribosyl transferase and NAD glycohydrolase activities by nicotinamide (5 and 10 mM) and 3-aminobenzamide (3AB) (1 and 10 mM).

ADP-Ribosylation in Cytoplasmic Ribonucleoprotein Particles Which Contain Silent mRNA. We demonstrated that ADP-ribosyl transferase activity is associated with cytoplasmic free mRNP isolated from a variety of organs and tumor cells: mouse plasmacytoma, rat liver, rat brain, Krebs II cell rat brain cultured neurons and astrocytes (25-28). Table 1 summarizes the mRNP poly(ADP-ribose) polymerase activity. On a protein basis, in contrast to the tumor cells, the activity associated with rat liver and whole rat brain free mRNP is very low. Nevertheless, it is 12 fold higher than the activity associated with the microsomal ribosomal fraction reported by Burizo *et al*. (19). In brain mRNP and mainly in neuronal mRNP the activity is much higher than in the rat liver mRNP on a DNA basis (not shown).

The maximum increase of ADP-ribosylation by plasmacytoma rat liver and brain mRNP when assayed with added DNA is about 20%. High concentrations of DNase hydrolyzing the small amount of DNA which might be present in the mRNP, provide an enzymatic activity close to the

35

Table 1. Poly(ADP-ribose) polymerase activity in free mRNP.

Tissue	nmol hr^{-1} mg^{-1} protein
Rat liver	2.2
Rat brain	1.3
Cultured chick embryo neurons	23.7
Cultured chick brain astrocytes	2.3
Plasmacytoma	36.0
Krebs II cells	23.4

For methods see (15, 25, 28).

control without DNase. This confirms that the mRNP ADP-ribosyl transferase is DNA independent (25-28). The ADP-ribosyl transferase activity estimated as nmole/hr/mg DNA present in the sample is much higher in free mRNP than in the nuclei differing by a factor of 75 and 12 for mouse plasmacytoma and rat liver, respectively. In brain mRNP the factor is even higher (28) since the DNA content was not measurable in these particles even employing the highly sensitive Labarca method (29). Only 4% in mouse plasmacytoma, 0.5% in rat liver, less than 0.005% in rat brain of the total cellular DNA was found in the free mRNP while the percentage of the overall ADP-ribosyl transferase activity present in these particles were 34%, 6% and 25% respectively (26-28).

At 2.5 mmolar concentration, thymidine, nicotinamide and 3-aminobenzamide were efficient inhibitors of the enzymatic activity, the apparent K_i with these inhibitors were 55 μM, 139 μM and 23 μM respectively in the case of plasmacytoma. When free mRNPs were incubated with snake venom phosphodiesterase or poly(ADP-ribose) glycohydrolase, but not with RNAses, a release of the labelled ADP-ribose in the incubation medium occurred (25-28).

When brain or plasmacytoma mRNP particles were incubated with pancreatic RNAse or T1 RNAse the ADP-ribosyl transferase activity was increased at least by a factor or 2. Such an activation was not observed for nuclear ADP-ribosyl transferase or with a purified calf thymus nuclear enzyme supplemented with or without RNA. The chain length of poly(ADP-ribose) in mRNP proteins also increased after RNAse treatment from 3.9 to 4.7 in plasmacytoma (26-28).

ADP-ribosylated proteins of plasmacytoma mRNP were examined by lithium dodecyl sulfate polyacrylamide slab gels and autoradiography. At least 5 main labeled bands of 116 kDa, 45 kDa, 38 kDa and 30 kDa were observed. The labeling increases with the increase of NAD concentration in the incubation medium parallel to a decrease of electrophoretic mobility of some bands. This may be attributed to a more potent ADP-ribosylation of these proteins as it has been observed with histones ADP-ribosylated by a

36

purified nuclear calf thymus poly(ADP-ribose) polymerase. When brain mRNP particles were used, 5 main ADP-ribosylated bands were observed: 120 kDa, 78 kDa,59 kDa, 43 kDa, and 37 kDa (28).

An aggregate was observed on the top of the autoradiogram of brain as well as of plasmacytoma mRNP lithium dodecyl sulfate polyacrylamide slab gels (26-28). It is probably due to the auto poly(ADP-ribosyl)ation of the enzyme as it was described for the nuclear ADP-ribosyl transferase. Using the protein blotting technique we have observed that the nuclear calf thymus poly(ADP-ribose) polymerase anti-serum reacts with the 116,000 molecular weight protein associated with the mRNP particles (26). These results strongly suggest some similarities between the nuclear and the mRNP poly(ADP-ribose) polymerase, although several differences could be demonstrated (DNA independence, RNAse effect). Finally, poly(ADP-ribose) glycohydrolase activity has been detected in free mRNP particles (30). Thus, one can conclude that enzymatic activities for synthesis and degradation of poly(ADP-ribose) are present in the mRNP particles.

In view of the mitochondrial autonomy and the existence of cell division independent neobiogenesis of these organelles it seems likely that in some respect mitochondrial ADP-ribosylation plays a role similar to that of nuclear ADP-ribosylation. There seems to exist certain links between ADP-ribosylation and mitochondrial replication. Our preliminary experiments showing an increase of ADP-ribosyl transferase activity during mitochondrial neobiogenesis are in favor of an involvement of ADP-ribosylation in mitochondrial replication. The topographical association of a part (25%) of protein ADP-ribosylation and mitochondrial DNA polymerase activities does also suggest such a relationship (11).

Although non-enzymatic ADP-ribosylation via glycohydrolase may occur, it seems unlikely that a basically hydrolytic event produced by NADase would regulate mitochondrial biogenesis and functions.

The existence of an ADP-ribosyl transferase in the free mRNP of several tissues was demonstrated; the enzyme can utilize some proteins of these particles as acceptors for mono- and poly(ADP-ribosyl)ation. Looking for a possible role of the ADP-ribosylation in free mRNP, one might hypothesize that post translational modification of mRNP proteins might act as a regulatory mechanism for the derepression followed by expression of mRNA. The addition of negatively charged polymers to protein acceptors of mRNP may induce structural changes and remove the translation repression. In fact, it has been shown that the chromatin associated poly(ADP-ribose) polymerase lost its DNA binding affinity once ADP-ribosylated (31) and that an increase of DNA polymerase activity followed polysome relaxation by ADP-ribosylation (32).

References

1. Chambon, P., Weill, J.D., Doly, J., Strosser, M.T., Mandel, P. (1966) Biochem Biophys Res Commun 25: 638-643
2. Reeder, R.H., Ueda, K., Honjo, T., Nishizuka, Y., Hayaishi, O. (1967) J Biol Chem 242: 3172-3179
3. Hasegawa, S., Fujimara, S., Shimizu, Y., Sugimura, T. (1967) Biochim Biophys Acta 149: 369-376
4. Doly, J., Mandel, P. (1967) CR Acad Sci 264: 2687-2690
5. Mandel, P., Okasaki, H., Niedergang, C. (1982) Progress in Nucleic Acid Research and Molecular Biology, 27,: 1-52 Academic Press, London, New York
6. Hayaishi, O., Ueda, K. (1982) ADP-Ribosylation Reactions Biology and Medicine, Academic Press, New York, London
7. Ueda, K., Hayaishi, O. (1985) Ann Rev Biochem 54: 73-100
8. Yaffe, M., Schatz, G. (1984) TIBS 9: 179-181
9. Kun, E., Zimber, P.H., Chang, A.C.Y. (1975) Proc Natl Acad Sci USA 72: 1436-1440
10. Burizo, L.O., Saez, L., Cornejo, R. (1980) Biochem Biophys Res Commun 103: 369-375
11. Kun, E., Kirsten, E. (1982) ADP-Ribosylation Reactions Biology and Medicine, Ed. Hayaishi, Academic Press, pp 193-205
12. Hilz, H., Koch, R., Fanick, W., Klapproth, K., Adamietz, P. (1984) Proc Natl Acad Sci USA 81: 3929-3933
13. Richter, C., Frei, B. This volume
14. Richter, C., Whirterhalter, K.M., Baumhuter, K., Lotscher, H.R., Moser, B. (1983) Proc Natl Acad Sci USA 80: 3188-3192
15. Masmoudi, A., Mandel, P. (1987) Biochemistry 26: 1965-1959
16. Madmoudi, A., Mandel, P. (1988) J. Neurochem 51: 188-193
17. Masmoudi, A., Mandel, P. This volume
18. Roberts, J.H., Stark, P., Giri, C.P., Smulson, M. (1975) Arch Biochem Biophys 171: 305-315
19. Burzio, L.O., Concha, I.I., Figueroa, J., Concha, M. (1983) ADP-Ribosylation, DNA Repair and Cancer, Miwa, M., Hayaishi, O., Shall, S., Smulson, M., Sugimura, T. (eds), Jpn Sci Soc Press, Tokyo/VNU Science Press, Utrecht, p. 141
20. Spirin, A.S. (1969) Eur J Biochem 10: 20-35
21. Spirin, A.S., Ajtkhozhin, M.A. (1985) Trends Biochem Sci 10: 162-165
22. Dreyfuss, G. (1986) Ann Rev Cell Biol 2: 459-498
23. Imaizumi-Scherrer, M.T., Maundrell, K., Civelli, O., Scherrer, K. (1982) Developmental Biology 93: 126-138
24. Kempf, J., Egly, J.M., Stricker, C.H., Schmitt, M., Mandel, P. (1972) FEBS Lett 26: 130-134
25. Elkaim, R., Thomassin, H., Niedergang, C., Egly, J.M., Kempf, J., Mandel, P. (1983) Biochimie 65: 653-659.
26. Thomassin, H., Niedergang, C., Mandel, P. (1985) Biochem Biophys Res Commun 133: 654-661
27. Thomassin, H., Gilbert, L., Niedergang, C., Mandel, P. (1985) ADP-Ribosylation of Proteins Althaus, FR., Hilz, H., Shall, S. (eds) Springer-Verlag, Berlin, Heidelberg, pp.148-153
28. Chypre, C., Maniez, C., Mandel, P. (1988) J Neurochem 51: 561-565
29. Labarca, C., Paigen, K. (1980) Anal Biochem 102: 344-352
30. Thomassin, H., Mandel, P. This volume.
31. De Murica, G., Johgstra-Bilen, J., Ittel, M.E., Mandel, P., Delain, E. (1983) EBMO 32: 543-548
32. Niedergang, C.P., de Murica, G., Ittel, M.E., Pouyet, J., Mandel, P. (1985) Eur J Biochem 146: 185-191

In Vitro Evidence for Poly(ADP-Ribosyl)ation of DNA Polymerase α-Primase and Phosphorylation of Poly(ADP-Ribose) Synthetase by Protein Kinase C

Koichiro Yoshihara, Yasuharu Tanaka, Asako Itaya, Tomoya Kamiya, Takashi Hironaka[1], Takeyoshi Minaga[1], and Samuel S. Koide[2]

Department of Biochemistry, Nara Medical University, Kashihara, Nara 634, Japan

Introduction

In a previous study (1-3), we found that terminal deoxyribonucleotidyl transferase (TdT), DNA polymerase α, DNA polymerase β, and DNA ligase II were markedly inhibited when incubated in a reconstituted poly(ADP-ribosyl)ating enzyme system. We have reported also the direct evidence for poly(ADP-ribosyl)ation of TdT (2) and DNA polymerase β (3). Based on these results, we proposed that a role of poly(ADP-ribose) synthetase in DNA repair is to cause an emergency halt of chromatin function at the damaged site to protect cells from abnormal metabolism in the chromatin (4).

In this report, we show that the inhibition of DNA polymerase α may be due to a direct modification of either the enzyme molecule itself or some protein cofactors required for enzyme activity. Recent studies of DNA polymerase α have revealed that the polymerase associates with a tightly bound primase (5, 6) and that the primase was markedly stimulated by a specific primase-stimulating factor (7). We also found that various basic proteins markedly stimulate the DNA polymerizing activity of DNA polymerase α-primase complex. We describe here the effect of poly(ADP-ribosyl)ation on the activities of the enzymes and the stimulating factors separately.

In an effort to study the possibility of the covalent modification of poly(ADP-ribose) synthetase [EC 2.4.2.30] itself, we found that the enzyme is phosphorylated by protein kinase C *in vitro*. The observed phosphorylation of the synthetase seems to be fairly specific to C-kinase species since the synthetase was a rather poor substrate for other protein kinases including cAMP-dependent, cofactor-independent, and Ca²⁺/calmodulin-dependent protein kinases.

[1]Cutter, Japan, Ltd, 6-9-1, Minatojima Nakamachi, Chuo-ku, Kobe

[2]The Population Council, The Rockefeller University, 1230 York Ave., New York, N.Y. 10021 U.S.A.

Results and Discussion

Effect of poly(ADP-ribosyl)ation on DNA-replicating enzyme system: Primase stimulating factor. When bovine thymus DNA polymerase α was purified by successive chromatography on phosphocellulose and Q-Sepharose columns, the enzymatic activity was separated into two major peaks upon the latter chromatography. One of the peaks showed so called DNA replicase activity (8) when assayed with single stranded poly (dT) as template and without the addition of primer, indicating that this DNA polymerase α fraction associates primase. After further purification of the enzyme by ssDNA-cellulose column chromatography, however, the DNA replicase activity of the DNA polymerase α-primase complex showed a marked decrease suggesting that a factor required for replicase activity may be separated from the enzyme complex. Supplementation of the inactive enzyme with various fractions eluted from ssDNA-cellulose revealed that a protein, which is tightly bound to ssDNA-cellulose and eluted at a relatively high salt concentration, possessed a high activity to restore the replicase activity of purified DNA polymerase α-primase. Thus, we purified the protein factor to near homogeneity. Details of the purification and properties of this factor (primase-stimulating factor; PSF) will appear elsewhere (7). Briefly, PSF showed a native molecular size of approximately 150 kDa as judged by Sephacryl S-200 column chromatography and was composed of two major polypeptides of 146 and 47 kDa judging from SDS-polyacrylamide gel electrophoresis. The factor did not affect the primer-dependent DNA polymerizing activity, which was assayed with either activated DNA or poly (dT)·oligo (rA) as template-primer.

PSF specifically stimulated the primase activity and, thus, the DNA replicase activity of purified DNA polymerase α-primase. As shown in Fig. 1, the primer synthesis by this enzyme was completely dependent on template, required appropriate concentration of deoxyribonucleoside triphosphate and was markedly stimulated by a very low concentration of PSF (more than 20-fold stimulation was observed at 10 ng/50 μl of PSF). The factor was also effective in stimulating the replicase activity supported by single stranded calf thymus or ØX 174 DNA as template, indicating that primer synthesis in these reactions was also stimulated. However, PSF is not primase itself since the factor neither shows any primer synthesis by itself nor does it stimulate the activity of *E. coli* DNA polymerase I with poly (dT) as template. A factor with properties similar to PSF has been found in mouse tissue by Yagura, Kozu, and Seno (8) but was not detected in other vertebrates tested by them (9). However, the molecular component of their factor was significantly smaller (63 kDa) than the one described here (8).

The effect of poly(ADP-ribosyl)ation on PSF was examined by monitoring the replicase stimulating activity. Purified PSF and DNA polymerase α-primase were separately incubated in a reconstituted poly(ADP-ribosyl)ating system and, after incubation, they were recombined with the untreated counterpart and replicase activity was examined. As shown in Fig. 2, incubation of PSF did not affect appreciably its replicase stimulating activity, while the treatment of DNA polymerase α-primase with increasing concentrations of NAD$^+$ markedly inhibited its replicase activity.

DNA polymerase α stimulating activity of basic proteins and poly(ADP-ribosyl)ation. When a highly purified DNA polymerase α was

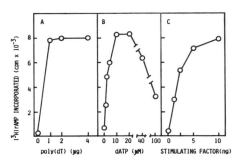 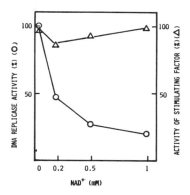

Fig. 1. (left) Requirements for primase activity. Complete assay mixture for primase assay contained 50mM Tris-HCl, pH 7.4, 10 mM MgCl$_2$, 1 mM DTT, 50 μg of BSA, 1 μg of poly (dT), 20 μM dATP, 100 μM [2,8-^{3}H]ATP (120 cpm/pmol), 10 ng of purified PSF and an appropriate amount of DNA polymerase α-primase in a total volume of 50 μl. The sample was incubated at 37°C for 60 min and the radioactivity of [^{3}H]AMP incorporated into acid-insoluble material was counted. The concentration of some components was varied as indicated.

Fig. 2. (right) Effect of poly(ADP-ribosyl)ation on replicase-stimulating activity of PSF and replicase activity of DNA polymerase α-primase. Purified PSF and DNA polymerase α-primase were separately incubated in poly(ADP-ribosyl)ating enzyme system (1) in the presence of the indicated concentration of NAD$^+$. Replicase assay was carried out principally as described by Yagura et al. (8) in the presence of 10 ng of purified PSF.

assayed with activated DNA as template-primer, the activity was stimulated approximately 10 to 15-fold by the addition of various basic proteins including histones, protamine, basic homopolyamino acids at an optimum protein/DNA ratio (w/w) of approximately 0.2 (10). The effect is considered to be through the blocking of non-productive DNA polymerase α-binding to a long single-stranded stretch of activated DNA just as is observed in eukaryotic single stranded DNA-specific binding protein (11). As shown in Fig. 3, histone H1 also markedly stimulated the reaction rate of DNA polymerase α. Since it is well known that histone H1 is a good acceptor of poly(ADP-ribose), we examined whether poly(ADP-ribosyl)ation of histone

Table 1. Effect of unmodified and poly(ADP-ribosyl)ated histone H1 on DNA polymerase α activity

Histone H1 added	Activity of DNA polymerase α (cpm/30 min)
none	428
unmodified histone H1 (0.4 µg)	6,233
ADP-ribosylated H1 (0.4 µg)	6,390
(10.3 ADP-ribose unit/molecule)	

H1 affected the DNA polymerase α-stimulating activity. As shown in Table 1, both unmodified and poly(ADP-ribosyl)ated histone H1 (carrying 10.3 ADP-ribose/molecule on an average) showed a similar extent of stimulation indicating that the amount of poly(ADP-ribose) bound to histone H1 does not affect the DNA polymerase α-stimulating ability.

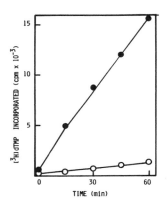 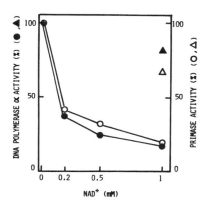

Fig. 3. (left) Stimulation of purified DNA polymerase α activity by histone H1. Assay for DNA polymerase α activity was carried out with activated DNA as template-primer as described (10) in the presence (●) and absence (○) of 4 µg/0.2 ml of histone H1.

Fig. 4. (right) Inhibition of DNA polymerase α and primase activities by poly(ADP-ribosyl)ation reaction. DNA polymerase α-primase was incubated in poly(ADP-ribosyl)ating enzyme system in the presence of the indicated concentration of NAD⁺ and the DNA polymerase α (●, ▲) and primase (○, △) activities were assayed as described in the legends of Fig. 3 and Fig. 1, respectively. ▲, △; the sample was incubated in the presence of 20 mM nicotinamide and 1 mM NAD⁺ in poly(ADP-ribosyl)ating reaction mixture.

Inhibition of primase and DNA polymerase α activities. Purified DNA polymerase α-primase complex was incubated in a reconstituted poly(ADP-ribosyl)ating enzyme system and, after the incubation, the DNA polymerizing activity with activated DNA as template-primer and the primer synthesizing activity with poly (dT) as template were separately assayed. Both activities decreased almost in a parallel manner with increasing concentration of NAD$^+$ and reached to approximately 20% of control at 1 mM NAD$^+$ (Fig. 4). An inhibition of poly(ADP-ribose) polymerase (20 mM nicotinamide) blocked the suppression of both activities.

All of these results indicate that the observed inhibition of DNA polymerase α by poly(ADP-ribosyl)ation is not due to the suppression of factors modulating the enzymatic activity but through the direct inhibition of DNA polymerase α-primase complex itself. Considering our previous observation (1) that the inhibited activity of DNA polymerase α by the incubation in a poly(ADP-ribosyl)ating system could be restored by a mild alkaline treatment, a procedure known to hydrolyze the ADP-ribosyl-protein linkage (12), the observed inhibition is probably due to the modification of either DNA polymerase α and/or primase subunit itself although direct evidence for covalent association of polymer to the enzyme subunit(s) remains to be proven.

Phosphorylation of poly(ADP-ribose) synthetase by protein kinase C. As shown in Fig. 5, the addition of purified poly(ADP-ribose) synthetase to a reaction mixture for protein kinase C in place of phosphate-accepting substrate markedly stimulated the incorporation of ^{32}P from [γ-^{32}P]ATP into acid-insoluble material in a dose-dependent manner.

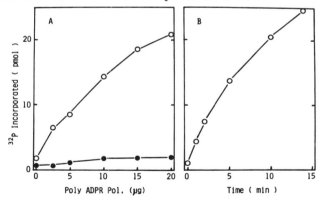

Fig. 5. Phosphorylation of poly(ADP-ribose) synthetase by purified rat brain protein kinase C. Enzyme reaction was performed principally as described by Kikkawa et al. (20) with 1-oleoyl-2-acetyl-rac-glycerol (OAG) and phosphatidyl serine as activator. A: The reaction was carried out with (O) or without (●) C-kinase activator (Ca^{2+}/OAG/phosphatidyl-serine); the indicated concentration of poly(ADP-ribose) synthetase was added to the reaction mixture in place of phosphate-accepting protein. B: Time course of the reaction.

Analysis of the reaction products by SDS polyacrylamide gel electrophoresis followed by autoradiography clearly demonstrated a radioactive band of product coincident with the protein band of poly(ADP-ribose) synthetase itself. Further, the synthesis of the product was dependent on both Ca^{2+} and diacylglycerol/phospholipid (Fig. 6). Two C-kinases from different sources, a purified rat brain enzyme and a crude frog liver enzyme, showed a quite similar mode of phosphorylation except that Ca^{2+}-dependency seemed to be more strict in the reaction of the frog liver enzyme. The observed phosphorylation of poly(ADP-ribose) synthetase seems to be specific to C-kinase species since the synthetase did not function as a substrate for cAMP-dependent and cofactor-independent protein kinases, although it may serve as a relatively poor substrate for Ca^{2+}/calmodulin-dependent kinase (Table 2).

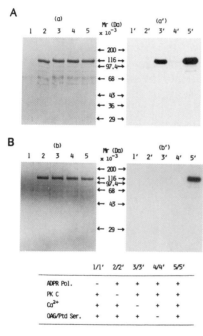

	1/1'	2/2'	3/3'	4/4'	5/5'
ADPR Pol.	−	+	+	+	+
PK C	+	−	+	+	+
Ca^{2+}	+	+	−	+	+
OAG/Ptd Ser.	+	+	+	−	+

Fig. 6. SDS-polyacrylamide gel electrophoresis-autoradiography of phosphorylated poly(ADP-ribose) synthetase. Poly(ADP-ribose) synthetase was phosphorylated by rat brain (A) and frog liver (B) C-kinases in the presence and absence of various components as indicated in the Table 2. The reaction products were analyzed by SDS-polyacrylamide gel electrophoresis-autoradiography. a, b; protein staining, a', b'; autoradiogram. These data are reprinted with permission from Biochem. Biophys. Res. Commun. 148, 709-711, 1987.

Since protein kinase C is localized mainly in the cytosol and translocated to the membrane when cells are stimulated (13), membrane-bound or cytosol proteins are considered to be the substrate for the enzyme. Recent studies, however, have revealed that some nuclear enzymes such as DNA methyltransferase (14), DNA topoisomerase II (15), and DNA polymerase α (16) serve as the target of the kinase *in vitro* and treatment of cells with carcinogenic phorbol ester enhances *in vivo* phosphorylation of histones (17). Furthermore, it is also reported that the treatment of cells with the phorbol ester causes proteolysis of C-kinase and redistribution of the cleaved enzyme (M-kinase) whose activity is independent of Ca^{2+} and phospholipid (18, 19). Thus, in spite of poor localization of C-kinase itself in nuclei, the occurrence of phosphorylation of poly(ADP-ribose) synthetase by C-kinase species *in vivo* and its biological significance may remain as inquiries worth studying further.

Table 2. Specific phosphorylation of poly(ADP-ribose) polymerase by protein kinase C.

Substrates	enzyme activity (^{32}P incorporated; pmol)[1]			
	C-kinase[2]	A-kinase	CaM-kinase	Cofactor-independent kinases [3]
poly(ADP-ribose) polymerase	23	1>	4	1>
histone	66	69	–	32
synapsin I	–	–	62	–
casein	–	–	–	29
phosvitin	–	–	–	20

[1]Kinase assays were carried out with 20 µg of acceptor proteins as indicated in the presence of various cofactors required for the respective kinase.

[2]C-Kinase; protein kinase C, A-kinase; cAMP-dependent protein kinase, CaM-kinase; Ca^{2+}/calmodulin-dependent protein kinase.

[3]The enzyme preparation is a mixture of histone-, phosphivtin-, and casein (type II)-kinases.

Acknowledgements: We thank Drs. Ushio Kikkawa and Yasutomi Nishizuka, Department of Biochemistry, Kobe University, for donation of purified rat brain C-kinase and useful discussion, and Miss Iyuko Matsuda for typing the manuscript.

References

1. Yoshihara, K., Itaya, A., Tanaka, Y., Ohashi, Y., Ito, K., Teraoka, H., Tsukada, K., Matsukage, A., Kamiya, T. (1985) Biochem Biophys Res Commun 128: 61-67
2. Tanaka, Y., Ito, K., Yoshihara, K., Kamiya, T. (1986) Eur J Biochem 155: 19-25
3. Ohashi, Y., Itaya, A., Tanaka, Y., Yoshihara, K., Kamiya, T., Matsukage, A. (1986) Biochem Biophys Res Commun 140: 666-673
4. Yoshihara, K., Itaya, A., Tanaka, Y., Ohashi, Y., Ito, K., Teraoka, H., Tsukada, K., Matsukage, A., Kamiya, T. (1985) ADP-Ribosylation of Proteins. Althaus, F.R., Hilz, H., Shall, S. (eds). Springer-Verlag Berlin Heidelberg pp 82-92
5. Suzuki, R., Masaki, S., Koiwai, O., Yoshida, S. (1986) J Biochem Tokyo 99: 1673-1679
6. Holmes, A.M., Cheriathundam, E., Bollum, F.J., Chang, L.M.S. (1986) J Biol Chem 261: 11924-11930
7. Itaya, A., Hironaka, T., Yoshihara, K., Kamiya,. T., (1988) Eur J Biochem 174: 261-266
8. Yagura, T., Kozu, T., Seno, T. (1982) J Biol Chem 257: 11121-11127

9. Yagura, T., Kozu, T., Seno, T., Saneyoshi, M., Hiraga, S., Nagano, H. (1983) J Biol Chem 258: 13070-13075
10. Hironaka, T., Itaya, A., Yoshihara, K., Minaga, T., Kamiya, T. (1987) Anal. Biochem 166: 361-367
11. Sapp, M., König, H., Riedel, H.D., Richter, A., Knippers, R. (1985) J Biol Chem 260: 1550-1556
12. Tanaka, Y., Yoshihara, K., Itaya, A., Kamiya, T., Koide, S.S. (1984) J Biol Chem 259: 6579-6585
13. Hirota, K. Hirota, A., Aguilera, G., Catt, K.J. (1985) J Biol Chem 260: 3243-3246
14. DePaoli-Roach, A., Roach, P.J., Zucker, K.E., Smith, S.S. (1986) FEBS Lett 197: 149-153
15. Sahyoun, N., Wolf, M., Besterman, J., Hsieh, T-S., Sander, M., Levine III, H., Chang K-J., Cuatrescasas, P. (1986) Proc Natl. Acad Sci USA 83: 1603-1607
16. Donaldson, R.W., Gerner, E.W. (1987) Proc Natl Acad Sci USA 84: 759-763
17. Patskan, G.J., Baxter, C.S. (1985) J Biol Chem 260: 12899-12903
18. Kishimoto, A., Najikawa, N., Shiota, M., Nishizuka, Y. (1985) J Biol Chem 258: 1156-1164
19. Guy, G.R., Gordon, J., Walker, L. Michell, R.H., Brown, G. (1986) Biochem Biophys Res Commun 135: 146-153
20. Kikkawa, U., Takai, Y., Minakuchi, R., Inohara, S., Nishizuka, Y. (1982) J Biol Chem 257: 13341-13348

Poly(ADP-Ribose) Glycohydrolase and ADP-Ribosyl Group Turnover

Kazuyuki Hatakeyama, Yasuo Nemoto[1], Kunihiro Ueda[1], and Osamu Hayaishi[2]

Department of Medical Chemistry, Osaka Medical College, Takatsuki, Osaka 569, Japan

Introduction

Poly(ADP-ribosyl)ation constitutes a novel type of posttranslational covalent modification of nuclear proteins. The synthesis of poly(ADP-ribose) is catalyzed by poly(ADP-ribose) synthetase (1). Recently, poly(ADP-ribose) groups on proteins were found to turn over rapidly *in vivo* (2, 3). The degradation of poly(ADP-ribose) is carried out by two different types of enzymes. ADP-ribosyl protein lyase cleaves the bond between mono(ADP-ribose) and protein (4). Another enzyme is poly(ADP-ribose) glycohydrolase, which is known to split the ribose-ribose glycosidic bond (5, 6). The glycohydrolase is the major enzyme responsible for polymer degradation *in vivo* (7). Therefore, the knowledge of its precise mode of action on poly(ADP-ribose) is important for elucidating not only the metabolism of poly(ADP-ribose) but also the biological functions of poly(ADP-ribosyl)ation. In the present study, we purified this glycohydrolase, and investigated its mode of action precisely.

Results and Discussion

Poly(ADP-ribose) glycohydrolase was purified to apparent homogeneity from a 0.4 M KCl extract of calf thymus approximately 70,000-fold with an overall yield of 3.2% through 11 steps (8). The enzyme in later steps of purification was very unstable. We overcame this problem by introducing Triton X-100 into the purification buffers. The molecular weight was estimated as 59,000 with SDS-polyacrylamide gel electrophoresis and 53,000 with Sephacryl S-200 gel filtration. Optimal pH for the reaction was around 7.0-7.5. The turnover number was 5,400 per min at 37°C. This turnover number is comparable to the value, 12,500, for a typical glycosidic hydrolase, β-galactosidase. The value, 5,400, is higher than the value of 200 determined for poly(ADP-ribose) synthetase of calf thymus (9) and 40 determined for ADP-ribosyl protein lyase of rat liver (4). The average

---

[1]Department of Clinical Science and Laboratory Medicine, Kyoto University Faculty of Medicine, Kyoto 606, Japan
[2]Osaka Medical College, Takatsuki, Osaka 569, Japan

number of glycohydrolase molecules per thymocyte was estimated to be about 2,000. This value is lower than the value of 50,000 determined for poly(ADP-ribose) synthetase of calf thymus (9) and 1,800,000 determined for ADP-ribosyl protein lyase of rat liver (4). Thus the intracellular content of glycohydrolase is relatively small, but the enzyme has a high molecular activity.

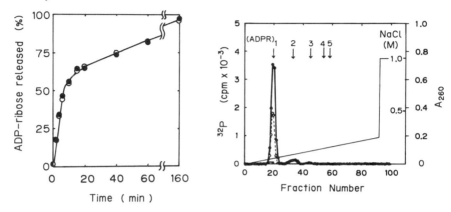

Fig. 1. (left) A release of monomeric ADP-ribose from free and poly(ADP-ribose) synthetase-bound poly(ADP-ribose) by the action of poly(ADP-ribose) glycohydrolase. Free poly(ADP-ribose) was prepared from synthetase-bound poly(ADP-ribose) with alkali treatment (Reprinted with permission from ref. 8).

Fig. 2. (right) Chain length analysis of poly(ADP-ribose) remaining bound to the synthetase after poly(ADP-ribose) glycohydrolase reaction. The standard reaction mixture contained 90 ng of glycohydrolase and 8.0 μM poly([^{32}P]ADP-ribose) (188 cpm/pmol) bound to the synthetase (1.45 ng). After incubation for 10 min, 20% Cl_3CCOOH-insoluble products were collected by centrifugation. After a release from protein with alkali treatment, the residual protein-bound (ADP-ribose) was subjected to chromatography on a DEAE-cellulose column. The elution positions of poly(ADP-ribose) (1-5 residues) were indicated by arrows (Reprinted with permission from ref. 8).

Degradation of poly(ADP-ribose) took place in a biphasic manner (Fig. 1). Initially, the reaction proceeded linearly at a rate of 85 μmol/min/mg of protein until 50% of the polymer was converted to ADP-ribose monomers, and then the reaction rate abruptly decreased to a rate of 4.0 μmol/min/mg. This marked change in the rate of reaction was observed at the same extent, 50%, of degradation, irrespective of the enzyme amount used or length of incubation. It was noticeable that the time course of degradation of the poly(ADP-ribose) synthetase-bound poly(ADP-ribose) was essentially the same as that of free poly(ADP-ribose). Subsequently, we examined the extent of the enzyme action on the synthetase-bound polymers by analyzing ADP-ribose remaining bound to the protein after extensive digestion. The ADP-ribose liberated from protein with alkali treatment was subjected to DEAE-cellulose column chromatography. Most of residual ADP-ribose was

ADP-ribose monomers (Fig. 2). This indicates that the glycohydrolase degraded the synthetase-bound poly(ADP-ribose) up to all but the last residue of ADP-ribose. Similar results were obtained with poly(ADP-ribose) bound to histone H1, histone H2B, or nonhistone proteins. Since the glycohydrolase degraded protein-bound poly(ADP-ribose) and produced exclusively ADP-ribose monomers, the enzyme was presumed to initiate hydrolysis at the adenosine terminus.

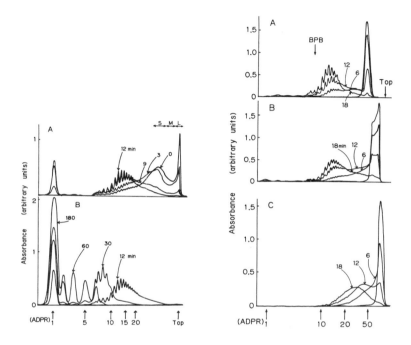

Fig. 3. (left) Polyacrylamide gel electrophoresis of degradation products of poly(ADP-ribose). The degradation products of poly([^{32}P]ADP-ribose) by poly(ADP-ribose) glycohydrolase were subjected to electrophoresis in a 20% polyacrylamide gel slab. The autoradiogram of the gel was scanned with a densitometer. A, the reaction products in a rapid phase (see Fig. 1). B, the reaction products in a slow phase. For L, M, and S, refer to legend to Fig. 4. The positions of poly(ADP-ribose) were shown by arrows (Reprinted with permission from ref. 8).

Fig. 4. (right) Polyacrylamide gel analysis of degradation products of ribose terminus-labeled poly(ADP-ribose) with different chain sizes. Three groups (referred to as L, M, and S, Fig. 3) of ribose terminus-labeled poly([^{32}P]ADP-ribose) were prepared by electroelution of polyacrylamide gel pieces from different areas. Enzymatic degradation products of these polymers were analyzed as described in the legend to Fig. 3. A, B, and C stand for polymers to S, M, and L, respectively (Reprinted with permission from ref. 8).

49

To determine the processivity of poly(ADP-ribose) glycohydrolase action, we analyzed the size distribution of degradation products of poly([^{32}P]ADP-ribose) at various time points using polyacrylamide gel electrophoresis. The densitometric profiles of autoradiograms are shown in Fig. 3. The original poly(ADP-ribose) preparation showed two peaks. With the progress of the reaction up to 12 min, both peaks became lower, the positions of the peaks being unchanged. This means that, while a portion of the polymers were digested, the remainder was left intact. These results indicate that the enzyme acted on polymers processively. The reaction rate of this early phase was 20 times higher than that of the following phase. During the processive digestion, small molecules of 8-20 ADP-ribose units accumulated in addition to the production of mono(ADP-ribose). This shows that, once an enzyme molecule acted on a polymer, the enzyme continued to degrade the same polymer processively up to a small size, and then transferred to another polymer. In the later phase (Fig. 3B), the position of the peaks of small polymers shifted gradually toward mono(ADP-ribose). This implies that the small polymers decreased in size evenly, and thus the glycohydrolase degraded small poly(ADP-ribose) nonprocessively, that is, the enzyme dissociated from a polymer molecule after every catalytic event.

Table 1. Kinetic parameters as a function of chain length.

Average Chain Length	K_m	V_{max}
	µM	µmol/min/mg protein
126	0.11	123
45	0.20	123
18	0.39	110
<10	~10	45

In order to further substantiate this view, we isolated polymers of defined sizes from the gel, as shown in Fig. 3A, and examined the action of enzyme against each of these polymers (Fig. 4). We employed ribose terminus-labeled polymers in this experiment, because the radioactivity of ribose terminus-labeled polymers was proportional to the number of polymers, irrespective of chain size. No radioactive mono(ADP-ribose) was initially produced, indicating that the enzyme did not attack the polymer from the ribose terminus. The profiles show that all these polymers were degraded processively by the enzyme. It is of interest that the size of accumulated small polymers in each class correlated closely with that of original polymers. The transition from a processive phase to a nonprocessive phase occurred at the same extent of degradation with these large polymers as with other sizes (data not shown). The foregoing view was further supported by the kinetic parameters of the enzyme for poly(ADP-ribose) of different sizes (Table 1). The K_m values for large polymers, for example, 0.11 μM for polymers of average chain size 126, were much lower than the K_m value, 10 μM, for small polymers. The average sizes of these large polymers were different, but their K_m values, were in the same order of magnitude. This may correspond to the fact that the polymers were different in total sizes, but similar in branch lengths (8).

Based on these results, we propose a model for the degradation of large poly(ADP-ribose) with varying sizes. The enzyme binds to the adenosine terminus of a branch of large poly(ADP-ribose) covering several ADP-ribose units from the terminus, trims the branch by liberating ADP-ribose monomers processively, dissociates from the trimmed branch, and moves to another branch. After several cycles of these events, a core structure consisting of short branches may accumulate, which is then degraded by the enzyme nonprocessively and slowly. Our view about degradation of poly(ADP-ribose) bound to protein is depicted in Fig. 5. Large poly(ADP-ribose) bound to protein is first degraded rapidly and processively to small polymer. The small polymer and other protein-bound small polymers are degraded slowly and nonprocessively. Finally, the bond between mono(ADP-ribose) and protein is split by ADP-ribosyl protein lyase (4).

MODE OF GLYCOHYDROLASE ACTION

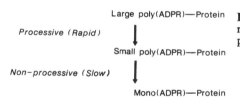

Fig. 5. A proposed model of poly(ADP-ribose) glycohydrolase action on protein-bound poly(ADP-ribose).

Similar results have been obtained in *in vivo* systems. A biphasic decay of poly(ADP-ribose) has been suggested in DNA-damaged cells (3,10,11). and the synthetase-bound polymers were reported to be degraded more preferentially than polymers bound to other acceptors *in vivo* (12). These *in vivo* observations could be explained by the different modes of action of poly(ADP-ribose) glycohydrolase on large and small poly(ADP-ribose), as revealed by our present study.

This inference provides some insight into functions of poly(ADP-ribose); the marked difference in the mode of degradation and in acceptors between large and small polymers may reflect the difference in their functions. The function of large poly(ADP-ribose) may be such as acting in a short time and terminating individually, while the function of small poly(ADP-ribose) may be such as acting in a long time and terminating concurrently.

Acknowledgements. This work was supported in part by grants-in-aid for Scientific Research and Cancer Research from the Ministry of Education, Science and Culture, Japan.

References

1. Ueda, K., Hayaishi, O. (1985) Ann Rev Biochem 46: 73-100
2. Juarez-Salinas, H., Sims, J.L., Jacobson, M.K. (1979) Nature 282: 740-741
3. Wielckens, K., George, E., Pless, T., Hilz, H. (1983) J Biol Chem 258: 4098-4104
4. Oka, J., Ueda, K., Hayaishi, O., Komura, H., Nakanishi, K. (1984) J Biol Chem 259: 986-995
5. Ueda, K. Oka, J., Narumiya, S., Miyakawa, N., Hayaishi, O. (1972) Biochem Biophys Res Commun 46: 516-523
6. Miwa, M., Tanaka, M., Matsushima, T., Sugimura, T. (1974) J Biol Chem 249: 3475-3482
7. Miwa, M., Nakatsugawa, K., Hara, K., Matsushima, T., Sugimura, T. (1975) Arch Biochem Biophys 167: 54-60
8. Hatakeyama, K., Nemoto, Y., Ueda, K., Hayaishi, O. (1986) J Biol Chem 261: 14902-14911
9. Ito, S., Shizuta, Y., Hayaishi, O. (1979) J Biol Chem 254: 3647-3651
10. Benjamin, R.C., Gill, D.M. (1980) J Biol Chem 255: 10493-10501
11. Wielchens, K., Schimidt, A., George, E., Bredehorst, R., Hilz, H. (1982) J Biol Chem 257: 12872-12877
12. Adametiz, P., Rudolph, A. (1984) J Biol Chem 259: 6841-6846

In Vitro ADP-Ribosylation Utilizing 2'Deoxy-NAD+ as a Substrate

Rafael Alvarez-Gonzalez, Joel Moss[1], Claude Niedergang[2], and Felix R. Althaus[3]

The Samuel Roberts Noble Foundation, Inc., Biomedical Division, Ardmore, Oklahoma 73402, USA

Introduction

The majority of the mono(ADP-ribosyl) transferases identified to date in animal tissues (1-5) are characterized by their specific modification of the guanidinium group of arginine residues. In contrast, poly(ADP-ribose) polymerase is known only to modify carboxylate groups on protein acceptors, i.e., glutamate (6-9), carboxy-terminal lysine (8) and aspartate (10). Both classes of enzymes have identical substrate stereospecificity in which the β-configuration of the anomeric carbon of NAD+ is converted to the α-configuration in the product (11, 12). No major differences between these two classes of enzymes in substrate structural requirements have been documented. The present study identifies a difference in behavior of an NAD+:arginine mono(ADP-ribosyl) transferase from turkey erythrocytes (2) and poly(ADP-ribose) polymerase from calf thymus (13) toward 2'dNAD+ as a substrate.

Results and Discussion

Synthesis and purification of 2'dNAD+. We used β-NMN+:adenyl transferase to synthesize [adenylate-^{32}P]2'dNAD+ from [α-^{32}p]2'dATP. Residual 2'dATP was removed by affinity chromatography on a boronate gel as described elsewhere (14). The elimination of contaminating β-NMN+ (Fig. 1A), which also bound to the boronate gel, was facilitated by its conversion to nicotinamide ribose by treatment with bacterial alkaline phosphatase (Fig. 1B). Figure 1C shows that following preparative HPLC, a single peak corresponding to pure 2'dNAD+ was detected. It is important to note that all of the ^{32}P radiolabel co-eluted with this peak.

---

[1] The Laboratory of Cellular Metabolism of the NHLBI of the National Institutes of Health, Bethesda, MD 20892, USA
[2] Centre de Neurochemie du CNRS, F-67084 Strasbourg, France
[3] The Institute of Pharmacology and Biochemistry, University of Zurich, Winterthurerstrasse 260, CH-8057, Zurich, Switzerland

53

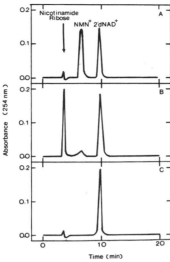

Fig. 1. Purification of 2'dNAD⁺ by strong anion exchange-HPLC. Retention time of boronate-purified compounds before (A) and after (B) treatment with bacterial alkaline phosphatase. (C) Retention time of pure 2'dNAD⁺ on a Partisil-10 SAX column of 250 mm x 4.6 mm I.D. utilizing a low salt buffer system at a flow rate of 1 ml/min.

Utilization of 2' dNAD⁺ as a substrate for purified poly(ADP-ribose) polymerase. We first determined if purified enzyme from calf thymus (13) contained endogenously bound polymeric ADP-ribose residues utilizing a highly specific and sensitive assay (15). Polymeric ADP-ribose residues were not detected in up to 17 pmol of total enzyme protein under conditions that allowed quantitative detection of 1 pmol of poly(ADP-ribose) (15). Pure DNA-independent poly(ADP-ribose) polymerase (13) catalyzed the formation of protein-bound [¹⁴C]poly(ADP-ribose) both in the presence or absence of histone H-1 at 10 µM [¹⁴C]β-NAD⁺. In contrast, no incorporation of high specific radioactivity [adenylate-³²P] 2'dNAD⁺ into acid-insoluble material was observed under identical conditions. These results indicated that 2'dNAD⁺ is not a substrate for the initiation reaction catalyzed by poly(ADP-ribose) polymerase. We have also examined 2'dNAD⁺ as a substrate for the elongation reaction. In this case, 40 µM [adenylate-³²P]2'dNAD⁺ was added to polymerase automodification reaction mixture after different times of incubation with 10 µM [¹⁴C]NAD⁺. No incorporation of ³²P radiolabel into acid insoluble material was observed. This indicated that 2'dNAD⁺ is not a substrate for the elongation reaction either. Interestingly, the addition of 2'dNAD⁺ to the enzyme automodification assay abruptly stopped the incorporation of [¹⁴C]NAD⁺ into acid insoluble material. This suggested an inhibitory effect of the 2'dNAD⁺ on the synthesis of poly(ADP-ribose). Further kinetic analysis by either the Dixon plot or Lineweaver-Burk reciprocal plot identified 2'dNAD⁺ as a noncompetitive inhibitor of NAD⁺ with an apparent K_i of 32 µM. It is important to note that similar results were observed with a DNA-dependent preparation of poly(ADP-ribose) polymerase purified from calf thymus by the procedure of Zahradka and Ebisuzaki (16).

In the late seventies it was reported that several chromatin proteins could be 2'dADP-ribosylated following incubation of a crude preparation of nuclei with radiolabeled 2'dNAD+ (17, 18). These observations have since been interpreted to indicate that 2'dNAD+ is a substrate for poly(ADP-ribose) polymerase (19). However, pure poly(ADP-ribose) polymerase does not utilize 2'dNAD+ as a substrate *vide supra*. Thus, the incorporation of radiolabeled 2'dNAD+ into acid insoluble material with crude rat liver nuclei may have resulted from either (i) the non-enzymatic ADP-ribosylation of chromatin proteins (20, 21) or (ii) the activity of a mono(ADP-ribosyl) transferase (4).

Utilization of 2'dNAD+ as a substrate for mono(ADP-ribosyl) transferases. The capability of mono(ADP-ribosyl) transferases to utilize 2'dNAD+ as a substrate was examined with a homogeneous preparation of an arginine-specific enzyme from turkey erythrocytes (2) using arginine-methyl ester as the substrate. The reaction was monitored by directly measuring the amount of product formed by HPLC on a Partisil 10-SAX column utilizing a low salt buffer system. The retention times of 2'dNAD+ and (mono-2'dADP-ribosylated)-(arginine methyl ester) corresponded to 9.28 and 6.88 min, respectively. A close quantitative correspondence between the amount of 2'dNAD+ consumed and the amount of the mono 2'dADP-ribosylated amino acid formed following incubation with mono(ADP-ribosyl) transferase was observed. Previous boiling of the enzyme totally prevented the reaction. Kinetic analysis of the initial rates indicated a K_m of the enzyme for 2'dNAD+ of 27.2 μM and a V_{max} of 36.4 μmol/min/mg of protein. In addition, to demonstrate that the product obtained corresponded to (mono-2'dADP-ribosylated)-(arginine methyl ester), we purified this putative material by HPLC and subjected it to treatment with a highly purified arginine-ADP-ribose specific hydrolase from turkey erythrocytes (22, 23). We utilized SAX-HPLC to monitor this reaction as indicated above. Incubation of (mono-2'ADP-ribosylated)-(arginine methyl ester) with the hydrolase at 30°C resulted in the formation of 2'dADP-ribose in a time-dependent fashion.

In conclusion, our data strongly support the notion that a mono(ADP-ribosyl) transferase rather than poly(ADP-ribose) polymerase was responsible for the previous observation that 2'dNAD+ is an ADP-ribosylation substrate for crude nuclei preparations (17-19).

Acknowledgement. This work was supported in part by the Swiss National Foundation for Scientific Research Grant 3.354.086. We are grateful to Mrs. Bettye Cox for typing the manuscript.

References

1. Moss, J., Vaughan, M. (1978) Proc Natl Acad Sci USA 75: 3621-3624
2. Moss, J., Stanley, S.J., Watkins, P.A. (1980) J Biol Chem 258: 5838-5840
3. Yost, D., Moss, J. (1983) J Biol Chem 258: 4946-4929

4. Tanigawa, Y., Tsuchiya, M., Imai, Y., Shimoyama, M. (1984) J Biol Chem 259: 2022-2029
5. Soman, G., Michelson, I.R., Louis, C.F., Graves, D.J. (1984) Biochem Biophys Res Commun 120: 973-980
6. Riquelme, P.T., Burzio, L.O., Koide, S.S. (1979) J Biol Chem 254: 3018-3028
7. Burzio, L.O., Riquelme, P.T., Koide, S.S. (1979) J Biol Chem 245: 3029-3037
8. Ogata, N., Ueda, K., Hayaishi, O. (1980) J Biol Chem 255: 7610-7615
9. Ogata, N., Ueda, K., Kagamiyama, H., Hayaishi, O. (1980) J Biol Chem 255: 7616-7620
10. Suzuki, H., Quesada, P., Farina, F., Leone, E. (1986) J Biol Chem 261: 6048-6055
11. Ferro, A.M., Oppenheimer, N.J. (1978) Proc Natl Acad Sci USA 75: 809-813
12. Moss, J., Stanley, S.J., Oppenheimer, N.J. (1979) J Biol Chem 254: 8891-8894
13. Niedergang, C.P., Okasaki, N., Mandel, M. (1979) Eur J Biochem 102: 43-57
14. Alvarez-Gonzalez, R., Juarez-Salinas, H., Jacobson, E.L., Jacobson, M.K. (1983) Anal Biochem 135: 69-77
15. Jacobson, M.K., Payne, D.M., Alvarez-Gonzalez, R., Juarez-Salinas, H., Sims, J.L., Jacobson, E.L. (1984) Methods Enzymol 106: 483-494
16. Zahradka, P., Ebisuzaki, K. (1984) Eur J Biochem 142: 503-509
17. Suhadolnik, R.J., Baur, R., Lichtenwalner, D.M., Uematsu, T., Roberts, J.H., Sudhakar, S., Smulson, M. (1977) J Biol Chem 252: 4134-4144
18. Lichtenwalner, D.M., Suhadolnik, R.J. (1979) Biochem 18: 3749-3755
19. Suhadolnik, R.J. (1982) ADP-ribosylation reactions: Biology and Medicine. O. Hayaishi, K. Uda (eds.), Academic Press, Inc. New York, pp. 65-75
20. Hilz, H., Koch, R., Fanick, W., Klapproth, K., Adamietz, P. (1984) Proc Natl. Acad Sci USA 81: 3929-3933
21. Ikejima, M., Gill, D.M. (1985) Biochem 24: 5039-5045
22. Moss, J., Jacobson, M.K., Stanley, S.J. (1985) Proc Natl Acad Sci USA 82: 5603-5607
23. Moss, J., Oppenheimer, N.J., West, R.E., Stanley, S.J. (1986) Biochem 25: 5408-5414

Reconstitution of an *In Vitro* Poly(ADP-Ribose) Turnover System

Luc Menard, Louis Thibault[1], Alain Gaudreau[1], and Guy G. Poirier[1]

Departement de Biologie, Faculte des Sciences, Universite de Sherbrooke, Sherbrooke, P.Q. J1K 2R1 Canada

Introduction

Poly(ADP-ribose) metabolism *in vivo* is carried out mainly by poly(ADP-ribose) polymerase and poly(ADP-ribose) glycohydrolase (1). Under normal conditions the level of poly(ADP-ribose) is stable in cells but if the cells are treated with a carcinogen (e.g. an alkylating agent) the level of poly(ADP-ribose) sharply increases in response to DNA damage (2). The short half-life measured *in vivo* after treatment of cells with alkylating agents (2) suggests that polymer degradation might occur during its accumulation thereby explaining the great discrepancy between the amount of NAD consumption and polymer accumulation. To better understand the relationship that is involved in poly(ADP-ribose) synthesis and degradation we have reconstituted *in vitro* a poly(ADP-ribose) turnover system.

Results and Discussion

The *in vitro* turnover system was designed to reflect the *in vivo* situation of poly(ADP-ribose) turnover. In this regard the NAD concentration used was 200 µM and short poly(ADP-ribose) half-lives were chosen (2). Poly(ADP-ribose) half-lives were determined by analyzing the degradation of poly(ADP-ribose) by poly(ADP-ribose) glycohydrolase after synthesis by poly(ADP-ribose) polymerase (Fig. 1A). We observed displacement of what may be non-covalently bound short polymer from the poly(ADP-ribose) polymerase using 3-AB. We then evaluated the accumulation of poly(ADP-ribose) by incubating both enzymes simultaneously to recreate a turnover condition (Fig. 1B). Three observations can be made: (i) the overall level of poly(ADP-ribose) was reduced under turnover conditions (compare Fig. 1A and 1B); (ii) the poly(ADP-ribose) level reached a plateau after 2 min; (iii) the poly(ADP-ribose) half-life changed from 55 sec to 15 min during turnover conditions. The reduction of poly(ADP-ribose) levels and the plateau that was generated did not result from an inhibition of either of the two enzymes because more than 85% of their activity was retained during the incubation period. However, the kinetics of the 2 enzymatic reactions were quite different;

---

[1]Centre de Recherches de l'Hotel-Dieu de Quebec, 11 cote du Palais, Quebec, P.Q., Canada G1R 2J6.

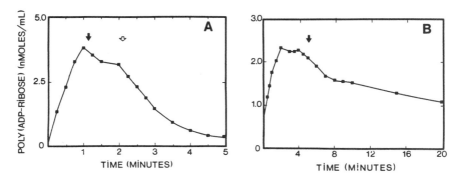

Fig. 1. Determination of the half-life of poly(ADP-ribose) using poly(ADP-ribose) polymerase and glycohydrolase. Poly(ADP-ribose) polymerase was purified up to the DNA-cellulose chromatography step (8) and poly(ADP-ribose) glycohydrolase was purified up to the poly-A cellulose chromatography step (6). A: poly(ADP-ribose) polymerase (3.8 units) was added to 50 mM Tris-Cl pH 8.0, 8 mM MgCl$_2$, 10 mM DTT, 200 μM NAD and [^{32}P]NAD (20 μCi) and incubated at 25°C. Aliquots were taken and poly(ADP-ribose) levels were determined by TCA precipitation (8). At 65 sec 3-AB was added to a final concentration of 10 mM (dark arrow) and at 125 sec poly(ADP-ribose) glycohydrolase (1.3 units) was added (open arrow). Poly(ADP-ribose) glycohydrolase activity was reduced by 50% under the incubation conditions described above. Poly(ADP-ribose) half-life after the addition of poly(ADP-ribose) glycohydrolase was 55 seconds. B: poly(ADP-ribose) accumulation during turnover conditions. The reaction was started by the simultaneous addition of poly(ADP-ribose) polymerase and glycohydrolase and poly(ADP-ribose) synthesis was monitored by TCA precipitation (8). At 5 min 3-AB was added (10 mM final concentration).

analysis by strong anion exchange HPLC of NAD consumed and ADP-ribose produced (3) showed that NAD consumption was biphasic with a rate of 15-20 nmol/ml consumed in the first 2 min and then almost no consumption during the next 3 min. This paralleled the kinetics of poly(ADP-ribose) accumulation. Nevertheless ADP-ribose production was constant during the 5 min of incubation at a rate of 2 nmol/ml/min. These results suggest that NAD consumption with synthesis of oligomers of ADP-ribose not covalently linked to proteins (not precipitable by TCA) but hydrolyzable by poly(ADP-ribose) glycohydrolase, may be produced by poly(ADP-ribose) polymerase under turnover conditions. The change in the half-life can be explained by analyzing the acceptor proteins on composite agarose-acrylamide gel electrophoresis. In addition to poly(ADP-ribose) polymerase, 2 other proteins were ADP-ribosylated; they showed apparent molecular weights of 30-40 kDa and 20-25 kDa (Fig. 2). However, the kinetics and levels of poly(ADP-ribose) accumulation and degradation were different for each acceptor. The lower molecular weight protein accumulated poly(ADP-ribose) at a constant rate, which did not affect its apparent molecular weight, although poly(ADP-ribose) polymerase accumulated high levels of modification (Fig. 2). The maximum level of modification of poly(ADP-ribose) polymerase occured when the plateau of polymer accumulation was reached (Fig. 1B) and when NAD consumption slowed down; the enzyme then stopped automodification but continued to

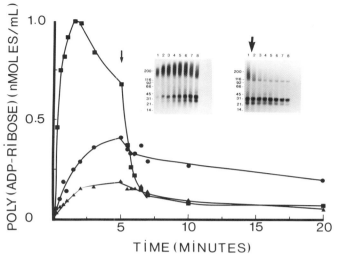

Fig. 2. Low pH gel electrophoresis of highly poly(ADP-ribosyl)ated protein. The low pH gel electrophoresis developed by Singh *et al.* (7) was modified in order to separate highly ADP-ribosylated proteins. Composite agarose-acrylamide gel contained 3% acrylamide, 0.05% bis-acrylamide, 0.5% agarose, 40 mM phosphate, pH 6.0 (pH adjusted with Tris base), 4.5 M urea, 0.1% SDS, 0.06% ammonium-persulfate and 0.1% (v/v) Temed. Aliquots were taken for each time point, treated with 25% TCA for 5 min and centrifuged at 12,000 x g for 30 min. The pellet was rinsed twice with TCA, twice with ether and dissolved in sample buffer. The gel was run at 4°C at 100 V for 5 hrs, stained and autoradiographed at -70°C using Kodak XAR-5 film. Polymer quantitation for each band was done by cutting the dried gel and counting it in a liquid scintillation counter. Time course of each ADP-ribosylated acceptor protein. (ß) poly(ADP-ribose) polymerase; (ç) 30-40 kDa acceptor protein; (') 20-25 kDa acceptor protein. Inset: Autoradiograms of the stained gels. Left: 1 to 8, 15 sec to 5 min, Right: 1 to 8, 5 to 20 min. Arrow indicates the time of addition of 3-AB. Molecular weight standards in kDa are shown on the left side of each gel.

modify other accepter proteins. Concomitantly, poly(ADP-ribose) glycohydrolase degraded more rapidly the polymer present on poly(ADP-ribose) polymerase. The preferential degradation of large polymers has been reported by others (4). The net effect of the differential degradation of poly(ADP-ribose) was the accumulation of oligo(ADP-ribosyl)ated proteins during turnover conditions.

We identified the 30-40 kDa protein as histone H1 which was present in low amounts in the poly(ADP-ribose) glycohydrolase preparation. The level of modification of histone H1 during turnover conditions was totally different from non-turnover conditions where it becomes hyper (ADP-ribosy)lated (5). The 20-25 kDa protein has not yet been identified but preliminary experiments indicate that it is not a histone or a degradation fragment of poly(ADP-ribose) polymerase.

To evaluate the effect of poly(ADP-ribose) glycohydrolase activity on the accumulation of poly(ADP-ribose) on the different acceptor proteins, we inhibited the enzyme with Tilorone, which is a competitive inhibitor (6). At the concentration used (100 μM), Tilorone had no effect on poly(ADP-ribose) polymerase activity but inhibited poly(ADP-ribose) glycohydrolase activity by 90% (Fig. 3). Tilorone drastically changed the kinetics of poly(ADP-ribose) accumulation and the profile of modification of the different acceptor proteins. Poly(ADP-ribose) polymerase became the major ADP-ribosylated protein. However, a fraction of the enzyme accumulated short polymers, a phenomenon that was not observed under turnover conditions when both enzymes were active (compare Figs. 2 and 3). The total accumulation of polymer on histone H1 was greatly reduced while no accumulation of polymer occurred on the 20-25 kDa protein. These results strongly suggest that poly(ADP-ribose) glycohydrolase could, through its polymer binding property, modulate poly(ADP-ribose) polymerase activity.

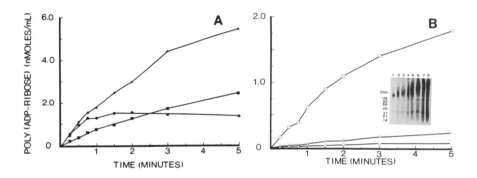

Fig. 3. Inhibition of poly(ADP-ribose) glycohydrolase effect on poly ADP-ribosylation of proteins. Incubation and determination of the levels of poly(ADP-ribose) were done as described in Fig. 1 but with the presence of 100 μM Tilorone R10,556 Da to inhibit poly(ADP-ribose) glycohydrolase. Gel electrophoresis was carried out as described in the legend of Fig. 2. A: Time course of poly ADP-ribosylation. (') poly(ADP-ribose) polymerase alone (2 units); (ç) poly(ADP-ribose) polymerase and glycohydrolase; (ß) poly(ADP-ribose) polymerase and glycohydrolase in the presence of Tilorone. B: Specific determination of poly(ADP-ribose) levels on each acceptor protein. (s) acceptor of MW > 116 kDa; (c) acceptor of 116 kDa; (t) acceptor of 30-40 kDa. Inset: autoradiogram of the stained gel, 1 to 8 represents the different time points. Molecular weight standards in kDa are shown on the left part of the gel.

In conclusion, under conditions of poly(ADP-ribose) turnover, short polymers accumulate on histone H1 whereas long polymers accumulate on poly(ADP-ribose) polymerase. If the synthesis of poly(ADP-ribose) is then inhibited, short polymers are more resistant to poly(ADP-ribose) glycohydrolase activity which results in their persistence. This phenomenon could lead to the accumulation of short polymers on, for example, histone H1 after repeated turnover events. Moreover, poly(ADP-ribose)

60

glycohydrolase reduces the activity of poly(ADP-ribose)polymerase and modulates the profile and level of poly(ADP-ribosyl)ated proteins by a structural (i.e. non-enzymatic) mechanism, probably by binding to poly(ADP-ribose).

Acknowledgements. We are grateful to Pierre de Grandpre for his technical advice for HPLC analysis. We wish to thank Dr. Alan Anderson for reading the manuscript.

References

1. Ueda, K., Hayaishi, O. (1985) Ann Rev Biochem 54: 73-100
2. Shall, S. (1984) Adv Rad Biol 11: 1-69
3. Alvarez-Gonzalez, R., Eichenberger, R., Loestscher, P., Althaus, F.R. (1986) Anal Biochem 156: 473-480
4. Hatekayama, K., Nemoto, Y., Ueda, K., Hayaishi, O. (1986) J Biol Chem 261: 14902-14911
5. Poirier, G.G., Niedergang, C., Champagne, M., Mazen, A., Mandel, P. (1982) Eur J Biochem 127: 437-442
6. Tavossoli, M., Tavassoli, M.H., Shall, S. (1985) Biochem Biophys Acta 827: 228-234
7. Singh, N., Leduc, Y., Poirier, G., Cerutti, P. (1985) Carcinogenesis 6: 1489-1494
8. Zahradka, P., Ebisuzaki, K. (1984) Eur J Biochem 142: 503-509

Abbreviations:

HPLC: high pressure liquid chromatography
PCA: perchloric acid
DTT: dithiothreitol
TCA: trichloroacetic acid
3-AB: 3-aminobenzamide

Effect of Selected Octadeoxyribonucleotides and 6-Amino-1,2 Benzopyrone on Poly(ADP-Ribose) Transferase Activity

Alaeddin Hakam, Jerome McLick, Kalman Buki, and Ernest Kun

Department of Pharmacology and The Cardiovascular Research Institute, School of Medicine, The University of California-San Francisco, San Francisco, California 94143 USA

Introduction

Polymerization of ADP-ribose from NAD$^+$ by poly(ADP-ribose) transferase is initiated by auto-mono-ADP-ribosylation of the enzyme followed by ADP-ribose transfer to other acceptor proteins and subsequent elongation (1). DNA is required as a coenzyme, and an enzyme-associated DNA, defined as sDNA (2), can be isolated which is more efficient than crude calf thymus DNA (3). It was reported that introduction of cleavage sites into double-stranded DNA coincides with an increased rate of poly(ADP-ribose) synthesis (4). However, only a few poly(ADP-ribosyl) transferase molecules (as located by electron microscopy) are at the ends of sDNA (2), and the existence of highly fragmented DNA does not inevitably coincide with a high rate of poly(ADP-ribosyl)ation (5). Furthermore, maximal activation of the enzyme in the nucleus may be catalyzed by "short DNA pieces" not detectable by methods which are suitable for assay of DNA fragmentation (6). We have observed that hormonal action *in vivo* and in cell cultures coincides with a decrease in poly(ADP-ribosyl)ation without a diminution of enzyme content (7). Since a specific interaction between hormone-receptor complexes and certain DNA sequences is well known [7], we assumed that the observed correlation between hormone action and decreased rates of poly(ADP-ribosyl)ation may suggest altered binding sites for the enzyme on DNA sequences (or structures) that are less efficient coenzymes for polyADP(ribose) transferase than the DNA sites prior to hormone action. Therefore, one may anticipate that the base sequence of oligo-DNA may be significant in determining coenzymic activity (8). Four oligo-DNA duplexes, all of which were octameric duplexes (octamers), were studied. Two of these were based on the consensus sequence identified by Yamamoto (9) to be required for hormone-dependent transcriptional enhancement *in vivo*. The third was identified in the regulatory region (203 SphI) of SV40 DNA (10) and the fourth was (dA:dT)$_8$. Application of synthetic octadeoxynucleotides, which are chemically defined coenzymes of polyADP(ribose) transferase, leads to the identification of a novel DNA related enzyme-inhibitory site. Drugs which inhibit tumorigenesis (11-13) bind to this site.

Results and Discussion

Only the initial phase of mono(ADP-ribosyl)ation by polyADP(ribose) transferase at very low concentrations of NAD$^+$ approximates Michaelis-Menten kinetics (14). At saturating concentrations of NAD$^+$ the rate of polymer formation, which follows a Poisson distribution (14), becomes important . Therefore the relationship between overall velocity and NAD$^+$ concentration cannot be described by a simple rate equation. Because of these complications we compare the catalytic efficiencies of octamers by correlating polyADP(ribose) transferase activity with the concentration of octamers varying over a very large range. Also binding constants (k$_D$) of the octamers and sDNA to polyADP(ribose) were calculated (Table 1).

Table 1. Binding of deoxyribonucleotides to poly(ADP-ribose) polymerase. Results are given as kD, determined by enzyme kinetics.

Deoxyribonucleotide species	Nature of enzymatic assay	
	Mono (ADP-ribosyl)ation[a]	Poly(ADP-ribosyl)ation[b]
Duplex A[c]	2×10^{-7}	1.4×10^{-7}
Duplex B[d]	2×10^{-7}	1.4×10^{-7}
Duplex C[e]	6×10^{-8}	7.0×10^{-8}
(dA-dT)$_8$	8×10^{-7}	4.0×10^{-7}
sDNA	1×10^{-9}	1.5×10^{-8}

Each value is an average of 3 determinations ($\pm$10% variation); kD values were calculated from enzymatic tests (Figs. 1 & 2) where deoxyribonucleotides were varied over a concentration range of 10^4.

[a]25 nM NAD$^+$
[b]200 μM NAD$^+$,
[c] 5'-A-G-A-T-C-A-G-T-3'
[d] 5'-A-G-A-A-C-A-G-A-3'
[e] 5'-G-C-A-T-G-C-A-T-3'
3'-T-C-T-A-G-T-C-A-5'
3'-T-C-T-T-G-T-C-T-5'
3'-C-G-T-A-C-G-T-A-5'

In Fig. 1, auto-mono-ADP-ribosylation of the enzyme was assayed at a constant substrate concentration of 25 nM NAD$^+$ as ADP-ribose donor (1). It should be noted that relatively small apparent differences in the saturation curves of the octamers actually correspond to significant differences because of the logarithmic abscissa. At or below an octamer/enzyme ratio of 1:1, duplex C was the most effective coenzyme, while (dA:dT)$_8$ was the poorest. The effect of sDNA on mono(ADP-ribosyl)ation was biphasic. Below an sDNA/enzyme ratio of 0.004:1 (first arrow) sDNA was hardly effective, whereas increasing the sDNA/enzyme ratio to 0.04:1 (second arrow), the coenzymatic effect of sDNA was maximal. A structural contribution of the much larger sDNA species, as compared to the synthetic octamers, apparently introduces macromolecular effects of sDNA on the enzyme which cannot occur with small oligo-DNAs.

63

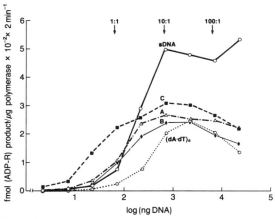

Fig. 1. Effect of sDNA and various octadeoxyribonucleotides on initiation activity of poly(ADP-ribose) transferase. NAD^+ concentration was 25 nM and V_{init} denotes rates in first 2 min at 24°C. A,B,C denote duplexes as defined in Table 1. sDNA is a mixture of double strands between 0.5 and 3.5 Kb. Ratios at the top of the figure refer to mol octamer/mol enzyme. In the case of sDNA, at the first arrow (1:1) the sDNA/enzyme ratio is 0.004:1, at the second arrow it is 0.04:1, and at the third 0.4:1. For additional experimental details see ref. 8. (Reprinted with permission from ref. 8).

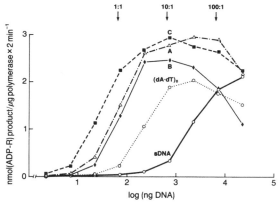

Fig. 2. Effect of various concentrations of sDNA and octadeoxyribonucleotides on rates of polymer formation by poly(ADP-ribose) transferase. NAD^+ concentration was 200 μM and 10 μg whole calf thymus histones were added per 0.10 ml incubation mixture. For other details see Fig. 1 and ref. 8. (Reprinted with permission from ref. 8).

Poly(ADP-ribosyl)ation at 200 μM NAD^+ as substrate with an excess of whole thymus histones as additional poly(ADP-ribose) acceptors, probably simulating conditions prevailing in chromatin, yielded a less complicated kinetics (Fig. 2). In this system duplex C WAS the best synthetic coenzyme, whereas the activation curve of a sDNA remained sigmoidal. In the cell nucleus, the concentration of DNA is far greater than that of the poly(ADP-ribose) transferase, therefore the observed inhibition in the *in vitro* model

64

(Fig. 2) may have biological relevance and it is possible that certain DNA sequences could be less effective coenzymes. Duplex C is significantly more effective than the other octamers, which argues against the exclusiveness of the "active termini" theory (4). Dissociation constants (k_D) calculated for the binding of each of the four octamers and sDNA to ADP(ribose) transferase are summarized in Table 1. Clearly sDNA exhibits the highest affinity for the enzyme, followed by octamer C and then A and B (being equal) and finally $(dA:dT)_8$ in decreasing order. The apparent differences in k_D as determined under conditions of mono or poly(ADP-ribosyl)ation are probably due to differing kinetics of the two processes (14). It may be assumed that certain base sequences present in the sDNA, being effective coenzymes like octamer C, are much more efficient when part of the macromolecular DNA structure. What the active base sequences in sDNA specifically are remains unknown, but the present results support a significant structural contribution of sDNA, as has been proposed earlier (2).

The availability of chemically well defined octamers (A and C) as sDNA substitutes provides a kinetic tool for the determination of the effects of enzyme inhibitors on the DNA binding site of the enzyme. The 6-amino derivative of 1,2-benzoypyrone, (6-aminocoumarin), competitively inhibits at the octamer duplex A or C sites with an apparent K_i of 28 μM. Without the inhibition at a fixed NAD^+ concentration, a Michaelis-Menten relationship exists between V_{init} and the concentration of the octamers (lowest curve in Fig. 3) with an apparent binding constant of 1 μM. The results shown in Fig. 3 identify a novel site of inhibitors, structurally unrelated to NAD^+ (12), which act at the DNA binding site of the enzyme.

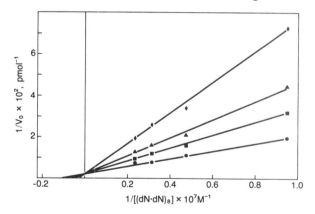

Fig. 3. The effect of 6-amino-1,2-benzopyrone on the initial velocity of auto-mono-ADP-ribosylation of the enzyme as a function of varying concen-trations of octadeoxyribonucleotide. No inhibitor present (●), 20 μM (■), 40 μM (▲) and 80 μM (◆) 6 amino-1,2-benzopyrone. NAD^+ concentration was constant at 25 nM. The K_i for 6-amino-1,2-benzo-pyrone is approx. 28 μM, and k_D, based on a 10-fold concentration range for the octadeoxyribonucleotide, is between 0.3 and 0.8 μM. $(dN-dN)_8$ denotes duplex A, defined in Table 1. Data points in Figs. 1-3 represent an average of 3 determinations each.

Acknowledgements. This research was supported by NIH grant HL-27317 and AFO-SR-85-0377 and 86-0064 (AFOSR). E.K. is a recipient of the Research Center Award of the USPHS. K.B. is a visiting scientist from Semmelweis Univ., Budapest, Hungary (Biochemistry II).

References

1. Bauer, P.I., Hakam, A., Kun, E. (1986) FEBS Lett 195: 331-338
2. Ittel, M.E., Jonstra-Bilen, J., Niedergang, C., Mandel, P., Delain, E. (1985) ADP-Ribosylation of Proteins, F.R. Althaus, H. Hilz, and S. Shall, (eds.), Springer-Verlag, Berlin pp. 60-68
3. Yoshihara, K., Kamiya, T. (1982) ADP-Ribosylation Reactions, O. Hayaishi, and K. Ueda, (eds.), Academic Press, New York pp. 157-171
4. Benjamin, R.C., Gill, D.M. (1980) J Biol Chem 255: 10502-10508
5. Skidmore, C.J., Jones, J., Oxberry, J.M., Chaudun, E., Counis, M.F. (1985) ADP-Ribosylation of Proteins, F.R. Althaus, H. Hilz, S. Shall, (eds.), Springer-Verlag, Berlin, pp. 116-123
6. Berger, N.A., Petzhold, S.J. (1985) Biochemistry 24: 4352-4355
7. Kun, E., Minaga, T., Kirsten, E., Hakam, A., Jackowski, G., Tseng, A., Brooks, M. (1986) Biochemical Action of Hormones, Vol. 13, J. Litwack, (ed.), Academic Press, New York, pp. 33-55
8. Hakam, A., McLick, J., Buki, K., Kun, E. (1987) FEBS Lett 212: 73-78
9. Yamamoto, K.R. (1985) 43rd Symposium of the Society for Development Biology, L. Bogorad, C. Adelman, (eds.), Alan Liss New York, pp. 3-20
10. Nordheim, A., Rich, A. (1983) Nature 303: 674-679
11. Kun, E., Kirsten, E., Milo, G.E., Kurian, P., Kumari, H.L. (1983) Proc Natl Acad Sci USA 80: 7219-7223
12. Milo, G.E., Kurian, F., Kirsten, E., Kun, E. (1986) FEBS Lett 179: 332-336
13. Tseng, A., Jr., Lee, W.M.F., Kirsten, E., Hakam, A., McLick, J., Buki, K., Kun, E. (1987) Proc Natl Acad Sci USA 84: 1107-1111
14. Kun, E., Minaga, T., Kirsten, E., Jackowski, G., McLick, J., Peller, L., Oredsson, S.M., Marton, L., Pattabiraman, N., Milo, G.E. (1983) Adv Enzyme Regul 21: 177-199

Two Tests of the Direction of Elongation of Poly(ADP-ribose): Residues are Added at the Polymerase-Proximal Terminus

Miyoko Ikejima, Gerald Marsischky, and D. Michael Gill

Department of Molecular Biology and Microbiology, Tufts University School of Medicine, Boston, Massachusetts 02111 USA

Introduction

Previous experimental approaches have suggested that poly(ADP-ribose) synthesis begins with the ADP-ribosylation of an "acceptor" protein, and that the poly(ADP-ribose) chain is elongated by the repetitive "distal" addition of further ADP-ribose residues to the 2'OH end (1). However the first product of poly(ADP-ribose) synthesis is an auto-modified polymerase (1-3), which is difficult to reconcile with distal addition. For this and other reasons we have considered the alternative possibility that a polymerase molecule would build a polymer of ADP-ribose upon itself (proximal addition). We describe here two tests of the direction of elongation of poly(ADP-ribose) both of which support proximal over distal addition.

Results and Discussion

Pulse-chase experiment. We developed a rapid pulse-chase method, which is illustrated in Fig. 1. Pulse-labeled polymerase was centrifuged through a small Sephadex G-50 column (4), which removed labeled NAD efficiently without much loss of enzyme activity, and was immediately incubated with unlabeled NAD. The products were released at high pH and digested with venom phosphodiesterase. If a chain were to grow by proximal extension, labeled AMP would be found in the digest with the same frequency as for the pulsed products. Alternatively, if new residues were added to the distal end, the terminal residue would become internal, and would generate PR-AMP, not AMP on digestion.

Poly(ADP-ribose) was purified almost to homogeneity from bull testes by the method of Agemori *et al.* (5). Purified polymerase was incubated briefly with [^{32}P]NAD (10°C, 30 sec, 5 μM NAD) in the presence of 10 μg/ml Hae III-digested calf thymus DNA. Incorporation was stopped by the addition of 3-aminobenzamide and unincorporated labeled NAD was removed. Pulse-labeled polymer was originally attached to the enzyme. After alkali treatment it showed a heterogeneous distribution of short chains on a 20% polyacrylamide gel (6). Venom phosphodiesterase digested these to [^{32}P]PR-AMP and [^{32}P]AMP, of which AMP represented 4.6% of the counts. After a chase with unlabelled NAD (200 μM NAD, 10 min, 25°C), pulse-chased chains are still enzyme attached but are much longer chains: at least 100 residues. After alkali-release and phosphodiesterase digestion [^{32}P]AMP was still found at the original frequency, 5.1% of the product. None would be expected had the terminal [^{32}P] residue been covered by the newly added cold ADP-ribose. Thus the pulse-labeled material still occupied a distal position on the chain and poly(ADP-ribose) must grow at the end proximal to the enzyme.

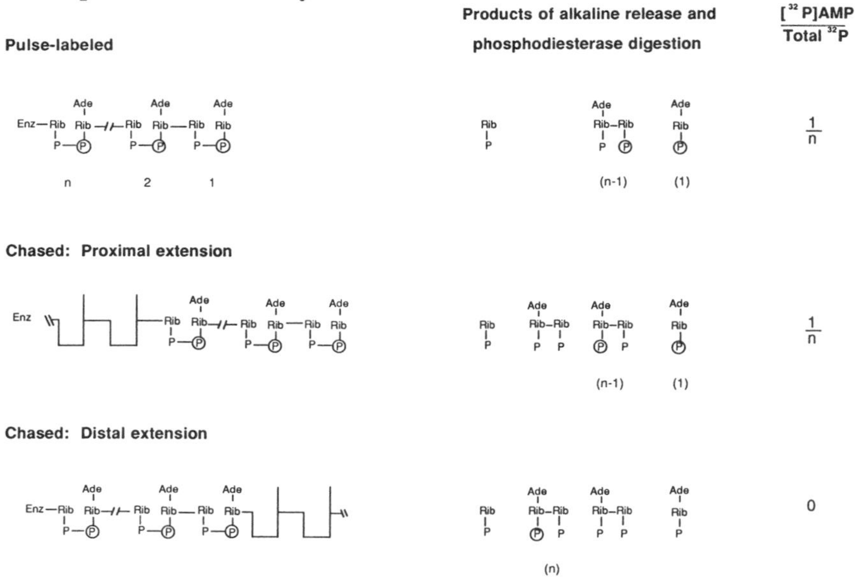

Fig. 1. Principle of the pulse-chase experiment. The pulse-labeled poly(ADP-ribose) chain (n residues) is shown elongated by proximal extension or distal extension. Labeled phosphate is circled. Digestion of poly(ADP-ribose) produces PR-AMP and AMP from the inner residues and the distal end, respectively. The expected numbers of labeled nucleotides among the total digests are written in parentheses.

Product formed with [^{32}P]2'-dNAD as a substrate. We also tested the direction of elongation with dNAD. Kinetic analysis showed that dNAD competitively inhibits the incorporation of NAD and has a lower affinity for the enzyme ($K_i/K_m = 4$). Since this nucleotide lacks the 2'OH group which is required for ribose-ribose linkages (1" → 2') between ADP-riboses, dNAD must either terminate chain growth by forming a cap or must not be incorporated at all. To determine the actual outcome we devised sensitive techniques to detect any incorporation of [^{32}P]dADP-ribose into poly(ADP-ribose) chains. Originally we performed this experiment in HeLa cell lysates and were unable to detect any dADP-ribose in the released oligomers (3). We have now repeated the experiment using purified polymerase to avoid the possibility that terminal dADP-ribose might have been removed by glycohydrolase in the HeLa lysate. The sensitivity of the experiment was maximized by using very highly labeled [^{32}P]dNAD (specific activity 540 times that of [32P]NAD whose incorporation is readily detected) and by removing any traces of [32P]NAD from the [32P]dNAD using borate column chromatography (7). The result was the same as before: dADP-ribose is not incorporated either into free oligomers or into enzyme-bound chains at anything like the level predicted by the distal addition model.

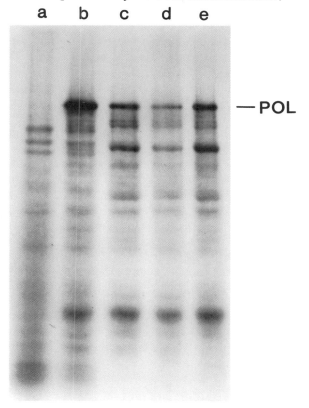

Fig. 2. Irreversible modification of the polymerase by dADP-ribose. Ten μl portions of 10% washed HeLa cell lysate were incubated with 10 μl of 25 μM [32P]NAD (63 Ci/mmol) and 100 μg/ml DNase I. 5 mM 3-aminobenzamide was included in sample (a). After incubation at 37°C for 5 min, 10% TCA was added to samples a and b. Samples c-e were chilled on ice for 5 min, and centrifuged. The pellets were resuspended in lysis buffer (40 mM Tris-Cl (pH 8), 10 mM Mg(O.CO.CH$_3$)$_2$, 5% Dextran, 0.05% Trition X-100 and 0.1 mM CaCl$_2$) plus DNase with (c): no nucleotide, (d): 5 mM unlabeled NAD or (e): 5 mM unlabeled dNAD, and were incubated again for 10 min. The incubation was stopped by the addition of TCA. The precipitates were dissolved in 40 μl of gel buffer containing 1% SDS, 1% 2-mercaptoethanol and were analyzed by 7%-15% polyacrylamide gel electrophoresis. Track (c) controls for losses in the wash. The bands labeled in the presence of 3-aminobenzamide are probably the result of non-enzymatic addition of [32P]dADP-ribose that was present in the precursor.

Although no [32P]dADP-ribose capped oligomer was formed, we did detect a slight covalent incorporation of dADP-ribose onto the enzyme itself.

69

In comparison with polymer synthesis, deoxyADP-ribosylation of polymerase was very slow, but it was DNA-dependent and was inhibited by 3-aminobenzamide, showing that some of the normal polymerase functions were employed. The dADP-ribose seemed to be added to the "automodification domain" (8), since the 67K fragment formed by partial digestion with chymotrypsin was labeled (3). The incorporated dADP-ribose was not removed by poly(ADP-ribose) glycohydrolase (9) showing that the residue was attached directly to the enzyme and was not part of a chain. Also the dADP-ribose residue was attached by an alkali-resistant bond whereas most poly(ADP-ribose) chains are readily removed by alkali. As shown in Fig. 2, neither excess unlabeled NAD nor excess unlabeled dNAD would release such [^{32}P]dADP-ribose from the enzyme. This irreversible modification caused inactivation.

We interpret these observations as follows. dNAD first enters the NAD binding site (8), and competes with the binding of NAD. It is unable to progress to the growing polymer chain, but at a slow rate can engage in a side reaction with a neighboring group. The latter is probably not a component of the active center but is sufficiently close to the active center to inhibit enzyme activity.

Our two approaches lead to the same conclusion. The persistence of terminal label on chased chains shows that these must grow at the enzyme-proximal end. Likewise the absence of dADP-ribose residues on completed chains implies that chains cannot accrete residues at the enzyme-distal (i.e. 2'OH) end. Enzyme proximal chain growth implies that the enzyme must act in some sense like a ribosome, alternating the growing chain between two attachment sites. Clearly these can be on the same enzyme molecule. The enzyme constructs poly(ADP-ribose) by itself, without a partner, and in particular without any other non-enzyme "acceptor".

References

1. Ueda, K., Hayaishi, O. (1985) Ann Rev Biochem 54: 73-100
2. Benjamin, R.C., Gill, D.M. (1980) J Biol Chem 255: 10493-10501
3. Ikejima, M., Sawada, Y., Dang, A.Q., Suhadolnik, R.J., Gill, D.M. (1983) ADP-Ribosylation, DNA Repair and Cancer, Miwa, M., Hayaishi, O., Shall, S., Smulson, M. and Sugimura, T. (eds), Japan Sci Soc Press, Tokyo/VNU Science Press, Utrecht, pp. 129-140
4. Maniatis, T., Aritseh, E., Sambrook, J. (1982) Molecular Cloning: A Laboratory Manual, Cold Spring Harbor Laboratory, Cold Spring Harbor, New York, pp. 466-467
5. Agemori, M., Kagamiyama, H., Nishikimi,M., Shizuta, Y. (1982) Arch Biochem Biophys 215: 621-627
6. Tanaka, M., Hayashi, K., Sakura, H., Miwa, M., Matsushima, T., Sugimura, T. (1978) Nucleic Acids Res 5: 3183-3194
7. Wielckens, K., Bredehorst, R., Adamietz, P., Hilz, H. (1981) Eur J Biochem 117: 69-74
8. Kameshita, I., Matsuda, Z., Taniguchi, T., Shizuta, Y. (1984) J Biol Chem 259: 4770-4776
9. Miwa, M., Tanaka, M., Matsushima, T., Sugimura, T. (1974) J Biol Chem 249: 3475-3482

Biosynthesis and Degradation of Poly(ADP-Ribose) Synthetase

Isamu Kameshita[1], Yasuhiro Mitsuuchi, Michiko Matsuda, and Yutaka Shizuta

Department of Medical Chemistry, Kochi Medical School, Okoh-cho, Nankoku-shi, Kochi 781-51, Japan

Introduction

Poly(ADP-ribose) synthetase is a chromatin-bound enzyme which produces a protein bound homopolymer of ADP-ribose using NAD as a substrate (1). The enzyme from calf thymus has been purified to homogeneity and extensively characterized in several laboratories (2-4). Using limited proteolysis, we recently demonstrated that the enzyme (Mr = 120,000) consists of three functionally different domains, the first (Mr = 46,000) for binding of DNA, the second (Mr = 22,000) for accepting poly(ADP-ribose) and the third (Mr = 54,000) for binding of the substrate, NAD (5, 6). We also demonstrated by immunoblotting that endogenous degradation products of the enzyme were present in calf thymus (7-9). Nevertheless, detailed processes of synthesis and degradation of this enzyme *in vivo* are not as yet fully understood.

In an effort to elucidate the degradation processes and the metabolic turnover rate of poly(ADP-ribose) synthetase in living cells, we prepared two types of polyclonal antibodies, one against the calf thymus native enzyme and another against its catalytic domain. These antibodies efficiently immunoprecipitated both calf and human enzymes in crude extracts. In this paper we present evidence to show that poly(ADP-ribose) synthetase is synthesized in HeLa cells without forming any precursor polypeptide and degraded with a half life of 18 hrs.

Results and Discussion

Antibodies against calf thymus poly(ADP-ribose) synthetase were prepared using the native enzyme (Mr = 120,000) and its NAD-binding domain (Mr = 54,000) as antigens. Fig. 1 shows the effect of these antibodies on the catalytic activity of the enzyme purified from calf thymus. The antibody against the native enzyme (a120K) as well as that against the catalytic domain (a54K) efficiently inhibited the enzyme activity, but preimmune IgG did not. The enzyme activity (0.25 unit) was neutralized to

---

[1]Present Address: Department of Molecular Genetics, Beckman Research Institute of the City of Hope, 1450 East Duarte Road, Duarte, CA 91010

50% with 3.4 µg of a120K or 0.7 µg of a54K, respectively. The reactivity of a120K and a54K with the enzyme in cell free extracts from calf thymus, HeLa cells, mouse testis, mouse lymphoma cells (L-1210), and chicken liver was also examined. As shown in Fig. 2, the inhibition patterns of the enzyme in HeLa cells both by a120K and a54K are quite similar to those of the calf enzyme. The enzymes in mouse testis and L-1210 were also significantly inhibited by a120K, but only slightly inhibited by a54K. The degrees of inhibition of the chicken enzyme both by a120K and a54K were considerably weaker than those of calf and human enzymes. In our earlier studies, we demonstrated that the major physicochemical characteristics of calf, human, and mouse enzymes are all similar to each other (7, 8) and that these enzymes are immunostained with the antibodies against the native enzyme on a nitrocellulose filter (7-9). The present results, taken together with the above data, suggest that the protein structures of these enzymes are very similar to each other and yet there are some differences in the fine structures, especially around the 54K catalytic portion of the enzyme.

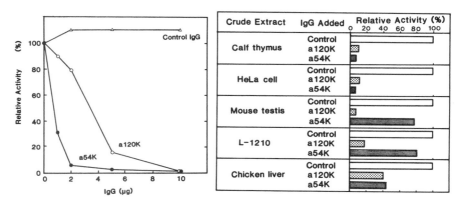

Fig. 1. (left) Effects of the antibodies on poly(ADP-ribose) synthetase activity. Antibodies were produced by immunizing BALB/c mice and IgG fractions were purified from antisera. The purified enzyme (0.25 unit) from calf thymus was preincubated in 50 mM Tris-HCl buffer (pH 8.0) containing 10% glycerol and 200 mM NaCl at 0°C for 30 min with varying amounts of a120K (O), a54K (●), or control (Δ) in a total volume of 20 µl, and the mixture assayed for the enzyme activity. The enzyme activity in the absence of IgG was taken as 100%.

Fig. 2. (right) Effects of the antibodies against calf thymus enzyme on poly(ADP-ribose) synthetase from various sources. A total of 10 µl of cell free extract (0.04-0.12 unit of the enzyme) in 50 mM Tris-HCl (pH 7.4), 300 mM NaCl, 50 mM NaHSO3, 1 mM EDTA, and 1 mM phenylmethylsulfonyl fluoride was preincubated at 0°C for 30 min with 10 µg of control IgG, a120K, or a54K, and the enzyme activity assayed.

To investigate whether these antibodies could be used for analyzing the metabolic turnover of poly(ADP-ribose) synthetase *in vivo*, we first examined the ability of these IgG fractions to immunoprecipitate the enzyme in a crude extract. Nuclei were isolated from calf thymus and incubated with 1.4 μM [^{14}C]NAD to ADP-ribosylate nuclear proteins. Nuclear extracts (Fig. 3, lane 1) were then incubated with antibody•protein A-sepharose complex, and immunoprecipitates formed were analyzed by SDS polyacrylamide gel electrophoresis and fluorography. A polypeptide of Mr = 120,000 was clearly detected when the nuclear extracts were incubated with a120K or a54K (Fig. 3, lanes 3 and 4), but not detected when they were incubated with preimmune IgG (Fig. 3, lane 2). Additional faint bands of Mr = 80,000 and Mr = 64,000 were precipitated by a120K (Fig. 3, lane 3), whereas a polypeptide of Mr = 57,000 was precipitated by a54K (Fig. 3, lane 4). In the previous study, we demonstrated that the polypeptides of Mr = 80,000 and Mr = 64,000 were immunostained with the antibody against DNA-binding domain of the enzyme (7). Furthermore, these fragments, like native enzyme, could be purified with DNA-affinity column and they served as efficient acceptors of poly(ADP-ribose) (9, 10). On the basis of these results, possible cleavage sites of poly(ADP-ribose) synthetase *in vivo* are schematically illustrated in Fig. 4.

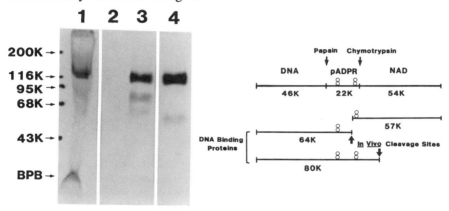

Fig. 3. (left) Fluorography after gel electrophoresis of proteins immunoprecipitated from the extracts of [^{14}C]NAD-labeled calf thymus nuclei. Calf thymus nuclei (5 x 10^6) were incubated with 1.4 μM [adenine-U- ^{14}C]NAD (423 dpm/pmol) at 0°C for 30 min and the reaction was terminated by the addition of 80 mM nicotinamide. The nuclei were disrupted in 0.5 ml of 50 mM Tris-HCl (pH 7.4) containing 300 mM NaCl, 50 mM NaHSO$_3$, 1 mM EDTA, 1 mM phenylmethylsulfonyl fluoride, and 0.1 unit/ml aprotinin (Sigma) by sonication and centrifuged at 10,000 x g for 10 min. The supernatant (lane 1) was incubated at 0°C for 1 hr with 20 μl of Protein A-Sepharose, previously reacted with 20 μg of control IgG (lane 2), a120K (lane 3), or a54K (lane 4). Protein A-Sepharose was then washed five times with phosphate buffered saline containing 0.1% sodium dodecyl sulfate, 0.5% deoxycholate, and 1% Triton X-100, and then washed twice with phosphate buffered saline. The final precipitate was suspended in 20 μl of 5 mM potassium phosphate (pH 7.0) containing 1% sodium dodecylsulfate, 50 mM dithiothreitol, 10% glycerol, and 0.004% bromophenol blue, and boiled for 90 sec. After centrifugation, the supernatant was analyzed by SDS polyacrylamide gel electrophoresis and fluorography as described previously (6).

Fig. 4. (right) Possible cleavage sites of poly(ADP-ribose) synthetase in calf thymus nuclei.

To determine the rate of degradation of poly(ADP-ribose) synthetase *in vivo*, HeLa cells were pulse-labeled with [³H]leucine and then chased in the same medium containing an excess amount of unlabeled leucine at different incubation times. Polypeptide of Mr = 120,000 was efficiently immunoprecipitated from the homogenates of HeLa cells by a120K and a54K, but not by preimmune IgG (data not shown). A decrease of radiolabeled poly(ADP-ribose) synthetase was measured by densitometric scanning of the polypeptide of Mr = 120,000 in a fluorogram. As shown in Fig. 5, the half life of this enzyme in HeLa cells was roughly estimated to be 18 hr.

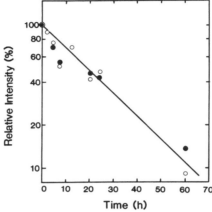

Fig. 5. Rate of degradation of poly(ADP-ribose) synthetase in HeLa cells. HeLa 3S cells (2.5 x 10⁷) in leucine-deprived Dulbecco's modified Eagle's medium (DIFCO Laboratories) containing 5% dialyzed fetal bovine serum were metabolically labeled with 50 µCi [³H]leucine (120 Ci/mmol) in a 5% CO_2 incubator for 1 hr at 37°C. After the addition of a 5,000 times excess amount of cold leucine (0.5 mM), cells were either harvested immediately or incubated further for chase. The radiolabeled cells were disrupted and immunoprecipitated using a120K (O) or a54K (●) as described in Fig. 3. Quantification of the radiolabeled poly(ADP-ribose) synthetase was carried out by densitometric trace of fluorogram using a chromatoscanner (Shimadzu CD-900).

Antibodies are very useful reagents for the detection of unstable precursor proteins which are easily degraded to molecular species (11). In our present study on poly(ADP-ribose) synthetase, however, we have not been able to detect any precursor polypeptide which is converted into a Mr = 120,000 polypeptide by posttranslational processing. This fact suggests that poly(ADP-ribose) synthetase is synthesized as a single polypeptide of Mr = 120,000 directly from its messenger RNA.

References

1. Hayaishi, O., Ueda, K. (1977) Annu Rev Biochem 46: 95-116
2. Mandel, P., Okazaki, H., Niedergang, C. (1977) FEBS Lett 84: 331-336
3. Yoshihara, K., Hashida, T., Tanaka, Y., Ohgushi, H., Yoshihara, H., Kamiya, T. (1978) J Biol Chem 253: 6459-6466

5. Nishikimi, M., Ogasawara, K., Kameshita, I., Taniguchi, T., Shizuta, Y. (1982) J Biol Chem 257: 6102-6105
6. Kameshita, I., Matsuda, Z., Taniguchi, T., Shizuta, Y. (1984) J Biol Chem 259: 4770-4776
7. Kameshita, I., Yamamoto, H., Fujimoto, S., Shizuta, Y. (1985) FEBS Lett 182: 393-397
8. Shizuta, Y., Kameshita, I., Ushiro, H., Matsuda, M., Suzuki, S., Mitsuuchi, Y., Yokoyama, Y., Kurosaki, T. (1986) Adv Enzyme Regul, Vol 25, G. Weber (ed.), Pergamon Press, New York, pp. 377-384
9. Kameshita, I., Matsuda, M., Nishikimi, M., Ushiro, H., Shizuta, Y. (1986) J Biol Chem 261: 3863-3868
10. Holtland, J., Jemtland, R., Kristensen, T. (1983) Eur J Biochem 130: 309-314
11. Teraoka, H., Tsukada, K. (1985) J Biol Chem 260: 2937-2940

Monoclonal Antibodies Against Poly(ADP-Ribose) Polymerase: Epitope Mapping, Inhibition of Activity and Interspecies Immunoreactivity

Daniel Lamarre, Gilbert de Murcia[1], Brian Talbot[2], Claude Laplante[2], Yvan Leduc, and Guy Poirier

Centre de Recherche en Cancérologie de l'Unversité Lavel à l'Hôtel-Dieu de Québec, Québec, G1R 2J6 Canada

Introduction

Poly(ADP-ribose) polymerase from different species appears to be structurally conserved in eucaryotes with regard to molecular mass, amino acid composition and enzyme activity as detected by gel electrophoresis (1-3). In addition, analysis of the antigenic structure by immunoblotting with polyclonal antibodies (4, 5) provides evidence that some epitopes are common to various animal cells. We have recently developed monoclonal antibodies against the main functional domains of calf thymus poly(ADP-ribose) polymerase (6). The purified monoclonal antibodies have been used to map three epitopes along the domains described by Kameshita *et al.* (7), to examine the presence of these epitopes in the enzyme from different species by immunoblot analysis, and to inhibit enzyme activity *in vitro*.

Epitope localization in poly(ADP-ribose) polymerase and interspecies analysis. We have further characterized the specificity of our monoclonal antibodies (6) by digesting the N-terminal 66 kDa poly(ADP-ribose) polymerase fragment (8) with trypsin to yield complementary fragments of 36 kDa and 29 kDa (Fig. 1A). The immunoreactivity of C-1-2, C-1-9 and C-2-10 antibodies on combined chymotryptic-tryptic digests is shown in Fig. 1B. C-1-9 reacted strongly with the 29 kDa fragment whereas C-2-10 reacted with the 36 kDa fragment. The epitope of the C-1-2 antibody specific for the NAD binding domain was mapped to the 17 kDa fragment generated by chymotryptic digestion of the 54 kDa C-terminal fragment. To further characterize the epitopes for antibodies C-1-9 and C-2-10, we performed enzymatic cleavage reactions on automodified enzyme. The digestion yielded 66 and 36 kDa ^{32}P-labelled fragments. As the 36 kDa fragment is detected by C-2-10 antibody, we concluded that this fragment is in the central position near of the automodification domain. The epitope of the C-1-9 antibody is thus situated on the N-terminal 29 kDa fragment. The tentative localization of epitopes along the 3 functional domains in shown in Fig. 2.

---

[1]IBMC-CNRS de Strasbourg 67000, France

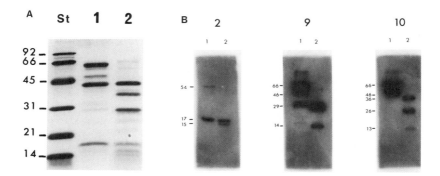

Fig. 1. Epitope mapping: Proteolytic cleavage of poly(ADP-ribose) polymerase. Stained gel (A) and immunoblot (B). The chymotryptic digestion (1) and the chymotryptic-tryptic digestion (2) were carried out at 25°C in 50 mM Tris pH 8, 0.2 M NaCl, 10 mM mercaptoethanol and 10% glycerol. The 30 min chymotryptic digestion was stopped with 1 mM PMSF and the subsequent 10 min tryptic digestion by treatment with 20% TCA. The digests were dissolved in SDS sample buffer and subjected to a 15% acrylamide SDS electrophoresis (6). The immunoblots were revealed with ^{125}I goat anti-mouse antibody (Amersham). Relative molecular masses of standards (St) and fragments as extrapolated from standards are indicated as kDa.

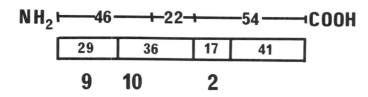

Fig. 2. Schematic drawing of the localization of the three epitopes along the functional domains. The orientation of the molecule was determined by Kameshita *et al*. (8).

The presence of the three epitopes in the poly(ADP-ribose) polymerase from chinese hamster ovary cells (CHO) and epidermoid human lung carcinoma cells (CALU-1) was also examined by immunoblotting. As shown in Fig. 3, an immunoblot performed on CHO subcellular fractions demonstrated a specific reaction at 116 kDa with C-2-10 antibody whereas C-1-2 and C-1-9 antibodies did not show any reaction. On the other hand,

77

the three epitopes are present in CALU-1 cells as demonstrated by a positive reaction with C-1-2, C-1-9 (Fig. 3) and C-2-10 antibodies. The central epitope of the enzyme that reacted with C-2-10 antibody has also been observed in extracts of rat cells (FR3T3), rat liver (data not shown) and SP 20 mouse plasmacytoma (Ph.D. thesis of H. Thomassin).

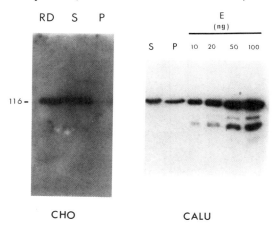

Fig. 3. Immunoblots of subcellular fractions of CHO and CALU-1 cells. In summary, cells (5 x 10⁶) permeabilized with 1% NP-40 (NP) were treated at 0° C with 250 µg/ml of DNase I and RNase A for 2 hr (RD), extracted for 1 hr with 2M NaCl buffer (S), and centrifuged to yield insoluble material (P). The different subcellular fractions (NP,RD,S,P), representing 100% of cellular protein content, were dissolved in a 4 M urea SDS sample buffer and subjected to western blot analyses. Poly(ADP-ribose) polymerase from CHO and CALU-1 cells were revealed in these cases with C-2-10 and C-1-9 antibodies respectively. Purified calf thymus enzyme (E) is used as reference.

Inhibition of enzyme activity by monoclonal antibodies. The antibodies that recognized the 29 kDa N-terminal fragment of poly(ADP-ribose) polymerase were the only group of monoclonal antibodies to inhibit enzyme activity. The effect of antibody concentration on enzyme activity is shown in Fig. 4. The highly purified C-1-23 (C-1-9 related antibody) was incubated for 4 hr at 4°C with 500 ng of purified enzyme. When the antibody to enzyme stochiometry was slightly more than 1, maximum inhibition (70%) was achieved without any degradation of the purified enzyme (Fig. 4).

Discussion

Immunological analysis of poly(ADP-ribose) polymerase has shown a high conservation of the central portion of the enzyme, whereas the C- and N-terminal portions of the molecule appear to be less conserved. The central 36 kDa epitope demonstrated a striking conservation as C-2-10 antibody reacted by immunoblot with the enzyme from every species tested (man, mouse rat, hamster and calf). The N-terminal 29 kDa epitope did not show a general distribution as suggested by the absence of immunoreactivity with the CHO enzyme. The binding of antibody on this epitope inhibited the activity of the purified calf thymus enzyme and in crude extracts. The antibody probably interfered with the formation of a DNA-enzyme complex essential for enzyme activity as suggested by the displacement of the antibody by DNA on the absorbed enzyme in an "ELISA" assay (data not shown). The C-1-2 antibody does not recognize the enzyme from most of the other species. However, the weak detection of the calf enzyme with C-1-2 antibody strongly suggests that this epitope is highly sensitive to denaturating agents or that it is not accessible after electrotransfer to nitrocellulose membrane. The binding of the C-1-2 antibody on the 17 kDa epitope has no effect on enzyme activity, suggesting that the antibody does not interfere with NAD binding or NADase activity. In fact, the presence of NAD (1 nM to 50 μM) in the ELISA assay did not alter the binding of the C-1-2 antibody to the enzyme while preincubation of enzyme with DNA considerably enhanced this binding (data not shown). These last results could be interpreted as an important modification of the enzyme in the presence of DNA which is detected with the C-1-2 antibody.

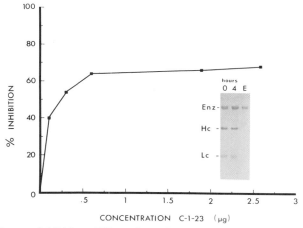

Fig. 4. Enzyme inhibition: Effect of an epitope specific antibody. Various amount of C-1-23 antibodies were incubated for 4 hr at 4°C with 500 ng of enzyme and subsequently assayed using standard conditions (6) for 5 min at 30° C in presence of 10 μg of DNA and total histone and 200 μM [^{32}P-]NAD (30,000 cpm/nmol). The percentage of inhibition is defined as (1 minus the ratio of the activity in the presence absence of antibodies) x 100. The specific activity of the purified enzyme after a 4 hr incubation was 1150 nmol/mg/min. Gel analysis of the mixture of an enzyme to antibody weight ratio of 1 is shown before (0) and after (4) incubation. E: enzyme, Lc: light chain and Hc: heavy chain of immunoglobulin.

The immunoblots on cell extracts demonstrated that the total cellular content of poly(ADP-ribose) polymerase is detected in a single band at 116 kDa. When cell extracts are freshly prepared in the presence of multiple protease inhibitors, no degradation fragments are detected with monoclonal or polyclonal antibodies but such fragments progressively appear upon storage of samples. Preliminary studies with irradiated CHO cells gave similar results and do not support the hypothesis that processing of the enzyme occurs during DNA repair as suggested by Berger *et al*.(9).

Acknowledgements. A special thanks to Dr. A. Anderson and Dr. M. Lalanne for editorial comments.

References

1. Holtlund, J., Kristensen, T., Ostvold, A.C., Laland, G. (1981) Eur J Biochem 119: 23-29
2. Agemori, M., Kagamiyama, H., Nishikimi, M., Shizuta, Y. (1982) Arch Biochem Biophys 215: 621-627
3. Scovassi, A.I., Izzo, R., Franchi, E., Bertazzoni, U. (1986) Eur J Biochem 159: 77-84
4. Jongstra-Bilen, J., Ittel, M.E., Jongstra, J., Mandel, P. (1981) Biochem Biophys Res Commun 103: 383-390
5. Kameshita, I., Yamamoto, H., Fujimoto, S., Shizuta, Y. (1985) FEBS Lett 182: 393-397
6. Lamarre, D., Talbot, B., Leduc, Y., Muller, S., Poirier, G. (1986) Biochem Cell Biol 64: 368-376
7. Kameshita, I., Matsuda, Z., Taniguchi, T., Shizuta, Y. (1984) J Biol Chem 259: 4770-4776
8. Kameshita, I., Matsuda, M, Ushiro, H., Shizuta, Y. (1986) J Biol Chem 261: 3863-3968
9. Surowy, C.S., Berger, N.A. (1985) Biochem Biophys Acta 832: 33-45

Poly(ADP-Ribose) Polymerase: Determination of Content from TCA-Precipitated Cells, Existence of Species-Specific Isoforms and Changes During the Growth Cycle

Andreas Ludwig and Helmuth Hilz

Institut für Physiologische Chemie, Universität Hamburg, D-2000 Hamburg 20, West Germany

Introduction

Poly(ADP-ribose) polymerase itself is one of the major acceptors of poly(ADP-ribose) residues under conditions of DNA fragmentation *in vitro* (1) and *in vivo* (2, 3). For a better understanding of the involvement of polymerase in alterations of chromatin structure and function it is important to know the absolute content of the enzyme under varying conditions. Because of the multiple factors influencing polymerase activity, determination of the content via activity is not reliable. We have developed a procedure for immunoquantitation of poly(ADP-ribose) polymerase from TCA precipitates. In this way artifactual degradation can be avoided. When applied in conjunction with Western blotting, the procedure also allows the analysis of isoforms of the enzyme present *in vivo*, and reactivation experiments on SDS gels from the same sample.

Results and Discussion

Immunoquantitation of poly(ADP-ribose) polymerase from TCA-insoluble fractions. Free and hemocyanin-bound poly(ADP-ribose) polymerase from pig thymus (4) was used to prepare polyclonal antibodies in rabbits. The highest titer was obtained with hemocyanin adducts. With the aid of these antibodies, we have developed a procedure for immunoquantitation of poly(ADP-ribose) polymerase. Amounts of antigen could be determined when the TCA precipitates from cells or tissues were dissolved and applied to nitrocellulose. Antibodies specifically binding to the absorbed antigen were then determined with (^{125}I) protein A. The procedure allowed quantitation of polymerase in the ng range (Fig. 1). Determination of cellular contents required only 10^5 cells. Recovery of added polymerase was almost quantitative.

Poly(ADP-ribose) polymerase content of mammalian cell lines. When antibodies purified to different degrees were applied to determine the poly(ADP-ribose) polymerase content of various mammalian cells, the values obtained were quite similar (Table 1). No significant differences were observed between tumor cells and non-transformed cells (Table 1). Nearly identical contents were also found in proliferating NIH 3T3 cells (3.26 ± 0.14 ng/μg DNA), and in SV40 transformed 3T3 cells (3.04 ± 0.24 ng/μg DNA).

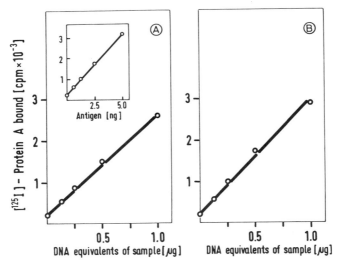

Fig. 1. Immunoquantitation of poly(ADP-ribose) polymerase extracted from the TCA insoluble fraction of rat liver. A: Reuber H35 hepatoma cells; B: rat liver; Inset in A: pure pig thymus polymerase.

Table 1. Determination of polymerase amounts with antibodies purified to different degrees

Cell Type	Antigen Equivalents (ng/μg DNA)		affinity purified antibodies[c]
	antiserum[a]	IgG fraction[b]	
HeLa	2.56 ± 0.06	2.09 ± 0.29	2.38 ± 0.19
3T3	2.51 ± 0.14	2.08 ± 0.55	2.11 ± 0.07

Mean values ± s.d. from four determinations. a: Rabbit antiserum against polymerase; b: IgG fraction of antiserum purified on protein A Sepharose column; c: Antibodies affinity purified against 116 K polypeptide.

Isoforms of polymerase in different mammalian cell lines. The patterns of antigenic polypeptides from TCA precipitated cells show characteristic species-specific differences. Human and bovine cell lines exhibit only a single 116 K band, while mouse and rat cells have an additional band of 98 K, independent of the growth cycle or transformation (Fig. 2). The 98 K isoform could not be renatured when subjected to reactivation (5) in SDS gels.

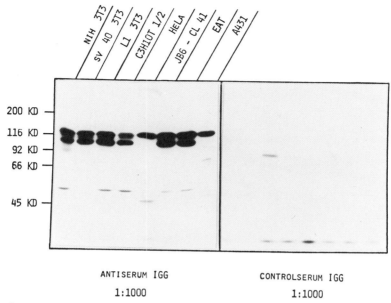

Fig. 2. Immunogenic patterns of mammalian cell lines. Western blot analysis with (^{125}I) Protein A after separating on 7.5% SDS-PAGE. Blotting was performed with IgG fraction of the antiserum.

Polymerase content in confluent cultures. In three cell lines, total antigenic material decreased when the cultures became confluent. The extent to which the decrease occurred was different in the different lines. In 3T3-L1 preadipocytes no significant change could be observed during 5 days in confluence. The phenomenon of decreasing polymerase content in non-dividing cells may relate to the observation of Jacobson *et al.* (6) that removal of potentially lethal damage in stationary cells does not require

Table 2. Polymerase contents of DNase stimulated cells in different phases of the growth cycle

Parameter	Growth Phase	Cell Line C3H1OT1/2	NH-3T3	3T3-L1
Polymerase content	log	3.77 ± 0.45	3.15 ± 0.03	2.93 ± 0.28
(ng/µg DNA ± s.d.)	confluent 1 D	2.18 ± 0.04	2.62 ± 0.31	2.69 ± 0.48
	confluent 5-6 D	1.40 ± 0.17	2.33 ± 0.12	3.72 ± 1.58
DNase stimulated	log	3.6 ± 1.2	14.8 ± 0.1	n.d.
activity	confluent 1 D	5.3 ± 2.2	6.8 ± 1.2	n.d.
(pmol/min/µg DNA ± s.d.)	confluent 5-6 D	3.8 ± 1.3	8.9 ± 0.4	n.d.

Mean values ± s.d. from three independent determinations except 3T3-L1 cells (duplicates). Polymerase activity was determined in permeabilized cells after addition of DNase.

poly(ADP-ribosyl)ation. The data in Fig. 2 and Table 2 also show that the activity as determined in permeabilized cells with the addition of DNase ("total activity") does not correlate with total immunogenic equivalents.

Acknowledgements. This work was supported by the Deutsche Forschungsgemeinschaft, SFB 232.

References

1. Ogata, N., Kawaichi, M., Ueda, K., Hayaishi, O. (1980) Biochem Int 1: 229-236
2. Kreimeyer, A., Wielckens, K., Adamietz, P., Hilz, H. (1984) J Biol Chem 259: 890-896
3. Adamietz, P., Rudolph, A. (1984) J Biol Chem 259: 6841-6846
4. Holtlund, J., Kristensen, T., Ostvold, A.C., Laland, S.G. (1981) Eur J Biochem 119: 23-29
5. Scovassi, A.L., Stefanini, M., Bertazzoni, U. (1984) J Biol Chem 259: 10973-10977
6. Jacobson, E.L., Smith, J.Y., Wielckens, K., Hilz, H., Jacobson, M.K. (1985) Carcinogenesis 6: 715-718

ADP-Ribosylation in Mitochondria: Enzymatic and Non Enzymatic Reactions

Ahmed Masmoudi and Paul Mandel

Centre de Neurochimie du CNRS, 67084 Strasbourg Cedex, France

Introduction

Mitochondria are known to be able to synthesize NAD⁺, the coenzyme involved in various oxidation systems (1, 2). Moreover it has been demonstrated that NAD⁺ is a substrate for enzymes such as ADP-ribosyl transferases and NAD glycohydrolases. ADP-ribosylation in mitochondria was first described by Kun and co-workers (3). Enzymatic transfer of ADP-ribose from NAD⁺ to an acceptor protein of 50 kDa has been observed in mitochondrial extracts. Richter *et al.* have reported a covalent modification by mono (ADP-ribosyl)ation of a 31 kDa protein present in submitochondrial particles (SMP) which seems to be involved in the regulation of Ca²⁺ release from mitochondria (4). According to Hilz *et al.* (5) non enzymatic ADP-ribosylation may occur in beef heart and liver mitochondria after enzymatic hydrolysis of NAD⁺ by NAD glycohydrolase and subsequent transfer of ADP-ribose to specific acceptors (5). We recently demonstrated (6) the existence of ADP-ribosyl transferase activity in rat liver mitochondria and mitoplasts. While NAD glycohydrolase activity was found to be inhibited by nicotinamide or 3-aminobenzamide, ADP-ribosyl transferase activity was found to be unaffected. ADP-ribose bound to mitochondrial proteins appears to be oligomeric. In this paper we demonstrate that removal of NAD glycohydrolase from solubilized mitochondrial proteins by hydroxyapatite chromatography does not affect the transfer of ADP-ribose from NAD⁺.

Results

ADP-ribosylation in rat liver mitochondria. Incubation of rat liver mitochondria or mitoplasts with radiolabeled NAD⁺ results in the incorporation of ADP-ribose into one major protein of 50 kDa and a minor one of 30 kDa. In order to confirm that the acid insoluble radioactive material transferred to these proteins is in fact due to ADP-ribosyl transferase activity and not to NAD glycohydrolase activity we investigated the effects of some inhibitors of these activities. Our results demonstrate that ADP-ribosyl transferase activity is involved in the process of mitochondrial ADP-ribosylation, because when mitochondrial NAD glycohydrolase activity was inhibited by 10 mM nicotinamide or 10 mM 3-aminobenzamide, ADP-ribose transfer still occurred (Fig. 1). The inhibition of the mitochondrial protein ADP-ribosylation by these inhibitors was only slight as compared to the effect on nuclear protein ADP-ribosylation.

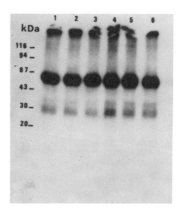 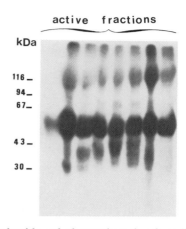

Fig. 1 (left) Autoradiogram of LDS polyacrylamide gel electrophoresis of rat liver mitochondria or mitoplasts depicting protein: ADP-ribosylation. Mitochondria or mitoplasts were incubated in the presence of 100 μM [³²P]NAD⁺. After 60 min of incubation the ADP-ribosylated proteins were precipitated with trichloroacetic acid at 0°C. Pellets were rinsed with cold ether and resuspended in LDS solubilization buffer (6). After electrophoresis gels were stained with Coomassie Brilliant Blue R-250. After destaining gels were dried and submitted to autoradiography with Amersham film RPN 6. Lane 1: mitochondria; Lane 2: mitochondria + 10 mM nicotinamide; Lane 3: mitochondria + 10 mM 3-aminobenzamide; Lane 4: mitoplasts (100 μg); Lane 5: mitoplasts + 10 mM nicotinamide; Lane 6: mitoplasts + 10 mM 3-aminobenzamide incubation with [³²P]NAD⁺ (2 μci) at NAD⁺ concentration of 100 μM.

Fig. 2. (right) Autoradiogram of LDS polyacrylamide gel electrophoresis (10%). Fractions eluted from hydroxylapatite column containing ADP-ribosyl transferase activity were incubated in the presence of 100 μM [³²P]NAD⁺ for 60 min. Gels were run for 16 hr at 15 mA using a gel thickness of 1 mm.

Partial separation of ADP-ribose transferase and NAD glycohydrolase activities in rat liver mitochondria. In order to demonstrate the involvement of ADP-ribosyl transferase in ADP-ribosylation of mitochondrial proteins, we have used one step of the purification procedure described by Moser and co-workers with slight modifications (7). Under these conditions, NAD-glycohydrolase activity is separated from solubilized submitochondrial proteins by hydroxylapatite chromatography. The proteins bound to hydroxyapatite were eluted using a potassium phosphate buffer gradient from 10 mM to 500 mM. The ADP-ribosyl transferase activity was eluted at a phosphate concentration of about 200 mM phosphate. The fractions containing ADP-ribosyl transferase activity were assayed for acceptor protein followed by lithium dodecyl sulfate (LDS) polyacrylamide gel electrophoresis. Fig. 2 shows three major ADP-ribosylated proteins ranging between 30 and 100 kDa in the presence of 100 µM NAD⁺ as substrate. ADP-ribosyl transferase enriched fractions were collected, pooled and concentrated. This material was dialyzed and applied to the DEAE-cellulose column. The column was washed with Tris/HCl, pH 8.1 buffer and eluted with a linear gradient of 0.05 M Tris/HCl, pH 8.1 and 0.05 M Tris/HCl, pH 8.1 containing 0.4 M sodium chloride. Fractions containing ADP-ribosyl transferase activity were collected and concentrated by ultrafiltration. All ADP-ribosyl transferase activity was recovered in the initial washes. However, the residual NAD glycohydrolase activity was eluted at about 200 mM sodium chloride. No ADP-ribosyl transferase activity was observed in eluted material.

These two fractions were assayed for ADP-ribosylated endogenous mitochondrial proteins. Fig. 3 shows that the fraction enriched in ADP-ribosyl transferase activity catalyzed the transfer of ADP-ribose from NAD⁺ to five protein acceptors. As shown after LDS polyacrylamide gel electrophoresis and autoradiography, the molecular weight of these proteins ranged from 30 kDa to 80 kDa (Fig. 3, lane A). The fraction enriched in NAD glycohydrolase activity was shown to ADP-ribosylate non-enzymatically one single protein of molecular weight 50 kDa (Fig. 3, lane B). Fig. 4 shows unambiguously that the protein which is modified in the fraction enriched in NAD glycohydrolase activity has a different mobility than the modified proteins in the fraction enriched in ADP-ribosyl transferase activity.

Discussion

In the present study we provide data in favor of the existence of an enzymatic transfer of ADP-ribose from NAD⁺ to mitochondrial proteins. The ADP-ribosyl transferase activity can be separated from NAD glycohydrolase activity using two-step preparative column chromatography. It is interesting to note that in intact mitochondria or mitoplasts (inner membrane devoid of outer membrane) only two proteins were shown to be ADP-ribosylated. When mitochondrial proteins were further purified some other proteins appeared to be ADP-ribosylated. This may be due either to dissociation of protein complexes and liberation of the new acceptor sites or to the presence of phosphodiesterase or nucleotide pyrophosphatase activities in rat liver mitochondria which can be separated from active fractions during the purification procedure. With regard to these results it is worth noting that mitochondria contain NAD⁺ at a concentration of 1-2 mM which is likely to be involved in ADP-ribosylation *in vivo* as endogenous substrate.

In our previous study we demonstrated the presence of an enzymatic ADP-ribosyl transferase activity in rat liver mitochondria (6). The use of a two-step purification of the ADP-ribosyl transferase described here supports our conclusion of this recent report. Non enzymatic reactions can occur *in vitro* in mitochondrial homogenates although only one single acceptor protein could be detected with partially purified NAD glycohydrolase. Our conclusion is in agreement with that of Payne *et al.* (8) who concluded that non-enzymatic reactions appear to be quantitatively non significant *in vivo*.

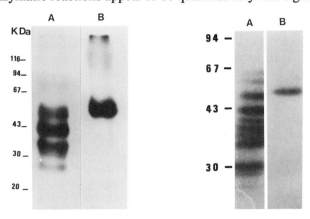

Fig. 3. (left) Autoradiogram of LDS polyacrylamide gel electrophoresis (10%). Fractions enriched in ADP-ribosyl transferase activity from hydroxylapatite column were loaded onto a DEAE-cellulose column: Lane A: fraction passing through the column and containing ADP-ribosyl transferase activity; Lane B: fraction eluted from the column with sodium chloride gradient and containing NAD glycohydrolase activity. Fractions were incubated in the presence of 100 μM [^{32}P]NAD$^+$ at 37°C.

Fig. 4. (right) Autoradiogram of SDS polyacrylamide gel electrophoresis (10%). Same fractions as Fig. 3.

References

1. Mandel, P., Okazaki, H., Niedergang, C. (1982) Prog Nucleic Acids Res Mol Biol 27: 1-51
2. Ueda, K., Hayashi, O. (1985) Ann Rev Biochem 54: 73-100
3. Kun, E., Zimber, P.M., Chang, A.C.Y., Puschendorf, B, Grunicke, M. (1975) Proc Natl Acad Sci USA 72: 1436-1440
4. Richter, C., Winterhalter, K.M., Baumhuter, S., Lotscher, H.R., Moser, B. (1983) Proc Natl Acad Sci USA 80: 3188-3192
5. Hilz, H., Koch, R., Fanick, W., Klapproth, K., Adamietz, P. (1984) Proc Natl Acad Sci USA 81: 3929-3933
6. Masmoudi, A., Mandel, P. (1987) Biochemistry USA 26: 1965-1969
7. Moser, B., Winterhalter, K.M., Richter, C. (1983) Arch Biochem Biophys 224: 358-364
8. Payne, D.M., Jacobson, E.L., Moss, J., Jacobson, M.K. (1985) Biochemistry USA 24: 7540-7549

Localization of the Zinc-Binding Sites in the DNA-Binding Domain of the Bovine Poly(ADP-Ribose) Polymerase

Alice Mazen, Daniel Lamarre[1], Guy Poirier[1], Gérard Gradwohl, and Gilbert de Murcia

IMBC du CNRS, 67084 Strasbourg Cédex, France

Introduction

The poly(ADP-ribose) polymerase is a chromatin associated enzyme of eukaryotic cell nuclei, which has an absolute requirement for DNA (1). *In vivo* (2, 3) and *in vitro* (4, 5) its activity is stimulated by DNA containing nicks or double stranded breaks. The binding of the enzyme to sites of strand breaks on DNA, is a prerequisite step to activation. Like other classes of proteins involved in nucleic acid binding poly(ADP-ribose) polymerase is a zinc metalloenzyme (6) and it is suggested that a metal-binding site is involved as part of the interaction of DNA and the enzyme. Kameshita *et al.* (7) elucidated the localization of three functional domains in the enzyme molecule. These domains are separable by mild proteolysis, the NH_2 fragment of 46 kDa corresponds to the DNA-binding domain, the COOH-terminal fragment of 54 K is the domain for the substrate NAD^+ binding, and the third one of 22 K contains the sites for accepting poly(ADP-ribose). We attempted to localize the zinc binding sites, having regard to the DNA-binding domain of the enzyme. Radioactive zinc (^{65}Zn) and ^{32}P-labelled nick translated DNA were used alternatively, to analyze electro-blots loaded with proteolytic fragments of the enzyme. The same blots were further immunostained with monoclonal antibodies. Our results showed that the radioactive zinc is only bound to the proteolytic fragments containing the DNA-binding domain of the enzyme.

Results and Discussion

The proteolytic fragments of the poly(ADP-ribose) polymerase, were obtained by mild digestion with papain and chymotrypsin as described by Kameshita *et al.* (7). Digestion of the enzyme with trypsin was performed at 20°C with a trypsin to protein ratio of 20:1 and the reaction was stopped with trichloroacetic acid at a final concentration of 20%. Poly(ADP-ribose) polymerase and the fragmens were separated electrophoretically on a 15%

1Centre de Recherches en Oncologie Moléculaire, Hôtel-Dieu de Québec, Québec, G1R 2J6 Canada polyacrylamide SDS gel. The mini-gels were run at 100 volts for one hour. Before transfer, the gels were incubated in a reduction buffer (25 mM Tris, 190 mM glycine, 0.2% SDS, 5% ß-mercaptoethanol pH 8.3) for one hour at 37°C (8). The electrophoretic transfer of the proteins from the gels to the nitrocellulose sheets was performed as described by Towbin (9), at 200 mA for one hour. The transferred fragments were characterized by immunodetection using monoclonal antibodies developed against poly(ADP-ribose) polymerase and analyzed for their reactivity against the NAD⁺ and DNA-binding fragments (10). The immunoreaction can be performed on blots which had already been treated for zinc-binding or DNA-binding and had been exposed to x-ray film. Zinc binding proteins were detected on the blots, after the nitrocellulose was incubated for one hour in buffer A (10 mM Tris-HCl pH 7.4) then for 15 min in a buffer containing ^{65}Zn (1 µCi ^{65}ZnCl$_2$, in 5 ml buffer containing 0.1 M KCl). The blots were washed twice for 10 min in buffer A containing 0.1 M KCl. The technique used to detect DNA-binding proteins has been described by Mellor et al. (11). The electroblots were placed in buffer B (20 mM Tris-HCl pH 7.8, 10 mM NaCl, 3 g/l Ficoll, 3 g/l polyvinylpyrrolidone) and washed twice for 15 min with shaking. The blots were then incubated for 1 hr at room temperature in a sealed plastic bag containing 0.2 µg ^{32}P-labelled nick translated DNA (0.6 x 108 cpm/µg) in 1 ml buffer B, were washed four times in 200 ml buffer B for 30 min, and dried. Autoradiographs were produced by exposure varying from 20 to 40 hrs or from 2 to 16 hrs for the detection of bound ^{65}Zn or bound ^{32}P-DNA, respectively.

Zinc-metalloproteins like anhydrase carbonic of M_r = 30000 and bovine serum albumin of M_r = 67000 (12) are present in the kit of low molecular mass standards used in gel electrophoresis. We first tested the binding of ^{65}Zn to these site markers and to the chymotrypsin and papaïn digestion fragments on the blots under a variety of ionic strength and pH conditions. We detected no bound radioactivity at pH 6.0; the exchange with ^{65}Zn occurred only at pH 7.4. Fig. 1 shows the binding of ^{65}Zn to the 30 K and 67 K markers (Fig. 1, blot 5) and exclusively to the 66K and 46 K fragments of the enzyme (Fig. 1, blot 3, 4) which have been reported to contain DNA-binding sites (7). Monoclonal antibodies C^I_9 also reacted exclusively with antigenic sites on the 46 K and the 66K fragments (Fig. 1, blot 1, 2). In order to correlate the ability of some domains to bind both zinc molecules and DNA, blots of the enzyme and papain digestion fragments were tested with three different specific probes: antibodies, ^{65}Zn and ^{32}P-DNA. In Fig. 2, it appears that exclusively the native enzyme and the 46 K fragment were visualized by their immunoreactivity with the monoclonal antibody C^I_9. The autoradiographs show that the same bands were labelled with ^{65}Zn and with ^{32}P-nick translated DNA (Fig. 2, blot 5, 6). Thus the two types of binding sites are localized on the same fragments.

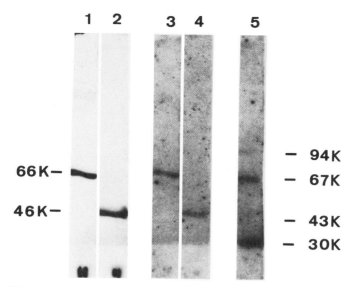

Fig.1. Electroblots of chymotrypsin and papain digestion fragments, resolved by a 15% SDS elecrophoresis. Fixation of ^{65}Zn. Chymotrypsin digestion (1, 3), papain digestion (2, 4), low molecular mass standards (5). Autoradiograms after incubation with ^{65}Zn (3, 4, 5). Immunoblots after incubation with monoclonal antibody CI$_9$ (1, 2).

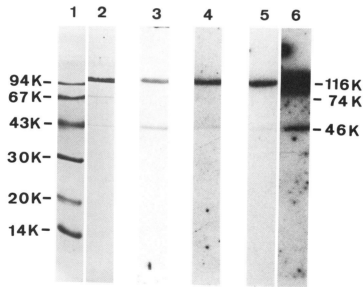

Fig. 2. Electroblots of papain digestion fragments, resolved by a 15% SDS electrophoresis. Stained gel: low molecular mass standards (1), papain digestion (2). Immunoblot incubated with monoclonal antibody CI$_9$ (3). Autoradiogram after incubation with ^{65}Zn (4), or with nick translated ^{32}P-DNA: 1 hr exposure (5) and 16 hr (6).

91

The 66 K fragment containing the DNA binding site, can be split by trypsin digestion in two smaller fragments (29 K and 36 K) which expressed different specificities with the monoclonal antibodies C^I_9 and C^{II}_{10}. Both antibodies reacted with the 66 K fragment. But C^{II}_{10} antibody bound to the 36 K fragment which possesses the site for accepting poly(ADP-ribose) (13) while C^I_9 antibody reacted with the 29 K fragment (see Fig. 4). Thus it can be concluded that the 29 K fragment is located in the N-terminal part of the enzyme.

Fig. 3 represents trypsin digestion fragments of the poly(ADP-ribose) polymerase. The C^I_9 antibody reacted with the native enzyme (116 K), the 46 K and 29 K fragments (Fig. 3, blot 4). ^{65}Zn and labelled DNA (in Fig. 3, blot 5 and blots 6, 7, respectively) bound to the exact same fragments but did not bind to the 36 K fragment.

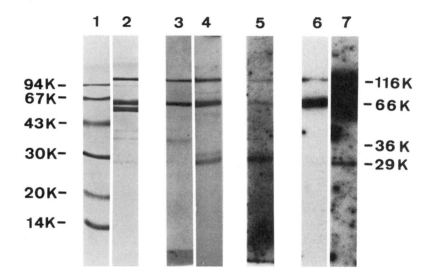

Fig. 3. Electroblots of trypsin digestion fragments. Stained gel: low molecular mass standards (1), trypsin digestion (2). Immunoblot incubated with monoclonal antibody C^{II}_{10} (3) or C^I_9 (4). Autoradiogram after incubation with ^{65}Zn (5) or with nick-translated [^{32}P]-DNA (6, 7).

These results taken together,. suggest that the zinc-binding sites and the DNA-binding domain are localized in the 29 K N-terminal region of the poly(ADP-ribose) polymerase. Metal-binding domains occur in a number of nucleic acid binding proteins and various eukaryotic regulatory proteins (14). Miller *et al.* (15) proposed a novel structural "finger protein model" of interaction between DNA and DNA binding protein. At present, we don't know if the N-terminal domain of the poly(ADP-ribose) polymerase contains sequences compatible with potential DNA-binding "fingers" in which four cysteines (or histidines) would be coordinated tetrahedrally with a Zn^{2+} ion. From the work of Zahradka and Ebisuzaki (6) we know that zinc ion participates in the interaction of the enzyme and DNA; our results show that zinc and DNA binding sites are both located in the N-terminal quarter of the enzyme. One can speculate that zinc stabilized DNA-binding fingers may be present in the poly(ADP-ribose) polymerase.

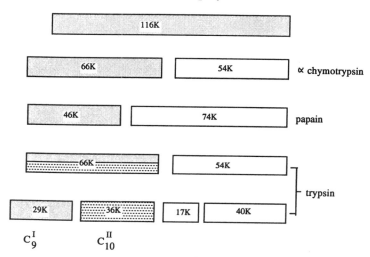

Fig. 4. Immunoreactivity of two monoclonal antibodies C^I_9 and C^{II}_{10} with the poly(ADP-ribose) polymerase and different proteolytic fragments.

References

1. Chambon, P., Weil, J.D., Doly, J., Strosser, M.T., Mandel, P. (1966) Biochem Biophys Res Commun 25: 638-643
2. Miller, E.G. (1975) Biochem Biophys Acta 395: 191-200
3. Berger, N.A., Weber, G., Kaichi, A.S. (1978) Biochim Biophys Acta 519: 87-104
4. Ohgushi, H., Yoshihara, K., Kamiya, T. (1979) J Biol Chem 255: 6205-6211
5. Benjamin, R., Gill, D.M. (1980) J Biol Chem 255: 10502-10508
6. Zahradkia, P., Ebisuzaki, K. (1984) Eur J Biochem 142: 503-509
7. Kameshita, I., Matsuda, Z., Taniguchi, T., Shizuta, Y. (1984) J Biol Chem 259: 4770-4776
8. Aoki, Y., Kunimoto, M., Shibata, Y., Suzuki, K.T. (1986) Anal Biochem 157: 117-122
9. Towbin, H., Staehelin, T. and Gordon, J. (1979) Proc Natl Acad Sci USA 76: 4350-4354
10. Lamarre, D., Talbot, B., Leduc, Y., Muller, S., Poirier, G. (1986) Biochem and Cell Biol 64: 368-376
11. Mellor, J., Fulton, A., Dobson, M., Roberts, N., Wilson, W., Kingsman, J., Kingsman, S.M. (1985) Nucl Acids Res 13: 6249-6263
12. Vallee, B.L., Walker, W. (1970) Proteins 5: 61-93
13. Lamarre, D., de Murcia, G., Talbot, B., Laplante, C., Leduc, Y., Poirier, G. (1987) This volume
14. Berg, J.M. (1986) Science 232: 485-487
15. Miller, J., McLachan, A.D., Klug, A. (1985) EMBO J 4: 1609-1614

Heterogeneity of Polyclonal Antisera to ADP-Ribose and Their Use as Probes for ADP-Ribosylation in Human Tumor Cells

Sarada C. Prasad, Peter J. Thraves, Jane Boyle, and Anatoly Dritschilo

Department of Radiation Medicine, Vincent T. Lombardi Cancer Research Center, Georgetown University Medical Center, Washington, D.C. 20007

Introduction

DNA strand breaks, induced by exposure to chemicals or ionizing radiation stimulate poly(ADP-ribose) polymerase activity (1, 2). The resultant protein modifications have been postulated to comprise an important step in the DNA repair process (3). Inhibitors of the polymerase have been shown to sensitize human fibroblasts (4) and certain tumor cells (5) to ionizing radiation and to inhibit the repair of potentially lethal radiation injury (6, 7). That the response of the tumor cell lines vary, with some showing sensitivity to inhibitors of poly(ADP-ribosyl)ation and irradiation while others do not, suggested a need for detailed investigation of the ADP-ribosylation process in these tumor cell lines. In the present study we report the quantitative variations in protein-bound mono(ADP-ribose) levels as well as poly(ADP-ribose) polymerase activities and cellular NAD levels of various tumor cells. To this end, we also describe the development and characterization of polyclonal antisera to mono(ADP-ribose) and its potential use as a probe for studies of ADP-ribosylation.

Results and Discussion

In the present report the immunogen used was ADP-ribose, conjugated to hemocyanin and the antisera generated in rabbits had affinity for ADP-ribose and NAD. Binding studies using [3H]ADP-ribose, [3H]NAD or [3H]5'AMP indicate that the two former ligands bound equally well (Fig. 1) while [3H]5'AMP did not bind at all. In addition, it was also observed that inclusion of Tween-20 (Fig. 1) in the buffer greatly enhanced the binding of [3H]NAD and [3H]ADP-ribose (data not shown). Cross reactivity studies with various related nucleotides (Table 1) revealed that many adenine ring-based compounds like ATP, 5'ADP, 3'5'ADP, N^6 5'AMP are good competitors indicating that the adenine ring is the major antigenic determinant (AD-1) (Fig. 2). However, the behavior of adenine (absence of ribose phosphate structure), deoxy-ATP (substitution of deoxyribose for ribose), 2',3' cAMP, 3',5' cAMP (cyclization of the nucleotide), 2'AMP, 3'AMP (position of the phosphate and ribose) in addition to poly A (polymerization of the adenine nucleotide) (Table 1) suggest some of the alterations on the adenine ring structure distort the recognition site for the antibody populations. These observations lead us to the conclusion that the antibodies exhibit a considerable amount of specificity to the ribose-phosphate extension of the adenine moiety (AD-2) (Fig. 2). α-NMN, β-NMN and nicotinic acid are not recognized even at several fold higher concentrations (Table 1) and therefore it is very clear that the nicotinamide ring (AD-4) (Fig. 2) is not part of the antigenic determinant. The fact that there is small distinction between the 50% displacement value of ADP-ribose (16.0 ng) and 5'AMP (10.0 ng) does not rule out the possibility that the second phosphate and ribose (AD-3) (Fig. 2) are also important part of the antigen antibody recognition site.

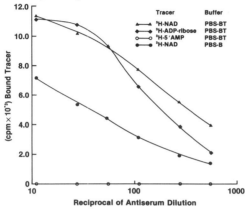

Fig. 1. Titration of mono(ADP-ribose) antiserum. Appropriately diluted serum was incubated overnight at 4°C with either [³H]NAD, [³H]ADP-ribose, or [³H]5'AMP (~ 30,000 cpm) in PBS-BSA buffer (final volume of 0.5 ml) either in the presence or absence of 0.05% Tween-20. Bound tracer was separated from free by DCC treatment. The bound fraction was counted in 10 ml of aquasol.

It is of importance to point out here that 5'AMP can compete with [³H]NAD (Table 1) or [³H]ADP-ribose (data not shown) for binding to the antiserum much more efficiently than the homologous ligands themselves. Also NAD is recognized by the anti-serum at two-fold higher concentration than ADP-ribose (Table 1). Therefore, it was necessary to determine the specificities of these antisera using 5'AMP-sepharose and NAD agarose affinity matrices. As shown in Table 2, most of the [³H]NAD binding activity is retarded by the affinity columns. This binding seems tight and rather specific because of the fact that there is no detectable activity in the unbound fraction or sodium chloride wash. Elution at pH 2.0 released nearly 80% of [³H]NAD binding activity. It is essential to note here that there was no [³H]-5'AMP binding activity seen in any fraction ruling out the interpretation that there was more than one set of antigenic determinant. Therefore, it is also evident that we are dealing with a homogeneous

95

antibody population to ADP-ribose extending the recognition to 5'AMP and NAD. Bredehorst *et al.* (8) hypothesized that ADP-ribose bound to protein in conjugated form is metabolized during immunization to 5'AMP and results in antisera specific to 5'AMP. However, this explanation holds only if [^{3}H]5'AMP can also directly bind to the antiserum.

Table 1. Specificity of mono(ADP-ribose) antiserum-cross reactivities of related nucleotides.

	ng Required for 50% Displacement		
	5'-Monophosphate	Triphosphates	Deoxytriphosphates
Adenosine	10.0	14.05	30.0
Guanosine	125.0	180.0	150.0
Cytosine	433.0	484.0	1000.0
Uridine	130.0	180.0	210.0
Thymidine	-	2350.0	-
Adenosine-Based Ligands			
Adenosine	6900.0		
3'5' cAMP	400.0		
2'5' ADP	42.6		
3'5' ADP	68.0		
5' ADP	62.0		
NAD-Analogues-Derivatives			
NAD	35.0		
ADP-ribose	14.0		
N^{6}NAD	362.0		
N^{6}5' AMP	10.1		
α-NMN	> 3320.0		
β-NMN	> 3320.0		
Nicotinic Acid	> 1220.0		
RNA-Hydrolytic Products			
2' AMP	> 6050.0		
3' AMP	> 6050.0		
2'3' cAMP	> 3520.0		
Nucleic Acids - Polymer			
Polymer	1310.0		
Poly A	1600.0		
DNA	> 35000.0		
RNA	1300.0		

Various concentrations of nonradioactive nucleotide analogues were incubated with 50 µl of antiserum (final dilution of 1:150) overnight at 0-4°C in volume of 0.4 ml PBS-BT. [^{3}H]NAD tracer was added (~ 30,000 cpm) in 50 µl and the incubation continued for another 24 hrs. Antibody bound tracer was separated from unbound fraction by DCC treatment. Controls were set up with no antiserum and with no competing ligands.

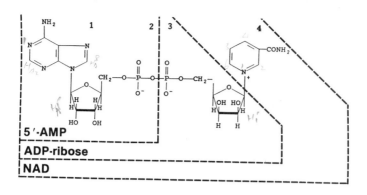

Fig. 2. Possible antigenic determinants in nicotinamide adenine dinucleotide (NAD). 1, 2, 3, 4 represent the likely antigenic determinants.

We observe that the radioimmunoassay displacement curves for NAD, ADP-ribose, 5'AMP using either [³H]NAD (Fig. 3) or [³H]ADP-ribose (data not shown) are extremely sensitive while the specificity is determined by the nature of the sample. The justification for using [³H]NAD or [³H]ADP-ribose as the radioactive tracer in the absence of any detectable binding to [³H]5'AMP is provided by the RIA-displacement curves for NAD, ADP-ribose, 5'AMP using [³H]NAD or [³H]ADP-ribose. The 50% displacement values for NAD, ADP-ribose, 5'AMP are 30-35 ng, 14-16 ng, and 9-11 ng with [³H]NAD or [³H]ADP-ribose as tracer. In the RIA for 5'AMP developed by Bredehorst *et al.* (9), the sample preparation protocol ensures that all of the mono ADP-ribose released by the alkali-treatment is converted to 5'AMP and at this point there is no free NAD or ADP-ribose remaining in the sample. Such a sample is rich in high concentrations of RNA hydrolytic products 2'AMP, 3'AMP, 2',3'AMP and intact DNA which do not interfere in the quantitation of 5'AMP using the present RIA. This RIA can also be used to quantitate cellular NAD pools after boronate chromatography of TCA-soluble fractions (studies in progress).

Table 3 reveals mono(ADP-ribose) levels determined as 5'AMP equivalents (after alkaline hydrolysis) in logarithmically growing human tumor cell lines using the currently described RIA. It is evident that all tumor cell lines show higher mono(ADP-ribosyl)ation as compared to normal human fibroblasts. The enzyme activity analysis by both the permeabilized cell system (10) and sonicated cell homogenates (11) brings out interesting differences in the various tumor cell lines tested. We make all comparisons for the various parameters tested in the different tumor cells to normal human fibroblasts. Firstly, the activity of the polymerase in fibroblasts (Table 3) is much higher in cell homogenates as compared to the permeabilized cells. Secondly, there is a striking increase in activity of

97

polymerase in Ewing's sarcoma (6-20 fold) and squamous cell carcinoma (1.4-4 fold), while lung carcinoma is almost identical to the fibroblast system by both the assay methods.

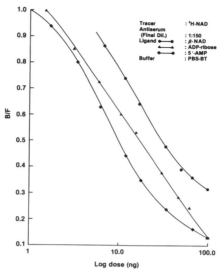

Fig. 3. Radioimmunoassay displacement curves. Various concentrations of nonradioactive nucleotide analogues were incubated with 50 µl of antiserum (final dilution of 1:150) overnight at 0-4°C in a volume of 0.4 ml of PBS-BT. [^{3}H]NAD tracer was added (~ 30,000 cpm) in 50 µl and the incubation continued for another 24 hrs. Antibody bound tracer was separated from unbound fraction by DCC treatment. Controls were set up with no antiserum and with no competing ligands.

Table 2. Binding characteristics of mono(ADP-ribose) antiserum.

Affinity Matrix	NAD Agarose		5' AMP - Sepharose	
		^{3}H-CPM Bound		
[^{3}H]Label	NAD	5' AMP	NAD	5' AMP
Antibody Fraction:				
Original Serum (1:100)	14774	ND	14774	ND
Unbound FRN	ND	ND	ND	ND
Salt Wash (0.5 M NaCl)	420	ND	ND	ND
pH 2.0 Eluate Frn #1	10742	ND	5543	ND
Frn #2	8768	ND	2805	ND
Frn #3	3884	ND	1617	ND

ND = Not detectable

The affinity matrices were (0.5 ml bed volume) equilibrated with PBS-BT buffer and 0.5 ml of antiserum fractionated on each column. The unbound material was washed with 20.0 ml loading buffer (PBS-BT) followed by a 0.5 M NaCl to remove nonspecific binding. Specific antibody fraction was eluted in the same buffer adjusted to pH 2.0 and the fraction (2.0 ml each) were immediately neutralized. Assayed 100 µls of each fraction with either [^{3}H]NAD (33,400 cpm = 0.90 ng) or ^{3}H-5'AMP (33, 610 cpm = 0.83 ng) in PBS-BT buffer.

The cellular content of NAD the substrate for poly(ADP-ribose) polymerase is probably one of the important factors regulating the enzyme activity. Several studies relate the significance of NAD metabolism to the poly(ADP-ribosyl)ation status of the acceptor proteins under a variety of experimental conditions (13). We find here (Table 3) that there is an inverse relationship between the levels of poly(ADP-ribose) polymerase activity and NAD levels of Ewing's sarcoma and laryngeal squamous cell carcinoma and lung carcinoma as compared to normal human fibroblasts.

Table 3. Comparisons in different cell lines

Cell Type	Radiation Survival Parameters*	Radiation Sensitization by 3AB or BZ*	Poly(ADP-ribose) Polymerase Activity Permeabilized	Sonicated	Protein-Bound Mono(ADP-Ribose	Cellular NAD
Ewing's Sarcoma (A4573)	D_o=2.40 Gy n=2.2	sensitized	20.20	6.26	2.20	0.33
Lung a Adeno-carcinoma (A549)	D_o=2.40 Gy n=3.4	not sensitized	1.34	1.17	1.76	1.88
Laryngeal Squamous Cell Carcinoma (SQ-20B)	D_o=2.40 Gy n=1.3	not sensitized	4.32	1.40	2.42	0.38
Normal Human Fibroblasts (NHF)	D_o=1.25 Gy n=1.0	sensitized	1.00 (0.51)	1.00 2.88)	1.00 (0.69)	1.00 (3.95)

Poly(ADP-ribose) polymerase activity was assayed according to Berger et al. (10) in permeabilized cells and Cherney et al. (11) cellular homogenates. Total (alkali-labile) protein-bound mono(ADP-ribose) levels were determined as described in this report while sample preparation for RIA was as reported by Bredehorst et al. (9). NAD assay was done by the cycling assay of Bernofsky et al. (12).

*Indicates data adapted from Thraves et al. (5). All data represents mean of 2-3 independent determinations. For reasons of clarity, data is normalized in comparison to normal human fibroblasts. In each case data for fibroblasts is identified in parenthesis as nmole/mg protein/min or nmole/mg protein.

The differences in the endogenous ADP-ribosylation status of proteins are not as striking as the polymerase activity itself in various cell lines. This observation could be interpreted to reflect a short half-life of the polymer or monomer and also the enzyme assay represents the potential of the cell to synthesize the polymer while RIA of protein-bound (endogenous levels) mono(ADP-ribose) evaluates the steady state levels.

In summary, we observe that (ADP-ribosyl)ation-status of a radiocurable tumor cell line (Ewing's sarcoma) markedly differs from other relatively radioincurable tumors (lung adenocarcinoma and squamous cell carcinoma). Therefore, the involvement of ADP-ribose in cellular radiosensitivity and DNA repair processes, may be at different levels or by different mechanisms.

Acknowledgements. This work was supported by grant number PDT-279, American Cancer Society, Inc. The authors thank Ms. Sandra Hawkins for her assistance in the preparation of this manuscript.

References

1. Skidmore, C.J., Davies, M.I., Goodwin, P.M., Halldorsson, M., Lewis, P., Shall, S., Ziaee, A. (1979) Eur J Biochem 101: 135-142
2. Nduka, N., Skidmore, C.J., Shall, S. (1980) Eur J Biochem 105: 525-530
3. Shall, S. (1984) Rad Biol 11: 1-69
4. Thraves, P.J., Mossman, K.L., Brennan, T., Dritschilo, A. (1985) Rad Res 104: 119-127
5. Thraves, P.J., Mossman, K.L., Brennan, T., Dritschilo, A. (1986) Int J Radiat Biol 50: 961-972
6. Ben-Hur, E., Utsumi, H., Elkind, M.M. (1984) Brit J Can 49(6): 39-42
7. Brown, D.M., Evans, J.W., Brown, J.M. (1984) Brit J Can 49(6): 27-31
8. Bredehorst, R., Ferro, A.M., Hilz, H. (1978) Eur J Biochem 82: 105-113
9. Bredehorst, R., Mila, M.S., Hilz, H. (1981) Biochem Biophys Acta 652: 16-28
10. Berger, A.N., Adams, J.W., Sikorski, G.W., Petzold, S.J., Shearer, W.T. (1978) J Clin Invest 62: 111-118
11. Cherney, B.W., Midura, R.J., Caplan, A.I. (1985) Dev Biol 112: 115-125
12. Bernofsky, C., Swan, M. (1973) Anal Biochem 53: 452-458
13. Olivera, B.M., Ferro, A.M. (1982) ADP-Ribosylation Reactions - Biology and Medicine, Academic Press, pp. 19-40

Abbreviations:

PBS-BT - Phosphate-buffered saline, 0.5% BSA, 0.05% Tween-20
DCC - Dextran-coated charcoal
3AB - 3-Aminobenzamide
BZ - Benzamide
AD - Antigenic determinant
RIA - Radioimmunoassay

ADP-Ribosylating Activity in *Sulfolobus solfataricus*

Piera Quesada, Maria Rosaria Faraone-Mennella, Mario De Rosa[1],
Agata Gambacorta[2], Barbara Nicolaus[2], and Benedetta Farina.

Dipartimento Chimica Org. e Biol., Universita di Napoli, 80134 Napoli, Italy

Introduction

The thermophilic microorganism *Sulfolobus solfataricus* is able to grow at low pH (3.5) and high temperature (87°C) and has been isolated from an acidic hot spring in Agnano (Napoli), Italy (1). This bacterium belongs to the archaebacteria, a phylogenetic group of microorganisms that can be distinguished from other bacteria and eukaryotes (2, 3). The archaebacteria group was initially defined on the basis of a partial 16S ribosomal RNA sequence comparison, which has been well supported by other biochemical features (2-4). More recent studies by Woese and collaborators, based on comparison of complete 16S ribosomal RNA sequences, confirmed and extended earlier archaebacteria phylogeny (3). Although archaebacteria taxonomy is complex and controversial, three main phenotypes have been described, characterized by their peculiar ecological niches, namely high salt environments (halophilic and haloalkaliphilic archaebacteria), high temperature environments (thermophilic and sulfur-dependent archaebacteria) and stringent anaerobic conditions (methanogenic archaebacteria). Several features indicate that the sulfur-dependent branch of archaebacteria, to which *Sulfolobus solfataricus* belongs, is more closely related to eukaryotes than eubacteria or other archaebacteria (3). One example is the diphtheria toxin reaction, which catalyzes the transfer of ADP-ribose to eukaryotic elongation factor 2. Kessel and Klink (5) demonstrated that the toxin is able to catalyze, in several archaebacteria including *Sulfolobus solfataricus*, the covalent binding of ADP-ribose to elongation factor, EF-2.

Many enzymes have been isolated and characterized from *Sulfolobus solfataricus*. All are thermophilic, thermostable and show resistance to denaturating agents and organic solvents (6-11). In this work we present preliminary results on an ADP-ribosyl transferase activity in *Sulfolobus solfataricus*.

Results and Discussion

Initial experiments revealed that ADP-ribosyl transferase activity was present in crude homogenates obtained from a *Sulfolobus solfataricus* cell-

---

[1]Istituto Chimica M.I.B. CNR, 80072 Arco Felice Napoli, Italy
[2]Istituto Biochimica Macromolecole, I Fac. Medicina, 80134 Napoli, Italy

free system. ADP-ribosyl transferase activity was determined, after incubation with [^{14}C]NAD$^+$ in crude homogenates of cells from the three different growth phases (Fig.1, A-C) as a function of temperature and pH. All three samples responded in a similar way. In each case the enzymatic activity appeared to be related to temperature more than to pH variations. In fact the activity did not vary over a broad pH range (5.0-8.0) although it decreased sharply at pH values of 4.0 and 9.0. This activity seemed to be related to the growth phase of the microorganism (Fig. 1, Table I). A higher concentration of the ADP-ribosyl transferase enzyme was present during the logarithmic growth phase (A) while in the late logarithmic phase (B) the specific activity (mU/mg) was reduced to about 50%. This value further decreased in the stationary phase (C).

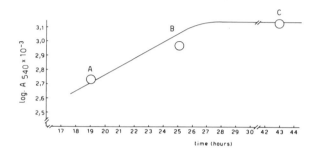

Fig. 1: Growth curve of *Sulfolobus solfataricus*. The organism was grown at 87°C in a 90 l fermenter (Terzano, Italy) with low mechanical agitation and aeration flux of 30 ml/min per liter of broth. The standard culture medium contained (g/l): KH$_2$PO$_4$ 3.1; (NH4)$_2$ SO$_4$, 2.5; MgSO$_4$ 7H$_2$O, 0.2; CaCl$_2$ 2H$_2$O, 0.25; yeast extract, 2.0. The pH of the culture medium was adjusted to 3.5 with 0.1 M H$_2$SO$_4$. Cell growth was quantified turbidimetically at 540 nm, absorbance of 0.6 corresponding to 138 mg of freeze dried cells/1. The doubling time of microorganisms under these conditions was 5.3 hr. The cells were then harvested, at different growth phases, by continuous-flow centrifugation in an Alfa-Laval model LKB 102 D-25 separator. Cells were washed with isotonic saline solution and lysed by freeze-thawing and suspended in Tris-HCl 20 mM, pH 7.5 (1:3 w/v) (crude homogenate).

Fig. 2 shows [^{14}C] incorporation as function of temperature obtained from a crude homogenate of *Sulfolobus solfataricus* harvested in the late logarithmic growth phase (Fig.1, B). ADP-ribosyl transferase activity increased quickly from 15°C, showed a transient plateau from 30°C to 50°C and reaches a maximum value at 80°C. To obtain information about cellular distribution of the ADP-ribosyl transferase activity, enzymatic activity was determined either on the soluble protein fraction (14) or on the nucleoprotein fraction obtained according to Searcy (15). The results showed a preferential association of the ADP-ribosyl transferase activity to the nucleoprotein fraction with a 45-fold increase of specific activity relative to crude homogenate. Identification of the reaction product as ADP-ribose was

further analysis on a boronate column (AffiGel 601, Bio Rad), modifying the procedure previously described by Jacobson *et al.* (16).

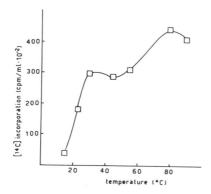

Fig 2: Effect of temperature on enzyme activity. Assay conditions as reported in Table 1.

Table 1. ADP-Ribosyl transferase activity in *Sulfolobus solfataricus* at different growth phases.

SAMPLE	mU/g of cells	mU/mg of protein	mU/mg of DNA
A LOGARITHMIC PHASE	7.75	0.15	1.03
B LATE LOGARITHMIC PHASE	2.14	0.08	0.36
C STATIONARY PHASE	1.40	0.06	0.25

ADP-ribosyl transferase activity assay contains, in a final volume of 125 µl, 80 mM Tris-HCl, pH 7.5, 5 mM NaF, 2 mM DTT, 11 nmoles of [U-^{14}C] NAD+ (4545 cpm/nmole) and enzyme, 5-10 µl. After 10 min at 80°C, 50% cold TCA (v/v) was added and the precipitate, collected on a Millipore filter, was counted in a liquid scintillation counter. One enzymatic unit corresponds to the incorporation of one µmole of NAD into acid insoluble material in one min at 80°C; for convenience, activity is expressed in mU. Protein and DNA content were determined respectively by the method of Bensadoun (12) and according to Richards (13).

In conclusion, these results indicate that *Sulfolobus solfataricus* cell-free system contains a significant ADP-ribosyl transferase activity. It seems of relevant interest that this archaebacterium is the most primitive organism in which ADP-ribosylation has been demonstrated. Furthermore the enzymatic activity appears to be thermophilic, a unique property never observed for the same kind of enzyme isolated from other sources (17). Further studies are in progress on the purification and characterization of the enzyme to investigate its biological role. As mentioned above, several biochemical properties demonstrated that sulfur-dependent archaebacteria are closer related to eukaryotes than to eubacteria (3). These findings support the hypothesis that the ADP-ribosyl transferase activity which we found associated primarily with the nucleoprotein fraction of *Sulfolobus solfataricus,* could play a role in any cellular event in which the enzyme is known to be involved in eukaryotic cells (17).

Acknowledgement: This work was supported by a grant from the MPJ (40%, 1985).

References

1. De Rosa, M., Gambacorta, A., Bu'Lock, J.D. (1975) J Gen Microbiol 86: 156-164
2. The Bacteria (1985), vol. VIII, (Woese, C.R., Wolfe, R.S., eds.) Acad. Press, N.Y.
3. Archaebacteria '85 (1986), (Kandler, O. & Zillig, W. eds.) G. Fischer Verlag, Stuttgard
4. De Rosa, M., Gambacorta, A., Gliozzi, A. (1986) Microbiol Rev 50: 70-80
5. Kessel, M., Klink, F. (1980) Nature 287: 250-251
6. Cacace, M.G., De Rosa, M., Gambacorta, A. (1976) Biochemistry 15: 1692-1696
7. Buonocore, V., Sgambati, O., De Rosa, M., Esposito, E., Gambacorta, A.J. (1981) Appl Biochem 3: 390-397
8. Carteni'-Farina, M., Oliva, A., Romeo,G., Napolitano, G., De Rosa, M., Gambacorta, A., Zappia, V. (1979) Eur J Biochem 101: 307-324
9. Giardina, P., De Biasi, M.G., De Rosa, M., Gambacorta, A., Buonocore, V. (1986) Biochem J 239: 517-522
10. Cacciapuoti, G., Porcelli, M., Carteni'-Farina, M., Gambacorta, A., Zappia, V. (1986) Eur J Biochem 161: 263-274
11. Rossi, M., Rella, R., Pensa, M., Bartolucci, S., De Rosa, M., Gambacorta, A. Raia, C.A., Dell'Aversano Orabona, N. (1986) System Appl Microbiol 7: 337-341
12. Bensadoun, A., Wenstein, D. (1976) Anal Biochem 70: 441-446
13. Richards, G.M. (1974) Anal Biochem 57: 369-376
14. Cammarano, P., Teichner, A., Londei, P., Acca., M., Nicolaus, B. Sanz, L., Amils, R. (1985) EMBO J 4 : 811-816
15. Searcy, D.G. (1975) Biochim. Biophys Acta 395: 535-547
16. Jacobson, M.K., Payne, D.M., Alvarez-Gonzales, R., Juarez-Salinas, H., Sims, J.L. Jacobson, E.L. (1984) Methods in Enzymology 106: 483-494
17. Ueda. K., Hayaishi, O. (1985) Ann Rev Biochem 54: 78-100

Alteration of Poly(ADP-Ribose) Metabolism By Ethanol: Kinetic Mechanism Of Action

James L. Sims and Robert C. Benjamin[1]

Department of Biochemistry, University of North Texas/Texas College of Osteopathic Medicine, Ft. Worth, Texas 76107 USA

Introduction

Ethanol, an experimental mimic of hyperthermia, potentiates the cytotoxicity of DNA damaging agents (1-3). Ethanol and hyperthermia have been suggested to act via similar or related mechanisms. This potentiation is an early event and occurs within the first few minutes of the DNA damage response, precisely the time when one expects to see the substantial but transient increase in poly(ADP-ribose) synthesis and accumulation (4). It has been reported that hyperthermia elicits an increase in poly(ADP-ribose) levels in cells which is potentiated by ethanol (5). We report here the first comprehensive study of the mechanism by which ethanol exerts its effect on poly(ADP-ribose) metabolism *in vivo* and *in vitro*. The data defines a mechanism by which ethanol has a net effect on polymer levels similar to that of hyperthermia, but via a uniquely different mechanism. This mechanism unifies some of the diverse effects attributed to ethanol treatment of cells via a common effect on poly(ADP-ribose) metabolism, explains the synergistic effects ethanol and hyperthermia, and implies that polymer metabolism may be intimately tied to ethanol pathology.

Results and Discussion

Fig. 1 shows the effect of 1% ethanol on poly(ADP-ribose) and NAD content following exposure of SV40 transformed mouse 3T3 cells to 20 µg/ml N-methyl-N'-nitro-N-nitrosoguanidine (MNNG). The maximum ethanol-dependent enhancement of MNNG-induced polymer accumulation is 2.5-fold and it occured at 20 min of MNNG treatment (Fig. 1A), corresponding to the peak of net polymer accumulation under these conditions (4). This increase was accompanied by a concomitant decrease in the NAD pool (Fig. 1B), being 20% lower at 15 min in ethanol-treated cells as compared to MNNG alone. Ethanol did not detectably alter polymer levels in cells not treated with MNNG (data not shown). Thus, ethanol stimulation of polymer accumulation *in vivo* appears to depend on simultaneous activation of the polymerase as provided by MNNG-induced DNA fragmentation.

---

[1]Department of Biological Science, University of North Texas, Denton, Texas 76202

105

Intracellular poly(ADP-ribose) levels are determined by the balance of synthesis and degradation. The ethanol-induced increase in polymer content could therefore be explained by either an elevated rate of synthesis or a reduced rate of degradation. Experiments in nucleotide-permeable cells possessing a chromatin-associated poly(ADP-ribose) polymerase but no functional polymer turnover system showed that ethanol has a direct effect on poly(ADP-ribose) polymerase (6). To determine if polymer degradation *in vivo* was affected by ethanol, polymer half-life was determined in the presence and absence of ethanol. MNNG-treated cells were transferred to medium containing 5 mM benzamide (to block further synthesis) at the peak period of poly(ADP-ribose) accumulation and polymer degradation was followed for 10 min. The polymer half-life was approximately 2.2 min in both the presence and absence of 1% ethanol (Fig. 2). Thus ethanol appears to stimulate polymerase activity in DNA-damage cells and does not appear to affect degradation.

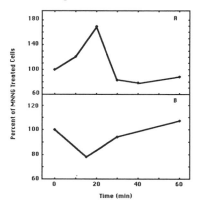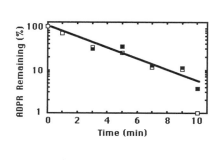

Fig. 1. (left) Effect of ethanol on poly(ADP-ribose) and NAD metabolism *in vivo*. SV40 transformed mouse 3T3 cells were grown as previously described (6) and treated with 20 μg/ml MNNG and 1% ethanol was added to some cultures. Poly(ADP-ribose) (panel A) and NAD (panel B) were extracted and quantified as described elsewhere (6). Zero time values were 4 pmol of poly(ADP-ribose) and 80 nmol of NAD per 10^8 cells. The data represent levels in ethanol-treated cells divided by levels in cells receiving MNNG alone at each time point multiplied by 100. (Reprinted with permission from ref. 6).

Fig. 2. (right) Degradation of poly(ADP-ribose) *in vivo*. Cell cultures were pretreated for 15 min with 50 μg/ml MNNG in the presence (closed symbols or absence (open symbols) of 1% ethanol. At zero time, culture medium was changed to medium containing 5 mM benzamide (closed symbols). Poly(ADP-ribose) was extracted and quantified as in Fig. 1. The polymer content at the time of benzamide addition was set to 100% and all other time points expressed as a percent of that value. (Reprinted with permission from ref. 6).

To determine how ethanol stimulates the polymerase, a kinetic analysis of NAD+ and DNA binding was performed in the presence and absence of

106

varying concentrations of ethanol. Results of initial velocity experiments carried out with calf thymus poly(ADP-ribose) polymerase indicate that K_{NAD} was unaffected by ethanol up to 8% (6). Three preparations of DNA, chosen for their varying abilities to activate the polymerase (K_{DNA}), were used to study the effect of ethanol on K_{DNA}. The results are shown in Fig. 3. K_{DNA} for Hpa II fragments of pBR322 (5'-phosphate extended end) was lowered nearly 15-fold by 6% ethanol (Fig. 3C). K_{DNA} values 5'-phosphorylated and dephosphorylated Hae III fragments of pBR322 (flush ends) were also lowered several fold (Fig. 3A,B). The magnitude of the ethanol induced lowering of K_{DNA} was inversely related to the K_{DNA} of the fragment in the absence of ethanol (7). In all cases V_{max} was unaffected by ethanol. Ethanol concentrations above 8% are inhibitory possibly due to denaturation of the enzyme (6). The results clearly show that ethanol alters the binding of the enzyme to the DNA component of chromatin and that this altered binding is responsible for the activation of the enzyme. This increased affinity for chromatin has the affect of shifting the kinetic mechanism of the polymerase, suggested to be a random sequential mechanism in which either NAD^+ or DNA can bind first (8), to a more ordered mechanism with DNA as the first substrate.

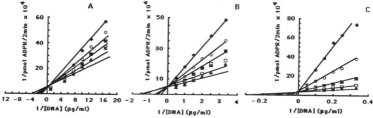

Fig. 3. Effect of ethanol on apparent K_{DNA}. calf thymus poly(ADP-ribose) polymerase was purified as described elsewhere (7) and initial velocities determined (8, 6) at saturating NAD^+ (2 mM) at varying fixed concentrations of ethanol while varying DNA about its K_m. Panel A shows data for alkaline phosphatase-treated Hae III fragments of pBR322, panel B shows data for Hae III fragments of pBR322 with phosphorylated 5'-termini, and panel C shows data for Hpa II fragments of pBR22. Symbols are: 0% ethanol (◆), 2% ethanol (◇), 4% ethanol (■), 6% ethanol (□), 8% ethanol (▲). (Reprinted with permission from ref. 6).

Cellular poly(ADP-ribose) levels reflect the collective impact of at least four factors. These are substrate NAD levels, the amount of DNA fragmentation, the affinity of the polymerase for the particular type of break(s) and the activity of the poly(ADP-ribose) glycohydrolase which degrades the polymer. The data we present here and elsewhere (6) collectively show that ethanol exerts its stimulatory effect on polymer metabolism both *in vivo* and *in vitro* by directly affecting the poly(ADP-ribose) polymerase itself, lowering K_{DNA} without affecting K_{DNA} or V_{max}. Cellular NAD levels and polymer degradation are unaltered by the presence of ethanol. DNA damage remains a requirement for polymerase activation under these conditions. Our findings are consistent with those reported by

Juarez-Salinas *et al.* (5) with regard to accumulation of polymer *in vivo* and agree with the observation made *in vitro* by Kristensen and Holtlund *et al.* (9) and Berger (10). It is of interest to note that two different cellular stresses, ethanol and heat shock, invoke similar cellular responses, but by apparently opposite mechanisms. Both lead to increased levels of poly(ADP-ribose); ethanol by directly increasing the activity of the poly(ADP-ribose) polymerase and heat shock by inhibiting the poly(ADP-ribose) glycohydrolase.

Acknowledgements: Grant support was provided by National Institutes of Health grants CA23994 and AG01274.

References

1. Westra, A., Dewy, W.C. (1971) Int J Radiat Biol 19: 467-477
2. Wallach, D.F.H. (1978) Cancer Therapy: By Hyperthermia and Ionizing Radiation (C.Steffer, ed), Urban and Schwarezenberg, Baltimore, pp. 19-28
3. Mondovi, B., Agro, A.F., Rotillio, G., Strom, R., Moricca, G., Fanelli, A.R. (1969) Eur J Cancer 5: 137-146
4 Juarez-Salinas, H., Sims, J.L., Jacobson, M.K. (1979) Nature 282: 740-741
5 Juarez-Salinas, H., Duran-Torres, G., Jacobson, M.K. (1984) Biochem Biophys Res Commun 122: 1381-1388
6. Sims, J.L., Benjamin, R.C. (1987) Arch Biochem Biophys 253: 357-366
7. Benjamin, R.C., Gill, D.M. (1980) J Biol Chem 255: 10502-10508
8. Benjamin, R.C., Cook, P.F., Jacobson, M.K. (1985) ADP-ribosylation of Proteins, FR Althaus, H Hilz, and S Shall (eds.), Springer-Verlag, Berlin,-Heidelberg, pp. 93-97
9. Kristensen, T., Holtlund, J. (1978) Eur J Biochem 88: 495-501
10. Berger, N.A., Weber, G., Kaichi, A.S. (1978) Biochem Biophys Acta 519: 87-104

The Effect of Benzamides on the Activity of Nuclear ADP-Ribosyl Transferase and the Accumulation of Poly(ADP-Ribose) *In Vivo*

Christopher J. Skidmore, Janet Jones, and Bhartiben N. Patel

Department of Physiology and Biochemistry, University of Reading, Reading RG6 2AJ, United Kingdom

Introduction

Investigation of the cellular role of the nuclear protein modification ADP-ribosylation has depended heavily on the use of benzamides as inhibitors of ADP-ribosyl transferase (EC 2.4.2.30) in intact cell studies. The substituted benzamides were first introduced by Whish (1) and have proved to be useful inhibitors of low toxicity and fairly high specificity. The role of ADP-ribosylation in DNA repair (2) has been strongly supported by the stimulation of the cytotoxicity of DNA-damaging agents by the benzamides.

In general, benzamides inhibit DNA repair-dependent processes however a number of anomalous effects have been observed. Benzamides and nicotinamides are not themselves mutagenic but have been reported to stimulate (3) and to inhibit (4) the mutagenic effects of alkylating agents in mammalian cells. Inhibitors of nuclear ADP-ribosyl transferase inhibit cellular transformation *in vitro* (5, 6) although they have been reported to increase transformation due to ethylating agents (6, 7). The ability of benzamides to affect transformation is dependent on a number of factors. When the effects on the cell cycle are taken into account, 3-methoxybenzamide stimulates transformation some ten-fold (8). We have studied the effect of the inhibitor 3-acetylamidobenzamide (3-aab) on the activity of nuclear ADP-ribosyl transferase in permeabilized cells in an attempt to discover the biochemical basis of these anomalies.

Results and Discussion

L1210 mouse leukemic cells were grown in RPMI 1640 with 10% horse serum and antibiotics and were harvested in mid-log phase. The cells were permeabilized by hypotonic cold shock by the method of Halldorsson *et al.* (9) and the nuclear ADP-ribosyl transferase activity assayed in the presence of a range of concentrations of 3-aab (Fig. 1). At concentrations greater than 10 μM the expected inhibition was obtained but at inhibitor concentrations in the nanomolar range, a pronounced increase in nuclear ADP-ribosyl transferase activity was observed. At 50 nM 3-aab the activity of the enzyme was over twice that found in control cells. We analysed the inhibitory effects of 3-aab further by product analysis. The products of the ADP-ribosylation reaction were separated into mono(ADP-ribose) and poly(ADP-ribose) by aminoethyl-cellulose column chromatography (10). At 100 μM NAD+ elongation (production of polymer) was the predominant reaction in control cells. When cells were treated with 25 μg/ml DNase I during the assay, the stimulated nuclear ADP-ribosyl transferase activity generated considerable amounts of mono(ADP-ribose). Whereas in unstimulated cells 3-aab largely inhibited polymer production, representing the elongation reaction, in stimulated cells both initiation and elongation were inhibited (Table 1). Further analysis (data not shown) demonstrated that the inhibition of monomer production was competitive.

Table 1. Product analysis of the ADP-ribose incorporated into permeabilized L1210 cells by aminoethyl-cellulose colum chromatography[a]

Treatment	ADP-ribose incorporated (pmol NAD+/min/10^6 cells)		
	monomer	polymer	total
Control activity[b]	0.36	1.14	1.42
Control + 50 nM 3-aab	1.70	0.79	2.50
Control + 50 μM 3-aab	0.36	0.20	0.48
Stimulated activity[c]	2.42	1.14	3.28
Stimulated + 50 μM 3-aab	0.68	0.32	1.01

[a]L1210 mouse leukemia cells (approx. 5 x 10^4 cells) were incubated for 5 min in NAD+ at a specific activity of 125 μCi/ml. TCA was added to stop the reaction at a final 30% and the precipitate washed with 20% TCA and then with ethanol. The precipitate was resuspended in 20 μl 9 M urea and digested in NaOH (final concentration 0.3 M) at 37°C for 16 hr to free ADP-ribose chains. The resulting solution was neutralized with HCl before further analysis. 100 μl of this material was counted to give total incorporation. Monomer and polymer fractions of ADP-ribose were separated on AE-cellulose (10). Columns of 1 ml bed volume were pre-equilibrated with 0.5 M acetic acid, 100 μl sample added and the column spun at 2000 rpm. The column was then washed again by centrifugation, with 100 μl aliquots of a stepwise gradient of 1-6 M acetic acid, water and then of a stepwise gradient of 1-5 M ethylamine. Fractions were placed in scintillation vials, dried and counted. (Reprinted with permission from ref. 13).
[b]Incubation with NAD+ alone at 100 μM.
[c]Incubation with 100 μM NAD+ and 25 μg/ml DNase I.

Investigation of the anomalous stimulatory effect of nanomolar concentrations of inhibitors showed that other benzamides and nicotinamides gave similar effects. When the degree of inhibition by each compound at 50 μM concentration was compared with the degree of stimulation observed at 50 nM a clear correlation emerged (correlation coefficient = -0.84) (Fig. 2). This suggested that the two effects derive from interaction at the same or similar sites. Product analysis showed that the stimulation caused by treatment with 50 nM 3-aab contained little change in the amount of elongation whereas the amount of initiation was increased some eight-fold (Table 1).

Recently we have asked whether the stimulatory effect of nanomolar 3-aab is purely an *in vitro* effect. Intact L1210 cells were incubated in varying concentrations of 3-aab for 10 min and then the poly(ADP-ribose) levels determined in a TCA precipitate as in Jacobson *et al.* (11). An inhibition of the steady-state polymer levels was observed below 100 nM 3-aab (Fig. 3). At 10 nM inhibitor the level of polymer was stimulated 5-fold.

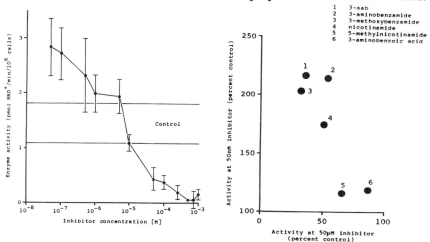

Fig. 1. (left) Effect of 3-acetylamidobenzamide on nuclear ADP-ribosyl transferase activity in permeabilized L1210 cells. Cells (7-9 x 10^6 per ml) (70 μl) in isotonic buffer were incubated with 10 μl of buffer or inhibitor solution and 20 μl 500 μM [^{3}H]NAD$^+$ at 26°C. The reaction was stopped by the addition of 50 μl 9 M urea, 48 mM nicotinamide and placed on ice. Assays were performed in triplicate for 0-5 min using NAD$^+$ of specific activity 12.5 μCi/ml. The stopped reaction mixture was transferred to dry Whatman GF/C discs (25 mm diameter) which had previously been soaked in 20% TCA in ether. The discs were stored at 4°C overnight and then washed as a batch five times in 5% TCA and twice in acetone. After drying, the radioactivity on the discs was counted by scintillation. (Reprinted with permission from ref. 13).

Fig. 2. (right) Correlation between the stimulatory and inhibitory effects of benzamides and nicotinamides. Nuclear ADP-ribosyl transferase activity was assayed as in the legend to Fig. 1 but in the presence of either 50 nM or 50 μM inhibitor. (Reprinted with permission from ref. 13).

We have observed a stimulation of ADP-ribosylation by nanomolar concentrations of benzamides both in intact and permeabilized cells. This stimulation is reminiscent of that found with competitive inhibitors of allosteric enzymes. At very low NAD$^+$ concentrations, Kun has reported finding a sigmoidal curve of velocity versus NAD$^+$ concentration for purified nuclear ADP-ribosyl transferase (12). The observed allosteric behavior is likely to arise from the interaction of NAD$^+$ binding sites either on the same or different enzyme molecules. There is no evidence that nuclear ADP-ribosyl transferase is an oligomeric protein. More than one nuclear ADP-ribosyl transferase molecule is involved however in the

111

modification of nuclear ADP-ribosyl transferase itself, which is the major acceptor in permeabilized cells and on treatment with DNA damaging agents. Such an interaction might give rise to the observed behavior.

These results suggest caution in attributing the effects of benzamides to the inhibition of nuclear ADP-ribosyl transferase unless that inhibition can be demonstrated by a decrease in the level of endogenous polymer synthesis. Intracellular concentrations of nuclear ADP-ribosyl transferase inhibitors have seldom been measured. In one case an extracellular benzamide concentration of 1 mM gave an intranuclear concentration of 4-8 μM (5). Nanomolar concentrations may well be achieved under other experimental conditions. The conflicting reports of the effects on whole cells with benzamides may be due to activation and inhibition of nuclear ADP-ribosyl transferase taking place under different experimental conditions.

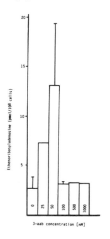

Fig. 3. Effect of nanomolar levels of 3-acetylamidobenzamide on the intracellular poly(ADP-ribose) levels in intact L1210 cells. Cells were incubated for 10 min at 30°C with the inhibitor. Estimation of the poly (ADP-ribose) level *in situ* in L1210 cells was performed on TCA precipitates as in Jacobson *et al*. (11). (Reprinted with permission from ref. 13).

References

1. Purnell, M.R., Whish J.D. (1980) Biochem J 185: 775-777
2. Shall, S. (1984) Adv Rad Biol 11: 1-69
3. Schwartz, J.L., Morgan, W.F., Brow-Lindquist, P., Afzal P., Weichselbaum, R.R., Wolff, S. (1985) Cancer Res 45: 1556-1559
4. Bhattacharyya, N., Bhattacharjee, S.B., (1985) ADP-Ribosylation of Proteins Althaus, F., Hilz, H., Shall, S., (eds) Springer-Verlag, Berlin
5. Kun, E., Kirsten, E., Milo, G.E., Kurian, P., Kumari, M.L. (1983) Proc Natl Acad Sci USA 80: 7219-7223
6. Borek, C., Ong, A., Cleaver, J.E., (1984) Carcinogenesis 5: 1573-1576
7. Lubet, R.A., McCarvill, J.T., Putman D.L., Schwartz, J.L., Schechtman, L.M., (1984) Carcinogenesis 5: 459-462
8. Jacobson, E.L., Smith, J.Y., Nunbhakdi, V., Smith, D.G., (1985) ADP-Ribosylation of proteins, Althaus, F., Hilz, H., Shall, S., (eds) Springer-Verlag Berlin
9. Halldorsson, H., Gray, D.A, Shall, S. (1978) FEBS Lett 85: 349-352
10. Stone, P.R., Purnell, M.R., Whish, W.J.D. (1981) Anal. Biochem. 110: 108-116
11. Jacobson, M.K., Payne, D.M., Alvarez-Gonzalez, R, Juarez-Salinas, H., Sims, J.L., and Jacobson, E.L. (1984) Methods Enzmol 106: 483-494
12. Bauer, P.I., Hakam, A., Kun, E. (1986) FEBS Lett 195: 331-338
13. Jones, J., Patel, B. N., Skidmore, C.J. (1988) Carcinogenesis 9: 2023-2026

Carba-Nicotinamide Adenine Dinucleotide: Synthesis and Enzymological Application of a Novel Inhibitor of ADP-ribosyl Transfer

James T. Slama and Anne M. Simmons

Department of Biochemistry, The University of Texas Health Science Center, San Antonio, Texas 78284 USA

Introduction

The covalent modification of proteins has been demonstrated to be an important and general *in vivo* mechanism for metabolic regulation (1). More recently it has been shown that nicotinamide adenine dinucleotide (NAD) can participate in such reactions as a donor of the adenosine diphosphate ribose (ADP-ribose) moiety (2, 3). Although the biological function of these reactions is still unknown, evidence has begun to accumulate suggesting that this class of modification constitutes an important regulatory mechanism.

One way to approach both the study of biological function and the chemistry of this important class of enzymes is to design and synthesize potent and specific inhibitors for individual enzymes in the family. Effective inhibitors can be used to elucidate active site structure (through covalent modification), as stable ligands for the development of affinity purifications, or to probe the regulation of the enzyme. They can also be applied to the study of the activity of the target enzyme *in vivo*.

The carbocyclic dinucleotide carba-NAD [1] is designed to function as an inhibitor of ADP-ribosyl transferases. In this compound, a cyclopentane ring replaces the furanose of the nicotinamide mononucleotide moiety of NAD. This modification will cause the C-N bond to be resistant to enzymatic cleavage. The analog will otherwise closely resemble NAD. We expect therefore that carba-NAD and its relatives will constitute an important set of mechanism-based inhibitors for NAD glycohydrolases and for ADP-ribosyl transferases.

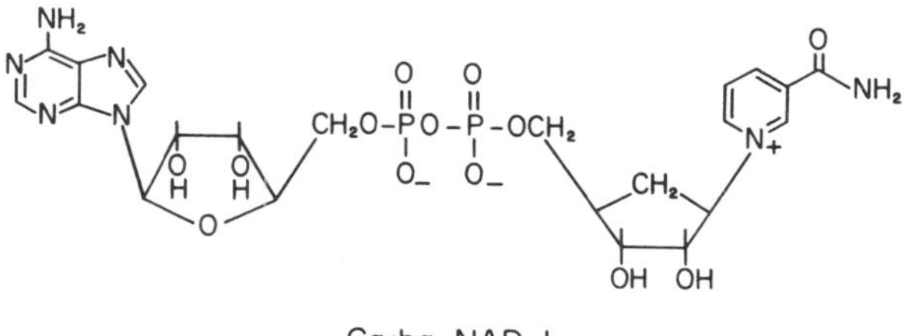

Carba-NAD, I

113

Results

The synthesis of carba-NAD beginning with the carbocyclic analog of 1-aminoribose (d,1-4β-amino-2α,3α-dihydroxy-1β-cyclopentanemethanol, [2]) (4,5) is shown in Scheme 1 and in Scheme 2. The primary amine [2] was converted to the nicotinamide monoucleoside analog [4] on treatment with 1-(2,4-dinitro-phenyl) pyridinium-3-carboxamide [3], using the Zincke reaction (6) . Selective monophosphorylation of [4] was accomplished in high yield using phosphorus oxychloride in trimethylphosphate (7), producing the carba-nicotinamide-5'-mononucleotide [5]. The product of phosphorylation [5] was purified by ion-exchange chromatography on DEAE cellulose. Purity of the product was verified by high pressure liquid chromatography, and its composition established using fast atom bombardment mass spectroscopy ([molecular ion + H]$^+$ was observed as a strong peak at m/z 333). The site of phosphorylation was established by ^{13}C NMR. The signal at 65.51 ppm assigned to the 5'-CH$_2$OH of [4] was shifted downfield 68.12 ppm and split into a doublet upon phosphorylation.

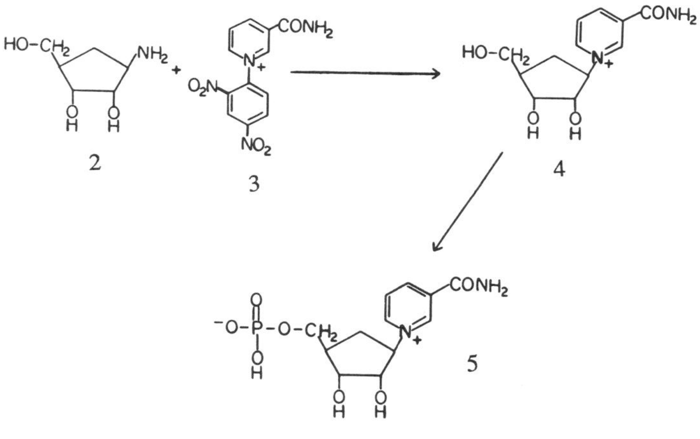

Scheme 1

Coupling of the nucleotide-5'-phosphate to an adenosine monophosphate derivative to produce the dinucleotide [1] was accomplished using the procedure of Furusawa *et al.* (8) (Scheme 2). In this reaction adenosine-5'-phosphoric di-n-butylphosphinothioic anhydride [6] is coupled to carba-NMN [5] after silver salt activation. We found that both the yield and the reproducibility of the coupling were improved by first protecting carba-NMN as its 2',3'-di-O-acetyl derivative. Acetylation increased the solubility of the nucleotide in mixtures of dimethyl formamide and pyridine, and allowed us to conduct the reaction using the more effective solvents containing higher percentages of pyridine.

114

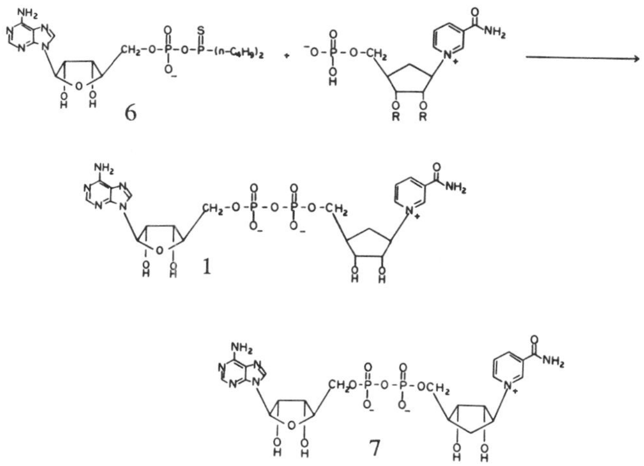

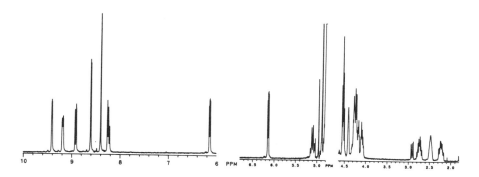

Scheme 2

After coupling, the acetyl groups are removed by brief treatment with methanolic ammonia. The product dinucleotides are purified by anion-exchange chromatography using DEAE-cellulose, and characterized spectroscopically. ^{1}H-NMR spectra of the product carba-NAD is shown in Figs. 1 and 2. The spectrum of the dinucleotide exhibits the expected contributions from the pyridine nucleotide portion and from the adenosine nucleotide.

Fig. 1. (left) ^{1}H-NMR (300 mHz, D2O, ref. to TSP) showing the downfield region of the mixture of carba-NAD [1] and its diastereomer, ψ-carba-NAD [7]. The signals at δ 8.40 and 8.61 are the adenosyl hydrogens. The adenosyl anomeric hydrogens appear as two doublets of δ 6.15 and 6.13. The remaining signals at δ 8.24, 8.92, 9.19, and 9.41 are assigned to protons on the pyridinium ring.

Fig. 2. (right) ^{1}H-NMR (300 mHz, D$_2$O, ref. to TSP) showing the upfield region of the mixture of carba-NAD [1] and its diastereomer, ψ-carba-NAD [7]. The signals at δ 2.20-2.27, 2.46-2.48, 2.69-2.80, and 5.10 are assigned to protons on the cyclopentane ring. The region from δ 4.05 to 4.9 ppm is complex and contains the ribosyl absorptions, the remaining carbocyclic absorptions, and an intense HDO signal.

115

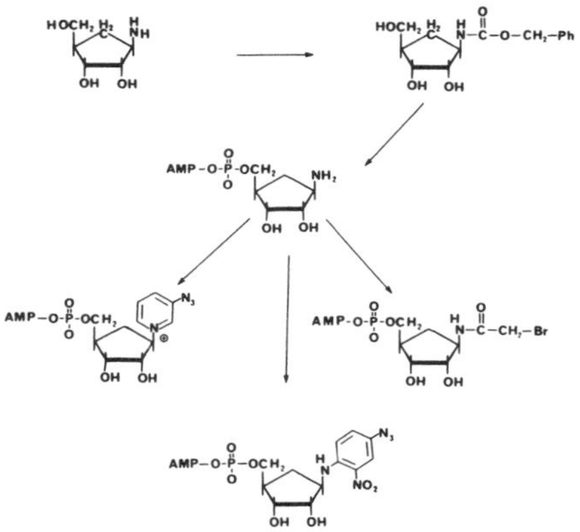

Scheme 3

Coupling of the racemic carba-NMN [5] to adenosine-5'-monophosphate leads to the production of two diastereomeric dinucleotides carba-NAD [1] and pseudocarba-NAD (Ψ-c-NAD, [7]) (Scheme 2). These are not separable by ion-exchange chromatography. The presence of two sets of anomeric signals at 6.13 and 6.15 ppm (Fig. 1) is the result of this mixture. The two dinucleotides are separable using reversed phase HPLC (Fig. 3). Only one of these diastereomers is a substrate for yeast alcohol dehydrogenase, and a configuration in the cyclopentane ring analogous to D-ribose is assigned to the carbocyclic sugar analog in this material.

The reduced dinucleotide, carba-NADH, has an electronic absorption spectrum with maxima at 260 nm and 360 nm. It exhibits a broad fluorescence emission at 456 nm, typical of a dihydropyridine. The fluorescence excitation spectrum of carba-NADH shows the expected maxima at 360 nm as well as a second, weaker excitation at 260 nm, due to intramolecular energy transfer from the adenosyl ring (9, 10). The similarities in the ratios of the 260 nm excitation to the longer wavelength excitation between carba-NADH and NADH indicate a similar amount of fluorescence energy transfer, and therefore a similar amount of stacked conformer present in solution. Thus, by this spectroscopic criteria, carba-NADH adopts a conformation in solution indistinguishable from that of NADH.

Carba-NAD is effectively recognized by pyridine nucleotide specific binding sites. Thus, carba-NAD is reduced by yeast alcohol dehydrogenase with a K_m of 1.7×10^{-3} M (K_m for NAD is 4.5×10^{-4} M) and with a maximal velocity 70% of that of NAD itself. Using equine liver alcohol dehydrogenase, a K_m of 7.6×10^{-6} M is obtained for carba-NAD (K_m for NAD is 24×10^{-6} M under identical conditions). Reduction occurs at a maximal velocity 25% of that measured for NAD itself. Carba-NAD is therefore demonstrated to be an effective substrate for this class of enzymes.

Low concentrations of carba-NAD have been demonstrated to inhibit the NAD glycohydrolase isolated from *Bungarus fasciatus* venom (11). The mechanism of inhibition was determined by Lineweaver-Burk analysis of the effect of fixed inhibitor concentration on the initial rate of the reaction (Fig. 4). The inhibition of the NAD glycohydrolase by mixtures of [1] and [7] was determined to be of the competitive type. Replots of the K_m^{app} versus inhibitor concentration (not shown) were linear, and a K_i of 74 μM determined. It is our expectation that the K_i for pure carba-NAD will be lower than this value. As it stands, the K_i for carba-NAD is close to the K_m of NAD, which we determine to be 33 μM.

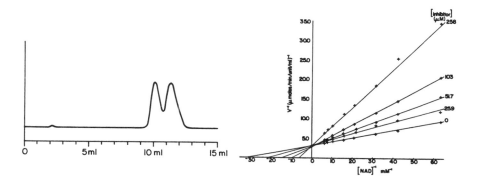

Fig. 3. (left) Separation of the diastereomeric c-NAD and ψ-c-NAD by reversed phase HPLC. A C-18 Dynamax scout column (4.6 mm x 25 cm; 8 micron packing; Rainin Instrument Co., Woburn, MA) was used, with an eluent of 20 mM NaH_2PO_4 with pH adjusted to 6.0 using tetrabutylammonium hydroxide, and run at a flow rate of 1 ml/min.

Fig. 4. (right) Competitive inhibition of NADase by a mixture of carba-NAD and its diastereomer. A typical assay contained 33 mM KH_2PO_4 (pH 7.5), 15-150 μM [carbonyl-^{14}C]NAD, 25-250 μM carba-NAD and ψ-carba-NAD, and 0.08 mU of enzyme in total volume of 0.3 mL. The assay was initiated by the addition of substrate, and the mixture incubated 30 min at 37°C. The reaction was quenched by withdrawing a 0.2 ml sample and applying it to a small (1 mL) column of Dowex 1x2, anion exchange resin, Cl- form. The product [carbonyl-^{14}C]-nicotinamide was specifically eluted from the column by washing 5 times with 1 mL portions of 20 mM Tris•HCl (pH 7.5) and the activity of the effluent determined by liquid scintillation counting. From this experiment a K_i of 74 μM for the mixture of carba-NAD and its diastereomer was determined.

Discussion

A synthesis of carba-NAD has been developed through which hundred milligram to gram quantities of this novel nucleotide analog can be produced. Spectroscopically, carba-NAD closely resembles its parent, NAD, indicating a similar conformation in solution. Finally, carba-NAD is recognized by pyridine nucleotide specific NAD binding sites. It is an efficient substrate for yeast and equine liver alcohol dehydrogenase, and an effective competitive inhibitor of the NAD glycohydrolase from *B. fasciatus* venom. The study of ADP-ribosylation next requires the production of a family of affinity and photoaffinity labels which incorporate the carbocyclic sugar. One such label would be the carba-NAD bearing an 8-azide on the adenosyl ring. Other affinity and photoaffinity labels will be available through a synthesis which diverges in its last step (Scheme 3). The synthesis of these target structures will produce a family of affinity and photoaffinity labels which can be used both to study the mechanism of ADP-ribosyl transfer and to elucidate the biological functions of these reactions.

Acknowledgements. This work was supported by grants from the National Institutes of Health as well as from the Robert A. Welch foundation.

References

1. Chock, P.B., Rhee, S.G., Stadtman, E.R. (1980) Ann Rev Biochem 49: 813-843
2. Ueda K., Hayaishi O. (1985) Ann Rev Biochem 54: 73-100
3. Hayaishi O., Ueda K. (1977) Ann Rev Biochem 46: 95-116
4. Kam B.L., Oppenheimer N.J. (1981) J Org Chem 46: 3268-3272
5. Cermak, R.C., Vince R. (1981) Tetrahedron Lett 22: 2331-2332
6. Lettre, H., Haede W., Ruhbaum E. (1952) Annalen der Chemie 579: 123-132
7. Yoshikawa, M., Kato T., Takenishi T. (1967) Tetrahedron Lett 50: 5065-5068
8. Furusawa, K., Sekine M., Hata T. (1976) J Chem Soc Perkin I: 1711-1716
9. Weber, G. (1957) Nature (London) 180: 1409
10. Shifrin, S., Kaplan, N.O. (1959) Nature (London) 183: 1259
11. Yost, D.A., Anderson, B.M. (1981) J Biol Chem 256: 3647-3653

Purification and Properties of Two Forms of Poly(ADP-Ribose) Glycohydrolase

Sei-ichi Tanuma

Department of Physiological Chemistry, Faculty of Pharmaceutical Sciences, Teikyo University, Sagamiko, Kanagawa 199-01, Japan

Introduction

In eukaryotic cells, chromatin function varies temporally and spatially with cell cycle and differentiation during development. One way a cell can control its chromatin function is thought to be by changing the configuration of chromatin. Post-translational modifications of chromosomal proteins have been proposed to be important in determining such conformational changes in chromatin structure. Since a minor but significant fraction of chromosomal proteins are reversibly poly(ADP-ribosyl)ated (1, 2) and the modification causes structural changes in chromatin (3, 4), it is assumed that the turnover of poly(ADP-ribose) on proteins has some biological role which is essential to expression of a nuclear function. Although a number of studies have shown that poly(ADP-ribosyl)ation may be involved in DNA repair and replication, gene expression, cell cycling, and differentiation, the physiological function of this modification has not yet been established (5).

Our approach to a better understanding of the biological function of poly(ADP-ribosyl)ation of chromosomal proteins has involved the extensive purification and characterization of the major enzymatic activities thought to be responsible for the degradation of poly(ADP-ribose) *in vivo* (6-8). The degradation of poly(ADP-ribose) has been thought to be catalyzed by two kinds of enzymes. One enzyme, poly(ADP-ribose) glycohydrolase (6-10), catalyzes hydrolysis of the glycosidic (1"-2')-linkages of poly(ADP-ribose). A second type is ADP-ribosyl-protein lyase (11), which has been purified from rat liver cytoplasm and is capable of splitting only mono(ADP-ribose)-protein linkages. Previously, we first purified poly(ADP-ribose) glycohydrolase to homogeneity from the nuclei of guinea pig liver (6). We also found that an poly(ADP-ribose) glycohydrolase activity was present in the cytosolic fraction. The present report deals with the purification and characterization of the two forms of poly(ADP-ribose) glycohydrolase from guinea pig liver. To distinguish the nuclear glycohydrolase from the cytosolic one, the nuclear and cytosolic enzymes are designated poly(ADP-ribose) glycohydrolase I and II, respectively.

Results

Enzyme purification. The poly(ADP-ribose) glycohydrolase activity present in the nuclei was solubilized only by sonication at high ionic strength and purified by sequential chromatographic steps on phosphocellulose, DEAE-cellulose, Blue Sepharose, and single-stranded DNA cellulose (6). Most of the poly(ADP-ribose) glycohydrolase activity detected in the cytoplasm was recovered in the cytosolic fraction and purified on phosphocellulose, DEAE-cellulose, Blue Sepharose, single-stranded DNA cellulose, and Red Sepharose chromatography. The nuclear and cytosolic poly(ADP-ribose) glycohydrolase (I and II) were purified more than 20,000 and 30,000-fold, respectively, with yields of 18 and 6%. The specific activities of the final preparations of poly(ADP-ribose) glycohydrolase I and II assayed with substrate poly(ADP-ribose) of average chain length of 15 and a concentration of 10 µM were 32.5 and 12.6 µmol·min^{-1}·mg protein^{-1}, respectively. The numbers of glycohydrolase I and II molecules per cell were calculated to be about 50,000 and 10,000, respectively.

Table 1. Effect of nucleotides on poly(ADP-ribose) glycohydrolase I and II

Compound	Glycohydrolase I K_i ($\pm$ S.D.)	Glycohydrolase II K_i ($\pm$ S.D.)
	µM	µM
ADP-ribose	16 ($\pm$ 1.5)	53 ($\pm$ 4.2)
cAMP	180 ($\pm$ 11)	230 ($\pm$ 18)
dibutyryl-cAMP	62 ($\pm$ 4.8)	----

Inhibitor constants (K_i) were determined using Dixon plots. Four different inhibitor concentrations were used for determination of each K_i value. At each inhibitor concentration, activity was determined using three different poly(ADP-ribose) concentrations (each point in duplicate). The best straight line was then determined using linear regression.

Catalytic properties. The optimum pH of glycohydrolase I and II activities were 6.9 and 7.4, respectively. In the experiments using an isoelectrofocusing gel, glycohydrolase I and II were found to be acidic proteins having pI values of 6.4 and 6.1, respectively. Analysis of degradation products from poly(ADP-ribose) with glycohydrolase I or II by cellulose thin layer chromatography indicated that both enzymes hydrolyzed poly(ADP-ribose) exoglycosidically to produce ADP-ribose. The K_m values of glycohydrolase I and II for poly(ADP-ribose) (n = 15) were estimated from double-reciprocal plots as 2.3 and 6.4 µM, respectively. The calculated V_{max} for hydrolysis of poly(ADP-ribose) was 36 and 15 µmol·min^{-1}·mg protein^{-1}. The optimal salt concentrations for the two enzymes were very different. Glycohydrolase I activity was stimulated about 1.5-fold at 50-100 mM KCl or NaCl. In contrast, glycohydrolase II activity was inhibited by the addition of NaCl or KCl; 50% of the activity remained at 100 mM. CaCl$_2$ had an inhibitory effect, but MgCl$_2$ had no effect on either enzyme. ADP-ribose and cAMP inhibited glycohydrolase I more strongly than glycohydrolase II (Fig. 1). The K_i values for ADP-ribose were calculated to be 16 and 53 µM for the reaction of glycohydrolase I and II, respectively

(Table 1). Interestingly, dibutyryl-cAMP was shown to possess a significant inhibitory effect on glycohydrolase I ($K_i = 62$ μM) but not on glycohydrolase II. ADP-ribose and cAMP were competitive inhibitors of both glycohydrolase I and II. These different sensitivities seemed to be a suitable parameter for identification of the two forms.

Molecular properties. The purified nuclear poly(ADP-ribose) glycohydrolase I exhibited one protein band on SDS-polyacrylamide gel electrophoresis (6). Based on the mobility of the glycohydrolase relative to that of protein standards, its molecular weight (Mr) was 75,500 (Fig. 2, lane f). On the other hand, Mr of the purified cytosolic glycohydrolase II was 59,300 (lane a). The native Mr of glycohydrolase I and II, determined by gel permeation on Sephadex G-100, were estimated to be 72,000 and 56,000, respectively. To detect the possible structural differences between the two glycohydrolases, proteolytic fragments produced by *S. aureus* V8 protease were compared. As shown in Fig. 2, peptides derived from glycohydrolase I appeared as three major bands (lane d), whereas the peptide map of glycohydrolase II indicated two major bands (lane c). All of these bands were almost distinguishable from one another, suggesting that glycohydrolase I and II are structurally different.

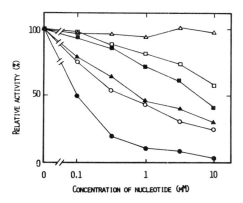

Fig. 1. Effect of ADP-ribose, cAMP, and dibutyryl-cAMP on poly(ADP-ribose) glycohydrolase I and II. The reaction was carried out under the standard conditions except that the various concentrations of ADP-ribose (●, ○), cAMP (■, □), or dibutyryl-cAMP (▲, △) as indicated were added to the reaction mixture. Values are normalized with respect to the activity of glycohydrolase I (●, ▲, ■) and glycohydrolase II (○, △, □) without these compounds, as 100%.

The results of amino acid compositions of glycohydrolase I and II are summarized in Table II. This analysis showed glycohydrolase I to have a relatively high proportion of basic (Lys + Arg) residues. In contrast, glycohydrolase II contained a low content of basic residues and a relatively high proportion of acidic amino acids. The numbers of other amino acid residues were similar to each other, although some differences such as proline were noted.

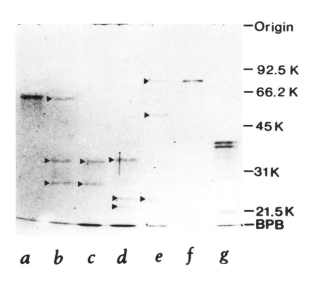

Fig. 2. (right) Determination of molecular weights of the purified glycohydrolase I and II and their peptide mapping with S. aureus V8 protease. About 5 µg of glycohydrolase II (lanes a - c) or I (lanes d - f) were treated at 100oC for 2 min in SDS-sample buffer (6) and then loaded onto 0.1% SDS and 12.5% acrylamide slab gel with 0 (a, f), 10 (b, e), and 100 (c, d) ng of the protease. Proteins were partially digested during electrophoresis. Lane g contains 5 µg of the protease alone. The molecular weight standards used are as follows: phosphorylase b (92,500), bovine serum albumin (66,200), ovalbumin (45,000), carbonic anhydrase (31,000), and soybean trypsin inhibitor (21,500). Positions of proteolytic fragments are indicated by the arrow (>).

Discussion

The present report clearly demonstrates that at least two forms of poly(ADP-ribose) glycohydrolase are present in eukaryotic cells: poly(ADP-ribose) glycohydrolase I (higher Mr) is tightly bound to some chromatin structure, presumably DNA in the nucleus, while poly(ADP-ribose) glycohydrolase II (lower Mr) is probably localized in the cytosol. This suggestion is supported by the observation that enucleated human erythrocytes contain an poly(ADP-ribose) glycohydrolase activity which has similar properties of poly(ADP-ribose) glycohydrolase II (8). However, we cannot rule out the possibility that poly(ADP-ribose) glycohydrolase II is originally present in the nucleus and easily leaks into the cytosolic fraction during the procedures of subcellular fractionation. In addition to their possibly different intracellular localizations, many of the general properties and the sensitivities to ADP-ribose and dibutyryl-cAMP are different. Several poly(ADP-ribose) glycohydrolase activities have been partially and extensively purified from post-nuclear fractions from several sources (6-10). These activities show some similarities to the poly(ADP-ribose) glycohydrolase II purified here (Table 3).

Table 2. Amino acid composition of poly(ADP-ribose) glycohydrolase I and II.

Compound	Glycohydrolase I	Glycohydrolase II
	mol/100 mol[a]	
Asx	8.98	9.51
Thr	4.01	4.2
Ser	10.2	11.0
Glx	11.2	13.2
Gly	12.6	13.8
Ala	9.41	9.86
Cys	T[b]	T
Val	6.02	5.82
Met	T	T
Ile	3.56	2.97
Leu	7.56	6.29
T	2.63	2.98
Phe	3.56	3.91
Lys	6.58	2.75
His	2.80	3.11
Arg	6.69	2.18
Pro	4.20	8.42
Trp	nd[c]	nd
Basics[d]	16.1	8.04
Acidics[e]	20.2	22.7
Acidic/Basic	1.26	2.83

[a]The results were from the analyses performed after 24 hr acidic hydrolysis.
[b]T = traces
[c]nd = not determined
[d]Basics = sum of Lys, His and Arg contents
[e]Acidics = sum of Asx and Glx

Table 3. Comparison of major characteristics of poly(ADP-ribose) glycohydrolases

Properties	Guinea Pig Liver(6) I	II	Pig Thymus(9)	Calf Thymus (10)
Localization	nuclei	cytosol	cytoplasm	cytoplasm
Mr[a]	75,500	59,300	61,500	59,000
			67,500	
Mode of hydrolysis	exo	exo	exo	exo
K_m for poly(ADP-Ribose) (μM)	2.3 (n=15)	6.4	1.8 (n=20)	0.1-10 (n=10-100)
V_{max} (μmol/min/mg)	36	15	19	50-100
Optimum pH	6.9	7.4	7.5	7.3
SH requirement	Yes	Yes	Yes	Yes
Effect of salt	S[b]	I[c]	I	S
Nucleotide Inhibitor	ADP-R cAMP db-cAMP[d] Ap4A[e]	ADP-R cAMP	ADP-R cAMP	ADP-R cAMP

[a]Mr was determined by SDS-polyacrylamide gel electrophoresis; [b]S = stimulation; [c]I = inhibition; [d]dibutyryl-cAMP; [e]diadenosine 5', 5'''-p[1],p[4]-tetraphosphate.

The findings that partial proteolytic degradation of the two forms gives rise to several discernible and distinguishable peptide bands and that the amino acid composition of glycohydrolase I is different from that of glycohydrolase II suggest that they may be structurally distinct polypeptides. Even though seemingly remote, there exists a possibility that glycohydrolase II may be a proteolytic product arising from glycohydrolase I. Proteolysis of this type may result in a change in the peptide mapping and amino acid analysis. This possibility cannot be discerned unless the polypeptide sequences and nucleotide sequences of the genes are determined.

In HeLa S3 cells, the cell cycle dependency of the two poly(ADP-ribose) glycohydrolase activities is different (7). Moreover, the two forms of poly(ADP-ribose) glycohydrolase present in human lymphocytes appear to be separately regulated during differentiation (unpublished results). Taken together, these results suggest that eukaryotic cells do contain two forms of poly(ADP-ribose) glycohydrolase. The presence of multiple forms of poly(ADP-ribose) glycohydrolase exhibiting differences in properties and subcellular localization may be related to diversity of function for the glycohydrolases. The nuclear glycohydrolase I is probably involved in chromatin associated events via depoly(ADP-ribosyl)ation of histones and nonhistone chromosomal proteins. On the other hand, the cytosolic glycohydrolase II may play an important role in extranuclear depoly(ADP-ribosyl)ation in mitochondria and ribosomes (5). However, we have obtained no clear evidence that would enable us to interpret the physiological significance of the family of poly(ADP-ribose) glycohydrolases. The further study of these purified poly(ADP-ribose) glycohydrolases may provide the information necessary to understand the relevance of the family of poly(ADP-ribose) glycohydrolases and the biological function of protein depoly(ADP-ribosyl)ation.

References

1. Tanuma, S., Johnson, G.S. (1983) J Biol Chem 258: 4067-4070
2. Tanuma, S., Johnson, L.D., Johnson, G.S. (1983) J Biol Chem 258: 15371-15375
3. Tanuma, S., Kanai, Y. (1982) J Biol Chem 258: 6565-6570
4. Poirier, G.G., de Murcia, G., Jongstra-Bilen, J., Niedergang, C., Mandel, P. (1982) Proc Natl Acad Sci USA 79: 3423-3427
5. Ueda, K., Hayaishi, O. (1985) Annu Rev Biochem 46: 73-100
6. Tanuma, S., Kawashima, K., Endo, H. (1986) J Biol Chem 261: 965-969
7. Tanuma, S., Kawashima, K., Endo, H. (1986) Biochem Biophys Res Commun 135: 979-986
8. Tanuma, S., Kawashima, K., Endo, H. (1986) Biochem Biophys Res Commun 136: 1110-1115
9. Tavassoli, M., Tavassoli, M.H., Shall, S. (1983) Eur J Biochem 135: 449-455
10. Hatakeyama, K., Nemoto, Y., Ueda, K., Hayaishi, O. (1986) J Biol Chem 261: 14902-14911
11. Oka, J., Ueda, K., Hayaishi, O., Komura, H., Nakanishi, K. (1984) J Biol Chem 259:986-995

Poly(ADP-Ribosyl)ation in Free Cytoplasmic mRNA-Protein Complexes

Hélène Thomassin[1] and Paul Mandel

Centre de Neurochimie du CNRS, 67084 Strasbourg Cédex, France

Introduction

Poly(ADP-ribosyl)ation is generally described as a nuclear event. However, extranuclear poly(ADP-ribose) polymerase activities have been detected in the cytosol of baby hamster kidney cells (1), the ribosomal fraction of HeLa cells (2) and rat testis (3). The activities looked like the nuclear enzyme in that they totally or partially depended on DNA. We have previously reported the association of a poly(ADP-ribose) polymerase with a specific ribonucleoprotein complex, namely free messenger ribonucleoprotein particles (free mRNP) (4-6). The enzyme has the particularity to be DNA-independent. In this paper, we have been interested in the proteins which are poly(ADP-ribosyl)ated in free mRNP. We also present evidence of the existence of a poly(ADP-ribose) glycohydrolase in free mRNP.

Results

Cytoplasmic poly(ADP-ribose) polymerase is associated with free mRNP. We have investigated the localization of poly(ADP-ribose) polymerase activity in the post-nuclear fraction of mouse plasmacytoma cells. We have not detected any poly(ADP-ribose) polymerase activity in the mitochondrial or in the polyribosomal fractions. However, cytoplasmic poly(ADP-ribose) polymerase activity has been found in the post-polyribosomal fraction, tightly associated with free mRNP (4-6). Free mRNP contain silent mRNA and have a different structure from polyribosomal mRNP which are actively engaged in protein synthesis (7, 8). The size of the particles depends on the size of the mRNA and free mRNP have sedimentation coefficients between 6S and 80S. Only free mRNP with sedimentation coefficients well below 40S can be efficiently separated from ribosomes by sedimentation velocity gradient centrifugation. However, free mRNP have a lower density than ribosomes and can be separated by isopycnic sedimentation. Thus we have isolated free mRNP from mouse plasmacytoma in D_2O-sucrose density gradients and have demonstrated that

[1]Present Address: Centre de Recherche en Cancérologie de l'Hôtel-Dieu de Québec, 11, côte du Palais, Québec, Québec, G1R 2J6, Canada

a poly(ADP-ribose) polymerase was present in these particles (4-6). In this report, we show that a poly(ADP-ribose) glycohydrolase is also associated with purified free mRNP.

Poly(ADP-ribose) protein acceptors in free mRNP. We have previously demonstrated that the poly(ADP-ribose) polymerase associated with free mRNP is able to modify free mRNP proteins of mol. wt. 21 kDa, 38 kDa, 45 kDa and 116 kDa (5, 6). Using the protein blotting technique and an antiserum directed against calf thymus poly(ADP-ribose) polymerase, we have shown that the 116 kDa protein corresponds to the enzyme. The poly(ADP-ribose) protein acceptors, including the poly(ADP-ribose) polymerase, can be dissociated from free mRNP by a 0.5 M KCl treatment (6). Several ribonucleoprotein particles have been isolated and characterized after a 0.5 M KCl treatment of free mRNP (7, 9). These complexes have been suggested to play a role in the structure of free mRNP or in the mRNA repression. In order to determine if the poly(ADP-ribosyl)ated proteins of mRNP belong to such ribonucleoprotein particles, we have treated free mRNP with 0.5 M KCl and fractionated them on 10-30% sucrose linear gradients. The poly(ADP-ribose) polymerase sedimented in the 6 S region and was always DNA-independent (Fig. 1A). The same experiment was performed with [^{32}P]poly(ADP-ribosyl)ated free mRNP. The TCA-insoluble radioactivity of each gradient fraction was measured. Poly(ADP-ribosyl)ated proteins sedimented in the 6S region or in a large zone around 12S (Fig. 1B). The analysis of poly(ADP-ribosyl)ated proteins by polyacrylamide gel electrophoresis as described in (6), has shown that the automodified poly(ADP-ribose) polymerase sedimented in the 6S region; the 38 kDa and the 45 kDa protein acceptors sedimented in the 12S region. The 21 kDa poly(ADP-ribosyl)ated protein sedimented in a large region from 16S to the bottom of the gradient and could stay partly associated with the mRNA (data not shown). We don't know whether the poly(ADP-ribose) polymerase is removed from the mRNP as a free protein or bound to a low mol. wt. RNA. However, different physiochemical analysis of purified nuclear poly(ADP-ribose) polymerase revealed a globular protein with a sedimentation coefficient between 4.6 and 5.8S (10). The other poly(ADP-ribosyl)ated proteins are associated with different ribonucleoprotein sub-particles of the free mRNP.

Poly(ADP-ribose) glycohydrolase in free mRNP. Because free mRNP contain poly(ADP-ribose) polymerase and poly(ADP-ribose) protein acceptors, we have looked for the presence of a poly(ADP-ribose) glycohydrolase in free mRNP. The particles were incubated in the presence of polymer with an average chain length of 28 ADP-ribose residues, synthesized by nuclear calf thymus poly(ADP-ribose) polymerase and purified by affinity chromatography as described in (11). A poly(ADP-ribose) glycohydrolase activity of 0.16 nmol ADP-ribose/min/mg protein has been detected, according to the procedure of Menard and Poirier (11). When free mRNP were first incubated in the presence of [^{32}P]NAD and then poly(ADP-ribosyl)ation inhibited by 3AB, a degradation of the poly(ADP-ribose) bound to the free mRNP proteins was observed. This degradation was inhibited by the ADP-ribose and corresponds to a poly(ADP-ribose) glycohydrolase activity of 1.2 nmol ADP-ribose/min/mg protein (Fig. 2). The difference between these two results could be explained by the structure of free mRNP. The catalytic site of the glycohydrolase might be positioned to hydrolyze more easily the poly(ADP-ribose) bound to the free mRNP proteins than exogenous long chains of polymer.

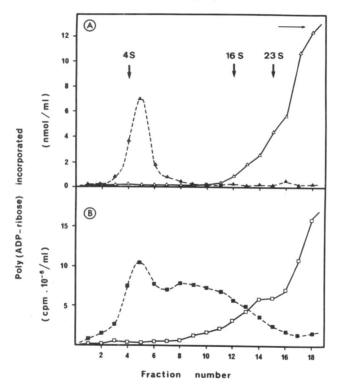

Fig. 1. Treatment of free mRNP with 0.5 M KCl. Mouse plasmacytoma free mRNP (150 μg protein) were layered on 10-30% (w/w) sucrose gradients over a 50% (w/w) sucrose cushion and centrifuged 15 hr at 39,000 rpm, 4°C in a Beckman SW 41 rotor. All sucrose solutions contained 10 mM TEACl (pH 7.4), 2 mM MgCl$_2$, 5 mM ß-mercaptoethanol and 0.1 mM PMSF. A, Aliquots were removed from each fraction of the gradients and tested for poly(ADP-ribose) polymerase activity as described in (6). Free mRNP treated with 0.05 M KCl (△), or with 0.5 M KCl (▲). B, free mRNP were [^{32}P]ADP-ribosylated with 1 μM NAD prior to analysis on 10-30% sucrose gradients. Protein aliquots from each fraction were precipitated with TCA and the radioactivity of the precipitates was measured. Free mRNP treated with 0.05 M KCl (□), or with 0.5 M KCl (■).

Discussion

It appears from our results that free mRNP possesses both the enzymatic activities of synthesis and degradation of poly(ADP-ribose). A cytoplasmic poly(ADP-ribose) glycohydrolase has also recently been described by Tanuma *et al.* (12). This suggests a balance between the ADP-ribosylation/de-ADP-ribosylation states of free mRNP proteins. In addition, free mRNP poly(ADP-ribosyl)ated proteins seem associated with different ribonucleoprotein particles which are removed from free mRNP by a 0.5 M KCl treatment. It is noteworthy that low mol.wt. RNA able to inhibit mRNA translation *in vitro* have been isolated from such ribonucleoprotein particles (13). Free mRNP proteins have also been suggested to play a role in the mechanisms and regulation of the translation of mRNA (7, 8). One could speculate a role of poly(ADP-ribosyl)ation in the structural changes that may occur in the free mRNP to permit the translation of the mRNA repressed in these particles.

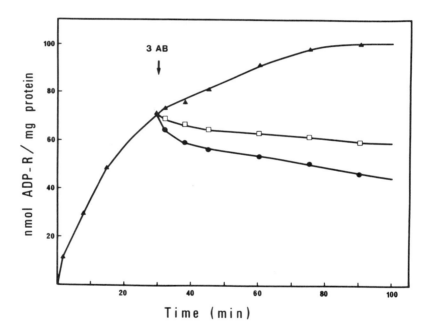

Fig. 2. Poly(ADP-ribose) glycohydrolase activity in free mRNP. Free mRNP isolated from mouse plasmacytoma cells were [^{32}P]ADP-ribosylated at 30°C with 200 μM NAD. 3AB was added after 30 min of reaction. Aliquots were taken at various times and precipitated in TCA (20% final concentration). The precipitates were collected on GFB Whatman filters and the radioactivity measured. (▲), 0 mM 3AB; (●), 6 mM 3AB; (□), 6 mM 3AB, 4 mM ADP-ribose.

128

Acknowledgements. This work was supported by the French CNRS and INSERM. HT was supported by a pre-doctoral fellowship from the Association de la Recherche pour le Cancer.

References

1. Furneaux, H.M. and Pearson, C.K. (1980) Biochem J 187: 91-103
2. Roberts, J.H., Stark, P., Giri, C.P. and Smulson, M. (1975) Arch Biochem Biophys 171: 305-315
3. Concha, I.I., Concha, M.I., Figueroa, J. and Burzio, L.O. (1985) ADP-Ribosylation of Proteins, F.R. Althaus, H. Hilz and S. Shall (eds.), Springer-Verlag, Berlin Heidelberg, pp. 139-146
4. Elkaim, R., Thomassin, H., Niedergang, C., Egly, J.M., Kempf, J. and Mandel, P. (1983) Biochimie 65: 653-659
5. Thomassin, H., Gilbert, L., Niedergang, C. and Mandel, P. (1985) ADP-Ribosylation of Proteins, F.R. Althaus, H. Hilz and S. Shall (eds.), Springer-Verlag, Berlin Heidelberg, pp. 148-153
6. Thomassin, H., Niedergang, C. and Mandel, P. (1985) Biochem Biophys Res Commun 133: 654-661
7. Vincent, A., Goldenberg, S., Standardt, N., Imaizumi-Scherrer, T., Maundrell, K. and Scherrer, K. (1981) Molec Biol Rep 7: 71-81
8. Spirin, A.S. and Ajtkhozhin, M.A. (1985) TIBS 10:162-165
9. Schmid, H.P., Akhayat, O., Martins de S.A.C., Puvion, F., Koehler, K. and Scherrer, K. (1984) EMBO J 3: 29-34
10. Althaus, F.R. and Richter, C. (1987) Molecular Biology Biochemistry and Biophysics, Vol. 37, Springer-Verlag, Berlin Heidelberg New York Tokyo, pp.3-226
11. Menard, L and Poirier, G.G. (1987) Biochem and Cell Biol 65: 668-73
12. Tanuma, S-I., Kawashima, K. and Endo, H. (1986) Biochem Biophys Res Commun 136: 1110-1115
13. Lorerboum, H., Digweed, M., Erdmann, V.A., Servadio, Y., Weinstein, D., De Groot, N. and Hochberg, A.A. (1986) Eur J Biochem 155: 279-287

Abbreviations:
mRNP - messenger ribonucleoprotein particles
TCA - trichloroacetic acid
3AB - 3-aminobenzamide
TEACl - triethanolamine chloride
PMSF - phenyl-methyl-sulfonyl-fluoride

New Inhibitors of Poly(ADP-Ribose) Synthetase

Marek Banasik, Hajime Komura[1], Isao Saito, Nazar A. N. Abed, and Kunihiro Ueda

Department of Clinical Science and Laboratory Medicine, Faculty of Medicine, Kyoto University, Shogoin, Sakyo-ku, Kyoto 606, Japan

Introduction

Despite the wealth of extensive studies on poly(ADP-ribose) synthetase, the biological function of this enzyme is not well understood (1). One effective approach to the biological function is to inhibit the enzyme activity *in vivo* using an inhibitor and to see cellular dysfunctions affected. Most of the inhibitors so far known, including nicotinamide, 3-aminobenzamide, and thymidine, have more or less side effects *in vivo* (1, 2), and the interpretation is accompanied by ambiguity. Furthermore, a comparison of various inhibitors on the enzyme activity *in vitro* would give us an insight into the nature of the active center of the enzyme. In this study, we made a survey of as many as 170 compounds for the inhibitory effect, and found many inhibitors with different types of structure.

Results and Discussion

Table 1 is a summary of representative data of our survey of inhibitors. Group 1 (nicotinamide derivatives), 2 (benzamide derivatives), 3 (acetophenone derivatives), and 4 (pyrimidine derivatives) are the expanded families of hitherto known inhibitors (3), while the other groups comprise entirely new types of compounds. It is evident that, in both groups, there are many new inhibitors with the potency comparable to or stronger than 3-aminobenzamide, the most widely used inhibitor (4). As judged by these data, structural features required for the inhibitory activity are summarized as follows: 1. At least one aromatic ring harboring conjugated double bonds (e.g. pyridine, benzene, pyrimidine) is necessary. Nipecotamide, a saturated analog of nicotinamide, was totally inactive. 2. At least one oxygen atom in a form of carbamyl, carboxyl or carbonyl group is necessary. In the presence of one such oxygen atom, a nitrogen atom is not always essential; for example, hydroxyacetophenone and naphthoquinone have no nitrogen atom. The oxygen atom appears to serve as an electron donor, and contribute to the interaction with enzyme (see below). 3. Among the benzamide derivatives, 3-substituted derivatives are generally, but not exclusively, more potent than 2- or 4-isomers.

---

[1]Suntory Institute for Bioorganic Research (SUNBOR), Wakayamadai, Shimamoto-cho, Mishima-gun, Osaka 618, Japan

Based on these data of benzamide derivatives, we examined a structure-activity relationship, i.e. which part of the molecule contributed most effectively to the inhibitory action. As shown in Fig. 1, the HOMO (highest occupied molecular orbital) of the oxygen atom of the carbonyl group exhibited a good correlation with inhibitory activity. This indicates that the ability of inhibitors to donate π-electrons to a positively charged residue or domain of the enzyme is one of the controlling factors. A significant contribution of LUMO (lowest unoccupied molecular orbital), which is a parameter of the ability to accept electrons at the carbonyl carbon, was also suggested. Biological effects of new inhibitors including cytotoxicity and side effects are currently under investigation.

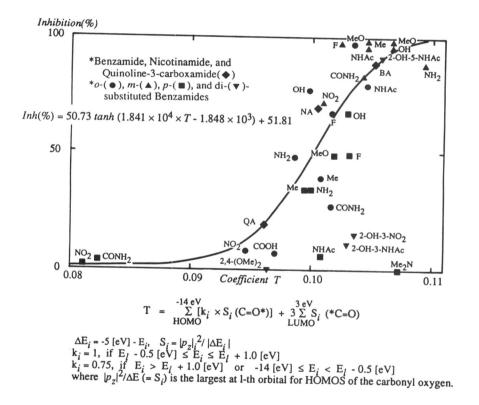

$$T = \sum_{HOMO}^{-14\,eV} [k_i \times S_i\,(C=O^*)] + 3\sum_{LUMO}^{3\,eV} S_i\,(^*C=O)$$

$\Delta E_i = -5\,[eV] - E_i, \quad S_i = |p_z|_i^2 / |\Delta E_i|$

$k_i = 1$, if $E_l - 0.5\,[eV] \leq E_i \leq E_l + 1.0\,[eV]$

$k_i = 0.75$, if $E_i > E_l + 1.0\,[eV]$ or $-14\,[eV] \leq E_i < E_l - 0.5\,[eV]$

where $|p_z|^2/\Delta E\,(= S_i)$ is the largest at l-th orbital for HOMOS of the carbonyl oxygen.

Fig. 1. Correlation between the inhibitory activity against poly(ADP-ribose) synthetase and coefficient $|p_z|^2/\Delta E$. The structure of benzamide for molecular orbital calculation was obtained by MM2 and MNDO optimizations. Molecular orbital calculation by a CNDO/2 program was performed on a simplification of fixing the dihedral angle between the benzene ring and the carbonyl group at 0°C.

131

Table 1. Inhibition of poly(ADP-ribose) synthetase activity by various compunds[a]

Compound	Inhibition at 1 mM	5 mM
1. Nicotinamide Derivatives		
α-Picolinamide	69	86
Nicotinamide	71	85
Isonicotinamide	51	71
1-Methylnicotinamide	27	52
cf.Nipecotamide	0	0
2. Benzamide Derivatives		
Benzamide	89	93
2-Aminobenzamide (Anthranilamide)	48	72
3-Aminobenzamide	88	96
4-Aminobenzamide	34	66
2-Chlorobenzamide	41	65
3-Chlorobenzamide	88	97
4-Chlorobenzamide	51	81
2-Fluorobenzamide	67	91
3-Fluorobenzamide	97	97
4-Fluorobenzamide	49	80
2-Hydroxybenzamide (Salicylamide)	78	90
3-Hydroxybenzamide	95	99
4-Hydroxybenzamide	67	86
2-Methoxybenzamide	97	100
3-Methoxybenzamide	98	100
4-Methoxybenzamide	49	82
2-Methylbenzamide	39	69
3-Methylbenzamide	98	99
4-Methylbenzamide	34	65
2-Nitrobenzamide	8	36
3-Nitrobenzamide	71	91
4-Nitrobenzamide	2	4
2-Acetamidobenzamide	79	92
3-Acetamidobenzamide	95	99
4-Acetamidobenzamide	6	22
3-Acetamidosalicylamide	11	31
5-Acetamidosalicylamide	91	100
2,4-Dimethoxybenzamide	0	30
3,5-Dimethoxybenzamide	46	71
4-N,N-Dimethylaminobenzamide	0	1
2-Hydroxy-3-nitrobenzamide	15	48
Phthalamide	27	55
Isophthalamide	83	95
Terephthalamide	4	0
Phthalamidic acid	7	23
cf.α-Phenylacetamide	0	0
3. Acetophenone Derivatives		
o-Acetamidoacetophenone	0	5
m-Acetamidoacetophenone	49	74
p-Acetamidoacetophenone	0	10
o-Aminoacetophenone	0	4
m-Aminoacetophenone	38	69
p-Aminoacetophenone	0	1
o-Hydroxyacetophenone	0	12
m-Hydroxyacetophenone	49	70
p-Hydroxyacetophenone	6	25

4. Pyrimidine Derivatives		
Uracil	3	33
Thymine	61	80
5-Aminouracil	4	23
5-Bromouracil	74	88
5-Fluorouracil	18	46
5-Nitrouracil	64	88
5. Phthalhydrazide Derivatives		
Phthalhydrazide	93	95
3-Aminophthalhydrazide (Luminol)	98	100
4-Aminophthalhydrazide (Isoluminol)	48	83
3-Nitrophthalhydrazide	96	100
4-Nitrophthalhydrazide	81	92
cf.Phthalimide	15	19
6. Naphthalene Derivatives		
1,4-Dihydroxynaphthalene	85	95
1-Hydroxy-2-methyl-4-aminonaphthalene (Vitamin K_5)	52	72
1,4-Naphthoquinone	73	100
2-Hydroxy-1,4-naphthoquinone	40	79
5-Hydroxy-1,4-naphthoquinone (Juglone)	81	94
2-Methyl-1,4-naphthoquinone (Vitamin K_3)	58	87
5,8-Dihydroxy-1,4-naphthoquinone	96	–
5-Hydroxy-2-methyl-1,4-naphthoquinone	66	88
7. Quinoline Derivatives		
4-Hydroxyquinoline	61	94
8-Hydroxyquinoline	9	42
Kynurenic acid	20	58
Xanthurenic acid	51	88
8. Carsalam Derivatives		
Carsalam	66	85
6-Acetamidocarsalam	4	5
8-Nitrocarsalam	7	31
Chlorthenoxazine	97	98
cf.Benzoyleneurea	95	98

[a]The reaction of poly(ADP-ribose) synthesis was carried out for 10 min at 37°C in the mixture (200 μl) of 100 mM Tris-HCl (pH 8.0), 10 mM $MgCl_2$, 5 mM dithiothreitol, 50 μg/ml of calf thymus DNA, 200 μM [^{14}C]NAD$^+$ (2.5 cpm/pmol), 10 μg/ml of poly(ADP-ribose) synthetase purified from calf thymus, and, if added, inhibitor as indicated. Cl_3CCOOH-insoluble ^{14}C was quantified using a liquid scintillation method.

Acknowledgements. The authors would like to thank Dr. Y. Ohashi (Nara Medical University) for a kind gift of purified poly(ADP-ribose) synthetase used in a part of this study. This work was supported in part by Grants-in-Aid for Scientific Research and Cancer Research from the Ministry of Education, Science, and Culture, Japan, and by the Suzuken Memorial Foundation.

References

1. Kunihiro, U., Hayaishi, O. (1985) Ann Rev Biochem 54: 73-100
2. Cleaver, J.E., Bodell, W.J., Borek, C., Morgan, W.F., Schwartz, J.L. (1983) ADP-Ribosylation, DNA Repair and Cancer, M. Miwa, O. Hayaishi, S. Shall, M. Smulson, T. Sugimura (eds.), Japan Scientific Societies Press, Tokyo, pp. 195-207
3. Ueda, K., Kawaichi, M., Hayaishi, O. (1982) ADP-Ribosylation Reactions: Biology and Medicine, O. Hayaishi, K. Ueda (eds.), Academic Press, New York, pp. 117-155
4. Purnell, M.R., Whish, W.J.D. (1980) Biochem J 185: 775-777

Monoclonal Antibodies to Human Placental Poly(ADP-Ribose) Synthetase

Hiroshi Ushiro and Yutaka Shizuta

Department of Medical Chemistry, Kochi Medical School, Nankoku, Kochi 781-51, Japan

Introduction

Poly(ADP-ribose) synthetase is a chromatin-bound enzyme which, in the presence of DNA, catalyzes polymerization of the ADP-ribose moiety of NAD to form a protein-bound homopolymer, poly(ADP-ribose) (1). The enzyme is automodified during the reaction. It has been shown that the enzyme consists of three functional domains for DNA-binding, automodification, and NAD-binding (2-4). The enzyme has recently been purified from human placenta and characterized (4, 5). To study the structure and function of the enzyme in more detail we have prepared monoclonal antibodies to the placental enzyme (6, 7). The present study defines three monoclonal antibodies to human poly(ADP-ribose) synthetase and shows that one of the antibodies which binds to the NAD binding domain inhibits the enzyme activity.

Results and Discussion

Immunization of BALB/c mice with human placental poly(ADP-ribose) synthetase yielded antisera with both neutralizing and binding activities (4). Fusion of spleen cells from the immunized mice with NS-1 murine myeloma cells yielded three positive hybridoma clones, which secreted monoclonal antibodies of IgG1 subclass (6, 7). Fig. 1 shows the effect of monoclonal antibodies on the enzyme activity. One of the antibodies, designated D4 and its Fab fragment inhibited the enzyme activity while the other two, designated G6 and X3, and their Fab fragments had no effect. From the ratio of the quantities of the enzyme to antibody D4 or its Fab fragment present in the neutralization assay, and from the extent of the inhibition of enzyme activity, one can estimate that approximately one mol of the enzyme is neutralized per mol of the antibody or its Fab fragment.

Antibody G6 bound to the native enzyme with an affinity similar to antibody D4 as shown by the indirect immunoprecipitation assay in Fig. 2. The immune complex of antibody G6 and the enzyme adsorbed to protein A-Sepharose exhibited enzyme activity: approximately 40% of the immunoprecipitated enzyme activity that was removed from the supernatant fraction was recovered in the pellet fraction. The immune complex of antibody D4 did not show enzyme activity. Antibody X3 did not remove any of the enzyme activity from the supernatant fraction. The reactivity of the antibodies was also assessed by immunoblot analysis (Fig. 3).

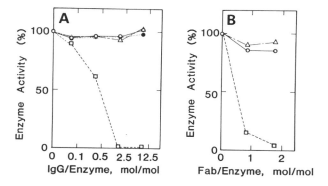

Fig. 1. Effect of monoclonal antibodies and Fab fragments on enzyme activity. A, purified poly(ADP-ribose) synthetase was preincubated with monoclonal antibodies at the molar ratios indicated. The enzyme activity was determined as described (4). Mab 137, a mouse monoclonal IgG1 to prostaglandin E2, was used as a negative control. O, antibody G6; □, antibody D4; △, antibody X3; ●, Mab 137. B, the enzyme was preincubated with Fab fragments. O, Fab from antibody G6; □, Fab from antibody D4; △, Fab from antibody X3. A hundred-percent activity refers to the value obtained from control enzyme without antibody or Fab fragment.

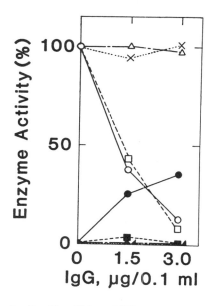

Fig. 2. (left) Indirect immunoprecipitation of poly(ADP-ribose) synthetase with monoclonal antibodies. Poly(ADP-ribose) synthetase (1.9 µg) was incubated with the indicated amounts of monoclonal antibodies and protein A-Sepharose (27% (v/v) suspension) in a final volume of 0.1 ml. The immune complex bound to protein A-Sepharose was sedimented by brief centrifugation. The enzyme activity in the supernatant and in the pellet was determined by the standard assay procedure. The enzyme activity in the supernatant: O, antibody G6; □, antibody D4; △, antibody X3; X, Mab 137. The activity in the pellet: ●, antibody G6; ■, antibody D4; ▲, antibody X3; ▶, Mab 137.

Antibodies G6 and X3 reacted with a single protein band corresponding to the enzyme. Antibody D4 yielded no positively stained band. These results indicate that antibody G6 binds to both the native and the denatured

135

enzyme, and its binding does not affect enzyme activity; antibody D4 binds only to the native enzyme and inhibits the enzyme activity; antibody X3 binds only to the denatured enzyme.

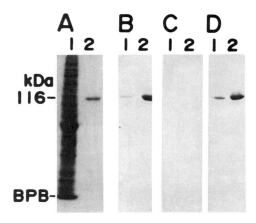

Fig. 3. Immunoblot of poly(ADP-ribose) synthetase recognized by monoclonal antibodies. Purified poly(ADP-ribose) synthetase and human placenta extract were resolved by SDS-polyacrylamide gel electrophoresis on a 7.5% acrylamide gel, and transferred to nitrocellulose membranes. A, gel stained with Coomassie blue; B-D, strips incubated with monoclonal antibodies G6, D4, and X3, respectively. Bound antibodies were detected by the peroxidase-antiperoxidase technique as described (4).

We previously showed that the placental enzyme, like the enzyme from calf thymus (2, 3), was cleaved by limited proteolysis into three functional domains for DNA-binding, automodification, and NAD-binding (4). To define the epitope specificity of the antibodies, we prepared the enzyme fragments by limited proteolysis with papain or chymotrypsin (Fig. 4A) and incubated with the antibodies (Fig. 4B). The immune complexes formed were adsorbed to protein A-Sepharose and resolved by SDS-polyacrylamide gel electrophoresis. Antibody G6 bound to the 72-kDa papain-digested fragment and the 62-kDa chymotryptic fragment. Antibody D4 bound to the 72-kDa fragment, the 54-kDa chymotryptic fragment, and a fragment of 40 kDa, which was derived from the 54-kDa fragment (4). Antibody X3 bound to none of the enzyme fragments in solution. The epitope specificity was also assessed by immunoblot analysis (Fig. 4C). The enzyme fragments, prepared as above, were resolved by SDS-polyacrylamide gel electrophoresis, transferred to nitrocellulose membranes, and analyzed for reactivity with the antibodies. Antibody G6 reacted with the 72-kDa and the 62-kDa fragments. Antibody X3 reacted with the 72-kDa, the 54-kDa, and the 40-kDa fragments. Antibody D4 yielded no positively stained band.

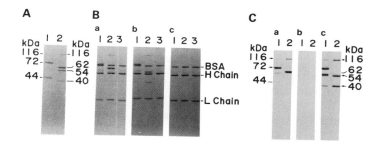

Fig. 4. Epitope mapping by indirect immunoprecipitation of enzyme fragments and by immunoblot analysis. The enzyme was partially digested with papain (lane 1) or chymotrypsin (lanes 2). A, the enzyme fragments were resolved by SDS-polyacrylamide gel electrophoresis on a 10% acrylamide gel, and stained with Coomassie blue. B, the antibodies G6 (a), D4 (b), and X3 (c) were incubated with the enzyme fragments. The immune complexes were adsorbed to protein A-Sepharose in the presence of 0.1% bovine serum albumin, and analyzed by SDS-polyacrylamide gel electrophoresis. Lane 3, antibodies incubated without the enzyme fragments. Protein bands of the heavy (H chain) and light (L chain) chains of IgG1 and nonspecifically absorbed albumin (BSA) are indicated on the right. C, the enzyme fragments were resolved by SDS-polyacrylamide gel electrophoresis, transferred to nitrocellulose membranes, and incubated with monoclonal antibodies G6 (a), D4 (b), and X3 (c). Bound antibodies were detected as described in the legend for Fig. 3.

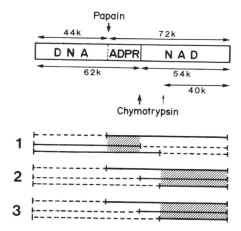

Fig. 5. Location of epitopes for monoclonal antibodies. DNA, ADPR, NAD represent domains for DNA-binding, automodification, and NAD-binding of the enzyme, respectively. Major cleavage sites attacked by papain and chymotrypsin are indicated. The solid lines represent the enzyme fragments that reacted with the antibody while the broken lines indicate the enzyme fragments that did not react with the antibody. 1-3, antibodies G6, D4, and X3, respectively.

Fig. 5 summarizes the epitope specificity of the antibodies. Antibody G6 recognizes an epitope located in the overlapping region of the 72-kDa and the 62-kDa fragments, which corresponds to the automodification domain. Antibodies D4 and X3 recognize epitopes located in the 40-kDa chymotryptic fragment, which constitutes the NAD-binding domain. It is reasonable to speculate that the epitopes for the latter two antibodies are different, since the reactivity of these antibodies with the native and the denatured enzyme are quite different: antibody D4 binds only to the native enzyme while antibody X3 binds only to the denatured enzyme. Furthermore, their interspecies cross-reactivity is different (7): antibody D4 cross-reacts with the enzyme from calf thymus but not with the enzyme from mouse testis, while antibody X3 cross-reacts with both the calf and the mouse enzymes. The two antibodies recognizing the native enzyme have been shown to be distinct from each other in their effect on the enzyme activity, and thus it is possible to use these antibodies to probe the function of poly(ADP-ribose) synthetase.

Acknowledgements. This work was supported in part by the Yamanouchi Foundation for Research on Metabolic Disorders and Grants-in-Aid for Science and Cancer Research from the Ministry of Education, Science, and Culture of Japan.

References

1. Ueda, K., Hayaishi, O. (1985) Annu Rev Biochem 54: 73-100
2. Nishikimi, M., Ogasawara, K., Kameshita, I., Taniguchi, T., Shizuta, Y. (1982) J Biol Chem 257: 6102-6105
3. Kameshita, I., Matsuda, Z., Taniguchi, T., Shizuta, Y. (1984) J Biol Chem 259: 4770-4776
4. Ushiro, H., Yokoyama, Y., Shizuta, Y. (1987) J Biol Chem 262: 2352-2357
5. Burtscher, H.J., Auer, B., Klocker, H., Schweiger, M., Hirsch-Kauffmann, M. (1986) Anal Biochem 152: 285-290
6. Ushiro, H., Yokoyama, Y., Shizuta, Y. (1985) Seikagaku 57: 894
7. Ushiro, H., Shizuta, Y. (1986) Seikagaku 58: 640

Characterization of the Human Autoantibody Response to Poly(ADP-Ribose) Polymerase

Hisashi Yamanaka, Erik H. Willis, Carol A. Penning, Carol L. Peebles, Eng M. Tan, and Dennis A. Carson

Department of Basic and Clinical Research, Research Institute of Scripps Clinic, La Jolla, California 92037 USA

Introduction

Poly(ADP-ribose) polymerase is a DNA-binding protein whose catalytic activity is stimulated strongly by DNA containing strand breaks (1). Recently, it has been proposed that a major function of the enzyme is to inactivate and eliminate cells with damaged DNA (2). DNA-binding proteins are established targets of autoimmunity in patients with systemic autoimmune diseases (3). Considering the possible functions of poly(ADP-ribose) polymerase in cells with DNA strand breaks, and the ability of the enzyme to undergo DNA-dependent automodification, it was conceivable that poly(ADP-ribose) polymerase could represent a potential autoantigen. For these reasons, we systematically searched for autoantibodies to poly(ADP-ribose) polymerase in human sera. Here we report the presence of autoantibodies to the poly(ADP-ribose) polymerase protein in patients with rheumatic diseases.

Poly(ADP-ribose) polymerase is a major acceptor of poly(ADP-ribosyl)ation in cells with damaged DNA (4). Utilizing this characteristic, we screened sera from patients with rheumatic diseases for the ability to immunoprecipitate [^{32}P]NAD labelled proteins of human CEM T lymphoblasts (5). Initially, 21 antinuclear antibody-positive sera were examined, of which, one specifically immunoprecipitated a 116 kDa poly(ADP-ribosyl)ated protein. Using this serum as a prototype, we surveyed by immunoblotting more than 200 sera from patients with autoimmune diseases and from control subjects. A total of six sera immunoblotted the same 116 kDa band in crude HeLa cell extracts, and also immunoprecipitated [^{32}P]NAD labelled CEM cell proteins (Fig. 1). These sera produced the same distinctive immunofluorescent pattern on HEp-2 cells, staining nuclei diffusely and reacting prominently with nucleoli and metaphase chromosomes (Fig. 2). As tested by immunoblotting, purified nucleoli were also enriched in the 116 kDa autoantigen.

To prove that the six sera specifically recognized the poly(ADP-ribose) polymerase protein, the enzyme was purified from calf thymus. Poly(ADP-ribose) polymerase and the autoantigenic activity co-eluted from both a DNA-cellulose column (Fig. 3) and a hydroxyapatite column. Incubation of the purified poly(ADP-ribose) polymerase with patients' IgG, coupled to

139

protein A-Sepharose, depleted the poly(ADP-ribose) polymerase activity in a dose-dependent manner. In control experiments, incubation of the enzyme with protein A-Sepharose beads coated with normal human serum or a control anti-nuclear antibody (anti-scl 70) did not alter the poly(ADP-ribose) polymerase activity. The purified enzyme yielded the same immunoblotting pattern as crude cell extracts. The immunoprecipitation of purified poly(ADP-ribose) polymerase, that had been previously [^{32}P]NAD labelled by the automodification reaction, similarly yielded a single band of 116 kDa. Poly(ADP-ribose) polymerase was also purified from human thymus. The human and calf enzymes bound the anti-poly(ADP-ribose) polymerase autoantibodies equivalently. Purified poly(ADP-ribose) polymerase coupled to Sepharose-4B, was used to isolate the anti-poly(ADP-ribose) polymerase autoantibodies from the serum of one patient. The affinity purified autoantibodies gave the same immunofluorescent pattern on HEp-2 cells, as whole serum, and also recognized the 116 kDa poly(ADP-ribose) polymerase protein in immunoblots.

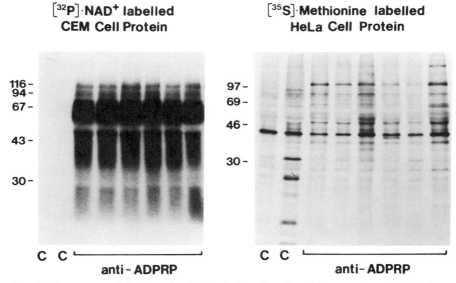

Fig. 1. Immunoprecipitation of poly(ADP-ribosyl)ated cellular protein. (A) Nuclear proteins of human T-lymphoblastoid line CEM cells were poly(ADP-ribosyl)ated with [^{32}P]NAD according to the method described before (4). Then, cells were lysed in NET-2 buffer (50 mM Tris-HCl, pH 7.4, 150 mM NaCl, 0.5% Nonidet P-40, 0.5% sodium deoxycholate, 0.1% sodium dodecyl sulfate [SDS], 5 mM EDTA and 0.5 mM PMSF). Solubilized proteins were reacted for 2 hr with human serum IgG, either from control (C) or from patients (anti-poly(ADP-ribose) polymerase), which had been immobilized onto protein-A Sepharose CL-4B beads. After centrifugation and washing of the beads with NET-2 buffer, antibody-bound protein was solubilized in Laemmli's sample buffer and boiled for 2 min. Immunoprecipitated proteins were fractionated on a 12.5% polyacrylamide gel containing 0.1% SDS, and autoradiographed. (B) HeLa cell proteins were labelled with [^{35}S]methionine overnight, and solubilized in NET-2 buffer. Immunoprecipitation was processed as described above.

To localize more precisely the autoantigenic epitopes on poly(ADP-ribose) polymerase, the enzyme was automodified with [^{32}P]NAD, partially proteolyzed, and then analyzed by immunoprecipitation. Previous experiments have shown that digestion of poly(ADP-ribose) polymerase with papain yields a 74 kDa fragment that contains the NAD binding and automodification domains. In contrast, digestion with α-chymotrypsin produces a 66 kDa fragment that includes the DNA binding and automodification domains (6). Besides reacting with the intact enzyme, one serum immunoprecipitated both the 74 kDa and 66 kDa fragments; the others recognized the 74 kDa fragment, but not the 66 kDa fragment. The interaction between purified poly(ADP-ribose) polymerase and [^{32}P]-labelled double stranded DNA was examined by a filter binding assay (7). The binding of purified poly(ADP-ribose) polymerase to DNA was perturbed *in vitro* by IgG autoantibody which recognized the DNA-binding domain but not by IgG which recognized only NAD-binding domain. These results define at least two autoantigenic epitopes on the poly(ADP-ribose) polymerase polypeptide, and indicate that the autoimmune response to the enzyme is heterogenous. However, none of the sera with anti-poly(ADP-ribose) polymerase antibodies contained other antibodies. Rather, each reacted monospecifically with poly(ADP-ribose) polymerase.

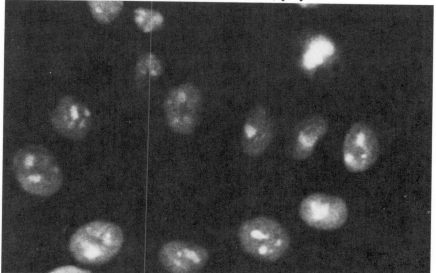

Fig. 2. Immunofluorescent detection of autoantibodies to poly(ADP-ribose) polymerase. Acetone-fixed HEp-2 cells reacted with a serum containing antibodies to poly(ADP-ribose) polymerase followed by fluorescein-conjugated goat anti-human IgG. The diffuse nuclear staining with nucleolar accentuation is apparent.

The patients with anti-poly(ADP-ribose) polymerase autoantibodies complained of persistent neuromuscular pain, principally affecting the lower extremities. One patient had Sjogren's syndrome: the others did not have any symptoms or signs of systemic lupus erythematosus, scleroderma, or other distinct connective tissue disease. Thus, the clinical significance of the autoantibodies is still unclear, and will require more extensive clinical correlative studies. Autoantibodies against DNA-binding proteins are common in rheumatic diseases. However, it is especially noteworthy that in the patient group reported here, the autoimmune response was directed exclusively against the poly(ADP-ribose) polymerase enzyme. In cells with damaged DNA, poly(ADP-ribose) polymerase is highly activated and also automodified. Thus, it is conceivable that the accumulation of DNA strand breaks in antigen presenting cells at sites of inflammation could lead to the modification of poly(ADP-ribose) polymerase, and might trigger an immune response against altered autoantigen.

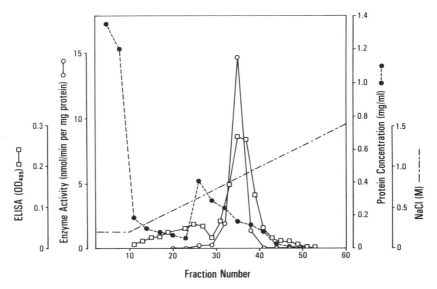

Fig. 3. Co-elution of poly(ADP-ribose) polymerase enzymatic and autoantigenic activity from a DNA cellulose column. Calf thymus was homogenized and a 40-80% ammonium sulfate fraction was collected and dialyzed into a buffer containing 50 mM Tris-HCl, pH 7.4, 0.2 M NaCl, 10 mM EDTA, 1 mM NaN$_3$, 1 mM glutathione, 50 mM NaHSO$_3$, 0.5 mM PMSF, 0.5 mM dithiothreitol, and 10% glycerol. The sample was loaded onto a DNA cellulose column (2.5 x 11 cm). After washing with the starting buffer, the DNA binding proteins were eluted with a linear 0.2-1.5 M NaCl gradient in the same buffer. The poly(ADP-ribose) polymerase catalytic activity in each 30 μl fraction was determined radiochemically (7). An ELISA was used to detect immunoreactive poly(ADP-ribose) polymerase. Fifty μl of each fraction was added to the wells of a polystyrene microtiter plate, and was incubated for two hr at 4°C. The coated wells were quenched with isotonic phosphate buffered saline, pH 7.4, containing 5% powdered milk. Then a 1:100 dilution of a serum containing anti-poly(ADP-ribose) polymerase autoantibodies was added to each well, and incubated for 1 hr. After washing, bound antibody was detected using affinity purified, alkaline phosphatase labelled goat anti-human IgG. Antibody binding is expressed as the OD$_{405}$ in each fraction.

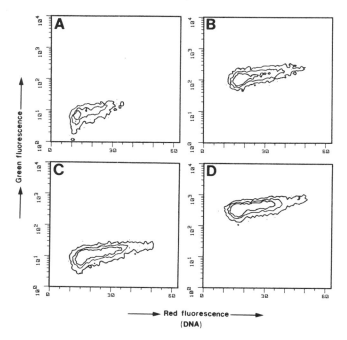

Fig. 4. Cellular content of poly(ADP-ribose) polymerase by two-color fluorescent measurement. Human CEM T lymphoblastoid cells were fixed with 45% ethanol for 30 min and stained with (A) FITC-labelled goat anti-human IgG, (B) FITC-labelled anti-poly(ADP-ribose) polymerase IgG, (C) normal human IgG followed by FITC-labelled goat anti-human IgG or (D) affinity purified anti-poly(ADP-ribose) polymerase IgG followed by FITC-labelled goat anti-human IgG (10 μg/ml in each experiment). Finally, cells were stained with propidium iodide (10 μg/ml) and two-color fluorescence was analyzed in a flow cytometer (11). Two-color contour plots shows DNA content (red fluorescence) on the X-axis, and FITC-antibody binding (green fluorescence) on the Y-axis.

In previous reports, rabbit antibodies against calf thymus poly(ADP-ribose) polymerase were suggested to have limited cross-reactivities with the enzyme from other species (8). In contrast, our preliminary results show that the human autoantibodies against poly(ADP-ribose) polymerase recognize poly(ADP-ribose) polymerase in many different species. Thus, these antibodies may be useful probes to investigate the biochemical role of poly(ADP-ribose) polymerase in different cell types. Recently, we have used the autoantibodies against poly(ADP-ribose) polymerase to measure the cellular content of poly(ADP-ribose) polymerase during the cell growth cycle, using flowcytometric analysis. Previous studies suggested the enzyme activity fluctuated markedly during the cell cycle (9). However, the fluorescent intensities of poly(ADP-ribose) polymerase in permeabilized ethanol-fixed human CEM T lymphoblasts were rather constant during the cell cycle. G2/M phase cells had approximately twice as much enzyme as

143

G1 phase cells, in accord with their increased content of DNA. Furthermore, the turnover rate of poly(ADP-ribose) polymerase in CEM cells was determined to be 12 hr (approximately half of the cell doubling time) by [^{35}S]methionine-labelling followed by immunoprecipitation. These data suggest that poly(ADP-ribose) polymerase stays bound to chromatin during the cell cycle, and does not turn over rapidly. It is likely that the reported changes in poly(ADP-ribose) polymerase activity during the cell cycle are related to the alterations in chromatin structure that modulate enzyme activity. Cell cycle dependent phenomenon, such as DNA synthesis and mitosis, are not tightly coupled to the synthesis of poly(ADP-ribose) polymerase proteins.

Acknowledgements. This work was supported in part by grants AR25443, GM23200, AR32063 and RR00833 from the National Institutes of Health. This is publication 4871BCR from the Research Institute of Scripps Clinic, La Jolla, California.

References

1. Pekala, P.H., Moss, J. (1983) Curr Top in Cellular Reg 22: 1-49
2. Berger, N.A. (1985) Rad Res 101: 4-15
3. Tan, E.M. (1982) Adv Immunol 33: 167-240
4. Surowy, C.S., Berger, N.A. (1983) Biochem Biophys Acta 740: 8-13
5. Yamanaka, H., Willis, E.H., Penning, C.A., Peebles, C.L., Tan, E.M., Carson, D.A. J Clin Invest, 80(3): 900-4
6. Kameshita, I., Matsuda, Z., Taniguchi, T., Shizuta, Y. (1984) J Biol Chem 259: 4770-4776
7. Mimori, T., Hardin, J.A. (1986) J Biol Chem 261: 10375-10379
8. Ushiro, H., Yokoyama, Y., Shizuta, Y. (1987) J Biol Chem 262: 2352-2357
9. Tanuma, S., Kanai, Y. (1982) J Biol Chem 257: 6565-6570
10. Carter, S.G., Berger, N.A. (1982) Biochemistry 21: 5475-5481
11. Kurki, P., Vanderlaan, M., Dolbeare, F., Gray, J., Tan, E.M. (1986) Exp Cell Res 166: 209-219

Abbreviations:
ELISA - enzyme-linked immunosorbent assay
SDS - sodium dodecyl sulfate
PMSF - phenylmethylsulfonyl fluoride

In Vivo Automodification of Poly(ADP-Ribose) Polymerase and DNA Repair

Jingyuan Zhang, Ramaswami Kalamegham, and Kaney Ebisuzaki

Cancer Research Laboratory, University of Western Ontario, London, Ontario N6A 5B7, Canada

Introduction

Poly(ADP-ribosyl)ation reactions have been implicated in various cellular functions including DNA repair, regulation of cell cycle, transformation and differentiation. Under *in vitro* conditions poly(ADP-ribose) polymerase, the enzyme catalyzing the synthetic reactions, is activated by DNA strand breaks (1) and in part controlled by automodification (2). However questions relating to the activity of the enzyme and the quantitive analysis of the extent of poly(ADP-ribosyl)ation of specific proteins under *in vivo* conditions are largely unanswered. The primary difficulties in such an analysis stem from the lability of the protein-(ADP-ribose) linkage and the lack of a specific radioactive labelling techniques. In order to simplify these problems, several investigators have utilized permeabilized cells and subcellular fractions in conjunction with labelling with NAD+ to identify poly(ADP-ribosyl)ated proteins. However these approaches are subject to artifacts including incidental DNA nicking, perturbations in the spatial relationships of macromolecules, as well as an arbitrary NAD concentration.

In this preliminary report, we describe the methodology we have used for the analysis of automodification in intact cells. The automodification of this enzyme during growth and in response to DNA damage reflects the dynamics of poly(ADP-ribosyl)ation and the fate of poly(ADP-ribose) polymerase in the cell.

Results and Discussion

Cell culture. BALB/c3T3 cells (clone A31, obtained from the American Type Culture Collection) were grown in Dulbecco's modified Eagle's medium supplemented with 10% calf serum (Colorado Serum Co.), penicillin and streptomycin. For measurement of poly(ADP-ribose) polymerase content and the degree of automodification, cells were labelled with [35S]methionine and [3H]adenosine, respectively. Calf thymus poly(ADP-ribose) polymerase, purified to apparent homogeneity (3, 4), was injected subcutaneously using Freund's complete adjuvant. A rabbit was injected 3 times at 2 week intervals (total protein, 2 mg) and bled 3 weeks after the last injection. Since the modified and unmodified forms of

145

poly(ADP-ribose) polymerase are susceptible to enzymatic attack and chemical degradation, we have used a number of inhibitors and chosen neutral or acidic conditions during the extraction of the enzyme (see flow chart). Under the conditions used, 94% of the total protein was extracted (data not shown). The enzyme was identified following immunoadsorption, polyacrylamide gel electrophoresis and autoradiography (Fig. 1). In a previous study, Ikai and Ueda (7) noted that the antibody used in their experiments recognized poly(ADP-ribose) polymerase containing relatively short poly(ADP-ribose) chains. In control experiments using purified poly(ADP-ribose) polymerase, we observed that highly modified enzyme was also immunoadsorbed (Fig. 2). Therefore the procedure used here would account for the total enzyme.

Flowchart.

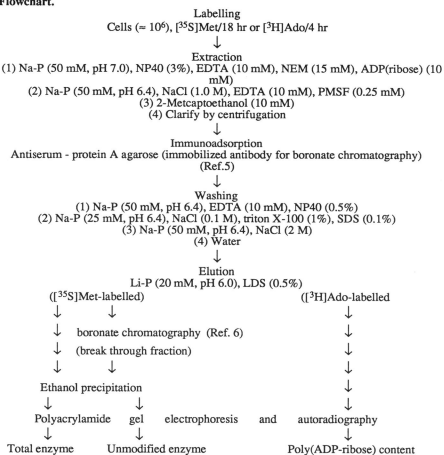

Labelling
Cells ($\approx 10^6$), [^{35}S]Met/18 hr or [^{3}H]Ado/4 hr
↓

Extraction
(1) Na-P (50 mM, pH 7.0), NP40 (3%), EDTA (10 mM), NEM (15 mM), ADP(ribose) (10 mM)
(2) Na-P (50 mM, pH 6.4), NaCl (1.0 M), EDTA (10 mM), PMSF (0.25 mM)
(3) 2-Metcaptoethanol (10 mM)
(4) Clarify by centrifugation
↓

Immunoadsorption
Antiserum - protein A agarose (immobilized antibody for boronate chromatography)
(Ref.5)
↓

Washing
(1) Na-P (50 mM, pH 6.4), EDTA (10 mM), NP40 (0.5%)
(2) Na-P (25 mM, pH 6.4), NaCl (0.1 M), triton X-100 (1%), SDS (0.1%)
(3) Na-P (50 mM, pH 6.4), NaCl (2 M)
(4) Water
↓

Elution
Li-P (20 mM, pH 6.0), LDS (0.5%)
([^{35}S]Met-labelled)   ([^{3}H]Ado-labelled
↓   ↓   ↓
↓   boronate chromatography (Ref. 6)   ↓
↓   (break through fraction)   ↓
↓   ↓   ↓
Ethanol precipitation   ↓
↓   ↓   ↓
Polyacrylamide   gel   electrophoresis   and   autoradiography
↓   ↓   ↓
Total enzyme   Unmodified enzyme   Poly(ADP-ribose) content

146

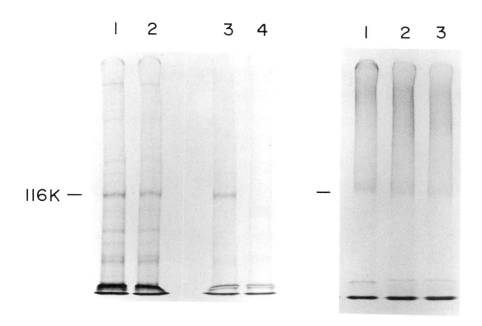

116K —

Fig. 1. (left) Identification of the immunoprecipitated enzyme after gel electrophoresis and autoradiography. This experiment was designed to test the specificity of the immuno-assay using either the free or immobilized antibody. Lane 1 shows total [^{35}S]-labelled enzyme obtained using the immobilized antibody; lane 2, unmodified enzyme obtained after boronate chromatography; lane 3, identical to lane 1 except that free antibody was used; and lane 4, pre-immune serum control for lane 3.

Fig. 2. (right) Immunoprecipitation of automodified enzyme. The automodified enzyme was prepared by incubation of sheep testis poly(ADP-ribose) polymerase (4) with [adenine-^{14}C]NAD. The autoradiography shows the automodified enzyme before immunoprecipitation (lane 1) and the enzyme immunoprecipitated with 25 µl antiserum (lane 2) and 50 µl antiserum (lane 3).

Enzyme-bound poly(ADP-ribose) content. Autoradiography and the accompanying densitometric scan of immunoprecipitates from normally growing [^{3}H]adenosine labelled cells indicate a single, automodified protein with an apparent Mr ranging from 116,000 to 144,000 (Fig. 3A, B). After treatment with MNNG (20 µg/ml, 15 min), the polymer content determined from [^{3}H]adenosine labelled enzyme increased by 267% (compared to the label in normal cells). However the poly(ADP-ribosyl)ated enzyme after DNA damage shared a similar molecular weight distribution, indicating that additional enzyme molecules were automodified. The addition of 5 mM 3-aminobenzamide (3AB) prior to and during DNA damage reduced the level of automodification and led to the appearance of a modified polypeptide, possibly an enzyme fragment which constituted 27% of the total poly(ADP-ribose). Control experiments using pre-immune serum (Fig. 3, lanes 1, 3,

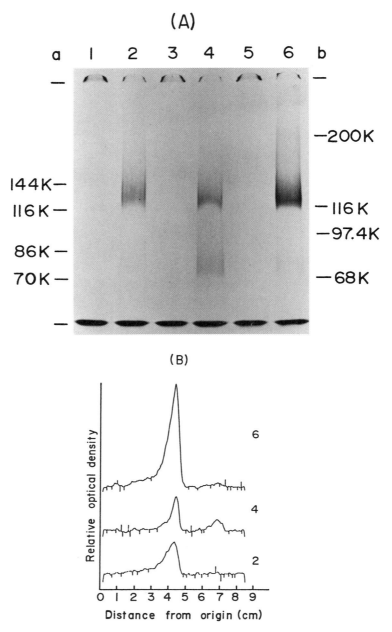

Fig. 3. Enzyme-bound poly(ADP-ribose) content. The autoradiography (A) and corresponding densitometric scan (B) were carried out with enzyme samples from [³H]adenosine labelled cells which were treated with MNNG (lane 6); 3AB and MNNG (lane 4); and without treatment (lane 2). The pre-immune serum controls for lanes 6, 4 and 2 are represented by lanes 5, 3, and 1, respectively.

5) attest to the high specificity of the immuno-assay for the modified poly(ADP-ribose) polymerase.

What fraction of the enzyme molecules are automodified? [^{35}S]Methionine labelled cells were used to determine the quantity of automodified enzyme relative to the total enzyme. As indicated in the flow chart, boronate-gel chromatography was used to distinguish the poly(ADP-ribosyl)ated enzyme. During normal growth, approximately 30% of the total enzyme was automodified (Fig 4.; see also Fig. 1). Treatment with MNNG resulted in the complete conversion of the enzyme to the automodified form without an appreciable change in total enzyme content. In other studies, Ogata *et al.* (8) noted that a modified protein of Mr 110,000-140,000, presumably poly(ADP-ribose) polymerase was the principal acceptor in MNNG-treated, permeabilized lymphocytes. Adamietz (9) noted that several proteins in addition to poly(ADP-ribose) polymerase were ADP-ribosylated in Yoshida hepatoma treated with dimethylsulfate.

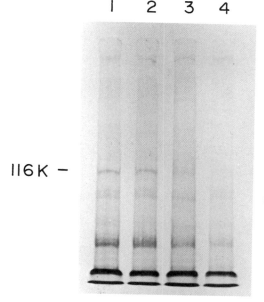

Fig. 4. The increase in automodified enzyme after MNNG treatment . Lanes 1 and 3 show total enzyme; lanes 2 and 4 unmodified enzyme. Lanes 1 and 2 show enzyme from normal cells; lanes 3 and 4, enzyme from cells treated with MNNG (20 µg/ml, 15 min).

Table 1. Relative content of poly(ADP-ribose) bound to the enzyme and its proteolytic fragment. The data were calculated from Fig. 3B.

| Treatment | Enzyme | Relative Polymer Content | | Fragment/total (%) |
		Fragment	Total	
Untreated Control	100 *	0	100	0
MNNG	261	6	267	2.2
3AB + MNNG	63	23	86	27.0

*Polymer content in the untreated control cells were normalized to 100.

The principal findings reported here involve the development of a quantative assay for automodification in the intact cell. By using specific antibodies, the method should be applicable for other poly(ADP-ribosyl)ated proteins as well. We note here that a substantial fraction (about 30%) of poly(ADP-ribose) polymerase molecules from growing cells was automodified. MNNG treatment resulted in the automodification of the total enzyme population. Significantly, treatment of these cells with the poly(ADP-ribose) polymerase inhibitor, 3AB, together with MNNG not only reduced automodification but also resulted in the appearance of a modified polypeptide.

References

1. Benjamin, R.C., Gill, D.M. (1980) J Biol Chem 255: 10502-10508
2. Zahradka, P., Ebisuzaki, K. (1982) Eur J Biochem 127: 579-585
3. Ito, S., Shizuta, Y., Hayaishi, O. (1979) J Biol Chem 254: 3647-3651
4. Zhang, J., Qiu, Z. (1986) Biochim Biophys Acta 882: 127-132
5. Schneider, C., Newman, R.A., Sutherland, D.R., Asser, U., Greaves, M.F. (1982) J Biol Chem 257: 10766-10769
6. Adamietz, P., Hilz, H. (1984) Methods Enzymol 106: 461-471
7. Ikai, K., Ueda, K. (1980) Biochem Biophys Res Commun 97: 279-286
8. Ogata, N., Kawaichi, M., Ueda, K., Hayaishi, O. (1980) Biochem Int 1: 229-236
9. Adamietz, P. (1985) ADP-Ribosylation of Proteins, F.R. Althaus, H. Hilz, S. Shall (eds.), Springer-Verlag, Berlin Heidelberg, pp. 264-271

Abbreviations:

Ado - adenosine
EDTA - ethylenediaminetetraacetate
LDS - lithium dodecyl sulfate
Li-P - lithium phosphate buffer
Met - methionine
MNNG - N-methyl-N'-nitro-N-nitrosoguanidine
Na-P - sodium phosphate buffer
NEM - N-ethylmaleimide
NP40 - nonidet P40
PMSF - phenylmethylsulfonyl fluoride
SDS -sodium dodecyl sulfate

ADP-Ribosylation and Chromatin Function

Felix R. Althaus, Georg Mathis, Rafael Alvarez-Gonzalez, Pius Loetscher, and Marianne Mattenberger

Institute of Pharmacology and Biochemistry, University of Zürich-Tierspital, CH-8057 Zürich, Switzerland

Introduction

Numerous lines of biological evidence converge on the concept that poly(ADP-ribosyl)ation may modulate various chromatin functions of higher eukaryotes, such as DNA repair, DNA replication, and gene expression in terminal differentiation (1). We have been intrigued by the question of how this process could be involved in such a diversity of biological processes. The observation that all of these chromatin functions are associated with changes of chromatin organization may hold an answer: poly(ADP-ribosyl)ation may primarily act to induce structural changes in chromatin and thereby modulate chromatin function. The present report conceptualizes the poly(ADP-ribosyl)ation system of higher eukaryotes as a polyanion generating system which induces structural changes in chromatin by modulating DNA-protein interactions at the nucleosomal level of chromatin organization. This concept extends previous models which rationalized the regulation of the DNA-binding protein poly(ADP-ribose) polymerase (2, 3).

Poly(ADP-ribose) in DNA excision repair: The inhibitor evidence and the role of polymer catabolism. The evidence linking poly(ADP-ribosyl)ation with DNA excision repair is heavily based on two types of evidence: i) *De novo* poly(ADP-ribosyl)ation of chromatin proteins is an obligatory postincisional event in DNA excision repair, and ii) Inhibition of the enzyme catalyzing this posttranslational modification somehow affects DNA excision repair and repair-dependent biological consequences of DNA damage such as cellular survival. However, these studies have not provided a clue as to which aspect of the postincisional formation of ADP-ribose polymers may be responsible for the modulation of repair functions. In addition, a critical examination of the available "inhibitor evidence" shows a puzzling contradiction of results (Fig. 1). At least 4 principally different inhibitor protocols have been used to probe the level of poly(ADP-ribose) involvement in DNA excision repair (Fig. 1A). The key difference is the time point of inhibitor addition relative to infliction of DNA damage and it is not known how this affects the poly(ADP-ribose) content of cells at the onset of incision and subsequent steps of repair (*vide infra*). If one then examines the more than 140 papers reporting effects of inhibitor treatments on DNA excision repair (1), one finds a wide spectrum of actions, the most consistent one being a potentiation of cytotoxicity (Fig. 1B). It is obvious that this diversity of results may be simply related to the different concentrations, sizes and complexities of ADP-ribose polymers which had been produced by different inhibitor protocols and various levels of DNA damage. Very recent data suggest that the size and complexity of ADP-ribose polymers *in vivo* are directly related to the overall nuclear polymer concentration (4).

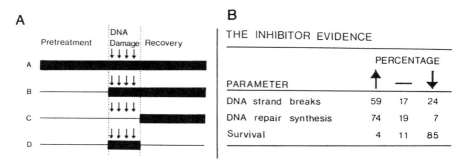

Fig. 1. Overview of various types of inhibitor protocols (A) and the diversity of effects observed on different repair parameters (B). The figures in B indicate the percentages of reports reporting identical effects on the repair parameters listed. The arrows symbolize an "increase" or "decrease", the horizontal line indicates "no effect" on the specified function. Data taken from ref. 1.

We have systematically studied the effects of DNA damage on poly(ADP-ribose) catabolism *in vivo* following inhibition of polymer biosynthesis with 3-aminobenzamide (5). The results can be summarized as follows (Fig. 2): i) Stimulation of poly(ADP-ribose) catabolism apparently is an obligatory postincisional event in DNA excision repair; ii) This stimulation is dependent on the dose of the DNA damaging agent and may be up to 680-fold. As a consequence, the $t_{1/2}$ of constitutive polymers, which is 7.7 hr in hepatocytes, was reduced to less than 41 sec in the presence of high levels of DNA strand breaks; iii) Enzymatic polymer breakdown is not directly stimulated by DNA strand breaks but is a function of polymer levels (sizes) (Fig. 2 and ref. 5).

Two generalizations emerge from these findings: i) Different inhibitor protocols (Fig. 1A) produce drastically different poly(ADP-ribose) concentrations dependent on the time of addition relative to infliction of DNA damage (Fig. 3); ii) Postincisional poly(ADP-ribose) turnover is coupled with the level of DNA damage. This suggests a requirement of dynamic polymer turnover on chromatin acceptor proteins in this and subsequent steps of DNA excision repair. We propose that this process is a reflection of DNA-protein shuttling as suggested by Ebisuzaki and associates (3) and that the velocity of this shuttling may be adjusted to the level of DNA damage by this mechanism. Thus poly(ADP-ribosyl)ation in eukaryotic chromatin may be regarded as a dynamic, strand break regulated polyanion generating mechanism, which operates to provide access for DNA repair enzymes to damaged templates. We have examined two aspects of this concept: i) Can ADP-ribosylation *in vitro* reduce DNA-protein interactions at the primary level of chromatin organization, i.e. the nucleosome? ii) What is the role of poly (ADP-ribose) in nucleosomal unfolding in DNA excision repair of mammalian cells *in vivo*?

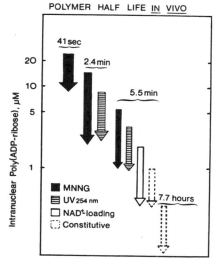

Fig. 2. Dependence of poly(ADP-ribose) catabolism on intracellular polymer concentrations. The arrows delineate the different ranges of polymer concentrations (expressed per nuclear volume of hepatocytes) for which identical half life values were obtained. Treatments used to elevate polymer concentrations included: MNNG, 200 μM (t1/2 = 41 sec), 50 μM (t1/2 = 2.4 min), and 20 μM (t1/2 = 5.5 min) N-methyl-N'-nitro-N-nitrosoguanidine; UV, 150 J/m^2 (t1/2 = 2.4 min), and 45 J/m^2 (t1/2 = 5.5 min) of ultraviolet light of 254 nm; NAD+-loading, increase of intracellular NAD+ in the absence of DNA damage as described in ref. 19.

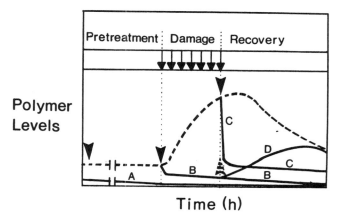

Time (h)

Fig. 3. Effects of the various inhibitor protocols (designated A - D in Fig. 1A) on intracellular polymer levels in mammalian cells. The scheme is based on the turnover data of Fig. 2 (5) and demonstrates the effect of the timing of inhibitor addition relative to infliction of DNA damage.. The dashed line shows the time course of polymer concentrations from constitutive to DNA strand break-induced levels, while lines marked A - D demonstrate the inhibitor effects in the respective protocols (arrowheads: time of inhibitor addition). For details see ref. 1.

153

DNA-protein interactions in poly(ADP-ribosyl)ated nucleosomal core particles. We have recently shown that nucleosomal core particles prepared by the method of Boulikas (6) retain high amounts of poly(ADP-ribose) polymerase activity (7). *In vitro* poly(ADP-ribosyl)ation caused a complete release of nucleosomal core DNA of 146 bp from its association with core particle proteins as examined by gel mobility shift analysis (Fig. 4). No release was observed in core particles which had been incubated in the presence of α-anomeric NAD, and which is not recognized as a substrate by poly(ADP-ribose) polymerase. Similarly, benzamide in the presence of β-NAD prevented the release of core DNA. A quantitative analysis of this phenomenon indicated that a separating force of 11 fN was sufficient to release core DNA from poly(ADP-ribosyl)ated core particles, while unmodified particles required at least 2-fold higher separating forces. These results then indicate that poly(ADP-ribosyl)ation may indeed reduce the binding of core DNA to core particle proteins. This system provides an attractive model to study the molecular details of the role of poly(ADP-ribosyl)ation in regulating DNA-protein interactions at the nucleosomal level of chromatin organization. It is important to note that the structural changes of chromatin in DNA excision repair in mammalian cells pertain also to this level of chromatin organization (*vide infra*). Therefore, detailed studies of the relevant poly(ADP-ribose) acceptor proteins involved in this *in vitro* phenomenon are in progress in our laboratory. It is interesting that free ADP-ribose polymers may release DNA-bound histone H4 *in vitro* as shown by Sauermann and associates (8), and it remains to be seen how this relates to our observations involving the model of nucleosomal core particles.

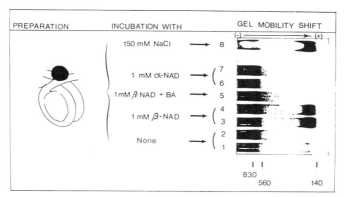

Fig. 4. DNA-protein interactions in isolated nucleosomal core particles following poly(ADP-ribose)-modification *in vitro*. Nucleosomal core particles were prepared as described (6, 7) and incubated as indicated. Subsequent gel mobility shift analysis was performed as described (7), using DNA restriction markers of 830, 560, and 140 bp.

Nucleosomal unfolding in poly(ADP-ribose)-depleted mammalian cells in DNA excision repair. In chromatin in mammalian cells, newly synthesized repair patches exhibit a transient micrococcal nuclease hypersensitivity (9). This hypersensitivity is thought to reflect local unfolding in the tightly packed nucleosomal organization of chromatin, causing exposure of "free" DNA domains. Studies by Smerdon and associates (10, 11) and others (10) have shown, that these domains refold into nucleosomally organized domains following completion of ligation (11); the $t_{1/2}$ of this process in UV excision repair has been estimated to be about 20 min (10). The sizes and functions of these domains are unknown and the mechanism(s) involved in their formation are yet to be elucidated. We have proposed earlier that postincisional poly(ADP-ribosyl)ation may be a key component of the mechanism(s) which induce unfolding of nucleosomal domains in DNA excision repair (12), as also suggested by our *in vitro* results (Fig. 4) (7). Based on our studies on polymer turnover (Fig. 3) (5), we were able to design protocols to completely deplete mammalian cells of chromatin-bound poly(ADP-ribose) (i.e. below 5 fmol polymeric ADP-ribosyl residues per million cells), and to determine whether depleted cells could still form unfolded chromatin domains in the course of DNA excision repair.

Isolation and characterization of unfolded ("free") DNA domains from intact mammalian cells: In conventional nuclease probing procedures for the study of chromatin organization in DNA excision repair, free DNA domains are preferentially degraded and they can only be characterized indirectly. We therefore developed a protocol to isolate the micrococcal nuclease-sensitive, free DNA fraction from bulk chromatin of intact cells (13). The key findings can be summarized as follows (13, 14): i) The proportion of unfolded DNA domains in DNA excision repair apparently changes periodically in early stages of excision repair. (13); ii) Unfolded domains may stretch over more than 10 nucleosomes in length. (13); iii) Excision of a model adduct (i.e. dG-adducts formed with the ultimate carcinogen N-acetoxy-2-acetylaminofluorene) occurred almost exclusively in these unfolded domains in the first 3 hr after onset of repair, suggesting that the establishment of these domains is required for efficient excision. (14); iv) The appearance of repair patches in unfolded domains is at least a two-step process: pulse-labelled repair patches initially exhibit a random distribution, and only during a subsequent chase period do they appear in the unfolded configuration (14). This suggests that the unfolding is not required for the synthesis of repair patches.

In poly(ADP-ribose)-depleted cells, this repositioning of repair patches was completely blocked (14), although overall repair patch synthesis was unaltered under these conditions. As a consequence, DNA adducts remained unexcised and gradually accumulated in free DNA domains (14). These results suggest a tight coupling of the excision step with the formation of unfolded DNA domains by a mechanism involving poly(ADP-ribosyl)ation of chromatin.

Model for the involvement of poly(ADP-ribosyl)ation in DNA excision repair. The data of this report may be conceptualized as shown in Fig. 5: Following incision, poly(ADP-ribose)polymerase translocates from its "resting" position in the nucleosomal core (7, 15) to adjacent high affinity binding sites (sites of incision) (Fig. 5A) and becomes activated (Fig. 5B; 2, 3, 16, 17). At this stage, repair patches are formed and, subsequently, patch-containing DNA domains become preferentially unfolded by a mechanism involving poly(ADP-ribosyl)ation Fig. 4; 14). Preliminary evidence from our laboratory indicates that the polymerase itself is the critical acceptor species modified under these conditions (H.P.Naegeli and F.R. Althaus, unpublished). Thus, the presence of the automodified polymerase may determine the sites of unfolding in damaged chromatin and coordinate unfolding with the incision reaction (14). Excision of adducts apparently requires the unfolded configuration (Fig. 5C). The evidence for this is derived from the observation, that excision is blocked when newly synthesized repair patches remain in the folded configuration, as observed in poly(ADP-ribose)-depleted cells (*vide supra*, and 14). Following adduct removal and ligation, repaired DNA domains refold (10, 11, 13). The velocity of this unfolding may be regulated at the level of polymer turnover, which we have shown *in vivo* to depend on the amount of strand break induced polymers (Fig. 2; 5). Thus, rapid turnover of polymers on poly(ADP-ribose) polymerase under high DNA damage conditions can rapidly restore the DNA binding properties as well as the activity of poly(ADP-ribose) polymerase (3), which then becomes involved in a new cycle of unfolding/refolding.

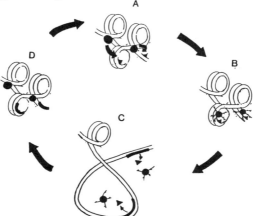

Fig. 5. Proposed model for the involvement of poly(ADP-ribose)polymerase in DNA excision repair. For explanations see text.

156

As far as our findings ascribe poly(ADP-ribosyl)ation a role in DNA excision repair, they contrast the hypotheses of Shall and associates (18) who envisioned poly(ADP-ribosyl)ation as a regulatory mechanism of ligation activity in the repair of alkylation damage in DNA. However, subsequent reports did not confirm such a specific role of poly(ADP-ribose) in DNA excision repair (1), although they failed to generate an alternative model. This contradictory phenomenology (1) may be reconciled on the premise that poly(ADP-ribosyl)ation affects local disruptions of chromatin structure and this secondarily affects DNA repair reactions.

Acknowledgements. This work was supported by grant 3.3540.86 (awarded to FRA) from the Swiss National Foundation for Scientific Research, the Sandoz Stiftung, and the Krebsliga des Kantons Zurich.

References

1. Althaus, F.R., Richter, C. (1987) ADP-ribosylation of proteins: Enzymology and biological significance. Molecular Biology, Biochemistry, and Biophysics (Springer Berlin Heidelberg New York Tokyo), Vol. 37
2. Ferro, A.M., Olivera, B.M. (1982) J Biol Chem 257: 7808-7813
3. Zahradka, P., Ebisuzaki, K. (1982) Eur J Biochem 127: 579-585
4. Alvarez-Gonzalez, R., Jacobson, M.K. (1987) Biochemistry 26: 3218-3224
5. Alvarez-Gonzalez, R. Althaus, F.R. (1988) J Biol Chem, submitted
6. Boulikas, T. (1984) Can J Biochem Cell Biol 63: 1022-1032
7. Mathis, G., Althaus, F.R. (1987) Biochem Biophys Res Commun 143: 1049-1054
8. Sauermann, G., Wesierska-Gadek, J. (1986) Biochem Biophys Res Commun 139: 523-529
9. Friedberg, E.C. (1984) DNA repair. W.H. Freeman, New York, pp.331-339
10. Nissen, K.A., Lan, S.Y., Smerdon, M.J. (1986) J Biol Chem 261: 8585-8588
11. Smerdon, M.J. (1986) J Biol Chem 261: 244-252
12. ADP-Ribosylation of Proteins (1985) Althaus, F.R., Hilz, H., Shall, S. (eds.). Springer Berlin Heidelberg New York Tokyo, pp. 235-241
13. Mathis, G., Althaus, F.R. (1986) J Biol Chem 261: 5758-5765
14. Mathis, G., Althaus, F.R. This volume
15. Leduc, Y., de Murcia, G., Lamarre, D., Poirier, G.G. (1986) Biochim Biophys Acta 875: 248-255
16. Benjamin, R.C., Gill, D.M. (1980) J Biol Chem 255: 10493-10501
17. Berger, N.A., Sikorski, G.W., Petzold, S.J., Kurohara, K.K. (1980) Biochemistry 19: 289-293
18. Shall, S. (1984) ADP-ribose in DNA repair: Adv Radiat Biol 11: 1-69
19. Loetscher, P., Alvarez-Gonzalez, R., Althaus, F.R. (1987) Proc Natl Acad Sci USA 84: 1286-1289

Modulation of Chromatin Structure by Poly(ADP-Ribosyl)ation

Ann Huletsky, Gilbert de Murcia[1], Sylviane Muller, Daniel Lamarre, Michel Hengartner, and Guy Poirier

Centre de recherche en cancérologie de l'Université Laval à L'Hôtel-Dieu de Québec, G1R 2J6 Québec, Canada

Introduction

The association of poly(ADP-ribose) polymerase with chromatin has been well described by Aubin *et al.* (1) and by Butt *et al.* (2) and we have found the enzyme to be associated mainly with tri- and tetranucleosomes (1). Poly(ADP-ribosyl)ation *in vitro* has been found to alter chromatin structure by decondensation of the 30 nm fiber (3). Poly(ADP-ribose) polymerase will alter chromatin structure by the modification of core histones, linker histone H1 and also by the interaction of automodified enzyme with chromatin. In this paper, we will review the different mechanisms by which poly(ADP-ribose) polymerase could modify specific nuclear proteins and alter chromatin structure. We present a model for the interaction of the enzyme with the various subnuclear components.

Results and Discussion

Modification of nuclear proteins by poly(ADP-ribosyl)ation. It has been shown by Okazaki *et al.* (4) that poly(ADP-ribose) polymerase could modify to the same extent nuclear proteins such as histone H1 and the core histones when the purified enzyme (DNA-independent activity) was incubated with individual histones. However, when purified poly(ADP-ribose) polymerase was incubated with chromatin, we found preferential modification of histone H1 which resulted in hyper(ADP-ribosylation) of this protein (5). This specific modification was associated with hypermodification of poly(ADP-ribose) polymerase at high NAD concentrations (200 μM) (5). However, lower concentrations of NAD were used, which lead to a lower level of poly(ADP-ribosyl)ation of the enzyme, modification of core histone H2A, protein A24 and mainly core histone H2B was observed (6).

In a previous study, we have demonstrated that in H1-depleted chromatin, core histones H2A, H2B and protein A24 became the major histone acceptors of poly(ADP-ribose) at a high concentration of NAD (200μM) (7).

[1]I.B.M.C. du C.N.R.S., Laboratoire de Biochimie II and Laboratoire d'immunologie, 15 rue René Descartes, 67084 Strasbourg Cedex, France

These modifications were also associated with a high level of poly(ADP-ribosyl)ation of the enzyme. Furthermore we have found that upon poly(ADP-ribose) glycohydrolase action on poly(ADP-ribosyl)ated chromatin, modification of histone H2B was more resistant than modification of other nucleosomal proteins (8). Menard *et al.* (9) have also shown, in a reconstituted *in vitro* poly(ADP-ribose) turnover system, that the half life on various acceptor proteins is quite different. These results suggest that the preferential distribution of poly(ADP-ribose) on nuclear proteins is related on the one hand to the localization of the enzyme on active (H1-depleted) and inactive (native) chromatin and on the other hand to the turnover rate of poly(ADP-ribose) on the enzyme which then determines the pattern of poly(ADP-ribosyl)ation of nuclear proteins. We will discuss in the last section the preferential distribution of the enzyme in the nucleus.

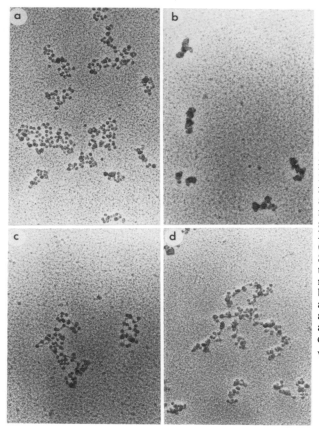

Fig. 1. Electron microscopic visualization of reconstituted poly(ADP-ribosyl)ated H1-depleted chromatin. H1-depleted polynucleosomes were incubated at 30°C with purified poly(ADP-ribose) in the presence of 200 µM NAD. The reaction was stopped by adding 1 mM 3-AB and nucleosomes were reconstituted by adding histone H1 slowly at a ratio (H1/nucleosome:1/1). Samples were diluted and treated for electron microscopy as described by Poirier *et al.* (5) (a and c) represent control and poly(ADP-ribosyl)-ated H1-depleted chromatin reconstituted with histone H1.

Modulation of chromatin structure by poly(ADP-ribosyl)ation: effect of poly(ADP-ribosyl)ation on chromatin superstructure. We have shown that poly(ADP-ribosyl)ation of native chromatin *in vitro* from endogenous and exogenous activity of the enzyme led to decondensation of the 30 nm chromatin fiber (3, 5, 10). This decondensation was associated with hypermodified forms of histone H1 at high NAD concentrations. We have thus studied the structural effect of poly(ADP-ribosyl)ation at a high NAD level on H1-depleted chromatin where modification of core histones was observed (7). This specific poly(ADP-ribosyl)ation prevented the recondensation of polynucleosomes reconstituted with native histone H1 (Fig. 1).

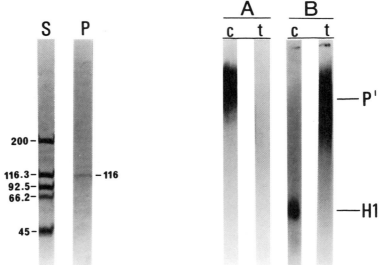

Fig. 2. Autoradiogram showing the interaction of the automodified poly(ADP-ribose) polymerase with native and H1-depleted chromatin. Following poly(ADP-ribosyl)ation at 200 μM NAD, H1-depleted chromatin and native chromatin were resolved on a 5-20% (w/v) isokinetic sucrose gradient containing 40 mM NaCl. Fractions corresponding to the chromatin peak (c) or to the top portion (t) of the gradients were pooled and precipitated with 25% TCA. Chromosomal proteins were then extracted, electrophoresed on a low porosity gel at pH 6.0 (9) and autoradiographed. Tracks S and P represent stained gel with high molecular weight standard and poly(ADP-ribose) polymerase respectively. A: H1-depleted chromatin, B: native chromatin. P': automodified poly(ADP-ribose) polymerase. H1: poly(ADP-ribosyl)ated histone H1. Numbers on each side of S and P represent molecular weight in kilodaltons.

Since the major poly(ADP-ribose) acceptor protein in chromatin and in intact cells is poly(ADP-ribose) polymerase (5, 11, 12), we have studied the interaction of the highly modified polymerase with native and H1-depleted chromatin in order to define its role as a potential modulator of chromatin structure. When poly(ADP-ribosyl)ated native chromatin was resolved on sucrose gradients, it was found that most of the modified enzyme was dissociated from chromatin and found on top of the gradients (Fig. 2B). The presence of the enzyme was confirmed by Western blot analysis of glycohydrolase treated preparations (data not shown). However, when H1-depleted chromatin was poly(ADP-ribosyl)ated, most of the modified enzyme remained associated with chromatin (Fig. 2A). Thus the presence of the modified enzyme in conjunction with hyper(ADP-ribosyl)ated histone H2B in H1-depleted chromatin might prevent the recondensation of this chromatin. Indeed, the modified enzyme remained bound to H1-depleted chromatin even in the presence of 0.35 M NaCl, which is usually used to dissociate the unmodified enzyme (data not shown). The presence of highly modified enzyme in specific regions of chromatin in the nucleus may thus have a role to play in the modulation of chromatin structure. Ueda *et al.* (13) have also suggested that the polymer synthesized on the enzyme bound in the vicinity of strand breaks during DNA repair might attract and bind DNA ligase and support the ligase activity. They thus suggest that polymer bound to the enzyme may also serve as a link between DNA damage and repair enzymes (13).

Effect of poly(ADP-ribosyl)ation on core nucleosome structure. To study the effect of poly(ADP-ribosyl)ation on nucleosome structure, we have used an immunological approach. We have analyzed the accessibility of core histone by ELISA to various monoclonal and polyclonal anti-core histone antibodies in poly(ADP-ribosyl)ated chromatin. We have found that the N-terminal domain of histone H2B, the C-terminal part of histone H3 (Fig. 3) and the N-terminal domain of histone H4 (data not shown) became accessible to the antibodies after poly(ADP-ribosyl)ation of native chromatin. These results suggest that poly(ADP-ribosyl)ation affected core histone-DNA interactions. Also, the histone H3/H4 complex was more accessible to an antibody directed against this complex (data not shown) suggesting a destabilization of the interaction between core histones within the octamer. Furthermore, this increased accessibility was proportional to the NAD concentrations used (data not shown) and was blocked by 3-aminobenzamide and nicotinamide (Fig. 3). These results strongly suggest that, in addition to affecting chromatin higher order structure, as we have demonstrated earlier (3), poly(ADP-ribosyl)ation would cause a destabilization of the core nucleosome structure. This might lead to the dissociation of DNA around the histone octamer such as shown by Huletsky *et al.* (7) and Mathis and Althaus (14) and to the dissociation of the histone octamer.

Nuclear distribution of poly(ADP-ribose) polymerase. Studies of Lamarre *et al.* (15) have shown the localization of unmodified poly(ADP-ribose) polymerase in chromatin and in nuclear matrix by Western blot analysis and that this localization depends on the cell types used. Also, automodified poly(ADP-ribose) polymerase identified by immunoprecipitation and autoradiography of [^{32}P]-labeled total proteins has been found to be mainly associated with chromatin and to a lesser extent in nuclear matrix (Lamarre *et al.*, unpublished observations). These results are in agreement with the recent studies of Wesierska-Gadek and Sauermann (16). Song and Adolph (17) have also shown the presence of automodified enzyme in nuclear matrix and in metaphase chromosomes. Finally, it has been suggested that lamins can be poly(ADP-ribosyl)ated *in vivo* in normal (16) and damaged cells (12, 18). These results provide further evidence for a localization of poly(ADP-ribose) polymerase in the vicinity of the lamina. Indeed, electron microscopic studies show a complete association of poly(ADP-ribose) polymerase with metaphase chromosome whereas in interphase cells there is a preferential association with heterochromatin (19).

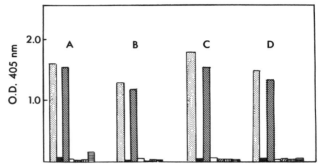

Fig. 3. Binding of polyclonal and monoclonal antibodies against histone to poly(ADP-ribosyl)ated native chromatin. ELISA were carried out on chromatin with rabbit antisera as described by Muller *et al.* (24) and with monoclonal antibodies by Muller *et al.* (25). Native chromatin directly diluted in PBS-T (▨), native chromatin which was incubated in the presence of Mg^{+2} and poly(ADP-ribose) polymerase (■), native chromatin incubated in the presence of Mg^{+2}, poly(ADP-ribose) polymerase and 200 μM NAD (▨), in the presence of Mg+2, poly(ADP-ribose) polymerase, 200 μM NAD and inhibitors, NAM 10 mM (□), 3-AB 5 mM (▨). Controls were also done with no enzyme in the presence of Mg^{+2}, 200 μM NAD and inhibitors, NAM (▥), 3-AB (▤). Antibodies used were: A: monoclonal antibody A21 supernatant (recognizes the N-terminal part of H2B in residues 1-25), B: monoclonal antibody C13 supernatant (recognizes region situated in residues 10-35 of H2B), C: antiserum to peptide 130-135 H3, D: antiserum against H2A.

As a model for the distribution of poly(ADP-ribose) polymerase in the nucleus of interphase cells, we propose from the studies of different groups and, from our studies, the presence of poly(ADP-ribose) polymerase in three nuclear compartments. The subnuclear structures are condensed chromatin, open chromatin and nuclear matrix (lamina) as depicted in Fig. 4.

In an undamaged cell poly(ADP-ribose) polymerase could be associated with chromatin and lamina. Indeed, Gradwohl *et al.* (20) have demonstrated that poly(ADP-ribose) polymerase can form looped structures in DNA *in vitro*. This observation suggests that the enzyme may be associated with DNA looped structures in chromatin in the absence of strand breaks (Fig. 4). Furthermore, the enzyme bound to the lamina may be associated with the base of chromatin looped domains in close contact with lamina (21). Upon DNA damage, the enzyme might redistribute according to strand breaks and excision repair taking place in these compartments. As the enzyme is being activated by strand breaks in H1-depleted chromatin or H1-containing chromatin, it poly(ADP-ribosyl)ates different acceptor proteins and has two modes of binding. In one case it dissociates from H1-containing chromatin and in the other case it binds with higher affinity to H1-depleted chromatin as shown in Fig. 2, possibly depending upon glycohydrolase activity in these two types of chromatin. In addition, the enzyme associated with lamina poly(ADP-ribosyl)ates lamins as suggested by Adamietz (12). This supports a localization of some repair events in the nuclear matrix as shown by McCready and Cook (22).

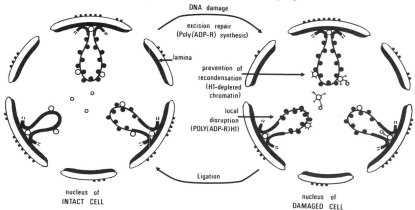

Fig. 4. Proposed model for nuclear localization of poly(ADP-ribose) polymerase and glycohydrolase and structural effects on chromatin during DNA repair. (■■) 30 nM chromatin fiber, (⚫•⚫) H1-depleted chromatin, (O) poly(ADP-ribose) polymerase, (o) poly(ADP-ribose) glycohydrolase, (-) oligo(ADP-ribose), (——⟨) poly(ADP-ribose), (·) histone H1.

The poly(ADP-ribosyl)ation of histones H1 and H2B and poly(ADP-ribose) polymerase *in vivo* during DNA repair (12, 18) destabilizes chromatin superstructure and core nucleosome structure. On the one hand, poly(ADP-ribosyl)ation of these histones causes local disruption of the 30 nm fiber of H1-containing chromatin (3) and prevents the recondensation of H1-depleted chromatin (Fig. 1 and 4). On the other hand, as shown in Fig. 3, core histones become accessible to core histone antibodies upon poly(ADP-ribosyl)ation of native and H1-depleted chromatin, thus suggesting a destabilization of core nucleosome structure. This effect would cause an increased accessibility of core DNA to various multienzymatic repair enzyme complexes in poly(ADP-ribosyl)ated chromatin. Indeed Althaus *et al.* (23), have shown that poly(ADP-ribosyl)ation is associated with the removal of bulky adducts from chromatin by repair enzymes. Upon resealing of DNA strand breaks, poly(ADP-ribose) polymerase redistributes to the three types of nuclear compartments (Fig. 4). It would re-equilibrate with open (H1-depleted chromatin) or condensed (H1-containing chromatin)

163

and nuclear matrix. In conclusion, we think that poly(ADP-ribosyl)ation is a key event in DNA repair by modulating the complex structure of eukaryotic chromatin. Thus poly(ADP-ribosyl)ation of chromatin components could depend upon the number of repair events in different nuclear compartments such as condensed chromatin, open chromatin or nuclear matrix.

Acknowledgements. We would like to thank Dr. A. Anderson for expert discussion on this paper. We also thank E. Lemay and M. Gagnon for excellent secretarial assistance and Hélène Voyer for expert art work.

References

1. Aubin, R.J., Dam, V.T., Miclette, J., Brousseau, Y., Poirier, G.G. (1982) Can J Biochem 60: 295-305
2. Butt, T.R., Brothers, J.F., Giri, C.P., Smulson, M. (1978) Nucleic Acids Res 5: 2775-2788
3. de Murcia, G., Huletsky, A., Lamarre, D., Gaudreau, A., Pouyet, J., Daune, M., Poirier, G.G. (1986) J Biol Chem 15: 7011-7017
4. Okazaki, H., Niedergang, C., Martinage, A., Couppez, M., Sautière, P., Mandel, P. (1980) FEBS Lett 110: 227-229
5. Poirier, G.G., de Murcia, G., Jongstra-Bilen, J., Niedergang, C., Mandel, P. (1982) Proc Natl Acad Sci USA 79: 3423-3427
6. Huletsky, A., Niedergand, C., Fréchette, A., Aubin, R., Gaudreau, A., Poirier, G.G. (1985) Eur J Biochem 146: 277-285
7. Huletsky, A., de Murcia, G., Mazen, A., Lewis, P., Chung, D.G., Lamarre, D., Aubin, R.J., Poirier, G.G. (1985) ADP-ribosylation of Proteins, F.R. Althaus, H. Hilz, S. Shall, (eds.) Springer-Verlag, Berlin Heidelberg, 180-189
8. Gadreau, A., Ménard, L., de Murcia, G., Poirier, G.G. (1986) Biochem Cell Biol 64: 146-153
9. Ménard, L., Thibault, L., Gadreau, A., Poirier, G.G. (1987) This volume
10. Aubin, R.J., Fréchette, A., de Murcia, G., Mandel, P., Grondin, G., Poirier, G.G. (1983) EMBO J 2: 1685-1693
11. Singh, N., Leduc, Y., Poirier, G., Cerutti, P. (1985) Carcinogenesis 6: 1489-1494
12. Adamietz, P. (1985) ADP-ribosylation of Proteins, F.R. Althaus, H. Hilz, S. Shall, (eds.) Springer-Verlag, Berlin Heidelberg, 264-271
13. Ueda, K., Ohashi, Y., Hatakeyama, K., Hayashi, O. (1983) ADP-ribosylation, DNA Repair and Cancer, M. Miwa, O. Hayashi, S. Shall, M. Smulson, T. Sugimura(eds.) Jpn Sci Soc Press, Tokyo, 175-182
14. Mathis, G., Althaus, F.R. (1987) Biochem Biophys Res Commun 143: 1049-1054
15. Lamarre, D., de Murcia, G., Talbot, B., Laplante, C., Leduc, Y., Poirier, G.G. (1987) This volume
16. Wesierka-Gadek, J., Sauermann, G. (1985) Eur J Biochem 193: 421-428
17. Song, M.K.H., Adolph, K.W. (1983) Biochem Biophys Res Commun 115: 938-945
18. Althaus, F.R., Richter, C. (1987) ADP-Ribosylation of Proteins: Enzymology and Biological Significance. Molecular Biology, Biochemistry and Biophysics, Vol. 37, Springer-Verlag, Berlin Heidelberg New York Tokyo
19. Fakan, S., Leduc, Y., Lamarre, D., Lord, A., Poirier, G.G. (1987) Proc 8th Int Symp on ADP-Ribosylation, Ft. Worth, Texas 1987, Abst 97
20. Gradwohl, G., Mazen, A., de Murcia, G., (1987) This volume
21. Weintraub, H. (1985) Cell 42: 705-711
22. McCready, S.J., Cook, P.R. (1984) J Cell Sci 80: 189-196
23. Althaus, F.R., Mathis, G., Alvarez-Gonzalez, R., Loetsher, P., Mattenberger, M. (1987) This volume
24. Muller, S., Plaue, S., Couppez, M., Van Regenmortel, M.H.V. (1986) Molec Immunol 23: 593-601
25. Muller, S., Isabey, A., Couppez, M., Plaue, S., Sommermeyer, G., Van Regenmortel, M.H.V. (1988) Molec Immunol, In press

Automodified Poly(ADP-Ribose) Synthetase Detected in Rat Liver Cells is Selectively Attached to a Nuclear Matrix-Like Structure

Martin Brauer, Jutta Witten, and Peter Adamietz

Institut für Physiologie Chemie der Universität Hamburg, 2000 Hamburg 20, West Germany

Introduction

The stimulation of poly(ADP-ribose) synthesis *in vivo* and *in vitro* appears to be generally accompanied by the automodification of poly(ADP-ribose) synthetase. It has been shown that this reaction is enhanced during recovery of cells from DNA damage (1), in response to the action of tumor promoting phorbol esters (2) and in isolated nuclei supplied with exogenous NAD (3). While there are reasonable doubts about a common molecular mechanism underlying the activation of the synthetase in these different cases (4), we were interested in the question of whether the automodification reaction is involved in the regulation of more than one biological function. Here we present evidence that the automodified poly(ADP-ribose) synthetase isolated from unstimulated normal rat liver exhibits distinct features that are not observed with poly(ADP-ribose) conjugates formed in response to stimulation by alkylating agents.

Results and Discussion

Poly(ADP-ribose) nonhistone conjugates of normal rat liver exhibit very large molecular sizes. The endogenous poly(ADP-ribose) nonhistone conjugates were purified from isolated rat liver nuclei using aminophenyl boronic acid agarose chromatography as described previously (5). A maximal yield of 290 pmol polymeric ADP-ribose residues per mg DNA was achieved with nuclei prepared at weak acidic pH. Application of neutral buffers was more effective with respect to removing contaminating proteins but caused losses of up to 90% of the poly(ADP-ribose) content, most probably as a consequence of less efficient inhibition of enzymatic degradiation (Fig. 1). Nevertheless, the electrophoretic patterns of both types of conjugate preparations looked very similar. The majority of the poly(ADP-ribose) groups appear to be excluded from the 7% polyacrylamide gel suggesting that the conjugates consisted of relatively large species. This result differs significantly from that observed when the endogenous synthesis of poly(ADP-ribose) was stimulated by DNA damage where the most prominent nonhistone conjugate migrated at 116 kDa and was identified as automodified poly(ADP-ribose) synthetase (1). A peak

165

with comparable mobility emerged only in preparations of basal nonhistone conjugates purified from neutral nuclei (closed circles in Fig. 1B). Since the ADP-ribose content and the protein stain at the corresponding gel region were enhanced by incubation with NAD *in vitro* (open circles in Fig. 1B), this also probably consists of automodified synthetase.

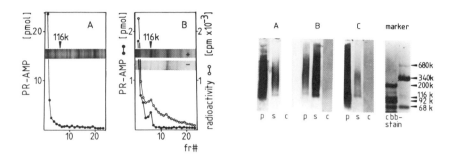

Fig. 1. (left) Electrophoretic analyses of poly(ADP-ribose) nonhistone proteins. Nuclei were prepared from the liver of male Wistar rats (220g) in the presence of citric acid (A) and at pH 7.5 (B) as described previously (10). Histones were extracted using 0.2 M sulfuric acid either before (closed circles) or after (open circles) a 5 min incubation with 0.5 mM [³H]NAD (1.2 Ci Mol⁻³) at 25° according to ref. 11. Purification of poly(ADP-ribose) nonhistone conjugates by boronate chromatography including preparation of samples for electrophoresis was done essentially as described elsewhere (5). Samples containing material equivalent to 22 μg DNA were applied per gel slot. The distribution of unlabelled poly(ADP-ribose) was determined by a PR-AMP specific radioimmuno assay according to ref. 5 (closed circles) while labelled poly(ADP-ribose) was analyzed by counting the acid insoluble radioactivity (open circles). Separate samples prepared with (+) or without (-) previous incubation with NAD were stained for protein using Coomassie Blue. The arrowheads mark the positions of bacterial β-galactosidase (116 kDa).

Fig. 2. (right) Immunoblot analyses of poly(ADP-ribose) nonhistone conjugates. Poly(ADP-ribose) nonhistone conjugates were purified from citric acid nuclei (A) or neutral nuclei without (B) and with (C) previous incubation with exogenous NAD. Electrophoretic separation in the presence of dodecyl sulfate was performed on 2% horizontal agarose gels using the sample buffer and running buffer as described by Laemmli (12). 15 μl samples containing material equivalent to 22, 200 and 2 μg DNA were applied to each lane in A, B and C, respectively. After transfer to nitrocellulose and treatment with 5% bovine serum albumin blots were incubated overnight with 2000-fold diluted rabbit anti-poly(ADP-ribose) antiserum (p), rabbit anti-poly(ADP-ribose) synthetase antiserum (s) (kindly provided by Prof. H. Hilz, Hamburg) and normal rabbit serum (c). Anti-rabbit-IgG alkaline phosphatase conjugate was used with a mixture of nitro blue phosphatase conjugate was used with a mixture of nitro blue tetrazolium chloride and 5-bromo-4-chloro-3-indolyl phosphate as color generating substrates. Marker proteins are bovine serum albumin (68 kDa), phosphorylase b (92 kD), β-galactosidase (116 kDa), rabbit muscle myosin (200 kDa), alpha-macroglobulin dimer (340 kDa) and tetramer (680 kDa). cbb-stain = Coomassie Blue stain.

The high molecular weight material resisted a 5 min period of boiling with 2% SDS and remained unchanged after treatment with reducing agents in the presence of 6 M guanidine hydrochloride, implying that the electrophoretic behaviour was not caused by simple aggregation phenomena. In order to find out whether this high molecular weight fraction contained poly(ADP-ribose) synthetase, which may have been shifted to the top of the gel as a result of extensive auto-ADP-ribosylation, samples of different preparations were separated on 2% agarose gels in the presence of dodecyl sulfate and indentified by immunoblotting. This technique allows a complete electrotransfer and provides more information about the size range of the conjugates as compared to analysis on polyacrylamide gels. As can be seen from Fig. 2A and B, only a very small part of the synthetase migrated to the position of the unmodified enzyme, as indicated by comparison with the mobilities of marker proteins. The bulk of the synthetase as well as poly(ADP-ribose) was found distributed over a broad region on the gel, corresponding to apparent molecular sizes ranging from 100 kDa to about 700 kDa. The ratio of poly(ADP-ribose) to poly(ADP-ribose) synthetase as judged from the relative staining intensities varied considerably but appeared to bear no relationship to the heterogeneity. Endogenous conjugates prepared from neutral nuclei (Fig. 2B) comprised a significantly lower relative amount of protein bound poly(ADP-ribose) residues as compared to those purified from citric acid nuclei (Fig. 2A) or after additional poly(ADP-ribose) synthesis *in vitro* (Fig. 2C). This raised the question of whether the retarded electrophoretic mobilities of the poly(ADP-ribose) synthetase can be explained by modification with poly(ADP-ribose) alone.

Evidence for the attachment of automodified poly(ADP-ribose) synthetase to a heterogenous population of carriers. When the mean chain length of the modifying poly(ADP-ribose) groups was determined by specific radioammuno assays (5, 6) a value of only 2.5 units ADP-ribose per chain was found for the endogenous conjugates while the *in vitro* conjugates contained 8.5 ADP-ribose moieties per residue. Furthermore, the presence of highly branched polymers or modification by very many short oligomers per molecule synthetase could be ruled out by using hydroxylapatite chromatography and cesium sulfate density gradient centrifugation. More than 80% of the endogenous poly(ADP-ribose) and synthetase of the two preparations with different extents of ADP-ribosylation eluted with 50 mM phosphate as determined by densitometric evaluation of immunoblots (hatched bars in Fig. 3). The elution of conjugates formed *in vitro* is shown for comparison (unhatched bars in Fig. 3). Fractions with mean chain lengths of 2.3, 6.5 and 10.5 needed 50 mM phosphate, 200 mM phosphate and 6 M guanidine hydrochloride for elution, respectively. On cesium sulfate gradients a slight shift toward higher densities was only observed with *in vitro* conjugates while the endogenous automodified synthetase co-banded with unmodified protein used for reference (data not shown). These results are consistent with the hypothesis that the automodified synthetase is bound to some kind of carrier that

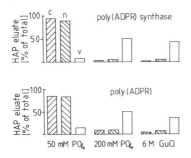

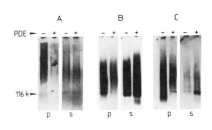

Fig. 3. (left) Hydroxylapatite chromatography of poly(ADP-ribose) nonhistone conjugates. Purified poly(ADP-ribose) nonhistone conjugates were applied to hydroxylapatite columns equilibrated with 2 M NaCl, 5 M urea, 10 mM sodium phosphate, pH 7.0. Fractions eluted after increase of the phosphate concentration to 50 and 200 mM as well as after applying 6 M guanidine hydrochloride (GuCl) were separated on agarose gels and the relative amounts of poly(ADP-ribose) and poly(ADP-ribose) synthetase determined by densitometric evaluation of immunoblots. hatched bars = endogenous conjugates obtained from citric acid nuclei (c) or neutral nuclei (n); unhatched bars = *in vitro* conjugates.

Fig. 4. (right) Effect of phosphodiesterase mediated degradation of poly(ADP-ribose) nonhistone conjugates formed *in vivo* and *in vitro*. 200 μl samples of poly(ADP-ribose) nonhistone conjugates either obtained from citric acid nuclei (A) or neutral nuclei before (B) and after (C) incubation with exogenous NAD were dialyzed against 1.5 M urea, 50 mM Tris/HCl, pH 7.0. For treatment with phosphodiesterase I (0.05 mg ml^{-1}, Boehringer) samples were adjusted to pH 8.0 and kept for 2 hr at 37°. The reaction products were dialyzed and analyzed by immunoblotting using anti-poly(ADP-ribose) antibodies (p) and anti-poly(ADP-ribose) synthetase antibodies (s).

The bond between the synthetase and the carrier is stable towards phosphodiesterase treatment and apparently not mediated by the protein bound ADP-ribose groups. As seen from Fig. 4A and 4B, the electrophoretic patterns of the synthetase in preparations with different degrees of ADP-ribosylation were not affected when the bulk of the modifying poly(ADP-ribose) residues was degraded by phosphodiesterase treatment. In contrast, the complex structure of the conjugates formed *in vitro* does appear to depend on the presence of poly(ADP-ribose) (Fig. 4C). Removal of poly(ADP-ribose) displaced most of the synthetase to the position of the unmodified enzyme, indicating a substantial difference in the properties of the endogenous conjugates.

The effect of hydroxylamine treatment shows that the bond linking the synthetase to the proposed carrier molecules is susceptible to nonenzymic base catalysis (Fig. 5A). Compared to the result obtained with *in vitro* conjugates; however, there were again qualitative differences. Whereas hydroxylamine removed the majority of the poly(ADP-ribose) residues from

168

both types of conjugates, a shift of a comparable percentage of synthetase to the 116 kDa position was only observed with *in vitro* material. The effect remained incomplete with the natural synthetase complex. In addition, treatment with hydroxylamine or sodium hydroxide under even milder conditions permitted the demonstration of an intermediate state where part of the endogenous synthetase had already been released from the unknown carrier before having lost its modifying poly(ADP-ribose) residues (Fig. 6). This seems the most reasonable explanation for the intermediate appearance of the band at the 116 kDa position stained using poly(ADP-ribose) antibodies.

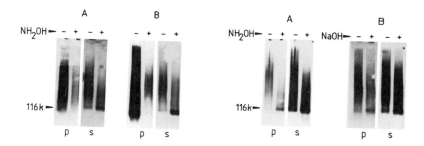

Fig. 5. (left) Cleavage of poly(ADP-ribose) nonhistone conjugates by treatment with hydroxylamine. 200 µl samples of poly(ADP-ribose) nonhistone conjugates either obtained from citric acid nuclei (A) or from neutral nuclei supplied with exogenous NAD (B) were made 2 M with respect to guanidine hydrochloride, and incubated for 1 hr at 37°. The reaction products were dialyzed and analyzed by immunoblotting using anti-poly(ADP-ribose) antibodies (p) and anti-poly(ADP-ribose) synthetase antibodies (s).

Fig. 6. (right) Partial cleavage of poly(ADP-ribose) nonhistone conjugates by mild treatment with hydroxylamine or sodium hydroxide. 200 µl samples of endogenous poly(ADP-ribose) nonhistone conjugates isolated from neutral nuclei were made 6 M with respect to guanidine hydrochloride and either treated with 0.5 M hydroxylamine hydrochloride pH 7.5 (A) or 10 mM sodium hydroxide for 10 min at 25°. The reaction products were dialyzed and analyzed by immunoblotting using anti-poly(ADP-ribose) antibodies (p) and anti-poly(ADP-ribose) synthetase antibodies (s).

Preparations of endogenous automodified poly(ADP-ribose) synthetase probably contain a component of the nuclear matrix. In order to provide more direct evidence for the existence and nature of the assumed carrier we applied several antisera developed against proteins present in nucleoids and the nuclear matrix. These proteins are known to form heterogenously sized populations of high molecular weight complexes. The results of a selection of these experiments are summarized in Fig. 7. Sera I and II that are obviously reactive with lamins A, B and C or lamin B, respectively, did not crossreact with the complex contained in the poly(ADP-ribose) nonhistone conjugate fraction. A positive reaction was only seen with serum III, which was probably specific to some kind of intermediate filament protein(s), though the precise antigen is not defined. It is probably a component of the nuclear matrix or closely related to it.

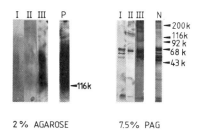

2% AGAROSE   7.5% PAG

Fig. 7. Immunoreaction of antibodies developed against nucleoids with components present in preparations of purified poly(ADP-ribose) nonhistone conjugates. Endogenous poly(ADP-ribose) nonhistone conjugates isolated from citric acid nuclei of rat liver were separated on a 2% agarose gel and analyzed by immunoblotting using either one of three different antisera developed against rat liver nucleoids designated as I, II and III, (kind gift of Prof. Jost, Gießen) or anti-poly(ADP-ribose) antibodies (p). The specificity of the anti-nucleoid antibodies is documented by immunoreaction with specific components of the nuclear matrix of rat liver (prepared according to ref. 7) using immunoblots transferred from 7.5% polyacrylamide gels. N = Coomassie Blue stain of rat liver nuclear matrix. See the legend to Fig. 1.

Automodified poly(ADP-ribose) synthetase formed in isolated nuclei in the absence of nucleases is associated with the DNA-nuclear matrix complex. Confirming other reports (7, 8) we have established that poly(ADP-ribose) synthetase is generally not bound to the nuclear matrix or nucleoids. As can be seen from Table 1, more than 99% of the nuclear content of the synthetase can be extracted by treatment with 2 M NaCl. The picture changes completely, however, when looking at the distribution of the poly(ADP-ribosyl)ated proteins including the automodified synthetase. After a brief incubation with tritiated NAD and subsequent dissociation with 2 M NaCl, 76% of the incorporated radioactive label co-sedimented with the DNA-matrix complex. The affect is abolished by the action of DNase I either applied before or after poly(ADP-ribose) synthesis. These findings are compatible with the assumption that poly(ADP-ribosyl)ation enhances the affinity of the acceptor molecules for DNA. This interpretation conflicts, however, with the hypothesis that an inverse relationship is responsible for inhibition of poly(ADP-ribose) synthetase activity as a result of auto-ADP-ribosylation (9). Alternatively, it may be that the small part of the nuclear content of poly(ADP-ribose) synthetase being selectively bound to a DNA matrix-complex represents the only active species of the enzyme when induction of unphysiological DNA breaks is prevented. The complex is apparently partially dissociated in the presence of 6 M guanidine hydrochloride or under nuclease attack followed by high salt treatment, to produce a heterogenously sized population of nucleoid fragments. The biological significance of the auto-ADP-ribosylation reaction is still unclear. It may be involved in the formation of the bond linking the synthetase to the DNA matrix-complex. We have shown that after stimulation by DNA damaging agents most of the automodified poly(ADP-ribose) synthetase formed does not appear to be bound to such a complex (5). This opens up the possibility that different biological responses might be mediated by the auto-ADP-ribosylation reaction. On the other hand, it cannot be excluded that in the alkylated cell the primary product of the auto-ADP-ribosylation reaction also consists of the high molecular weight complex, but this has not been observed so far because of the rapid release of the 116 kDa enzyme.

Table 1. Association of poly(ADP-ribose) protein conjugates with nucleoids.

Treatment	Amount of synthetase % of total	Amount of protein bound poly(ADP-ribose) (acid insoluble radioactivity) % of total
2 M NaCl		
pellet	1	76 ± 19[1]
supernatant	99	24 ± 15
DNase I + 2 M NaCl		
pellet	1	7 ± 6
supernatant	99	93 ± 6

[1] Average from 4 experiments

Nuclei were prepared at neutral pH as described previously (10) and briefly incubated with NAD (for details see legend of Fig. 1) either in the absence or presence of 0.1 mg ml[-1] DN'ase I. Incorporation of labelled NAD was stopped by low speed sedimentation of the nuclei through a 0.5 M sucrose buffer containing 10 mM benzamide. The resuspended nuclei were then treated for 15 min at 0° with 2 M NaCl, 10 mM Tris/Cl, pH 7.5 and subjected to ultracentrifugation (1 h at 150.000 x g, 4°). Poly(ADP-ribose) was determined by counting the acid insoluble radioactivity whereas the relative amounts of poly(ADP-ribose) synthetase were evaluated from immunoblots using a densitometer.

References

1. Adamietz, P. (1985) F.R. Althaus, H. Hilz, S. Shall (eds.) (264-271) ADP-ribosylation of Proteins. Springer, Berlin Heidelberg New York
2. Singh, N., Leduc, Y., Poirier, G., Cerutti, P. (1985) Carcinogenesis 6: 1489-1494
3. Jump, D.B., Smulson, M. (1980) Biochemistry 19: 1024-1030
4. Singh, N., Poirier, G., Cerutti, P. (1985) EMBO J 4: 1491-1494
5. Adamietz, P., Rudolph, A. (1984) J Biol Chem 259: 6841-6846
6. Wielckens, K., Bredehorst,R., Adamietz, P., Hilz, H. (1981) Eur J Biochem 117: 69-74

7. Lewis, C.D., Lewkowski, J.S., Daly, A.K., Laemmli, U.K. (1984) J Cell Sci Suppl 1: 103-122
8. Wesierska-Gadek, J., Sauermann, G. (1985) Eur J Biochem 153: 421-428
9. Ferro, A.M., Olivera, B.M. (1982) J Biol Chem 257: 7808-7813
10. Adamietz, P., Wielckens, K., Bredehorst, R., Lengyel, H., Hilz, H. (1981) Biochem Biophys Res Commun 101: 96-103
11. Adamietz, P., Bradehorst, R., Hilz, H. (1978) Biochem Biophys Res Commun 81: 1377-1383
12. Laemmli, U.K. (1970) Nature 227: 680-685

Studies on the Nuclear Distribution of ADP-Ribose Polymers

Maria E. Cardenas-Corona[1], Elaine L. Jacobson, and Myron K. Jacobson

Departments of Biochemistry and Medicine, Texas College of Osteopathic Medicine, University of North Texas, Fort Worth, Texas 76107 USA

Introduction

Nuclear compartmentalization of chromatin appears to play an important role in the regulation of chromatin functions such as *de novo* DNA replication, DNA repair replication, carcinogenesis and gene expression (1). Likewise, regulation of these chromatin functions by poly(ADP-ribosyl)ation of nuclear proteins, due to the potential of this polymer to reversibly bring about changes in chromatin conformation *in vitro*, has been proposed. Furthermore, there is the postulation that these changes could be mediated by covalent and non-covalent interactions of poly(ADP-ribose) with chromatin components as well as by the fast turnover of these polymers (2-4). Hence, this proposal implies that close proximity of poly(ADP-ribose) with chromatin regions engaged in these functions is essential. Thus, an analysis of the nuclear distribution of ADP-ribose polymers will further help in the support of this proposal or in the formulation of new ones.

Results and Discussion

In these studies a simple method which allows the *in vivo* radiolabeling and detection of poly(ADP-ribose) (5, 6) was combined with nuclear fractionation techniques to analyze the distribution of ADP-ribose polymers. Due to the instability of the covalent protein-poly(ADP-ribosyl) bonds at pH values above 7.0 (7), it was important to conduct these experiments at pH values below 7.0 to avoid disruption of these linkages. We have shown that the isolation of nuclei at pH 3.1 and at pH 6.5 resulted in a good recovery of polymer (8). In the present studies all nuclear fractionation experiments were performed at pH 6.5 since it represented a good compromise between physiological conditions and preservation of protein-poly(ADP-ribose) linkages. Except where otherwise indicated, the data presented in this study was obtained from cells that have been treated with MNNG following a brief period of hyperthermia. Previous studies have shown that MNNG results in the production of DNA strand breaks, which leads to a rapid activation of

---

[1]Present address: Swiss Institute for Experimental Cancer Research, 1066 Epalinges, Switzerland

173

poly(ADP-ribose) polymerase (9, 10) while hyperthermic treatment results in increased polymer levels due to a decreased activity of poly(ADP-ribose) glycohydrolase (11).

Are ADP-ribose polymers randomly or non randomly distributed in chromatin? Isolated nuclei from hyperthermia-MNNG-treated cells were digested with MNase and the DNA and poly(ADP-ribose) released into the digestion supernatant were determined. While 80% of the total DNA was rendered soluble by the nuclease, 90% of the total ADP-ribose residues remained with the MNase resistant chromatin (Fig 1). These data show that ADP-ribose polymers are not randomly distributed in chromatin but rather they exhibit a predominant association with the chromatin fraction which was resistant to MNase digestion. However the possibility that these polymers confer resistance to the nuclease digestion can not be ruled out at this point. The MNase resistant chromatin fraction isolated in these experiments contained approximately 20% of the total DNA and proteins that isolate with the nuclear matrix. Thus, the possibility of the association of ADP-ribose polymers with the nuclear matrix was examined.

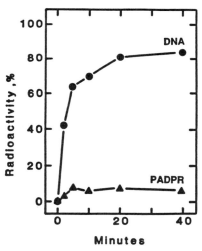

Fig. 1. DNA and poly(ADP-ribose) released during the digestion of isolated nuclei with MNase. Isolated nuclei from hyperthermia-MNNG-treated cells were digested at 30°C with 250 U/mg DNA of MNase. At the indicated times, aliquots from the digestion mixtures were removed and extracted twice for nucleosomes (8,18) . The content of DNA and poly(ADP-ribose) in the supernatant was determined as previously described (5, 6) (For experimental details see 8).

About 50% of the total Poly(ADP-ribose) residues are tightly associated with the nuclear matrix. Nuclear matrices were prepared and the poly(ADP-ribose) content of these preparations was examined. About 50% of the total polymer residues were associated with the nuclear matrices (Table 1). The association of poly(ADP-ribose) to the nuclear matrix could be explained in two ways, either the polymers are associated *in vivo* or this association occurred *in vitro* during the cell fractionation. Two independent lines of evidence support an *in vivo* association: i) exogenous polymers added to isolated nuclei prior to the preparation of nuclear matrices did not bind significantly to these matrices; ii) nuclear matrices isolated by

174

extraction of nuclei with buffers containing 2 M NaCl, 0.3 M $(NH_4)_2SO_4$ and 25 mM LIS contained from 50 to 70% of the total poly(ADP-ribose) (Table 1). This rules out an artifactual association of poly(ADP-ribose) with the nuclear matrix promoted by hypertonic salt extraction.

Table 1. Distribution of poly(ADP-ribose) in nuclear fractions and the effect of salt extraction.

Fraction	Treatment			
	None		DNAase I	
	DNA	Poly(ADP-ribose) Percent	DNA	Poly(ADP-ribose) Percent
DNAase I supernatant	3	4	88	5
2 M NaCl extract	9	43	12	48
Nuclear Matrix	88	53	<1	47
DNAase I supernatant	<1	3	88	5
0.3 M $(NH_4)_2SO_4$ extract	2	16	10	22
Nuclear Matrix	98	80	1	73
25 mM LIS extract	0	42	<1	43
DNAase supernatant	2	<1	99	<1
Nuclear Matrix	98	58	1	57

Nuclear matrices from hyperthermia-MNNG-treated cells were isolated by extraction of DNAase I-digested nuclei with buffers containing the indicated salt concentration. The DNA and poly(ADP-ribose) content were determined as described in legend to Fig. 1 (for experimental details see 8). Results are expressed as % of the total nuclear content.

In contrast to the high poly(ADP-ribose) content of the nuclear matrix, its DNA content was less than 1% of the total (Table 1). This result raises the interesting possibility that 50% of the total poly(ADP-ribose) is associated with a chromatin fraction which is clustered and tightly bound to the nuclear matrix. However, this possibility has to be reconciled with the observation that the fraction of poly(ADP-ribose) released from the nuclei by salt extraction was not dependent on prior nuclease digestion, since approximately the same fraction of polymers was released by salt extraction of undigested nuclei (Table 1).

The nuclear matrix-associated poly(ADP-ribose) represent short polymers while complex polymers are released from the nucleus by the salt extraction. Molecular sieve chromatography analysis revealed that the nuclear matrix-associated ADP-ribose polymers, which represented 50% of the total residues are short polymers, while the polymers extracted by salt contain both short and complex chains (Fig. 2). These results demonstrate that there are considerably more polymer molecules in the nuclear matrix as compared to the salt extractable fraction. Therefore, it is very likely that in the nuclear matrix there are more molecules of protein modified by these polymers. However, modification of proteins at multiple sites is also consistent with these data.

In these studies we have detected a small increase in the fraction of nuclear matrix-associated polymers generated following hyperthermia (not shown), where polymer half-life is 10 min, as compared to polymers synthesized in response to MNNG treatment, where polymer half-life is less than 2 min (11). These results and the observation that only short ADP-ribose polymers (Fig. 2) were detected in the nuclear matrix suggest that turnover for nuclear matrix-associated polymers is faster than for non matrix-associated polymers. The fact that practically no endogenous poly(ADP-ribose) glycohydrolase activity could be detected with the nuclear matrix in the experiment shown in Table 2 does not conflict with these results since previous studies have shown that this enzyme has a low affinity for short polymers (12, 13). These observations also agree with other reports (5, 13) which have suggested that *in vivo* ADP-ribose polymer chain length seems to be tightly regulated by poly(ADP-ribose) glycohydrolase.

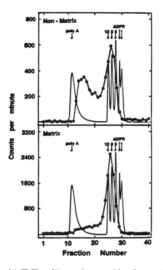

Fig. 2. Molecular sieve chromatography of non-matrix and matrix-associated ADP-ribose polymers. Radiolabeled ADP-ribose polymers isolated from the nuclear matrix (upper panel) and salt extract (lower panel). The peaks for the positions of non radiolabeled standards of polyA, $(Ap)_{11}A$ (12), $(Ap)_7A$ (8), $(Ap)_4A$ (5), ADP-ribose and AMP (1) are indicated. (—) absorbance at 254 nm and (●) radioactivity.

Is poly(ADP-ribose) mediating the association of some proteins with the nuclear matrix? Digestion of isolated nuclear matrices with poly(ADP-ribose) glycohydrolase resulted in the release of 80% of ADP-ribose polymers. Fig. 3 shows glycohydrolase treatment also resulted in the release of 2% of the total protein (Table 2). Both poly(ADP-ribose) degradation and protein release were inhibited by ethacridine, an inhibitor of poly(ADP-ribose) glycohydrolase (14) (not shown). This suggests that the release of proteins was due to glycohydrolase activity. Among the proteins released from the nuclear matrix by poly(ADP-ribose) glycohydrolase were prominent polypeptides at 170 kDa, 116 kDa, 100 kDa, 70 kDa, 67 kDa, 55 kDa and a set at 20 to 36 kDa. It is clear that some of these polypeptides

176

were highly enriched in the glycohydrolase-released fraction, relative to the nuclear matrix. This result raises the interesting possibility that the association of these proteins to the nuclear matrix is mediated by poly(ADP-ribose). The nature of this association remains to be determined. Our results are consistent with studies by other workers that have identified several nuclear matrix proteins as covalent acceptors for poly(ADP-ribose) both *in vivo* (15) and *in vitro* (16, 17).

Table 2. Probing of nuclear matrices with poly(ADP-ribose) glycohydrolase.

Condition	Poly(ADP-ribose)	Protein
	% Released	
Control	8	0.1
Poly((ADP-ribose) Glycohydrolase	80	2.0

Isolated nuclear matrices from [^{35}S]methionine labeled cells were digested with 5 Uml of poly(ADP-ribose) glycohydrolase for 1 hr at 30°C. Following centrifugation of digestion mixtures at 3000 x g for 10 min at 4°C the poly(ADP-ribose) content in the pellets was determined. The protein fraction released into the digestion supernatant was estimated by liquid scintillation counting.

Altogether, these observations suggest the occurrence poly(ADP-ribose) metabolism in two compartments of chromatin; one that is nuclear matrix-associated and one that is not. The elucidation of the biological significance of this compartmentalization awaits further information concerning the molecules which interact with the polymers in both compartments.

References

1. Nelson, W.G., Pienta, K.J,. Barrack, E.R., Coffey, D.S. (1986) Ann Rev Biophys Chem 15: 457-475
2. Althaus, F.R., Mathis, G., Alvarez-Gonzalez, R., Loetscher, P., Mattenberger, M. (1988) This volume
3. de Murcia, G., Huletsky, A., Lamarre, D., Gaudreau, A., Pouyet, J., Daune, M., Poirier, G.G. (1986) J Biol Chem 15: 7011-7017
4. Huletsky, A., de Murcia, G., Muller, S., Lamarre, D., Hengartner, M., Poirier, G.G. (1988) This volume
5. Alvarez-Gonzalez, R., Jacobson, M.K. (1987) Biochemistry 26: 3218-3224
6. Aboul-Ela, N., Jacobson, E.L., Jacobson, M.K. (1988) Anal Biochem In press
7. Riquelme, P.T., Burzio, L.O., Koide, S.S. (1979) J Biol Chem 254: 3018-3028
8. Cardenas-Corona, M.E., Jacobson, E.L., Jacobson, M.K. (1987) J Biol Chem 262: 14863-14866
9. Juarez-Salinas, H., Sims, J.L., Jacobson, M.K. (1979) Nature 282: 740-741
10. Jacobson, M.K., Levi, V., Juarez-Salinas, H., Barton, R.A., Jacobson, E.L. (1980) Cancer Res 40: 1797-1802
11. Jonsson, G.G., Jacobson, E.L., Jacobson, M.K. (1988) Cancer Res 48: 4233-4239
12. Hatekayama, K., Nemoto, Y., Ueda, K., Hayaishi, O. (1986) J Biol Chem 261: 14902-14911
13. Menard, L., Thibault, L., Gaudreau, A., Poirier, G.G. (1988) This volume
14. Tavassoli, M., Tavassoli, M.H., Shall, S. (1985) Biochem Biophys Acta 827: 228-234

15. Adolph, K.W. (1987) Biochem Biophys Acta 909: 222-230
16. Adolph, K.W., Song, M.K.H. (1985) Biochem Biophys Res Commun 126: 840-847
17. Wesierska-Gadek, J., Sauermann, G. (1985) Eur J Biochem 193: 421-428
18. Aubin, R.J., Dam, V.T., Miclette, J., Brousseau, Y., Poirier, G.G. (1982) Can J Biochem 60: 295-305

Abbreviations:
MNNG - N-methyl-N'-nitro-N-nitrosoguanidine
MNase - micrococcal nuclease
LIS - lithium diiodosalicylate
Ethacridine - ethoxyacridine lactate
SDS-PAGE - sodium dodecyl sulfate polyacrylamide gel electrophoresis

A Proposed Molecular Mechanism of Nucleosome Unfolding. Effect of Poly(ADP-Ribose) on DNA-Histone H4 Interaction

Georg Sauermann and Jozefa Wesierska-Gadek

Institute of Tumorbiology-Cancer Research, University of Vienna, A-1090 Vienna, Austria

Introduction

Poly(ADP-ribosyl)ation of nuclear proteins has been reported to occur during the process of DNA repair, DNA replication and DNA transcription (1). Interestingly, the primary structure of chromatin is considered to be altered in the course of these events (2). In the present study, the effect of poly(ADP-ribose) chains *per se* on histone-DNA interaction was investigated. The data show that poly(ADP-ribose) is affecting histone-DNA binding and that the observed effects are histone-specific. While poly(ADP-ribose) effectively competes with DNA for histone H4 binding, it equally competes with DNA for histone H3 binding and only weakly affects histone H1-DNA interactions. On account of our previous finding that histone H4 has a high affinity for poly(ADP-ribose), we have proposed the hypothesis that poly(ADP-ribose) chains, transiently and locally formed by activated nuclear ADP-ribosyltransferase, may loosen DNA-core histone H4 interactions and thus cause partial or complete unfolding of nucleosome core DNA from the core particle (3). By this molecular mechanism nucleosome DNA could become accessible to enzymes and other factors under certain biological conditions. Our concept is different from that of other authors who previously proposed that poly(ADP-ribosyl)ation of histones is affecting another level of chromatin organization, namely the higher order chromatin structure (4, 5).

In this communication, data on the interaction between poly(ADP-ribose), DNA and individual histones and on the effect of poly(ADP-ribose) on isolated nucleosomes are presented. Considering known facts about nucleosome structure and the kinetics of DNA excision repair, the proposed molecular mechanism of poly(ADP-ribose) action on nucleosome structure is discussed.

Results

Using a membrane filter binding assay, the competition between poly(ADP-ribose) and DNA for binding of individual histones was measured (3). The size of the poly(ADP-ribose) molecules from rat liver nuclei ranged from 10 to 60 monomer units (3). Fig. 1 shows the effect of unlabeled

179

poly(ADP-ribose) on the formation of complexes between [³²P]DNA and individual histones. While low poly(ADP-ribose) concentrations sufficed to inhibit DNA-histone H4 complex formation, higher poly(ADP-ribose) concentrations were required to comparably inhibit DNA-histone H1 complex formation. The data of the reciprocal experiments depicted in Fig. 2 are in accordance with the above results. Relatively low DNA concentrations affected [³²P]poly(ADP-ribose)-histone H1 complex formation, while higher DNA concentrations were needed to lower the amount of poly(ADP-ribose)-histone H4 complexes.

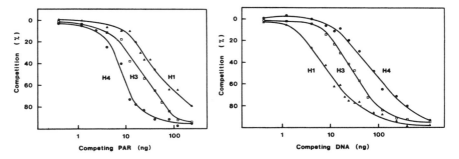

Fig. 1. (left) Effect of poly(ADP-ribose) on DNA-histone complex formation. 40 ng samples of rat liver [³²P]DNA were incubated with increasing concentrations of unlabeled poly(ADP-ribose) in 10 mM Tris-HCl (pH 7.0), 150 mM NaCl, 3 mM EDTA at 22°C for 10 min. After addition of 20 ng histone H1, H3 or H4, respectively, and 10 min incubation in a final volume of 150 μl, the mixtures were placed on cellulose nitrate membrane filters, washed, and the amount of retained complexes measured. Other conditions as in (3).

Fig. 2. (right) Effect of DNA on poly(ADP-ribose)-histone complex formation. Conditions as in Fig. 1 except that [³²P]poly(ADP-ribose) and increasing concentrations of unlabeled DNA were used.

It thus appears that the effect of poly(ADP-ribose) on DNA-histone interactions is histone-specific. Poly(ADP-ribose) is competing with DNA effectively for binding of histone H4, equally for binding of histone H3 and only ineffectively for binding of histone H1. Since we considered that the high affinity of histone H4 for poly(ADP-ribose) might have biological implications, the formation of histone H4 complexes was further studied. As histone H4 plays a major role in the organization of the nucleosome (6, 7), it was of interest to examine in which manner preexisting DNA-histone H4 complexes would be affected by poly(ADP-ribose). In the experiment of Fig. 3, preformed complexes containing labeled or unlabeled polymers were treated with the respective opposing poly(ADP-ribose) or DNA during a second incubation period. It can be seen that in a reversible process, poly(ADP-ribose)-histone H4 complex formation was favored in agreement with the data of the direct competition experiments of Figs. 1 and 2. In order

180

to test the specificity of poly(ADP-ribose) action, the effect of the polyanion poly(A) on DNA-histone H4 binding was examined. As depicted in Fig. 4, poly(A) is not capable of exerting an effect comparable to that of poly(ADP-ribose).

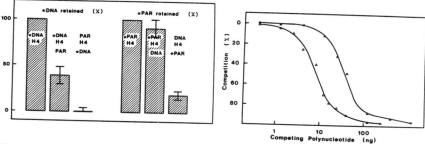

Fig. 3. (left) Effect of poly(ADP-ribose) and of DNA on preformed polynucleotide-histone H4 complexes. Mixtures of the components indicated in the first 2 lines were preincubated. Then the third component was added. 16 ng of [^{32}P]polynucleotide, 48 ng of the unlabeled polymer and 26 ng histone H4 were used. Other conditions as in Fig. 1.

Fig. 4. (right) Effect of poly(A) and of poly(ADP-ribose) on [^{32}P]DNA histone H4 binding. Conditions as in Fig. 1. Increasing concentrations of unlabeled (▲) poly(A) and (△) poly(ADP-ribose).

In comparison to DNA, the isolated poly(ADP-ribose) is of low molecular weight. In order to test whether the competition behaviour depended on DNA length, experiments were performed with DNA restriction fragments. Fig. 5 demonstrates that the capability of DNA to compete with [^{32}P]poly(ADP-ribose) for histone H4 binding did not depend on DNA size. Even DNA molecules comparable in length to those of poly(ADP-ribose) reacted in the same manner as high molecular weight DNA.

In the experiment of Fig. 6 we examined whether poly(ADP-ribose) would induce complete dissociation of DNA from isolated nucleosome particles. Fig. 6 suggests that, after addition of poly(ADP-ribose), part of the core DNA molecules were released as they moved in front of the mononucleosome zone. We are, however, aware that more detailed analyses are required in order to determine the type and degree of dissociation.

Discussion

The present data show that poly(ADP-ribose) containing two phosphate groups per monomer unit, does not generally and unspecifically displace less negatively charged polyanions such as DNA from complexes with basic proteins. The observed specificities of poly(ADP-ribose) action may, however, give a clue as to the potential role of the polymer. The structural organization of the nucleosome core particle is mainly determined by interaction of DNA with the histones H3 and H4 which contact the central turn of the superhelix (6, 7). It can, therefore, be envisaged that, by action of poly(ADP-ribose), the DNA-core histone binding is loosened, especially at the site of histone H4-DNA interaction. This may lead to partial or complete unfolding of nucleosomes core DNA from the core particle, thereby making DNA accessible to reactants. Poly(ADP-ribose) chains may be formed locally after stimulation of nuclear ADP-ribosyltransferase at specific sites. The enzyme may be activated by DNA strand breaks or by some other mechanism during certain phases of DNA repair, replication, or transcription. The energy consuming formation and degradation of poly(ADP-ribose) would be useful if the molecule had such a temporary local function. The above scheme would also explain the reported short biological half life of poly(ADP-ribose) (8, 9). Since one polymer molecule may interact with more than one core histone molecule, poly(ADP-ribose) may also have the additional function of keeping the core histone available and of maintaining local structures, until the process will be completed for which the dissociation of DNA and core histone is required.

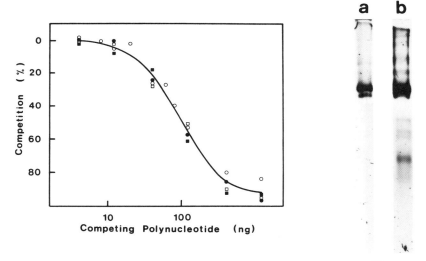

Fig. 5. (left) Competition between DNA restriction fragments and [^{32}P]poly(ADP-ribose) for histone H4 complex formation. Conditions as in Fig. 2. Size of Sau III DNA restriction fragments: (○), <125 bp; (●), 125-560 bp; (■) , 560-2000 bp; (□), undigested DNA.

Fig. 6. (right) Effect of poly(ADP-ribose) on isolated nucleosome particles. Soluble long chromatin was extracted from micrococcal nuclease-digested rat liver nuclei. After stripping from histone H1 with 500 mM NaCl and additional nuclease digestion, the material was submitted to sucrose gradient centrifugation. Electrophoresis in 5% native polyacrylamide gel. A, control nucleosomes (4.5 μg DNA in main mononucleosome zone; 146 bp). B, sample plus 1.5 μg poly(ADP-ribose). Ethidium bromide staining.

182

Lan *et al.* (10) reported that newly formed repair patches are nonuniformly distributed along nucleosome core DNA. Their finding that the distribution of the newly formed repair patches correlated with the arrangement of histone H4 binding sites along both strands of core DNA (7) may, therefore, be cited in support of our hypothesis (3). One aspect concerning DNA excision repair remains to be considered further. Histones H2A and H2B, which contact the end of the nucleosome core DNA, can readily exchange in and out of nucleosomal structures (11). Therefore, the DNA sequences contacting the histones H2A and H2B, as well as the DNA sequences of the linker region, may be accessible for DNA repair even in the absence of poly(ADP-ribose). Accordingly, inhibition of poly(ADP-ribose) synthesis may impair only part of the total DNA repair activity.

References

1. Ueda, K., Hayaishi, O. (1985) Ann Rev Biochem 4: 73-100
2. Eissenberg, J.C., Cartwright, I.L., Thomas, G.H., Elgin, S.C.R. (1985) Ann Rev Genet 19: 485-536
3. Sauermann, G., Wesierska-Gadek, J. (1986) Biochem Biophys Res Comm 139: 523-529
4. Wong, M., Malik, N., Smulson, M. (1982) Eur J Biochem 128: 209-213
5. Poirier, G.G., de Murcia, G., Jongstra-Bilen, J., Niedergang, C., Mandel, P. (1982) Proc Natl Acad Sci USA 79: 3423-3427
6. Klug, A., Rhodes, D., Smith, J., Finch, J.T., Thomas, J.O. (1980) Nature 287: 509-516
7. Shick, V.V., Belyavsky, A.V., Bavykin, S.G., Mirzabekov, A.D. (1980) J Mol Biol 139: 491-517
8. Wielckens, K., Schmidt, A.A., George, E., Bredehorst, R., Hilz, H. (1982) J Biol Chem 257:12872-12877
9. Jacobson, E.L., Antol, K.M., Juarez-Salinas, H., Jacobson, M.K. (1983) J Biol Chem 258: 103-107
10. Lan, S.Y., Smerdon, M.J. (1985) Biochemistry 24: 7771-7783
11. Louters, L., Chalkley, R. (1985) Biochemistry 24: 3080-3085

On The Nature of the Interaction Between Poly(ADP-Ribose) Polymerase and DNA

Srinivas S. Sastry and Ernest Kun

Cardiovascular Research Institute, University of California-San Francisco, San Francisco, California 94143 USA

Introduction

We have employed nuclease protection experiments to directly probe the nature of the contacts between poly(ADP-ribose) polymerase and defined DNA sequences. The data suggest that the polymerase exhibits a broad preference for A + T rich regions and may recognize certain unusual conformation(s) on the DNA molecule.

Results and Discussion

DNase I footprinting analysis. A 209 bp *Eco*RI-*Pst*I fragment of SV40 DNA was labelled at the 5' *Eco*RI site and DNase I protection assays were performed with purified calf thymus poly(ADP-ribose) polymerase. The large region of protection from nuclease attack (approximately 66 bp, regions I+II, Fig. 1A, 1B) and the pattern of cleavage [a 9-13 bp periodicity coinciding with the helical screw as expected for DNase I cutting (1)] suggest that the DNA lies outside the protein core. These data confirm earlier reports using a different technique (2). This result is reminiscent of the DNase I protection data of gyrase-DNA and histone-DNA complexes. At a low mass ratio of protein/DNA only a small region, (region I, about one turn of the helix), is protected from DNase I (lanes 8, 9, 11, Figs. 1A and B). At a higher protein/DNA ratio increased protection occurs, region I plus region II, approximately 66 bp, 6-7 helical turns, lane 10, Figs. 1A and B). Region II is contiguous with region I, suggesting cooperative interactions between the polymerase molecules; binding of the polymerase to region I preceeds this cooperativity. Assuming that all polymerase molecules have full binding activity, approximately seven protein monomers bind to each DNA molecule at this concentration (lane 10, Fig. 1A). Similar results were obtained when only the bottom strand was labelled (data not shown). The data from Fig. 1A do not indicate whether the polymerase makes any contacts upstream of region I because bands smaller than 17 bp are not resolved on this gel. Experiments with Iron (II)-EDTA, which cuts uniformly at almost every base on the DNA helix, might clarify this. At the high protein/DNA mass ratios polymerase molecules seem to bind along the helical screw of the DNA, protection occurring in blocks equal to about one turn of the helix and the most pronounced sites on naked DNA are the least protected in the presence of the polymerase (lane 10, Fig. 1A). Experiments with several other fragments of comparable lengths but with low overall A + T content did not reveal DNase I footprints (data not shown) indicating for the first time that poly(ADP-ribose) polymerase shows preferential binding to regions of high overall A + T content. A + T rich regions are known to occur in biologically important contexts such as the 5' and 3' ends of genes, origins of replication, regulatory signals, etc., (4, 5). Whether the polymerase protein participates in any of these regulatory processes remains to be determined. Lanes 4, 5, 6 (Fig. 1A) indicate that increasing amounts of poly(dI-dC) can compete out the footprint, possibly because the polymerase also binds to the ends of DNA (see below). It appears from these and other

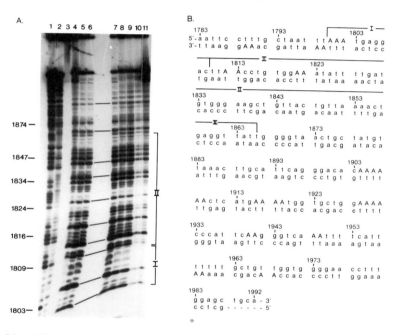

Fig. 1A. DNase I footprinting. SV40 DNA, restricted with *Eco*RI, 5' ^{32}P-labelled, cut with *Pst*I, and the 209 bp fragment was gel purified, by conventional procedures. Purified poly(ADP-ribose) polymerase was incubated (25°C, 15 min) with ^{32}P-DNA (50 ng, 2 x 10^4 CPM) in 25 mM TRIS-Cl (pH 8.0), 10 mM MgCl$_2$, 100 mM NaCl, 0.5 mM DTT, 50 mM Na-Cacodylate (pH 8.0) in a final volume of 50 μl. DNase I (10 u/ml) was added to the mix and transferred to 37°C for 30 sec. Reactions terminated by heating in EDTA (10 mM) + SDS (0.1%) followed by phenol extraction and EtOH precipitation. Samples were loaded on an 8% sequencing gel. Lanes 1, 2: C + T, C chemical cleavage markers; lane 3: undigested ^{32}P-DNA; lanes 4-6: ^{32}P-DNA + 100 ng, 5000 ng, 1000 ng of poly (dI-dC) respectively + 300 ng of the polymerase; lanes 7-11: ^{32}P-DNA + 0, 5, 20, 300, 60 ng of the polymerase, respectively.

Fig. 1B. Sequence of the SV40 *Eco*RI-*Pst*I fragment. Regions I & II indicate protection from DNase I. Capitalized ApA "wedges" (9) and A tracts (10) indicate bending loci.

185

experiments (data not shown) that the polymerase has the ability to bind non-specifically to the ends of DNA and more specifically to the internal length of some DNAs.

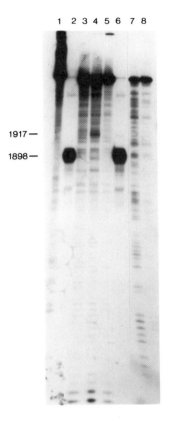

1917—
1898—

Fig. 2. (left) λ exo protection of the 3'-^{32}P-EcoRI-PstI fragment. Binding assays and gel separation are described in Fig 1A except that the buffer was 67 mM glycine-KOH (pH 8.1) + 3.5 mM $MgCl_2$ + 3 mM β-mercaptoethanol at 25°C for 30 min. Lane 1: undigested ^{32}P-DNA (50 ng); lane 2: ^{32}P-DNA (50 ng) + λ exo (15 u); lane 3: ^{32}P-DNA + polymerase (60 ng) + λ exo (15 u); lane 4: ^{32}P-DNA + polymerase (300 ng) + λ exo (15 u); lane 5: as in lane 4 + 50 ng of unlabelled EcoRI-PstI fragment; lane 6: as in lane 5 + 500 ng of unlabelled fragment; lanes 7, 8: C + T, C markers.

Poly(ADP-ribose) polymerase binds to the ends of DNA. Fig. 2 shows the results of λ exonuclease digestion of the EcoRI-PstI fragment in the presence of the polymerase. Over 90% of the DNA fragments are digested uniformly to approximately 1898 bases (lane 2, Fig. 2). Polymerase binding renders most of the DNA resistant to exonucleolytic attack (lanes 3, 4; Fig. 2) by blocking the 5' end. Experiments using exonuclease III also revealed binding of the polymerase to 3' ends of DNA (data not shown). A fraction of the DNA molecules are digested to various lengths even at high concentrations of polymerase (lanes 3, 4; Fig. 2). Lane 4 (Fig. 2) shows a prominent pause site probably indicating the 5' boundary (1917 bp) of a region where a polymerase binds specifically to the bottom strand. A specific stop site for the λ exonuclease (1898 bp) exists in the absence of

186

polymerase protein, suggesting a conformational impediment to exonucleolytic progression. However, in the presence of the polymerase this stop site is not observed indicating a polymerase-induced change in the conformation of the DNA (lanes 3, 4; Fig. 2).

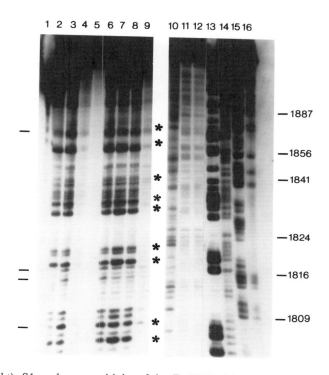

Fig. 3. (right) S1 nuclease sensitivity of the *Eco*RI-*Pst*I fragment. S1 and Mung bean nuclease digestions (25°C, 30 min) were performed in their respective buffers (7) except 50 mM NaCl was also present. Lane 1: undigested ^{32}P-DNA (100 ng); lane 2: ^{32}P-DNA (100 ng) + S1 (10 u/ml); lane 3: ^{32}P-DNA + S1 (40 u/ml); lane 4: ^{32}P-DNA + S1 (40 u/ml) + polymerase (200 ng); lane 5: as in lane 4 + polymerase (600 ng); lane 6: as in lane 5 + 1 μg of unlabelled DNA fragment ; lane 7: ^{32}P-DNA + S1 at 37°C; lane 8: DNA solution heated to 65°C for 10 min and cooled to 25°C for 15 min before S1 was added; lane 9: ^{32}P-DNA + 60% formamide + S1 at 25°C; lane 10: ^{32}P-DNA + Mung bean nuclease (30 u/ml); lane 11: as in lane 10 + polymerase (200 ng); lane 12: as in lane 11 + polymerase (600 ng); lanes 13-16: G, A, T, C markers, respectively.

The unusual conformation of the isolated *Eco*RI-*Pst*I SV40 fragment. The findings reported above could reveal new information about the mechanism of binding of poly(ADP-ribose) polymerase to unusual secondary structures on DNA. DNA structure is known to be polymorphic (5, 6). Chemical and enzymatic probes (e.g. S1 nuclease) are available to

187

probe non-B DNA conformations (7). Fig. 3 (lanes 2, 3) shows that S1 nuclease digestion of the top strand yielded discrete bands (asterisks, Fig. 3) and these sites are specifically rendered inaccessible by the polymerase (lanes 4, 5; Fig. 3) suggesting binding of the polymerase to distorted non-B DNA regions recognized by S1. It is determined that these conformational anamolies are stable at 37°C (lane 7), reform rapidly at 25°C after partial denaturation at 65°C (lane 8), and are susceptible to denaturation by 60% formamide (lane 9). Lane 10 (Fig. 3) demonstrates that Mung bean nuclease recognizes different structures as compared to S1. The Mung bean nuclease-specific sites are not protected by the polymerase, unlike the S1 cutting sites (lanes 11, 13; Fig. 3), suggesting structure-specific binding by the polymerase. One possibility is that Mung bean nuclease dissociates the polymerase from the DNA before cleavage occurs. This appears unlikely because partial protection by the polymerase is seen at some sites. The striking sensitivity of this SV40 fragment to S1 and several other structure-specific probes (data not shown) appears to be a novel phenomenon since sensitivity of double-stranded DNA to these reagents has been reported to be completely dependent on negative superhelicity (7, 8). Finally, the sequence of the *Eco*RI-*Pst*I fragment contains several putative bending loci (capital letters, Fig. 2B) (9, 10). It is of interest to find out to what extent DNA bending influences the positioning of the DNA along the protein core of the polymerase as in the case of the nucleosome octamer (1, 11).

Acknowledgements. This research was supported by grants HL-27317 (NIH), AFO-SR-85-0377 and AFO-SR-86-0064 (Air Force Office of Scientific Research), Washington, D.C.. Ernest Kun is a recipient of the Research Career Award of U.S.P.H.. We thank Kulim Buki for preparing the polymerase.

References

1. Drew, H.R., Travers, A.A. (1985) J Mol Biol 186: 773-790
2. de Murcia, G., Jongstra-Bilen, J., Ittel, M.E., Mandel, P., Delain, E. (1983) EMBO J 2: 543-548
3. Moreau, J., Matyash-Smirniaguina, L., Schierrer, K. (1981) Proc Natl Acad Sci USA 78: 1341-1345
4. Gerard, R., Gluzman, Y. (1986) Mol and Cell Biol 6: 4570-4577
5. Patel, D.J., Pardi, A., Itakura, K. (1982) Science 216: 581-590
6. Wells, R.D., Goodman, T.C., Hillen, W., Horn, G.T., Klain, R.D., Larson, J.E., Muller, U.R., Neuendorf, S.K., Panayotatos, N., Stirdivant, S.M. (1980) Prog Nucl Acid Res Mol Biol 24: 167-267
7. Singleton, C.K., Kilpatrick, M.W., Wells, R.D. (1984) J Biol Chem 259: 1963-1967
8. Singleton, C.K., Klysick, J., Wells, R.D. (1983) Proc Natl Acad Sci USA 80: 2447-2451
9. Ulanovsky, L.E., Trifonov, E.N. (1987) Nature 326: 720-723
10. Koo, H-S., Wu, H-M., Crothers, D.M. (1986) Nature 320: 501-507
11. Travers, A.A. (1987) Trends in Biochem Sci 12: 108-112

Inhibition of ADP-Ribosyl Transferase Activity Blocks the Stable Integeration of Transfected DNA into the Mammalian Cell Genome

Farzin Farzaneh, George N. Panayotou[1], Lucas D. Bowler[1], Timothy Broom[1], and Sydney Shall[1]

Molecular Genetics Unit, Department of Obstetrics & Gynecology, King's College School of Medicine and Dentistry, Denmark Hill, London SE5 8RX, England

Introduction

The transfer of genetic information into animal cells by the DNA transfection method is a powerful and widely used technique. The most commonly used DNA transfection method, which employs the co-precipitation of the donor DNA with calcium phosphate, involves several discrete steps (1, 2). These include the uptake of the transfected DNA by the recipient cell, transport of the DNA to the nucleus, concatenation and stable integration of DNA into the host cell genome and finally the expression of the stably-integrated genes (2, 3). The stable expression of a transduced DNA molecule which is unable to maintain its own autonomous replication requires its stable integration into the host cell genome. Both the concatenation and the stable integration of the host genome involve the formation and ligation of DNA strand-breaks. DNA strand-breaks are an essential activator for the nuclear adenosine diphosphoribosyl transferase (4), the activity of which is required in a number of circumstances for the efficient ligation of DNA strand-breaks in eukaryotic cells (5-7), possibly because it regulates DNA ligase activity (8, 9). Previous studies have demonstrated the involvement of nuclear ADP-ribosyl transferase activity in a variety of cellular functions that require the ligation of DNA strand-breaks. These include DNA excision repair (10), sister chromatid exchanges (11, 12), mitogen activation of quiescent lymphocytes (13-15), antigenic variation in *Trypanosoma brucei* (16, 17) and a number of specific examples of cytodifferentiation (18-28). In this study we have investigated the possible involvement of ADP-ribosyl transferase activity in DNA mediated gene transfer, by studying the effect of ADP-ribosyl transferase inhibition on the transfection of NIH/3T3 cells and a human fibroblast cell line (XP12RO-SV40) with four different plasmids. Results demonstrate that the stable integration into the host genome of several different DNA molecules (transduced by the calcium phosphate co-precipitation method) is blocked by inhibition of ADP-ribosyl transferase activity in a number of

---

[1]Cell and Molecular Biology Laboratory, Biology Building, University of Sussex, Brighton BN1 9QG, England

different mammalian cells. However, inhibition of ADP-ribosyl transferase activity affects neither the uptake into the cell nor the expression of the transduced genes either before or subsequent to their stable integration into the host genome.

Results and Discussion

The plasmid, pRSV.cat contains the chloramphenicol acetyl transferase (CAT) gene inserted 3' into the Rous sarcoma virus terminal repeat (29). The expression of the CAT activity can be readily assayed as the acetylation of radio-labelled chloramphenicol measured by thin layer chromatography (29). In the plasmid, pSV2.neo, the aminoglycoside phosphotransferase (*neo*) gene is inserted 3' into the simian virus 40 (SV40) origin (30). Stable integration and expression of the *neo* gene makes the recipient cells resistant to the antibiotic Geniticin (G418), and allows *neo* colonies to form in the presence of Geniticin (30). The plasmid, pSVO.1 was constructed by the insertion of full length SV40 DNA at the *Bam*H1 site of pAT153. The plasmid pSV3.gpt codes for bacterial guanine-phosphoribosyl transferase and permits the selection of cells resistant to mycophenolic acid (31). This plasmid also contains the whole early region of SV40. The stable integration and expression of this plasmid or of pSVO.1 leads to the transformation of and focus formation in NIH/3T3 cells. We have used pRSV.cat to study the possible role of ADP-ribosyl transferase activity in the uptake of the transfected DNA into the cell and in its transient expression prior to integration. The plasmids pSV2.neo, pSVO.1 and pSV3.gpt were used to study the role of ADP-ribosyl transferase in the stable integration and expression of the transfected DNA.

The stable expression of transfected genes in proliferating cells requires both the ligation (concatenation) of the transfected DNA fragments and their stable integration into the host genome (3). In the absence of concatenation and stable integration, which both require the ligation of DNA strands, the transfected DNA fragments are lost by dilution (and presumably by degradation) during cell replication, with first order kinetics (32, 33). However, the transient expression of the DNA sequences, which can be detected soon after transfection, requires neither concatenation nor integration into the host genome.

Inhibition of nuclear ADP-ribosyl transferase activity, by the competitive inhibitor 3-methoxybenzamide (34), has no detectable effect on the transient expression of CAT activity in NIH/3T3 cells transfected with either pRSV.cat, or co-transfected with both pRSV.cat and pSV2.neo. Comparable levels of CAT activity were found in cell extracts from cultures exposed to 2 mM 3-methoxybenzamide, either from 2 hr prior to transfection or from 12 hr after transfection up to the time of assay at 42 hr. Therefore, inhibition of ADP-ribosyl transferase activity in NIH3T3 cells either during the uptake of

190

transfected DNA into the cells, or during the subsequent period of expression prior to stable integration into the genome, had no effect on the detectable levels of CAT activity. These observations demonstrate that ADP-ribosyl transferase activity is not required for processes involved either in the uptake of the transfected DNA into the cell or in the transient expression of this DNA, including transcription and translation.

By contrast, ADP-ribosyl transferase activity is required for the stabilization of transfected DNA in the genome. The function of ADP-ribosyl transferase activity in the stable integration of transfected DNA and its subsequent expression is demonstrated by the effect of ADP-ribosyl transferase inhibitors on the formation of *neo* colonies in NIH/3T3 cells transfected with pSV2.neo (Table 1). In control cultures transfected with pSV2.neo, resistant colonies were formed at a frequency of 665 ± 98 per ug per 10^6 cells. The continuous presence of 2mM 3-methoxybenzamide almost completely abolished the formation of resistant colonies. Inhibition of ADP-ribosyl transferase activity for limited periods following transfection reduced the number of resistant colonies depending on the period of inhibition. When 3-methoxybenzamide was present only during the first two days after transfection, it slightly inhibited the formation of *neo* colonies. Similarly, presence of 3-methoxybenzamide later than the seventh day after transfection reduced the number of *neo* colonies only marginally. Maximum inhibition of the formation of *neo* colonies was obtained when 3-methoxybenzamide was present either continuously, or during the third to seventh day after transfection. The continuous presence of 2 mM 3-methoxybenzoate, which does not inhibit ADP-ribosyl transferase activity (34), did not reduce the number of *neo* colonies (Table 1).

Transformation of NIH/3T3 cells with pSVO.1 was inhibited about 97% by the continuous presence of 4 mM 3-aminobenzamide (Table 2), which is also a competitive inhibitor of nuclear ADP-ribosyl transferase activity (34). By contrast, 4 mM 3-aminobenzoate (the non-inhibitory analogue of 3-aminobenzamide) had no effect on the transformation frequency. Similarly, complete inhibition of cell transformation was observed when the plasmid pSV3.gpt, which contains the early region of SV40, was used in the presence of 3-aminobenzamide. The addition of 4 mM 3-aminobenzamide to cells which had already been transformed did not inhibit the growth of these cells, demonstrating that inhibition of ADP-ribosyl transferase activity does not block the expression of the transformed phenotype. Neither 4 mM nor 10 mM 3-aminobenzamide affected the cloning efficiency of NIH/3T3 cells which was about 32% both in the presence and absence of the inhibitors.

The same inhibition of successful transformation was also observed when a human fibroblast cell line (XP12RO-SV40) was transfected with 20 µg of the plasmid pSV3.gpt and then selected for the bacterial Eco.gpt gene in the presence of mycophenolic acid. In 9 plates there was an average of 114 colonies per plate, while in 7 plates to which 4 mM 3-aminobenzamide had also been added, there was an average of 15 colonies per plate;   a

191

suppression of 87% of the transformation frequency. This difference is significant with p ≤ 0.01. No spontaneous mycophenolic acid-resistant colonies were observed in these studies when the cells were grown for 14 days and no colonies were seen when cells were transfected with the carrier salmon-sperm DNA only.

Table 1. Effect of inhibition of ADP-ribosyl transferase activity on the formulation of neo^R colonies in NIH/3T3 cells transfected with pSV2.neo.

Treatment	Duration	Number of neo^R Colonies (per ug Plasmid DNA per 10^6 Cells ± S.D.)
Control		665 ± 98
2 mM 3-Methoxybenzamide	0-------------14	4
" "	0--2	420 ± 85
" "	1--3	240 ± 62
" "	3-----7	138 ± 43
" "	7------14	560 ± 81
2 mM 3-Methoxybenzoate	0-------------14	593 ± 74

Nearly confluent (~75%) NIH/3T3 cells were transfected with pSV2.neo DNA as described in legend to Fig 1. The medium was changed 24 hr after transfection and at 48 hr the cultures were trypsinised from each plate and seeded into 10 plates in the presence of 500 µg/ml of Geniticin. For ADP-ribosyl transferase inhibition, 2 mM 3-methoxybenzamide, or 2 mM methoxybenzoate was added to each culture for the indicated period. The growing neo^R colonies were identified twelve days later after fixing and staining the plates. Colonies smaller than 1 mm in diameter, probably the result of a transient resistance to geniticin, were not included in the quantitation of neo^R colonies. (Reprinted with permission from ref. 35).

Table 2. Effect of inhibition of ADP-ribosyl transferase activity on transformation of NIH/3T3 cells by pSVA.1.

Drug present	Number of plates	Mean number of transformed foci per plate	Foci/µg DNA
None	4	41.5	20.8
3-aminobenzoate	52	37.5	18.8
3-aminobenzamide	22	1.1	0.6

Sub-confluent NIH/3T3 cells were transfected with 2 µg of pSVO.1 DNA and 28 µg of salmon sperm DNA essentially as described in legend to Fig. 1. Where indicated 3-aminobenzoate, or 3-aminobenzamide, were added to cells at a final concentration of 4 mM, 30 min prior to transfection. Four to 6 hours after transfection the medium with the DNA precipitate was replaced with fresh mediun, which also contained the inhibitors or analogues where appropriate. The medium was again replaced at 24 hr and every three days thereafter. Foci of transformed cells were visible after 7 to 14 days and were counted at 10-14 days. The difference between plates containing 3-aminobenzamide and either of the other two sets of plates was statistically significant with p<0.001 (student t test). (Reprinted with permission from ref. 35).

These studies demonstrate that ADP-ribosyl transferase activity is involved in DNA-mediated gene transfer, because it is required for the stable integration of transfected DNA into the host genome. ADP-ribosyl transferase activity is not required for the uptake or the expression of transfected DNA either prior to or subsequent to its stable integration. The observations reported here also provide a potentially useful approach for the inhibition of DNA transfection.

Acknowledgements. This work was supported by grants from the British Medical Research Council and Cancer Research Campaign. We thank Dr. Cornelia Gorman for the pRSV.cat and Dr. Paul Berg for the pSVO.neo and pSV3.gpt plasmids.

References

1. Graham, F.L.,van der Eb, A. (1973) Virology 52: 456
2. Graham, F.L. *et al.* (1980) Introduction of Macromolecules into Viable Mammalian Cells, R. Baserga, C.Croce, G. Rovera (eds.) Liss, New York, pp. 3-25
3. Scangos, G., Ruddle, F.H. (1981) Gene. 14: 1
4. Benjamin, R.G., Gill, D.M. (1980) J Biol Chem 255:10493
5. Shall, S., *et al.* (1984) Genes and Cancer, J.M. Bishop, J.D. Rowley, M. Greaves (eds.) Liss, New York, pp. 175-183
6. Ueda, K., Hayaishi, O. (1985) Annu Rev Biochem 54: 78
7. Gaal, J.C., Pearson, C.K. (1986) Trends in Biochemical Sciences 11: 171
8. Creissen, D., Shall, S. (1982) Nature 296: 271
9. Ohashi, Y., *et al.* (1983) Proc Natl Acad Sci USA 80: 3604
10. Durkacz, B.W., *et al.* (1980) Nature 283: 593
11. Natarajan, A.T., *et al.* (1982) Prog Mut Res 4: 47
12. Morgan, W.F., Cleaver, J.E. (1982) Mut Res 104: 361
13. Johnstone, A.P., Williams, G.T. (1982) Nature 300: 368
14. Johnstone, A.P. (1984) Eur J Biochem 140: 401
15. Greer, W.L., Kaplan, J.G. (1984) Biochem Biophys Res Commun 115: 834
16. Farzaneh, F., *et al.* (1985) Mol Biochem Parasitol 14: 251
17. Cornelissen, A.W.C.A., *et al.* (1985) Biochem Pharmacol 34: 4151
18. Farzaneh, F., Shall, S., Zalin, R. (1980) Novel ADP-ribosylation of Regulatory Enzymes and Proteins, M.E. Smulson, T. Sugimura (eds.), Elsevier, Amsterdam, pp. 217-225
19. Farzaneh, F., *et al.* (1982) Nature 300: 362
20. Farzaneh, F., *et al.* (1985) ADP-Ribosylation Reactions, F.R. Althaus, H. Hilz, S. Shall (eds.), Springer-Verlag, Berlin, Heidelberg, pp. 433-439
21. Farzaneh, F., Shall, S., Johnstone, A.P. (1985) FEBS Lett 200: 107
22. Farzaneh, F., *et al.* (1987) Nucleic Acids Research 15: 3493
23. Farzaneh, F., *et al.* (1987) Nucleic Acids Research 15: 3503
24. Althaus, F.R. (1982) Nature 300: 366
25. Francis, G.E., *et al.* (1983) Blood 62: 1055
26. Francis, G.E., *et al.* (1984) Leukemia Res 8: 407
27. Williams, G.T. (1983) Exp Parasitol 56: 409
28. Williams, G.T. (1984) J Cell Biol 99: 79
29. Gorman, C.M., *et al.* (1982) Proc Natl Acad Sci USA 79: 6777
30. Southern, P.J., Berg, P. (1982) J Mol Appl Genet 1: 327
31. Mulligan, R.C., Berg, P.(1980) Science 209: 1422
32. Klobutcher, L.A., Miller, C.L., Ruddle, F.H. (1980) Proc Natl Acad Sci USA 77: 3610
33. Scangos, G., *et al.* (1981) Mol Cell Biol 1: 111
34. Purnell, M.R., Whish, W.J.D. (1980) Biochem J 185: 775
35. Farzaneh, F. *et al.* (1988) Nucleic Acids Res. 16: 11319-11326

Role of Poly(ADP-Ribose) Polymerase in Ribosomal RNA Gene Transcription

Samson T. Jacob, Jane F. Mealey, and Rabinder N. Kurl[1]

Department of Pharmacology and Cell and Molecular Biology Center, The Pennsylvania State University College of Medicine, Hershey, Pennsylvania 17033 USA

Introduction

Our laboratory has been involved in the identification and characterization of the transcription factors for RNA polymerase I (Pol I)-directed transcription of ribosomal RNA genes (rDNA). To this end, we initially isolated and partially purified a transcriptionally active protein complex which contains RNA polymerase I and the essential Pol I transcription factors (1). Such a fraction was obtained from whole cell extracts (1) or from nuclear extracts (2). Subsequently, we demonstrated that a fraction obtained during chromatography of the cell extract on a heparin sepharose column could prevent nonrandom transcription of cloned rat rDNA in an *in vitro* system (3). The major protein in this fraction exhibited characteristics of purified poly(ADP-ribose) polymerase. The present report summarizes the properties of this protein, and describes experiments showing the dramatic appearance of accurately initiated transcript in an unfractionated whole cell extract or nuclear extract from a tissue following addition of the protein factor.

Results

Fractionation of whole cell or nuclear extract. The cell extract and nuclear extract were prepared from rat mammary adenocarcinoma cells (1) and rat liver nuclei (2). The extracts were fractionated by chromatography on a DEAE-Sephadex column and the fractions eluting with 175 mM $(NH_4)_2SO_4$ were further chromatographed on a heparin Sepharose column using a linear NH_4Cl or KCl gradient. Fractions containing RNA polymerase I (fraction HS-B) and fractions eluting between 0.5 and 1 M were pooled (fraction HS-C). The major protein in this pooled fraction was then analyzed.

Properties of the major protein in the HS-C fraction. When fraction HS-C was subjected to SDS-PAGE, followed by staining with Coomassie

---

[1]Present address: Department of Pathology, George Washington University Medical Center, Washington, D.C. 20037

blue, only one band was visible indicating a high degree of purity at this stage of protein fractionation. It exhibited a molecular weight of 116,000 (3).

We first determined the effect of this protein on RNA polymerase I activity assayed with linearized rDNA fragment (spanning from -167 bp to 2.0 kb) cloned in pBR322 (pB 7-2.0) (4). Preliminary studies indicated that the HS-C protein had an insignificant effect on RNA polymerase I (assayed in the presence of alpha-amanitin to block RNA polymerases II and III) when DNA with negligible nicks was used as the template. However, when DNA was nicked by treatment with DNase I, it acted as an excellent template in the reaction. At a fixed concentration of the protein, RNA polymerase I activity increased with increasing amounts of nicked DNA whereas at a fixed concentration of DNA, nonspecific transcription of the nicked template by RNA polymerase I decreased with increasing amounts of the HS-C protein (Tables 1 and 2). At the highest concentration of the protein, as much as 60% inhibition of transcription was observed relative to the control sample.

Table 1. Effect of fraction HS-C on transcription of nicked DNA.

Amount of DNA (ng)	Transcription of nicked DNA (% of control)
100	18
200	38
300	55
400	63
100 (untreated DNA)	100

Varying amounts of rDNA nicked with DNase I were transcribed using 2 µg of fraction HS-B (which contains RNA polymerase I and transcription factors (1) and 1 µg of fraction HS-C (which contains poly(ADP-ribose) polymerase). Transcription was measured by the incorporation of $[\alpha^{32}P]UTP$ into TCA-insoluble precipitate. α-amanitin was included at a concentration of 130 µg/ml.

Table 2. Effect of varying amounts of fraction HS-C on transcription of nicked DNA.

Amount of fraction of HS-C (ng)	Transcription of nicked DNA (% of control)
200	75
400	67
600	59
800	46

Varying amounts of fraction HS-C were used in the reaction containing 2 µg of fraction HS-B and 200 ng of nicked DNA. Transcription was measured as described in the legend to Table 1.

ADP-ribosylation of HS-C protein. The requirement of nicked DNA for the action of HS-C protein, and the similarity in molecular weight of HS-C to that of highly purified poly(ADP-ribose) polymerase (5) suggested that

HS-C protein is poly(ADP-ribose) polymerase. To test this possibility further, the protein was subjected to ADP-ribosylation with [^{14}C]NAD and nicked DNA. The reaction product was subjected to SDS-PAGE. Most of the radioactivity was located near the origin or at a position that approximately corresponded to the Coomassie blue-stained protein band in a parallel track (data not shown). Since poly(ADP-ribose) polymerase is capable of auto-ADP-ribosylation and this modification can alter electrophoretic mobility (6), the two radioactive peaks observed after ADP-ribosylation of HS-C protein probably correspond to the different levels of ADP-ribosylation occurring in the enzyme molecule. That the radioactive peaks were indeed generated by a specific enzymatic reaction was shown by the lack of incorporation of [^{14}C]NAD at 4°C or in the absence of nicked DNA (data not shown).

Specific transcription of cloned rat rDNA in the presence of fraction HS-C. The above experiments suggested that the protein present in fraction HS-C can prevent random transcription of DNA. Slattery *et al.* (7) have purified a factor which eliminates nonspecific transcription by RNA polymerase II and is identical with poly(ADP-ribose) polymerase. In view of these observations we argued that HS-C factor, which exhibits the characteristics of poly(ADP-ribose) polymerase, could promote specific transcription of rDNA by RNA polymerase I. For this purpose, we used cloned rat rDNA (1, 2) which yields a 635 nucleotide-long transcript if transcription occurs at the correct site. Although fraction HS-B derived from rat mammary adenocarcinoma cells can yield a distinct 635 nucleotide-long transcript in most experiments (1), some preparations of HS-B were devoid of a factor(s) that prevented specific transcription of rDNA. The latter preparations were used to determine the efficacy of fraction HS-C. Increasing amounts of the latter fraction reduced the random transcription progressively and a distinct transcript was visible at the optimal concentration of HS-C (Fig. 1). The change from a dark smear on the autoradiogram to a clear radioactive profile was significant and, in some cases, quite dramatic. A reduced incorporation at higher concentrations of HS-C may be due to a shift from an optimal HS-B/HS-C ratio to a less favorable ratio of these components and/or due to an increase in an unknown inhibitor associated with fraction HS-C.

We then tested whether HS-C protein can promote specific and accurate transcription of rat rDNA in an unfractionated extract. Previous studies in our laboratory (2) have shown that unfractionated nuclear extracts from either rat hepatoma or rat liver cannot support rDNA transcription. If the lack of transcription in such extracts is predominantly due to nicks in the template caused by the action of DNase, addition of fraction HS-C should prevent random transcription and yield a distinct band corresponding to the expected size of the transcript. That this is indeed the case was proven by the addition of exogenous HS-C; addition of 45 ng to approximately 18 µg of unfractionated net hepatoma nuclear extract revealed the accurate transcript (Fig. 2, lanes 1 and 2).

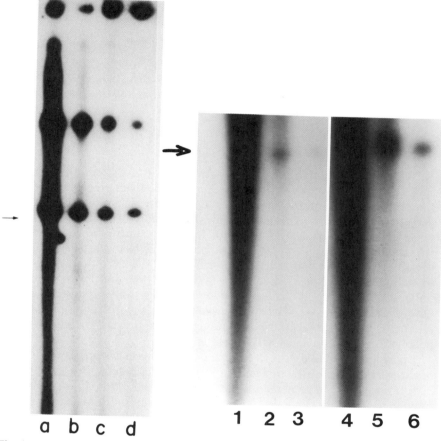

Fig. 1. (left) Transcription of rat rDNA in the presence of poly(ADP-ribose) polymerase associated with factor HS-C. The plasmid (pB 7-2.0) containing -167 bp to 2.0 kb of rat rDNA (1) was cleaved with *Xho* I and used in a run-off transcription assay. If transcription occurs at the correct site (nucleotide + 1), the transcript must be 635 nucleotides-long (1). Lane a, transcription using fraction HS-B (10 µg); Lanes b-d correspond to transcription using 10 µg of fraction HS-B and 0.6 µg, 0.8 µg and 1µg of fraction HS-C, respectively (3). The arrow corresponds to a 635 nucleotides-long transcript.

Fig. 2. (right) Effect of poly(ADP-ribose) polymerase on transcription of cloned rat rDNA in unfractionated whole cell extract and unfractionated nuclear extract derived from Morris hepatoma 3924A. Whole cell and nuclear extracts were prepared as described (1). HS-C was added at two different concentrations and *Xho* I-cleaved rDNA was used as a template in a run-off assay (Fig. 1). Lanes 1, 2 and 3 correspond to transcription in 18 µg unfractionated hepatoma nuclear extracts in the presence of 45 ng and 90 ng of HS-C, respectively. Lanes 4, 5 and 6 represent transcription in unfractionated whole cell extract (18 µg) in the absence or in the presence of 45 ng and 90 ng of HS-C, respectively.

197

Similar experiments were also performed with unfractionated whole cell extracts. Unlike unfractionated tissue extracts (2), unfractionated whole cell extracts usually yield the correct transcript without addition of exogenous HS-C. However, some preparations of unfractionated cell extracts are unable to support transcription of rDNA. When such preparations were used for rat rDNA transcription, a smeared autoradiogram with no distinct bands was observed. However, addition of HS-C in these preparations prevented random transcription and resulted in the production of the correct transcript (Fig. 2, lanes 3-5). Higher concentrations of HS-C had an inhibitory effect on the transcription, as observed even in reconstituted systems consisting of relatively pure fractions (Fig. 1).

Discussion

The present studies have demonstrated that a protein factor which exhibits properties of poly(ADP-ribose) polymerase can prevent nonspecific transcription of rat rDNA in either whole cell extracts or nuclear extracts derived from tissues. This factor is analogous to that described for RNA polymerase II-directed transcription (7). Highly purified fractions do not require this factor for accurate transcription of rDNA (Mealey and Jacob, unpublished data) probably due to removal of DNA-nicking enzyme(s) following purification. This observation does not preclude the importance of poly(ADP-ribose) polymerase *in vivo* in rDNA transcription following DNA damage. Clearly, rDNA can be nicked under these conditions and the requirement for poly(ADP-ribose) polymerase to prevent binding of RNA polymerase I to these nicks will be essential.

The appearance of a distinct band corresponding to the accurate transcript of rat rDNA in an unfractionated hepatoma nuclear extract in response to exogenous HS-C is of considerable interest. This observation raises the possibility of using such a system to study rDNA transcription in response to a variety of physiological stimuli that are known to alter rRNA synthesis. Unfractionated samples have certain advantages over the fractionated samples since purification procedures could result in differential recovery of the transcription factors from control and test samples.

The remarkable increase in poly(ADP-ribose) polymerase by DNA-damaging agents including carcinogens (8), suggests that this enzyme is in some way involved in DNA repair (9-11). Although the exact mechanism by which this enzyme induces the repair has not been fully established, it is clear that it has an important role in rDNA transcription particularly when rDNA is nicked. Since as little as 45 ng of the HS-C factor could prevent random transcription of rDNA in the reconstituted system containing approximately 18 μg of total protein, the effect of the poly(ADP-ribose) polymerase must be specific rather than a simple masking effect of large amounts of protein.

In the course of these studies, we observed that RNA polymerase I can be ADP-ribosylated and that this modification can partially inactivate the enzyme (Mealey and Jacob, unpublished data). We are now testing whether ADP-ribosylation of RNA polymerase I can adversely affect rDNA transcription. We have failed to ADP-ribosylate RNA polymerase I completely in the *in vitro* system and are in the process of developing the optimal conditions for this reaction. Only then can we do meaningful experiments to determine the precise role of ADP-ribosylation of RNA polymerase I or of other essential transcription factor(s) in the transcription of ribosomal RNA genes.

Acknowledgements. The authors thank Susan DiAngelo for expert technical assistance. This work was supported by the United States Public Health Service Grants CA31894 and CA25078 (STJ). Some of the data have been reprinted with permission from Nucleic Acids Research 13, 89-101, 1988.

References

1. Kurl, R.N., Rothblum, L.I., Jacob, S.T. (1984) Proc Natl Acad Sci USA 81: 6672-6675
2. Kurl, R.N., Jacob, S.T. (1985) Proc Natl Acad Sci USA 82: 1059-1063
3. Kurl, R.N., Jacob, S.T. (1985) Nucleic Acids Res 13: 89-101
4. Rothblum, L.I., Reddy, R., Cassidy, B. (1982) Nucleic Acids Res 10: 7345-7362
5. Purnell, M.R., Stone, P.R., Whish, J.D. (1980) Biochem Soc Trans 8: 215-227
6. Ogata, N., Veda, K., Kawaichi, M., Hayaishi, O. (1981) J Biol Chem 256: 4135-4137
7. Slattery, E., Dignam, J.D., Matsui, T., Roeder, R.G. (1983) J Biol Chem 258: 5955-5959
8. Mandel, D., Okazaki, H., Niedergang, C. (1982) Progr Nucleic Acid Res Mol Biol 27: 1-51
9. Veda, K., Ogata, N., Kawaichi, M., Inada, S., Hayaishi, O. (1982) Curr Top Cell Reg 21: 175-187
10. James, M., Lehman, A. (1982) Biochemistry 21: 4007-4013
11. Cleaver, J., Bodell, W., Morgan, W., Zelle, B. (1983) J Biol Chem 258: 9059-9068

ADP-Ribosylation of HeLa Chromosomal and Nuclear Proteins: Response to DNA Damage Caused by Chemical Mutagens

Kenneth W. Adolph

Department of Biochemistry, University of Minnesota Medical School, Minneapolis, Minnesota 55455 USA

Introduction

Recent experimental evidence has indicated a central role for ADP-ribosylation of chromosomal proteins as a response to DNA damage produced by chemicals and radiation (1-3). ADP-ribosylation is involved in the initial response to DNA damage, as this paper demonstrates, and in the final step in DNA repair, which is the ligation of broken DNA ends to produce a continuous double helix. Repair of DNA is accompanied by an increase in DNA ligase activity, which apparently results from the attachment of ADP-ribose to the enzyme. ADP-ribosylation could have other functions connected with DNA repair. Since the histones are major acceptors of ADP-ribose units, the structure of nucleosomes and the 30 nm chromatin fibers could be altered by the modification. A number of other nuclear proteins are also substrates for the modification, including the lamins of the peripheral lamina.

This paper describes experiments concerned with the response of nuclei and chromosomes to DNA damage caused by the chemical mutagen, dimethyl sulfate. The experiments demonstrate this response to be a substantial increase in the level of ADP-ribosylation of both nonhistones and histones. The study was additionally concerned with examining the nature of the ADP-ribosylated proteins, including identifying the primary substrates.

Results

Effect of dimethyl sulfate on ADP-ribosylation of HeLa nuclear proteins and chromosomal proteins. The autoradiograms included in Fig. 1 reveal the influence of dimethyl sulfate (DMS) on [^{32}P]NAD labeling of the nonhistones and histones associated with nuclei and chromosomes. The experiments were performed by incubating permeabilized cells with radioactive isotope before nuclei or chromosomes were isolated (4, 5). Cells were pretreated with DMS in serum-free medium (Fig. 1B) or isolation buffer (Fig. 1C). Labeling of nuclear proteins is shown in Fig. 1B to be greatly stimulated by treatment with the DNA-damaging agent. The optimal concentration of DMS under the conditions used was 1.0 mM. The protein compositions of the samples were largely unaffected by the treatment (Fig.

1A). A similar situation of ADP-ribosylation was observed (Fig. 1C) for metaphase chromosome proteins. For these samples, the optimal DMS concentration was 3.0 mM.

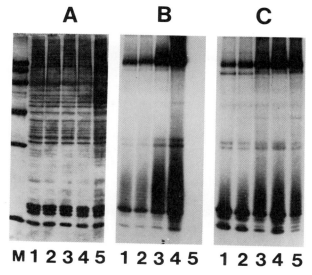

Fig. 1. ADP-ribosylation of HeLa nuclear and chromosomal proteins following treatment with DMS. Interphase cells were incubated with DMS in growth medium, metaphase-arrested cells in chromosome isolation buffer. After labeling the proteins with [^{32}P]NAD, nuclei and chromosomes were isolated, and the proteins were separated on SDS-polyacrylamide gels (12.5%) (6, 7). A gel of nuclear proteins stained with Coomassie blue is in (A), and the corresponding autoradiogram is in (B). The concentrations of DMS were: lane 1, 0 mM; lane 2, 0.01 mM; lane 3, 0.1 mM; lane 4, 1.0 mM; lane 5, 10.0 mM. An autoradiogram showing [^{32}P]NAD-labeled chromosomal proteins is in (C). In this case, the DMS concentrations were: lane 1, 0 mM; lane 2, 0.3 mM; lane 3, 1.0 mM; lane 4, 3.0 mM; lane 5, 10.0 mM. The stained gel contains molecular weight markers in lane M of, from top to bottom, β-galactosidase (116,000 daltons), phosphorylase b (97,000), bovine serum albumin (66,000), chicken egg albumin (45,000), carbonic anhydrase (30,000), and lysozyme (14, 000).

Except for bands in the histone region, the most conspicuous acceptor of label for both untreated and treated samples was a species of 116,000 daltons. This protein is probably poly(ADP-ribose) polymerase. The histone region, which also shows a substantial increase in ^{32}P-labeling, is more complicated. Mutagen treatment resulted in the increased labeling of a series of equally spaced bands. The ladder of bands could represent modified histones with poly(ADP-ribose) chains of different lengths or free chains of poly(ADP-ribose).

The effect of incubation with the mutagen was quantified by densitometry of the autoradiograms (Fig. 2). In each panel, the increase in ^{32}P-labeling is plotted for total proteins, total nonhistones, and for the 116,000 dalton species. The results are presented for nuclei from interphase cells incubated with DMS in serum-free medium (A) and in nuclei isolation buffer (B), as well as for chromosomes from mitotic cells incubated in chromosome isolation buffer (C). Panel B indicates that incubation of intact cells in growth medium is much more effective than incubation of permeabilized cells in isolation buffer.

Two-dimensional gel electrophoresis of histones from interphase and mitotic cells treated with dimethyl sulfate. 2-D gel electrophoresis was carried out with proteins separated by nonequilibrium pH gradient electrophoresis in the first dimension and on 8% polyacrylamide slab gels in the second (6, 7). Separation of histones was enhanced by use of a urea solubilization buffer containing protamines to displace histones from DNA (8). The resulting autoradiograms for histones of both nuclei and chromosomes are in Fig. 3. Incubation of mitotic cells with [^{32}P]NAD before chromosomes were isolated led to prominent labeling of histones H2B and H4 (A, B), and spots are also visible at the positions of H1A and H1B. ^{32}P-incorporation is also seen at the positions of histones H2A and H3. In Fig. 3B, a similar pattern is observed for chromosomal histones from cells treated with DMS. The primary acceptors of isotope are H2B and H4, with radioactivity also present at the positions of H1A, H1B, H2A, and H3. A major change in the relative distribution of ^{32}P between species is clearly not a feature of DMS treatment.

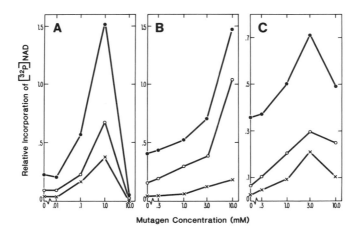

Fig. 2. Quantification of the increase in [^{32}P]NAD labeling of nuclear and chromosomal proteins resulting from incubation with DMS. Autoradiograms, such as those in Fig. 1, were scanned with a Cary spectrophotometer equipped with a gel scanning stage. The relative weights of the tracings were measured for incorporation by total proteins (●), nonhistones (○), and the 116,000 dalton protein (x). Panel A, nuclear proteins from cells treated in growth medium; panel B, cells treated in isolation buffer; panel C, metaphase chromosome proteins from cells incubated with mutagen in chromosome isolation buffer.

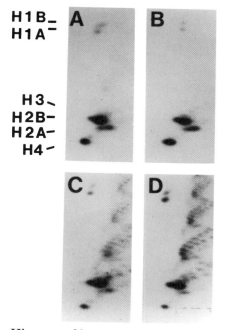

H1B—
H1A—

H3—
H2B—
H2A—
H4—

Fig. 3. Histones, resolved by two-dimensional gel electrophoresis, from cells treated with DMS. [^{32}P]NAD-labeled nuclei and chromosomes were isolated, and the histones were solubilized in protamine-containing buffer (8). Autoradiograms (A) of control chromosomes and (B) from cells treated with 1.0 mM DMS; (C) labeled histones of control nuclei and (D) nuclei from DMS-treated (1.0 mM) cells. The isoelectric focusing dimension (horizontal) used nonequilibrium pH gradient electrophoresis. SDS-polyacrylamide slab gels (12.5%) were run in the second dimension (vertical).

Histones of interphase nuclei were also resolved on two-dimensional gels (panels C and D). The distribution of label associated with the histones is not unlike that found for chromosomes, but the overall pattern is more complex due to the presence of other spots. This radioactivity represents the ladder of bands on autoradiograms of 1-D gels.

The results imply that the accessibility of the histones to poly(ADP-ribose) polymerase and the nature of the modification do not change

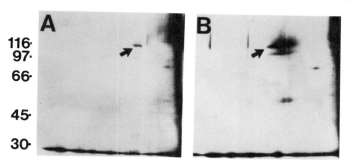

116-
97-
66-
45-
30-

Fig. 4. Chromosomal nonhistones, resolved by 2-D gel electrophoresis, from metaphase cells incubated with DMS. Cells in chromosome buffer were treated with 10.0 mM DMS. In the second dimension, the slab gels were of 8% acrylamide. The autoradiogram in (A) displays the labeled proteins of untreated chromosomes, while (B) shows the radioactive species of mutagen-treated cells. The arrow indicates the 116,000 dalton protein.

dramatically due to mutagen treatment. Thus, a greater occupancy of modification sites or a lengthening of existing ADP-ribose chains probably causes the increased [^{32}P]NAD labeling.

Two-dimensional gel electrophoresis of nonhistones from mutagen-treated nuclei and chromosomes. The differences between the modification of nuclear and chromosomal nonhistone proteins are revealed most strikingly on autoradiograms of 2-D gels. Fig. 4A shows the characteristic distribution of radioactive species associated with isolated metaphase chromosomes (untreated). The pattern is extremely simple, with a single ^{32}P-labeled spot, indicated by an arrow, predominating. This is the 116,000 dalton species, identified as poly(ADP-ribose) polymerase. Upon pre-treating mitotic cells in chromosome isolation buffer with DMS, the 116,000 dalton species, as well as one of 100,000 daltons, became highly labeled (Fig. 4B).

Autoradiograms of 2-D gels of ADP-ribosylated nuclear proteins from cells treated with DMS are presented in Fig. 5. In the experiment shown, cells were treated with the mutagen in growth medium. The concentration of DMS was varied for different samples to include (A) 0 mM, (B) 0.1 mM, (C) 1.0 mM, and (D) 10.0 mM. A higher level of ADP-ribosylation is seen for all the major acceptors, which includes species with molecular weights of 60,000-70,000. These include the lamins of the peripheral nuclear lamina. As with metaphase chromosomes, species of 116,000 and 100,000 daltons are highly modified. A number of additional spots and streaks can be observed. These also display a greater level of ^{32}P-labeling. The relative distribution of label between species undergoes some changes not seen on 1-D gels, so that, for example, the 100,000 dalton protein has more ^{32}P-label bound than the 116,000 dalton protein after 1.0 mM DMS treatment. The increase in the modification is abruptly reversed at 10.0 mM DMS, which may reflect a harmful effect of extreme DNA damage or may be a consequence of the effect of DMS on cell physiology.

Discussion

Autoradiograms revealed that the dominant acceptor of radioactive label from [^{32}P]NAD was a 116,000 dalton band, poly(ADP-ribose) polymerase. Since the activity of the enzyme depends upon nicked DNA, an increase in the automodification of the enzyme as a result of damaging DNA with mutagens is compatible with its known properties. The other major ADP-ribosylated proteins were the histones, particularly H2B, and a ladder of bands extending to higher molecular weights. The bands appear to represent ADP-ribose chains of up to 20 units that are probably covalently attached to the histone polypeptides.

The alterations in DNA structure produced by mutagen treatment which most effectively increase ADP-ribosylation could include not only breaks in the DNA double helix, but also the loss of guanine bases and methylation of bases. The finding of an optimal concentration of mutagen complicates interpretation of the effect of mutagen treatment. One explanation is that excessive DNA damage can inhibit the activity of the polymerase. Proper binding of the enzyme to DNA or movement along the DNA strands could be hindered. Another explanation is that the mutagens disturb the physiology of the cells. Previous studies from this laboratory have demonstrated that major changes in ADP-ribosylation occur during the cell cycle and as a result of various *in vitro* treatments (9-13). The results of these studies, along with the present results, indicate the broad and dynamic significance of ADP-ribosylation of chromosomal and nuclear proteins.

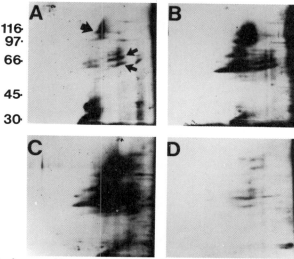

Fig. 5. Nuclear nonhistones, resolved by 2-D electrophoresis, from mutagen-treated interphase cells. Cells in growth medium were incubated with DMS at concentrations of 0 mM (panel A), 0.1 mM (panel B), 1.0 mM (panel C), and 10.0 mM (panel D). The large arrow indicates the 116,000 dalton species, and the small arrows indicate lamins A and C.

References

1. Hayaishi, O., Ueda, K. (eds.) (1982) ADP-ribosylation Reactions. Academic Press, New York
2. Purnell, M.R., Stone, P.R., Whish, W.J.D. (1980) Biochem Soc Trans 8: 215-227
3. Shall, S. (1984) Adv Radiat Biol 11: 1-69
4. Adolph, K.W. (1980) J Cell Sci 42: 291-304
5. Adolph, K.W. (1980) Chromosoma 76: 23-33
6. Laemmli, U.K. (1970) Nature 227: 680-685
7. O'Farrell, P.Z., Goodman, H.M., O'Farrell, P.H. (1977) Cell 12: 1133-1142
8. Sanders, M.M., Groppi, V.E., Browning, E.T. (1980) Anal Biochem 103: 157-165
9. Adolph, K.W., Song, M.-K. H. (1985) Biochemistry 24: 345-352
10. Song, M.-K. H., Adolph, K.W. (1983) Biochem Biophys Res Commun 115: 938-945
11. Adolph, K.W., Song, M.-K. H. (1985) FEBS Lett 182: 158-162
12. Adolph, K.W. (1985) Arch Biochem Biophys 243: 427-438
13. Adolph, K.W. (1987) Arch Biochem Biophys 253: 176-188

Poly(ADP-Ribose) Polymerase: DNA Complexes Visualized as Looped Structures by Electron Microscopy

Gérard Gradwohl, Josette Dunand, Alice Mazen, and Gilbert de Murcia

I.B.M.C. du C.N.R.S., Laboratoire de Biochimie II, 67084 Strasbourg, France

Introduction

Poly(ADP-ribose) polymerase is a chromatin bound enzyme which catalyzes the covalent attachment of ADP-ribose units from the coenzyme NAD to various nuclear proteins (1-3). Poly(ADP-ribosyl)ation is a posttranslational modification which appears to be involved in DNA excision repair, cellular proliferation and differentiation (1-3) and in modulation of chromatin structure (4, 5). This DNA dependent enzyme is inactive unless stimulated by DNA strand breaks. Although the DNA structures which activate the enzyme have been identified (6, 7) the basis for the DNA requirement as well as the stimulation of the enzyme activity are not yet understood. We have studied the interaction between poly(ADP-ribose) polymerase and different DNAs using electron microscopy and gel retardation electrophoresis.

Results

Procedures. Highly purified calf thymus poly(ADP-ribose) polymerase was used to avoid DNA relaxation due to topoisomerase I contaminants (8, 9). Several types of DNA were purified from a pBR322 form I stock: form III (*Eco*RI linearized), form IIa (topoisomerase I relaxed), form IIc (DNase I nicked). Alternatively the stock DNA was irradiated with 250 J/m^2 at 254 nm and subsequently digested with *M. luteus* UV endonuclease (form IIb). Purified enzyme and DNA were mixed at a given protein/DNA ratio (mole/mole) in binding buffer containing 25 mM Tris HCl pH 8, 2 mM MgCl$_2$, 0.4 mM DTT, 100 mM NaCl, and incubated during 30 min at 25°C. The complexes were then processed for electron microscopy as previously described (10).

For agarose gel retardation electrophoresis, aliquots corresponding to 0.5 μg of DNA in the complex, were mixed with 5 μl of sample buffer containing 25% glycerol and bromophenol blue, deposited onto a 1% agarose gel and run in Tris-phosphate-EDTA buffer at 30 V for 14-16 h.

Gel retardation experiments. When increasing quantities of enzyme was mixed with form I pBR322 DNA in binding buffer, form I DNA was progressively retarded whereas electrophoretic migration of form II DNA was practically unchanged (Fig. 1). The same result was obtained with a naturally occurring dimer of form I pBR322 DNA (form ID) which was also preferentially complexed by increasing concentrations of poly(ADP-ribose) polymerase. In order to compare the binding affinity of the enzyme to different DNA conformations, the enzyme was incubated with form I and form IIa DNA in equal proportions, or form IIb or form IIc. The results of these competition experiments are shown in Fig. 2. One can see that in the absence of nicks in the DNA (panel A), poly(ADP-ribose) polymerase was preferentially bound to form I DNA as in the previous experiment. Conversely, when nicks were created in DNA by *M. luteus* UV endonuclease (panel B) the enzyme seemed to be equally distributed between form I and form IIb or to bind preferentially to form IIc (panel C).

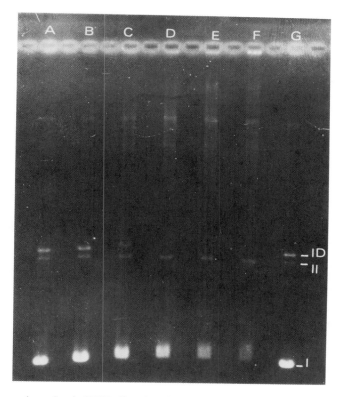

Fig. 1. Formation of poly(ADP-ribose) polymerase • DNA complexes studied by gel retardation electrophoresis. pBR322 form I DNA (containing about 10% form II DNA) was mixed to purified calf thymus poly(ADP-ribose) polymerase at different ratios r (mole/mole).
(A, G: r=0; B: r=15; C: r=30; D: r=45; E: r=60; F: r=75).

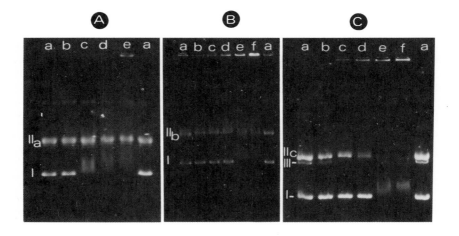

Fig. 2. Competition binding experiments between purified poly(ADP-ribose) polymerase and different topological forms of pBR322 DNA.
Equal proportions of form I and form IIa (A) or form IIb (B) or form IIc (C) were incubated with poly(ADP-ribose) polymerase at different ratios (a: r=0; b: r=15; c: r=30; d: r=45; e: r=60; f: r=75).

Electron microscopy. Similar experiments were done in order to visualize the enzyme-DNA interaction by electron microscopy. Fig. 3a illustrates the competition between form I and form IIa DNAs. One can see that the enzyme was mostly bound to form I DNA. Conversely the competition between form I and form IIc DNA shows that most of the poly(ADP-ribose) polymerase molecules were bound to form IIc pBR322 DNA as shown in Fig. 3b.

For each competition experiment enzyme-bound DNA and free DNA (form I and form II) were scored. The results presented in Table I indicate that in the absence of breaks in DNA, poly(ADP-ribose) polymerase preferentially binds to form I DNA, whereas in the presence of nicks most of the enzyme molecules are bound to DNase I relaxed circular DNA. These results are thus in full agreement with the gel retardation experiments.

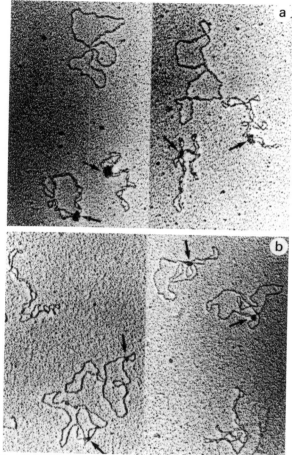

Fig. 3. Visualization of poly(ADP-ribose) polymerase • DNA complexes by electron microscopy. Competition for enzyme binding to (a) form I and form IIa pBR322 DNA, (b) form I and form IIc pBR322 DNA. The arrows indicate a preferential location of poly(ADP-ribose) polymerase bound to form I DNA in panel (a); and a preferential location of the enzyme on form IIc DNA in panel (b) (r=30).

Careful examination of the enzyme DNA complexes displayed in Fig. 3b revealed the presence of DNA loops which result from the interaction between DNA and one or more enzyme molecules. The enzyme DNA interaction was further investigated by the use of different morphological forms of pBR322 DNA. Fig. 4 depicts several examples of enzyme DNA complexes. In practically all the cases, DNA in the complex was frequently seen as a multi-looped structure with the poly(ADP-ribose) polymerase molecule located at the cross-over point of the DNA strands. These DNA loops are in fact produced by the binding of the enzyme to two distinct DNA molecules (intermolecular binding) (Fig. 4a) or by intramolecular binding of the enzyme to two strands of the same DNA molecule (Fig. 4c, 4d). In some cases two distinct enzyme molecules can be observed at the same DNA cross-over point (Fig. 4a, 4c), indicating probable protein-protein interactions. The micrographs 4e to 4i depict some characteristic features of the interaction between poly(ADP-ribose) polymerase and *Eco*RI linearized pBR322 DNA (form III). One can observe enzyme binding either to one *Eco*RI extremity (Fig. 4e) or to the two extremities (Fig. 4g) resulting in the DNA molecule being re-circularized. In addition, DNA loops are also visible with linear DNA (Fig. 4 f-i).

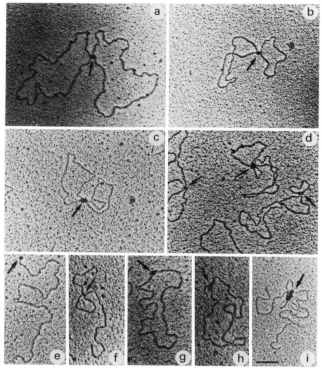

Fig. 4. Visualization of DNA loops obtained after binding of poly(ADP-ribose) polymerase to: (a, b) form IIa pBR322 DNA, (c, d) form IIc, (c-i) form III in all cases (r=30).

Discussion

Both the electron microscopy and gel retardation electrophoresis results indicate that in the absence of single or double strand breaks, purified poly(ADP-ribose) polymerase preferentially binds to superhelical DNA. By electron microscopy we have found a high frequency of enzyme molecules at intersections of DNA duplexes, independently of the type of DNA used.

The higher affinity of the enzyme to form I DNA can be explained by the higher number of cross-over points of the DNA strands in a form I DNA molecule as compared to a relaxed covalently closed circular DNA molecule. We note that such a high affinity of form I DNA for the cooperative binding of histone H1 to supercoiled DNA (14) has already been observed. Interestingly, on chromatin super-structure this linker histone has been found to be the major poly(ADP-ribose) acceptor (4, 11, 12) inducing *in vitro* the relaxation of the 25 nm chromatin fiber. Thus, in the absence of nicks poly(ADP-ribose) polymerase could compete with histone H1 for the same DNA binding sites whereas in the presence of nicks H1 could be preferentially poly(ADP-ribosyl)ated.

For high enzyme DNA ratios, poly(ADP-ribose) polymerase appears to react preferentially and cooperatively in the vicinity of poly(ADP-ribose) polymerase molecules already bound to a DNA intersection, leading to the formation of large protein aggregates in interaction with multi-looped DNA duplexes (data not shown). In the presence of single and/or double strand breaks in DNA, poly(ADP-ribose) polymerase preferentially binds to form II DNA. Under these conditions the enzyme activity is strongly stimulated (Table 1); this is in agreement with previously published work (6, 7). Direct comparison by electron microscopy between active (Fig. 4c, 4d) and inactive (Fig. 4a, 4b) poly(ADP-ribose) polymerase-DNA complexes does not explain the mechanism by which the enzyme activity is switched on. However, by this technique we have found that loop formation is a common feature observed with all the DNAs used.

We do not know whether in the case of active complexes the enzyme is interacting directly with the strand break or if the loop formation allows the poly(ADP-ribose) polymerase to act at a distance. This last hypothesis would require either the presence of two DNA binding sites on the same molecule or the possibility for the enzyme to form dimers as in the case of gyrase-DNA complexes (13) or for proteins involved in gene regulation where DNA bending and loop formation was observed by electron microscopy (15, 16).

Table 1

DNA	Relative enzyme activating efficiency	DNA bound enzyme	
		Form I	Form II
Calf thymus activated DNA	1		
Form I	0.04		
Form IIa	0.07		
Form IIb	0.14		
Form IIc	0.57		
Form III	0.12		
(Form I + Form IIa)		73%	30%
(Form I + Form IIc)		20%	68%

Acknowledgements. The authors wish to thank Dr. D.B. Windsor for a careful revision of the text. Some of the data has been reprinted with permission from Gradwohl, *et al.*, Biochem. Biophys. Res. Commun. 148: 913-919, 1987.

References

1. Mandel, P., Okazaki, H., Niedergang, C. (1982) Prog Nucl Acids Res 27: 1-51
2. Ueda, K., Hayaishi, O. (1985) Ann Rev Biochem 54: 73-100
3. Gaal, J.C., Pearson, C.K. (1985) Biochem J 230: 1-18
4. Poirier, G., de Murcia, G., Jongstra-Bilen, J., Niedergang, C., Mandel, P. (1982) Proc Natl Acad Sci USA 79: 3423-3427
5. de Murcia, G., Huletsky, A., Lamarre, D., Gaudreau, A., Pouyet, J., Daune, M., Poirier, G. (1986) J Biol Chem 261: 7011-7017
6. Benjamin, R.C., Gill, D.M. (1980) J Biol Chem 255: 10502-10508
7. Ohgushi, H., Yoshihara, K., Kamiya, T. (1980) J Biol Chem 255: 6205-6211
8. Ferro, A.M., Higgins, N.P., Olivera, B.M. (1983) J Biol Chem 258: 6000-6003
9. Mandel, P., Jongstra-Bilen, J., Ittel, M.E., de Murcia, G., Delain, E., Niedergang, C., Vosberg, H.P. (1983) ADP-Ribosylation, DNA Repair and Cancer, M. Miwa et al. (eds.), Japan Sci. Soc. Press, Tokyo/VNU Science Press, Utrecht, pp. 71-81
10. de Murcia,G., Koller, T. (1981) Biol Cell 40: 165-174
11. Aubin, R.J., Dam, V.T., Miclette, J., Brousseau, Y., Huletsky, A., Poirier, G. (1982) Canadian J Biochem, 60: 1085-1094
12. Niedergang, C, de Murcia, G., Ittel, M.E., Pouyet, J., Mandel, P. (1985) Eur J Biochem 146: 185-191
13. Moore, C.L., Klevan, L., Wang, J.C., Griffith, J.D. (1983) J Biol Chem 258: 4612-4617
14. De Bernardin, W., Losa, R., Koller, T. (1986) J Mol Biol 189: 503-517
15. Ptashne, M. (1986) Nature 322: 697-701
16. Krämer, H., Niemöller, M., Amouyal, M., Revet, B., Von Wilcken-Bergmann, B., Müller-Hill, B. (1987) EMBO J. 6: 1481-1491

Poly(ADP-Ribose) May Signal Changing Metabolic Conditions to the Chromatin of Mammalian Cells

Pius Loetscher, Rafael Alvarez-Gonzalez[1], and Felix R. Althaus

Institute of Pharmacology and Biochemistry, University of Zürich (Tierspital), CH-8057 Zürich, Switzerland

Introduction

The concept of "suicidal NAD+ depletion" proposed by Berger and associates (1) emphasizes the aspect that carcinogen-inflicted DNA damage in mammalian cells dramatically stimulates the utilization of NAD+ for poly(ADP-ribose) biosynthesis and concomitantly depletes the cellular NAD+ pool(s). Thus, the chromatin-associated enzyme poly(ADP-ribose) polymerase may serve as part of a metabolic "shut-off" mechanism under conditions of excessive DNA damage (1-3). Here we examined the hypothesis that metabolic conditions affecting the availability of NAD+ may be signaled to chromatin by the intermediacy of poly(ADP-ribose). Therefore, experimental conditions were set up to determine whether an altered level of posttranslational poly(ADP-ribose) modification could be forced upon the chromatin of mammalian cells by directly manipulating the intracellular NAD+ pool with exogenously supplied NAD+ (4).

Results

Fig. 1A shows that within a 60 min incubation of cultured hepatocytes with 500 µM NAD+, the intracellular NAD+ concentration increased by 50% and reached almost twice the concentration of control cells after prolonged incubation. This increase also occured under conditions in which the exogenously supplied NAD+ concentration was significantly below the endogenous NAD+ concentrations of these cells (Fig. 1B).

The results of Table 1 demonstrate that following a 1 or 2 hr incubation of hepatocytes with 3 different concentrations of exogenously supplied NAD+, the absolute intracellular increases were nearly identical to the membrane transfer of an equivalent amount of radiolabeled material, identified as intact NAD+ by affinity chromatography in conjunction with high pressure liquid chromatography analysis.

The results of Table 2 confirmed the translocation of intact NAD+ across the plasma membrane. Using NAD+ preparations radiolabeled in two different positions of the molecule, i.e. [adenine-2, 8-3H]- and [carbonyl-

[1]Present Address: Biomedical Division, The Samuel Roberts Noble Foundation, Inc., P.O. Box 2180, Ardmore, OK 73402, USA

^{14}C]NAD$^+$, we were able to directly compare the isotope ratio of the radiolabeled NAD$^+$ exogenously supplied with the isotope ratio of the NAD$^+$ transferred across the membrane.

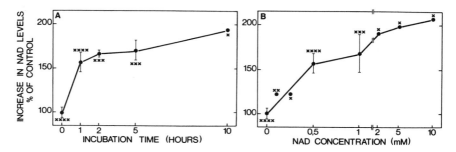

Fig. 1. Time course (A) and dose dependence (B) of NAD$^+$ uptake by adult rat hepatocytes in primary monolayer culture (4). A, the hepatocyte monolayers were incubated in the presence of 500 μM NAD$^+$ in the medium (5 ml/dish) and the intracellular NAD$^+$ concentration was determined at the times indicated (4). B, hepatocyte monolayers were incubated with various concentrations of NAD$^+$ in the medium, and the intracellular NAD$^+$ concentrations were determined (4) after 60 min. The values are expressed as percent of untreated cells not incubated with NAD$^+$. Control levels (= 100%): 1.3 ± 0.082 nmol NAD$^+$/mg protein (mean ± SEM, n = 8), which corresponds to an intracellular NAD$^+$ concentration of 400 ± 25 μM (mean ± SEM, n = 8). The values represent: *, single determination, **, average of two determinations, *** mean ± SEM of 3 independent determinations, or ****, mean ± SEM derived from 8 independent determinations, all involving separate cell preparations.

Table 1. Quantitative analysis of the intracellular NAD$^+$ increase[a].

Incubation Time (hr)	NAD$^+$ in Medium (μM)	Intracellular NAD$^+$ (μM)	
		Total	Increase Due To Transmembrane Transfer
1	100	113	81
1	500	169	128
1	1000	194	142
2	100	189	158
2	500	232	226

[a]determined as described (12)

At least 81% of the NAD$^+$ taken up by cultured hepatocytes eqilibrates freely with the intracellular NAD$^+$ pool, which is utilized for poly(ADP-ribosyl)ation of nuclear proteins (data not shown). The data in Table 3 show that a 70% increase of intracellular NAD$^+$ caused a five-fold increase of chromatin-associated poly(ADP-ribose). Concomitantly, the ratio of ADP-ribosyl residues in poly(ADP-ribose) over those contained in NAD$^+$ rose three fold (Table 3). This indicates a new equilibrium between these two pools of ADP-ribosyl residues. In this new steady state, the catabolic rate of chromatin-bound poly(ADP-ribose) was unaltered (data not shown). This demonstrates that a significant increase of chromatin-bound poly(ADP-ribose) could be forced upon hepatocytes by raising the intracellular NAD$^+$ pool.

Table 2. Evidence for membrane transfer of intact NAD$^+$, radiolabeled in two different positions, in cultured hepatocytes concomitant with the increase of intracellular NAD$^+$.

Incubation Time (hr)	NAD$^+$ in Medium (μM)	[adenine-2,8-^{3}H]NAD$^+$/[carbonyl-^{14}C]NAD$^+$ isotope ratio		
		Extracellular	Intracellular	Extra: Intra
1	100	0.71	0.82	0.87
1	500	0.68	0.69	0.98
2	100	0.68	0.80	0.85

Table 3. Consequences of the elevation of intracellular NAD$^+$ concentrations on the level of poly(ADP-ribose) modification of hepatocellular chromatin and the ratio between ADP-ribosyl residues contained in NAD$^+$ and poly(ADP-ribose)

	Intracellular NAD$^+$ (μM)	Poly(ADP-ribose) (nm)	Poly(ADP-ribose)/NAD$^+$
Control Cells	458.0 ± 34.6	66.0 ± 2.8	1.4 x 10^{-4}
NAD+-loaded Cells	786.0 ± 57.1	340.0 ± 60.8	4.3 x 10^{-4}

Intracellular NAD and poly(ADP-ribose) were determined according to ref. 12.

Discussion

We propose that conditions altering the availability of NAD$^+$ may be signaled to chromatin by the intermediacy of poly(ADP-ribose) (Fig. 2). The question is whether this signal operates under physiological conditions. Some reports have provided direct evidence that NAD$^+$ depletion forced upon cells by incubation in nicotinamide-free medium, arrests poly(ADP-ribose) synthesis (5) and drastically alters chromatin functions (5-7). For example, DNA repair in response to a chemical carcinogen was arrested under these conditions, but could be reestablished with nicotinamide, which restored normal intracellular NAD$^+$ and poly(ADP-ribose) levels (5). Similarly, myoblast differentiation was reversibly inhibited either by nicotinamide starvation or by selective inhibition of poly(ADP-ribose) polymerase (7). These results illustrate that nutritional manipulation of

215

intracellular NAD⁺ pools in living cells may be signaled to chromatin and may cause an adjustment of chromatin functions to altered metabolic requirements (Fig. 2, C, D). Under physiological conditions, small changes in the posttranslational modification of proteins secondary to fluctuations in the availability of NAD⁺ for poly(ADP-ribose) biosynthesis can be expected to effect only minor, barely detectable adjustments in chromatin functions. These minor changes may serve to keep the phenotypic expression of chromatin functions in tune with the metabolic requirements of mammalian cells. This adds a new element to the suicide concept by Berger and associates (1, Fig. 2, A, B). Thus poly(ADP-ribose) becomes part of a regulatory cycle (Fig. 2).

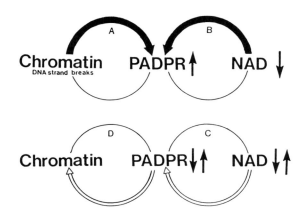

Fig. 2. Bidirectional signaling through poly(ADP-ribose). A and B, suicidal NAD depletion in the presence of excessive amounts of DNA strand breaks as proposed by Berger and associates; C and D, reverse signaling of NAD availability through poly(ADP-ribose) (this report, 4). For further discussion see text.

Numerous reports have emphasized the dominant role of DNA strand breaks in regulating poly(ADP-ribose) biosynthesis *in vivo,* and the observation that DNA strand breaks stimulate purified poly(ADP-ribose) polymerase *in vitro* provides convincing support of this view (1, 3, 6). However, in several examples of cytodifferentation, drastic changes in constitutive poly(ADP-ribosyl)ation activity could not be ascribed to DNA strand breakage, and the signal prompting these changes has not been identified (8-11). It is possible that these alterations in constitutive poly(ADP-ribosyl)ation activity reflect the varying availability of NAD⁺ in cytodifferentiation.

216

Acknowledgements. This work was supported by the Swiss National Foundation for Scientific Research (Grant 3.354.086), the Sandoz Foundation, and the Krebsilga des Kantons, Zürich.

References

1. Berger, N.A. (1985) Radiat Res 101: 4-15
2. Alvarez-Gonzalez, R., Eichenberger, R., Althaus, F.R. (1986) Biochem Biophys Res Commun 138: 1051-1057
3. Althaus, F.R., Richter, C. (1987) ADP-Ribosylation of Proteins: Enzymology and Biological Significance. Springer Berlin Heidelberg New York Tokyo, pp. 245
4. Loetscher, P., Alvarez-Gonzalez, R., Althaus, F.R. (1987) Proc Natl Acad Sci. USA 84: 1286-1289
5. Jacobson, E.L., Juarez, D., Sims, J.L. (1980) Fed Proc 39: 1739 (Abstract)
6. Shall, S. (1984) Adv Radiat Biol 11: 1-69
7. Farzaneh, F., Zalin, R., Brill, D., Shall, S. (1982) Nature 300: 362-366
8. Jackowski, G., Kun, E. (1981) J Biol Chem 256: 3667-3670
9. Althaus, F.R., Lawrence, S.D., He, Y., Sattler, G.L., Tsukada, Y., Pitot, H.C. (1982) Nature 300: 366-368
10. Sugimura, M., Fram, R., Munroe, D., Kufe, D. (1984) Dev Biol 104: 484-488
11. Brac, T., Ebisuzaki, K. (1985) ADP-Ribosylation of Proteins, F.R. Althaus, H. Hilz, S.Shall (eds.), Springer Berlin Heidelberg New York Tokyo, pp. 446-452
12. Alvarez-Gonzalez, R., Eichenberger, R., Loetscher, P., Althaus, F.R. (1986) Anal Biochem 156: 473-480

217

Uncoupling of DNA Excision Repair and Chromatin Rearrangements in Poly(ADP-ribose)-Depleted Cells

Georg Mathis[1] and Felix R. Althaus

Institute of Pharmacology and Biochemistry, University of Zürich (Tierspital), CH-8057 Zürich, Switzerland

Introduction

The repair of DNA damage in mammalian cells is accompanied by alterations in chromatin organization such that newly synthesized repair patches become transiently accessible to enzymatic and chemical probes (1, 2). We have previously demonstrated that during early stages of DNA excision repair, these alterations are reflected by periodic changes of the relative amount of free, 8-methoxypsoralen-accessible DNA domains in chromatin (Fig. 1) (2,3).

We have isolated these domains from hepatocytes in different stages of DNA excision repair (2, 3) and studied the distribution of newly synthesized repair patches as well as the position of bulky DNA adducts in these sites. The advantage of this approach as compared to micrococcal nuclease probing of chromatin is that 8-methoxypsoralen can be covalently bound to free DNA domains *in situ,* and subsequent fractionation of chromatin does not effect its position (2-7).

We have reported earlier that the coordinate development of periodic changes in chromatin organization in DNA excision repair is significantly impaired in poly(ADP-ribose)-depleted cells (Fig. 1) (3). This disturbance at the chromatin organizational level apparently represents the earliest poly(ADP-ribosyl)ation-dependent event in the course of DNA excision repair observed so far (8). Here we present evidence suggesting the rearrangement of newly synthesized repair patches into free DNA domains is a prerequisite for efficient excision of a bulky adduct and that this rearrangement process involves poly(ADP-ribosyl)ation of chromosomal proteins.

Results and Discussion

Primary cultures of rat hepatocytes were incubated for 20 min with the radioactively labeled ultimate carcinogen N-acetoxy-2-acetylaminofluorene (AAAF), the carcinogen was removed, and the cells were allowed to repair DNA damage for up to 3 hr. Within this repair period, excision of the bulky

---

[1]Present Address: Medical College of Virginia, Department of Pathology, P.O. Box 662, Richmond, VA 23298

AAAF deoxyguanosine adducts occurred almost exclusively in free DNA domains as is shown in Table 1. Concomitant with the excision of DNA damage, rearrangements of newly synthesized repair patches into free DNA could be observed (Fig. 2). Virtually complete depletion of the cells of chromatin-bound poly(ADP-ribose) was achieved by a 24 hr preincubation in the presence of benzamide which resulted in less than 10 femtomoles of ADP-ribosyl residues per million hepatocytes as determined by the method of Jacobson et al. (9, 10). This treatment caused an almost complete block of both adduct excision (Table 1) and rearrangements of repair patches (Fig. 2).

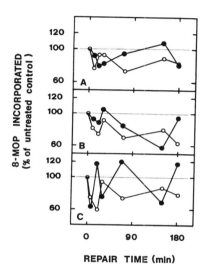

Fig. 1. Repair-associated fluctuations of the relative amount of free DNA in chromatin: effect of poly(ADP-ribose) depletion. Hepatocytes were cultured for 24 hr in the presence or absence of 8 mM 3-aminobenzamide [poly(ADP-ribose) depletion, see text], exposed to 5 (A), 50 (B), or 500 (C) μM AAAF for 20 min and allowed to continue repair in the presence of 3-aminobenzamide at 37°C after removal of the carcinogen. At the times indicated, the portion of free DNA in chromatin was determined using [8-methoxyl-^{3}H] methoxypsoralen: the amount of radioactivity covalently bound to DNA is directly proportional to the relative amount of free DNA in the chromatin of intact cells (2, 4). The results are expressed in percent of the values of untreated control cells. A, 5 μM AAAF; B, 50 μM AAAF; C, 500 μM AAAF. O---O, control; ●---●, poly(ADP-ribose) depletion (n = 2).

Table 1. Excision of AAAF-DNA adducts in total and free DNA of poly(ADP-ribose)-depleted and control hepatocytes.

| Repair Interval [a] | % of Initial Adducts Excised | | | |
| | Control | | Poly(ADP-ribose) Depletion[b] | |
	Total DNA	Free DNA	Total DNA	Free DNA
20 min Pulse	0	0	0	0
100 min Chase	39.3 ± 9.2	39.2 ± 4.2	5.5 ± 2.5	9.9 ± 0.5
160 min Chase	53.0 ± 4.0	52.3 ± 4.1	11.5 ± 2.4	7.6 ± 4.5

[a]Primary cultures of hepatocytes were incubated for 20 min in the presence of 50 μM [G-^{3}H]AAAF (pulse), washed extensively and allowed to continue repair for the intervals indicated (chase). The amount of DNA-bound AAAF radioactivity was determined in total and 8-methoxypsoralen-cross-linked, free DNA (2; n = 3, mean ± SEM, p < 0.05).

[b]Cells were preincubated for 24 hr with 8 mM benzamide.

219

These observations indicate a tight coupling of the excision step with the establishment of free DNA domains by a mechanism involving ADP-ribosylation of chromatin proteins. Interestingly, the overall repair synthesis during the 20 min pulse labeling period (Fig. 2) was not significantly affected by the poly(ADP-ribose) depletion: thymidine incorporation per mol AAAF-DNA adduct was 511 ± 85 mmol in control and 408 ± 95 mmol in poly(ADP-ribose)-depleted hepatocytes (n = 3). The consequence of the inhibited excision was a gradual increase of AAAF adducts in free DNA domains of poly(ADP-ribose)-depleted cells (Table 2).

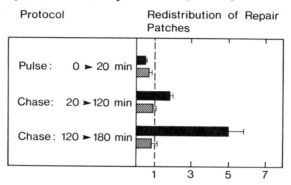

Fig. 2. Redistribution of repair patches relative to free DNA domains in poly(ADP-ribose)-depleted and control hepatocytes. After a 24 hr adaptation period in the presence or absence of 8 mM benzamide [poly(ADP-ribose) depletion, see text], primary cultures of rat hepatocytes were exposed to 50 μM AAAF and newly synthesized repair patches were simultaneously pulse-labeled with [methyl-³H] thymidine for 20 min. Incorporation of radioactivity was stopped by extensive washes with medium containing unlabeled thymidine, followed by a chase in the continuous presence or absence of benzamide. Then free DNA domains were isolated and analyzed for their content of radioactively labeled repair patches (2). The values (Redistribution of Repair Patches) reflect the relative accumulation of repair patches in free DNA (value = 1: no accumulation; value > 1: accumulation; 2). Black bars, control; shaded bars, poly(ADP-ribose) depletion. n = 4, mean ± SEM, p < 0.01.

Table 2. Relative distribution of unexcised AAAF-DNA adducts in chromatin of poly(ADP-ribose)-depleted and control hepatocytes.

Repair Interval[a]	$Ratio = \dfrac{AAAF\ Adducts\ (free\ DNA)}{AAAF\ Adducts\ (total\ DNA)}$	
	Control	Poly(ADP-Ribose) Depletion[b]
20 min Pulse	1.09 ± 0.06	1.19 ± 0.13
100 min Chase	1.03 ± 0.16	1.73 ± 0.13
160 min Chase	1.28 ± 0.15	1.92 ± 0.05

a; b; see. footnotes of Table 1.

Our experiments identify a tight functional linkage between structural chromatin rearrangements and the excision step, and dissect the process of repair patch synthesis from the subsequent patch repositioning relative to nucleosome organized chromatin regions. In addition, our results suggest that chromatin-associated poly(ADP-ribosyl)ation acts to coordinate DNA excision repair with structural rearrangements of chromatin (11).

Acknowledgements. This work was supported by Grant 3.354.086 from the Swiss National Foundation for Scientific Research, the Sandoz Stiftung, and the Krebsliga des Kantons, Zürich.

References

1. Friedberg, E.C. (1985) DNA Repair, W.H. Freeman & Co., New York
2. Mathis, G., Althaus, F.R. (1986) J Biol Chem 261: 5758-5765
3. Althaus, F.R., Mathis, G. (1985) ADP-Ribosylation of Proteins. F.R. Althaus, H. Hilz, S.Shall, (eds) Springer, Berlin, Heidelberg, New York, pp. 235-241
4. Mathis, G., Althaus, F.R., Ibid., pp. 230-234
5. Hearst, J.E. (1981) J Invest Dermatol 77: 39-44
6. Ben Hur, E., Prager, A., Riklis, E. (1979) Photochem Photobiol 29: 921-924
7. Reeves, R. (1984) Biochim Biophys Acta 782: 343-393
8. Althaus, F.R., Richter, C. (1987) ADP-Ribosylation of Proteins: Enzymology and Biological Significance. Springer, Berlin, Heidelberg, New York.
9. Jacobson, M.K., Payne, D.M., Alvarez-Gonzalez, R., Juarez-Salinas, H., Sims, J.L., Jacobson, E.L. (1984) Methods in Enzymology 106: 483-494
10. Alvarez-Gonzalez, R., Eichenberger, R., Althaus, F.R. (1986) Anal Biochem 156: 473-480
11. Althaus, F.R., Mathis, G., Alvarez-Gonzalez, R., Loetscher, P., Mattenberger, M. (1987) This volume

Highly Sensitive Method for Immunohistochemical Detection of Poly(ADP-Ribose) in Various Animal Species

Kazuhiko Nakagawa, Prince Masahito, Takatoshi Ishikawa, Haruo Sugano, Hisae Kawamitsu[1], Masanao Miwa[1], and Takashi Sugimura[1]

Cancer Institute (Japanese Foundation for Cancer Research), 1-37-1 Kami-Ikebukuro, Toshima-ku, Tokyo 170, Japan

Introduction

Poly(ADP-ribose) has been expected to play some regulatory roles in nuclear functions. However, its amount and distribution *in vivo* have not yet been fully elucidated. An immunohistochemical method should be useful for the study of the natural occurrence of poly(ADP-ribose) at the cellular level. We have developed a highly sensitive immunohistochemical method for detecting poly(ADP-ribose) in conventional histological sections (1, 2).

Results and Discussion

In this method, samples of freshly excised tissues were fixed by boiling them in a 10% neutral formaldehyde solution for 5 min. The fixed tissues were then embedded in paraffin and cut into 3 μm thick sections. The sections on glass slides were deparaffinized and dehydrated by passage through xylene and a graded alcohol series. Then they were incubated with a primary antibody against poly(ADP-ribose). The antibody used, IgG monoclonal antibody 10H, recognized the linear structure of poly(ADP-ribose) described previously (3). Bound antibody was stained with the avidin biotin peroxidase complex or by an indirect immunofluorescence technique.

By this method the nuclei of the boiled tissues were strongly stained. No staining was observed in tissues fixed in formaldehyde solution at room temperature. The reason why prompt boiling of the tissue resulted in intense staining is unknown. Possibly this treatment inactivates an enzyme(s) involved in poly(ADP-ribose) degradation or causes a structural change in the chromatin rendering the poly(ADP-ribose) accessible to antibody. Some positive staining was also detected in tissues boiled in other fixatives. However, a 10% neutral formaldehyde solution gave the stronger staining

---

[1]National Cancer Center Research Institute, 1-1, Tsukiji 5-chome, Chuo-ku, Tokyo 104, Japan

with the lowest nonspecific staining, and resulted in the best preservation of the integrity of the tissues.

Several examinations were carried out to confirm the specificity of the immunohistochemical staining. Pretreatment of the sections with phosphodiesterase markedly decreased the staining in the nuclei, while digestion of the sections with DNase or RNase had no effect on the immunostaining. Incubation with antibody that had been absorbed with poly(ADP-ribose) did not result in any positive staining reaction, while incubation with antibody that had been preincubated with related compounds did not affect the staining. These results indicated that this histological staining method was specific for poly(ADP-ribose).

Poly(ADP-ribose) polymerase or glycohydrolase activities have been found in various animals. The present method was used to investigate the phylogenetic distribution of poly(ADP-ribose) in vertebrates [(mammalia, aves, reptilia, amphibia, osteichthyes (12 species)] and invertebrates (9 phyla; 24 species) (2). Poly(ADP-ribose) was detected in almost all organs of all 36 animal species. But the staining intensity varied considerably in different organs and in different cell types within the organs. In general, the whole nuclear area stained homogeneously, whereas the cytoplasm showed only background staining. However, as seen in spermatocytes in rat testis and regenerating rat liver cells, the staining appeared in clusters in the nuclei, indicating positive poly(ADP-ribose) staining on chromatin condensation areas. The intensity of poly(ADP-ribose) staining also differed at different stages of differentiation of the cells (e.g., during cytodifferentiation during spermatogenesis).

In previous attempts to demonstrate poly(ADP-ribose) histocytologically by an autoradiographic method with [^{3}H]NAD or immunohistochemically (4, 5), the grains observed in autoradiographs did not seem unequivocally to be poly(ADP-ribose). Moreover, in most immunohistochemical methods used so far, frozen sections or cultured cells fixed in ethanol or acetone were preincubated with NAD to intensify the staining. But the physiological distribution and fluctuation of poly(ADP-ribose) cannot be determined accurately if the fixed sections or cells still contain poly(ADP-ribose) synthesizing or decomposing enzyme activity. Our immunohistochemical method in which these enzyme activities were probably promptly inhibited by boiling the tissue permits study of the actual amount and location of poly(ADP-ribose) *in vivo*. Furthermore our method for detecting poly(ADP-ribose) in conventional histological sections is simple and seems to be highly sensitive. The present findings suggest that further studies on the cellular levels of poly(ADP-ribose) in various systems will be useful for more thorough understanding of the physiological role of poly(ADP-ribose) in the cell.

References

1. Ishikawa, T., Nakagawa, K., Kawamitsu, H., Miwa, M., Sugimura, T. (1986) Proc Jap Cancer Assoc 45th Annu Meeting; 42
2. Prince Masahito, Ishikawa, H., Nakagawa, K., Sugano, H., Kawamitsu, H., Miwa, M., Sugimura, T. (1986) Proc Jap Cancer Assoc 45th Annu Meeting; 68
3. Kawamitsu, H., Hoshino, H., Okada, H., Miwa, M., Momoi, H., Sugimura, T. (1984) Biochemistry 23: 3771-3777
4. Ikai, K., Ueda, K., Hayaishi, O. (1980) J Histochem Cytochem 28: 670-676
5. Kanai, Y., Tanuma, S., Sugimura, T. (1981) Proc Natl Acad Sci USA 78: 2801-2804

Poly(ADP-Ribosyl)ation of Nuclear Proteins by Oxidant Tumor Promoters

P. Cerutti, G. Krupitza, and D. Mühlematter

Department of Carcinogenesis, Swiss Institute for Experimental Cancer Research, 1066 Epalinges, Switzerland

Oxidant tumor promoters. In the promotion phase of carcinogenesis, initiated cells which have undergone a genetic change as a consequence of the exposure to a carcinogen are allowed to expand preferentially at the cost of the surrounding tissue. The pathways which can lead to this result are expected to vary for different types of promoters and tissues. A major mechanism involves the differential modulation of the expression of growth- and differentiation- related genes in initiated and non-initiated epithelial cells and surrounding mesenchyme. Several scenarios are possible. For example in mouse skin promotion by the phorbolester promoter, TPA, a combination of general growth stimulation and preferential terminal differentiation of non-initiated cells may allow papilloma formation. Selective toxicity of a promoter to the tissue which surrounds initiated cells represents another important mechanism which facilitates the formation of a tumor nodule. Among the several classes of physiological and xenobiotic agents with promotional activity, oxidants are of particular interest (1-3). Because they are ubiquitous they may represent "natural" promoters. Traces of active oxygen (AO) are present in the air, while large amounts are released by stimulated inflammatory leukocytes and there are multiple metabolic reactions which produce AO intracellularly (1). Indeed in many instances tissue inflammation may be a necessary step in carcinogenesis (4-6). Numerous xenbiotic oxidants have been shown to possess promotional activity on initiated mouse skin and in tissue culture systems. Among them are H_2O_2 (7), active oxygen released by xanthine/xanthine oxidase (8), ozone, peroxyacetic acid, chlorobenzoic acid, benzoylperoxide, decanoylperoxide, cumene-hydroperoxide, p-nitro-perbenzoic acid and periodate (7,9). Oxidants cause DNA strand breakage and it was expected that they also induce the poly(ADP-ribosyl)ation of nuclear proteins. This was first demonstrated for TPA-stimulated human monocytes (10) and benzoylperoxide treated mouse 3T3 cells (11) and in the present report for extracellularly generated AO in mouse epidermal cells JB6. It should be noted that promotable clone 41 of JB6 cells, but not the promotion resistant clone 30, is stimulated by AO to grow in soft agar and monolayer cultures (R. Larsson, I. Zbinden and P. Cerutti, unpublished and reference 9).

The functional significance of constitutive and induced poly(ADP-ribosyl)ation still remains unclear. Poly(ADP-ribosyl)ation of structural chromosomal proteins such as the histones is expected to affect chromatin conformation and there is support for this notion from *in vitro* experiments. Poly(ADP-ribosyl)ation may also change the activity and specificity of functional chromosomal proteins involved in DNA and RNA metabolism. Examples are the participation of poly(ADP-ribose) in repair, recombinational events (12) and gene amplification (13). Poly(ADP-ribose) could also participate in the regulation of gene expression similar to post-translational phosphorylation reactions. It has recently been recognized that certain carcinogens such as UV and mitomycin C, besides causing DNA damage, modulate the expression of the same families of genes which are induced by serum factors, certain polypeptide growth factors and phorbolester-type promoters. For the understanding of the promotional action of AO our recent observation that it very potently induces *c-fos* and *c-myc* in JB6 cells is particularly intriguing (D. Crawford, I. Zbinden and P. Cerutti, unpublished). The post-translational modification of transcription factors and proteins which affect mRNA stability could play a role. Upstream enhancer sequences (serum responsive element, SRE) and DNA binding proteins have been identified which regulate the expression of "immediate-early" genes (14) which include actin and the proto-oncogene *c-fos* (15-17). Since *de novo* protein synthesis is not required for the induction of "immediate-early" genes (18) the post-translational modification of regulatory proteins is a distinct possibility. A second family of genes is induced later and involves the binding of transcription factor AP1 to an oligonucleotide consensus sequence (19). Posttranscriptional mechanisms further regulate the expression of *c-fos* and *c-myc*. As yet unidentified protein factors destabilize the mRNAs of these genes. The activity of these factors could be regulated by carcinogen-induced post-translational modification (20). Support for an involvement of poly(ADP-ribose) in regulation of gene expression derives from the observation that poly(ADP-ribose) transferase serves as a factor for the accurate transcription of ribosomal-DNA genes (21) and that topoisomerase I (22) and II (23, 24) are poly(ADP-ribose) acceptors. It will be an exciting challenge to explore the possible role of poly(ADP-ribose) in carcinogen-induced processes of genetic regulation.

We report here on the mechanism of DNA strand breakage and poly(ADP-ribosyl)ation by AO in comparison to the methylating agent N-methyl-N'-nitro-N-nitrosoguanidine (MNNG) in promotable JB6 clone 41 cells. We also summarize first results on the identity of the nuclear proteins which serve as poly(ADP-ribose) acceptors in response to AO. We chose xanthine/xanthine oxidase (X/XO) as a source of extracellular AO because it produces a mixture of superoxide and H_2O_2 which resembles the oxidants released by macrophages and mouse epidermal cells JB6 because promotable (clone 41) and non-promotable (clone 30) variants are available (9).

Active oxygen induces DNA breakage in a reaction which requires Ca^{2+} and Fe. Monolayer cultures of JB6 (clone 41) were exposed to X/XO (15 µg X plus 1.5 µg XO/ml) for 30 min at 37° in Ca^{2+}-free medium 199 supplemented with 8% fetal calf serum. Under these conditions a maximal O_2^- concentration of 2 to 3 x 10^{-5} M is reached after 15-20 min. Extracellularly generated AO efficiently induced DNA breakage as determined by the alkaline elution method (25). In previous work we had found that the intracellular Ca^{2+}-chelator Quin 2 partially suppressed DNA breakage by arachidonic acid hydroperoxides (26). Quin 2 (2.5 µM) completely inhibited AO induced strand breakage indicating a stringent requirement for intracellular Ca^{2+}. The need for Ca^{2+} suggests that AO triggers a complex chain of metabolic reactions and it appears unlikely that DNA breakage results from the direct attack by "primary" radicals contained in AO or hydroxylradicals formed in Fenton reactions. However, Fenton-type reactions do play a role at some step of the pathway leading to strand breakage. This is indicated by our observation that the diffusible Fe-chelator o-phenanthroline (100 µM) (27) in Fe-free medium 199 completely suppressed strand breakage. The observation that Ca^{2+}- and Fe-chelators independently completely suppressed AO-induced DNA breakage may indicate that these ions are required for consecutive steps. The following sequence of events could result in strand breakage: Fe-dependent Fenton-reactions damage the plasma membrane, activate phospholipase A_2, stimulate the arachidonic acid cascade and consequently the formation of arachidonic acid hydroperoxides. These and other lipid-hydroperoxides might induce DNA damage in Fe-catalyzed radical reactions (28). Alternatively, AO and secondary lipidperoxides may induce changes in the distribution and concentration of intracellular Ca^{2+} (3, 29, 30). This could conceivably result in DNA fragmentation because of the activation of a Ca^{2+}, Mg^{2+}-dependent endonuclease. The Ca^{2+}- and Fe-dependent mechanism of DNA strand breakage is unique for oxidants, neither Quin 2 or o-phenanthroline affected DNA breakage by MNNG (D. Mühlematter and P. Cerutti, unpublished).

Exposure to active oxygen results in a transient increase in poly(ADP-ribose) concentrations. We measured the AO induced increase in poly(ADP-ribose) in monolayer cultures of JB6 clone 41 by the fluorimetric method of Jacobson *et al.* (31) which avoids cell permeabilization. At a concentration of 50 µg X/ml plus 5 µg XO/ml, which generates within 30 min a concentration of 5 x 10^{-5} M O_2, the level of poly(ADP-ribose) reached 70 pmoles εRAdo per 10^8 cells. After this maximum, poly(ADP-ribose) concentrations slowly decreased and had not returned to control levels after 2 hrs. Because of their inhibitory action on AO induced DNA breakage we studied the effects of the Ca^{2+}-chelator Quin 2 and the Fe-chelator o-phenanthroline on poly(ADP-ribose). Both drugs completely prevented the increase in poly(ADP-ribose) by AO at relatively low doses. In contrast, experiments carried out in parallel indicated that

Quin 2 had only a minor effect on poly(ADP-ribose) in response to MNNG (20 μg/ml). The data is summarized in Table 1. Therefore, it appears very likely that the primary requirement for Ca^{2+} and Fe is for DNA strand breakage by AO. We conclude that the suppression of poly(ADP-ribose) by Ca^{2+} and Fe-complexers is a consequence of the suppression of DNA strand breakage (D. Mühlematter and P. Cerutti, unpublished).

Table 1. Suppression of poly(ADP-ribosyl)ation in response to active oxygen by Ca^{2+}-complexer Quin 2 (JB6 CI 41)

	pmol ε RADO/10^8 cells
Control	8.7 ± 0.1
AO (50/5 μg/ml X/XO)	70.1 ± 7.0
AO (50/5 μg/ml X/XO) + 5 μM Quin 2	22.9 ± 4.6
MNNG (20 μg/ml)	77.7 ± 2.3
MNNG (20 μg/ml) + 5 μM Quin 2	66.9 ± 5.8

Histone- and non-histone poly(ADP-ribose)-acceptors following treatment with active oxygen or MNNG. For the elucidation of the functional role of poly(ADP-ribose) it is important that the nuclear acceptor proteins are identified under physiological conditions in the intact cell. Progress in this regard has been slow because it has not been possible, in general, to label the cellular NAD pools with high specific radioactivity. While *in vitro* studies with permeabilized cells or isolated nuclei are useful there is no guarantee that they accurately reflect the *in vivo* situation. In the present study we have combined boronate affinity chromatography with the immunoblot technique for the identification of the major poly(ADP-ribose)-acceptors in AO and MNNG-treated JB6 clone 41 cells. Histones were extracted from nuclear preparations with 0.25 M H_2SO_4. The pellet remaining after the treatment of nuclei with 5% PCA was analyzed for its content of non-histone proteins. Poly(ADP-ribosyl)ated proteins contained in either fraction were purified on aminophenyl-boronate columns (32, 33). The histone fractions were electrophoresed on 15% polyacrylamide gels according to Zweidler (34) while the non-histone proteins were separated on 7% SDS-polyacrylamide stacking gels (35) in a pH 6 phosphate-tris buffer (36). Histone immunoblots were carried out with polyclonal rabbit antibodies against the calf thymus histones H2A, H2B, and H3 (37) and reacted with ^{125}I-labeled donkey anti-rabbit IgG. Fig. 1 shows an autoradiogram for H3 from clone 41 cells which had been treated with 5 μg/ml MNNG for 20 min or with AO released by 50 μg X/ml plus 5 μg XO/ml for 30 min, respectively. Slightly retarded bands relative to the histone standards are visible in the untreated control and the treated samples. These bands are tentatively assigned to poly(ADP-ribosyl)ated H3.1 and H3.2/H3.3. The bands are darker in the AO and MNNG treated samples than in controls indicating increases in poly(ADP-ribosyl)ated H3. Densitometer tracings from this (see Fig. 1) and analogous autoradiograms of immunoblots for H2A and H2B allow estimates of the amounts of poly(ADP-ribosyl)ated proteins which are contained in a particular band relative to untreated controls. The data is listed in Table 2. It is evident that H3 represents the major poly(ADP-ribose) acceptor in response to AO and MNNG in JB6 clone 41 cells. Aminophenyl boronate chromatography retains all ADP-ribosylated proteins regardless of poly(ADP-ribose) chain length and number of substituted sites. Therefore, our data give no information about

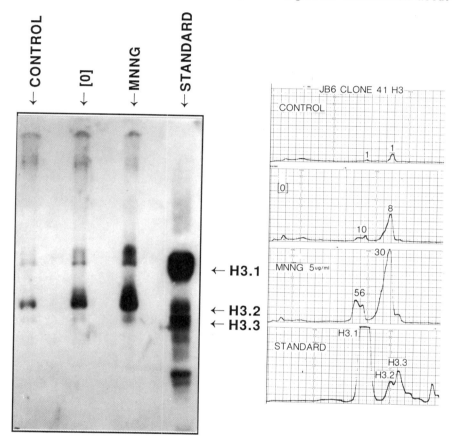

Fig. 1 Immunoblots of acid extracts of nuclear preparations from mouse epidermal cells JB6 (clone 41) which had been treated with active oxygen produced by xanthine/xanthine oxidase (50 μg/ml xanthine plus 5 μg/ml xanthine oxidase) for 30 min or 5 μg/ml MNNG for 20 min. The extracts were purified on a boronate affinity column and the proteins separated on 15% polyacrylamide gels (34). A polyclonal rabbit antibody against histone H3 (37) was used for the immunoblot and the blot was reacted with [^{125}I]-labeled donkey anti-rabbit IgG. Densitometer scannings are shown at the right side of the autoradiogram and were obtained with a Zeineh "soft laser" scanning densitometer.

the extent of poly(ADP-ribosyl)ation of a specific histone. Unfortunately,the monoclonal antibody against poly(ADP-ribose) available to us (39) gave high backgrounds on the immunoblots (in contrast to the blots for non-histone proteins discussed below) and could not be used to approach this question.

We had shown previously that the phorbolester promoter TPA increased poly(ADP-ribosyl)ation of the same histones in mouse embryo fibroblasts C3H1OT1/2 but in this case H2B was the major acceptor (38). H2B was also the major poly(ADP-ribose) acceptor in Yoshida hepatoma cells following exposure to the methylating agent dimethylsulfate (32). However, it should be recognized that no immunoblots were used for the identification of the individual histones in the previous work.

Table 2. Enhancement factors for poly(ADP-ribosyl)ation of histones in response to active oxygen and MNNG (JB6 CI 41)

Histone	Control	AO[1]	MNNG[1]
H2A	1	3.7	18.3
H2B.1 + H2B.2	1	2.1	21.2
H3.1	1	10.5	56.3
H3.2 + H3.3	1	8.2	30.2

[1]From densitometer readings of autoradiograms of immunoblots of acid nuclear extracts from cells which were treated with 50 μg xanthine/ml plus 5 μg/ml xanthine oxidase for 30 min (AO) or 5 μg/ml MNNG for 20 min.

The gels of the non-histone protein fractions were analyzed by the immunoblot technique using monoclonal antibodies against poly(ADP-ribose) (39) and polyclonal rabbit antibodies against calf thymus poly(ADP-ribose) transferase (40) and topoisomerase I. Fig. 2 shows representative autoradiograms for AO treated cells and untreated controls and Table 3 lists enhancement factors relative to untreated controls from densitometer scannings. From the immunoblot with poly(ADP-ribose) antibody on the right side panel of Fig 2 it is evident that AO strongly induced the poly(ADP-ribosyl)ation of a multitude of non-histone proteins. Most proteins which serve as acceptors are already present in small amounts in the untreated control. The strongest bands are found at 28, 32, 50 and 120 kDa. The immunoblots with topoisomerase I antibody in the middle panel show 3 major bands at approximately 70, 76-78 and 100 kDa. According to densitometer tracings AO treatment increased the intensities of the latter two bands 7 and 8 fold (see Table 3). The molecular weights of these major bands agree with published data for topoisomerase I variants and fragments from Hela cells (41) and calf thymus (42). The left side panel shows the immunoblots with poly(ADP-ribose) transferase antibody. Major bands are found at 32, 50, 70, 88 and 120 kDa. While the 120 kDa band corresponds to the intact protein (43), proteolytic fragments of the transferase from 32 to 96 kDa have been observed previously in preparations from several animal species (44-46). AO treatment induced a moderate increase in the intensities of all bands (1.4 - 5 fold, see Table 3). The fact that the 32 kDa band increased only 2.5 fold on the poly(ADP-ribose) transferase blot but 37.5 fold on the poly(ADP-ribose) blot in response to AO suggests that the poly(ADP-ribose) chains are of considerable length and/or that the transferase is substituted at multiple sites. It is conceivable that highly modified transferase molecules are preferentially degraded by proteolytic enzymes. Preliminary results suggest that besides being phosphorylated the proto-oncogene protein c-*fos* may serve as poly(ADP-ribose) acceptor. Evidence for the extensive posttranslational modification of *c-fos* has been reported (47).

Table 3. Enhancement factor for poly(ADP-ribosyl)ation of non-histone proteins in response to active oxygen (JB6 CI 41)

kDa	Poly(ADP-ribose)	Topo I	poly(ADP-ribose) transferase
120	3.5		5
100	induced	7	
88	0.6		3
76/78		8	
70	induced	1	1.4
50	10.5		2.7
32	37.6		2.5
28	10.3		
22	5.2		

Ratio of densitometer readings of autoradiograms of immunoblots of non-histone proteins from control cultures and cultures treated with 50 μg xanthine/ml plus 5 μg/ml xanthine oxidase for 30 min.

Concluding Remarks. A role for poly(ADP-ribosyl)ation in carcinogenesis is suggested by the observation that certain inhibitors of poly(ADP-ribose) transferase are anticarcinogenic in model systems (48-50). On the molecular level it has been demonstrated that poly(ADP-ribose) participates in important reactions of DNA metabolism including DNA repair, transcription initiation, integration into the genome of transfected DNA and gene amplification. On many occasions it has been speculated that the disturbance of one or several of these crucial genetic functions may play a role in carcinogenesis. Therefore, it is reasonable to postulate that, in response to carcinogen exposure, poly(ADP-ribose) is involved at some stage of malignant transformation. In relationship to tumor promotion previous work has shown that the oxidant benzoylperoxide (11) and the phorbolester promoter TPA (10, 46, 51) moderately increased poly(ADP-ribose) levels in cultured mouse fibroblasts. Our present work demonstrates that extracellular AO of the type produced by inflammatory leukocytes promotes mouse epidermal cells JB6 clone 41 in culture and efficiently increases the poly(ADP-ribose) substitution of histones and non-histone proteins. While

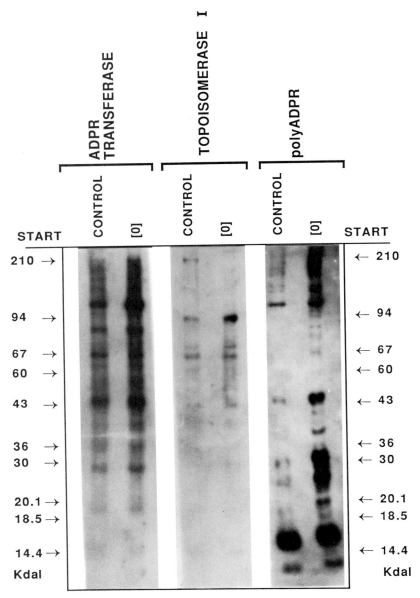

Fig. 2. Immunoblots of acid insoluble pellets obtained from nuclear preparations of mouse epidermal cells JB6 (clone 41) which had been treated with active oxygen as described in the legend to Fig 1. After purification on a boronate affinity column the proteins were separated on 7% SDS-polyacrylamide stacking gels (35) in a pH 6 phosphate-Tris buffer (36). The left side panel shows autoradiograms of immunoblots with polyclonal rabbit antibody against calf thymus poly(ADP-ribose) transferase (40), the middle panel immunoblots with a polyclonal rabbit antibody against calf thymus topoisomerase I and the right side panel immunoblots with monoclonal antibody against poly(ADP-ribose) (39). Appropriate [125I]-anti IgG antibodies were used for radioactive labeling.

232

we identified poly(ADP-ribose) transferase and topoisomerase I as major non-histone acceptors many additional bands on the poly(ADP-ribose) immunoblots remain unidentified and many poly(ADP-ribosyl)ated proteins are probably below the level of detection by our technology. As is the case for protein phosphorylation reactions in carcinogenesis, which receive much attention at present, progress will depend on the biochemical characterization of additional poly(ADP-ribose) acceptor proteins and the definition of their function in normal and carcinogen disturbed cells. Of particular interest will be the elucidation of the role of poly(ADP-ribose) in the posttranslational modification of transcription factors, regulatory proteins affecting mRNA stability and of proteins involved in recombinational events and gene amplification.

Acknowledgements. This work was supported by the Swiss National Science Foundation, the Swiss Association of Cigarette Manufacturers and the Association for International Cancer Research. We are grateful to several researchers for gifts of antibodies: Dr. M. Bustin for the antibodies against histones, Dr. K. Ueda for the antibody against poly(ADP-ribose) transferase, Drs. M. Darby and H-P. Vosberg for antibody against topoisomerase I and Drs. H. Kawamitsu and T. Sugimura for the antibody against poly(ADP-ribose). Without their generosity this work would not have been possible.

References

1. Cerutti, P. (1985) Science 227: 375-381
2. Kozumbo, W., Trush, M., Kensler, T. (1985) Chem Biol Interactions 54: 199-207
3. Cerutti, P. (1987) Growth Factors, Tumor Promoters and Cancer Genes N. Colburn, H. Moses and E. Stanbridge (eds) , Alan R. Liss, N.Y.
4. Slaga, T., Fisher, S., Viaje, A., Berry, D., Bracken, W., LeClerc, S., Miller, D. (1978) Carcinogenesis 2: Mechanisms of Tumor Promotion and Co-carcinogenesis T. Slaga, A. Sivak and R. Boutwell (eds) Raven Press, N.Y. p. 173
5. See Carcinogenesis: A Comprehensive Survey Vol. 8: (1985) M. Mass, D. Kaufman, J. Siegfried, V. Steele, S. Nesnow (eds), Raven Press, N.Y.
6. DiGiovanni, J., Chenicek, K., Ewing, M. (1986) J Cell Biochem Suppl 10C, 163, Abstr L189
7. Slaga, T., Solanki, V., Logani, M. (1983) Radioprotectors and Anticarcinogens Nygaard and M. Simic, (eds) Academic Press, N.Y. pp 471-485
8. Zimmerman, R., Cerutti, R. (1984) Proc Natl Acad Sci USA 81: 2085-2087
9. Gindhart, T., Nakamura, Y., Stevens, L., Hegameyer, G., West, M., Smith, B., Colburn, N. (1985) Carcinogenesis: A Comprehensive Survey Vol. 8: M. Mass, D. Kaufman, J. Siegfried, V. Steele, S. (eds), Nesnow, Raven Press, N.Y.
10. Singh, N., Poirier, G., Cerutti, P. (1985) Biochem Biophys Res Commun 126: 1208-1214
11. Reported by H. Hilz at the 7th International Symposium on ADP-Ribosyaltion Reactions Vitznau, Switzerland (1984)
12. Shall, S. Personal communciation
13. Bürkle, A., This volume
14. Lan, L., Nathans, D. (1987) Proc Natl Acad Sci USA 84: 1182-1186
15. Prywes, R., Roeder, R. (1986) Cell 47: 778-784
16. Greenberg, M., Siegfried, Z., Ziff, E. (1987) Mol Cell Biol 7: 1217-1225
17. Treisman, R. (1986) Cell 46: 567-574
18. Greenberg, M., Hermanowski, A., Ziff, E. (1986) Mol Cell Biol 6: 1050-1057
19. Angel, P., Imagawa, M., Chin, R., Stein, B., Imbra, J. and Rahmsdorf, H., Jonat, C., Herrlich, P., Karin, M. (1987) Cell 49: 729-739
20. Suzuki, H., Quesda, P., Farina, B., Leone, E. (1986) J Biol Chem 261: 6048-6055

21. Slattery, E., Dignam, J., Matsui, T. Roeder, R., (1983) J Biol Chem 288: 5955-5959
22. Jongstra-Bilen, J., Ittel, M-E., Niedergang, C., Vosberg, H-P-, Mandel, P. (1983) Eur J Biochem 136: 391-396
23. Gasser, S., Laroche, T., Falquet, J. de la Tour E., Laemmli, U. (1986) J Mol Biol 188: 613-629
24. Darby, M., Schmitt, B., Jongstra-Bilen, J., Vosberg, H.P. (1985) EMBO J 4: 2129-2134
25. Kohn, K. Erickson, L., Ewig, R., Friedman, C. (1976) Biochemistry 15: 4629-4637
26. Ochi, T., Cerutti, P. (1987) Proc Natl Acad Sci USA 84: 990-994
27. Mello Filho, A., Hoffman, M., Meneghini, R. (1984) Biochem J 218: 273-275
28. Kozumbo, W., Mühlematter, D., Jorg, A., Emerit, I., Cerutti, P. (1987) Carcinogenesis 8: 521-526
29. Richter, C., Frei, B., Cerutti, P. (1987) Biochem Biophys Res Comm 146: 253-257
30. Bellomo, G., Jewel, S., Orrenius, S. (1982) J Biol Chem 257:11558-11562
31. Jacobson, M., Payne, D., Juarez-Salinas, H., Sims, J., Jacobson, E. (1984) Methods in Enzymology Vol. 106 F. Wold and K. Moldave (eds), Academic Press, N.Y. pp 483-494
32. Adamietz, P., Rudolph A. (1984) J Biol Chem 259: 6841-6846
33. Kreimeyer, A., Wielckens, K., Adametz, P., Hilz, H. (1984) J Biol Chem 259: 890-896
34. Zweidler, A. (1978) Methods in Cell Biology Vol. 17 G. Stein, J. Stein, L. Kleinsmith, (eds), Academic Press, N.Y. pp 223-232
35. Laemmli, U. (1970) Nature 270: 680-685
36. Holtland, J., Jemtland, R., Kristensen, J. (1983) Eur J Biochem 130: 309-314
37. Bustin, M., (1978) The Cell Nuclues Vol. IV H. Busch, (ed), Academic Press, N.Y. pp 196-238
38. Singh, N., Cerutti, P. (1985) Biochem Biophys Res Comm 132: 811-819
39. Kawamitsu, H., Hoshino, H., Okada, H., Miwa, M., Momoi, H., Sugimura, T. (1984) Biochemistry 23: 3771-3777
40. Ikai, K., Ueda, K. (1983) J Histochem Cytochem 31: 1261-1264
41. Liu, L., Miller, K. (1981) Proc Natl Acad Sci USA 78: 3481-3491
42. Schmitt, B., Buhre, U., Vosberg, H-P. (1984) Eur J Biochem 144: 127-134
43. Shizuta, Y., Kameshita, I., Agemori, M., Ushiro, H., Taniguchi, T., Otsuki, M., Sekimizu, K., Natori, S. (1985) ADP-Ribosylation of Proteins, F. Althaus, H. Hilz, S. Shall (eds), Springer-Verlag, Berlin, pp 52-59
44. Zahradka, P., Ebisuzaki, K. (1984) Eur J Biochem 142: 503-509
45. Berger, N., Surowy, C. and Petzold, S. (1985) ADP-Ribosylation of Proteins F. Althaus, H. Hilz, S. Shall (eds), Springer-Verlag, Berlin, pp 129-138
46. Singh, N., Leduc, Y., Poirier, G., Cerutti, P. (1985) Carcinogenesis 6: 1489-1494
47. Curran, T., Miller, A., Zokas, L., Verma, I. (1984) Cell 36: 259-268
48. Kun, E., Kirsten, E., Milo, G., Kurian, P., Kumari, H. (1983) Proc Natl Acad Sci USA 80: 7219-7223
49. Borek, C., Morgan, W., Ong, A., Cleaver, J. (1984) Proc Natl Acad Sci USA 81: 243-249
50. Tseng, A., Lee, W., Kirsten, E., Hakam, A., McLick, J. Buki, K., Kun, E. (1987) Proc Natl Acad Sci USA 84: 1107-1111
51. Singh, N., Poirier, G., Cerutti, P. (1985) EMBO J 4: 1491-1494

The Cytogenetic Effects of an Inhibition of Poly(ADP-Ribose) Polymerase

William F. Morgan, James E. Cleaver, Jeff D. Shadley, John K. Wiencke, and Sheldon Wolff

Laboratory of Radiobiology and Environmental Health, University of California, San Francisco, San Francisco, California 94143 USA

Introduction

Poly(ADP-ribose) polymerase is a ubiquitous enzyme found in high concentrations in the nuclei of most eukaryotic cells. DNA is an essential cofactor for enzyme activity. When cells are exposed to agents that result in DNA damage that can be expressed as direct or enzyme-induced DNA breaks, poly(ADP-ribose) is rapidly synthesized from nicotinamide adenine dinucleotide (NAD). Polymer synthesis and its subsequent covalent binding results in the modification of a large number of intracellular proteins. In general, all enzymes whose functions have been analyzed after poly(ADP-ribosyl)ation appear to be inhibited (1). Although poly(ADP-ribose) synthesis occurs extensively in the vicinity of DNA breaks, its physiological significance for chromatin structure and function, and its possible role in break repair, have not yet been demonstrated.

Inhibitors of poly(ADP-ribose) synthetase, in particular, 3-aminobenzamide (3AB), have been extensively used to elucidate possible biological functions of poly(ADP-ribosyl)ation. Inevitably, then, inhibitors occupy a prominent place in the cytogenetic investigation reported here, and care must be exercised to ensure that the pharmacological properties of an inhibitor are not equated or confused with the physiological function of poly(ADP-ribose). In this chapter, we will review the cytogenetic consequences of inhibiting poly(ADP-ribose) polymerase.

Results and Discussion

Chromosome aberrations. Ionizing radiations break chromosomes presumably through DNA double-strand breakage (2, 3) and are capable of inducing aberrations in all phases of the cell cycle. 3AB potentiates the induction of chromosome aberrations by sparsely ionizing radiations, such as X- and γ-rays (4-7), but does not affect the yields induced by densely ionizing fast neutrons (4). This indicates that poly(ADP-ribose) plays a different role in response to these types of radiation.

The effect of 3AB on radiation-induced aberrations is also cell cycle-dependent (Fig. 1). Although Natarajan et al. (6) originally found that 3AB increased aberrations in human lymphocytes X-irradiated in G_0, three subsequent investigations have failed to substantiate this (4, 7, 8). Furthermore, no effect of 3AB was found when radiation-induced aberrations were analyzed in prematurely condensed chromosomes from unstimulated G_0 lymphocytes (8). Several studies have confirmed that lymphocytes cultured with 3AB and irradiated during G_1, show dramatically increased numbers of radiation-induced dicentric and ring chromosomes. There is, however, some discrepancy regarding the effect of the poly(ADP-ribose) inhibitors during the DNA synthesis (S) period and G_2. Kihlman and Andersson (5) found that 3AB did not affect the aberration frequency when human lymphocytes were X-irradiated in S or G_2. We found that lymphocytes exposed to X-rays during early S or late G_2 showed more aberrations when 3AB was present, whereas aberration yields in lymphocytes irradiated in late S or early G_2 were unaffected by the presence of 3AB (7). Since 3AB-modulated cytogenetic damage appears to be cell cycle dependent, minor differences in experimental conditions that affect lymphocyte cycle progression or the times of X-ray treatments relative to the position in the cell cycle could lead to disparate results.

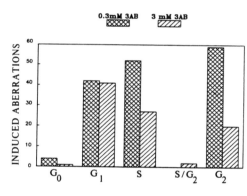

Fig. 1. The effect of 3AB (0.3 and 3.0 mM) on aberrations induced by 2 Gy of X-rays at various stages of the lymphocyte cell cycle. Irradiation in G_0 was before mitogen stimulation; G_1, 21 hr after addition of PHA; S, 10 hr before terminating cultures; S/G_2, 6 hr before terminating cultures, and G_2, 3 hr before termination. A lower radiation dose of 0.4 Gy was used for G_2, lymphocytes because of their radiosensitivity. 3AB was present for the total culture time. Aberration yields in the unirradiated controls and 3AB-treated cultures have been subtracted for simplicity. From Wiencke and Morgan (7).

At relatively high X-ray dose rates, the increase in aberration yield, namely dicentric and ring chromosomes (2 hit aberrations) is proportional to the square of the dose. Repair of chromosome breaks occurs within 4 hr after X-irradiation as measured by dose fractionation experiments (9). When a single radiation dose is fractionated into two smaller doses given 5 hr apart, there is time for breaks induced by the first dose to rejoin before the second dose is given. The yield of 2 hit aberrations observed in this case is the sum of the yields of each of the half doses rather than proportional to the total dose, i.e., four times the half dose (10). When, on the other hand, lymphocytes were cultured with 3AB (10), or in nicotinamide deficient medium (11) (thereby eliminating an essential precursor for polymer synthesis), the yield of chromosome dicentrics was proportional to the square of the dose, i.e., equal to four times the effect of the half dose rather than that expected from the sum of the doses (11). These results indicate that chromosome repair is inhibited by metabolic conditions known to suppress poly(ADP-ribosyl)ation.

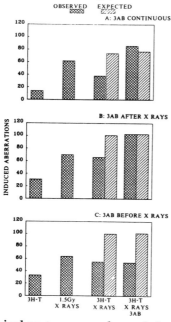

Fig. 2. The effects of continuous (A), pre- (B) or post-irradiation (C) exposure to 2 mM 3AB on the frequency of chromatid aberrations in human lymphocytes treated with 0.1 μ Ci/ml [³H]dThd given 24 hr after mitogen stimulation (adapting dose) and 1.5 Gy of X-rays as the challenge dose given 6 hr before termination of cultures. Aberration yields in the unirradiated controls and 3AB-treated cultures have been subtracted for simplicity. From Wiencke *et al.* (10).

Chemical mutagens, such as ethyl methanesulfonate (EMS), in contrast to ionizing radiation, produce long-lived lesions in DNA that give rise to cytogenetic damage only when the cell passes through the S phase of the cell cycle. Combined treatments of EMS and 3AB markedly increased the frequency of chromatid aberrations (12, 13). Although the exact stage of the cell cycle at the time of EMS treatment was not determined, the results suggest that 3AB was most effective in potentiating EMS-induced aberrations during the S phase, but might potentiate aberrations throughout the cell cycle. An investigation on the influence of 3AB in G2 on the frequency of chromosomal aberrations induced by thiotepa and N-methyl-N-nitro-N'-nitrosoguanidine led to the conclusion that when 3AB treatments were limited to G2, 3AB had no effect on aberration frequency induced by those compounds (14).

The Adaptive Response to Ionizing Radiation. Human lymphocytes treated with a low dose of ionizing radiation from either [³H]thymidine or X-rays, and subsequently challenged with a higher dose of X-rays have fewer chromatid aberrations than those not receiving the low dose pretreatment. This phenomenon has been termed the adaptive response to ionizing radiation; it is thought that the low dose pretreatment induces a

237

chromosomal repair system that is present at the time of the high dose challenge and can act on that damage (10, 15, 16).

Since poly(ADP-ribosyl)ation is somehow involved in the cytogenetic response to ionizing radiation, the relationship between the enzyme and the adaptive response to ionizing radiation was investigated. When 3AB was present during the period between the challenge dose and termination of the culture, no adaptive response was observed. If the inhibitor was removed from the culture medium before the challenge dose, the adaptive response was preserved (Fig. 2). Thus, poly(ADP-ribose) activity after the challenge dose exposure is required for expression of the adaptive response.

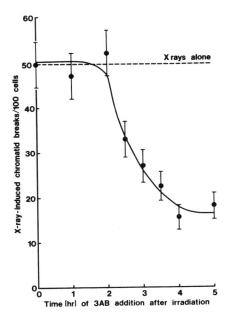

Fig. 3. The effect of adding 2 mM 3AB at various times after 1.5 Gy of X-rays on the induction of chromatid breaks in human lymphocytes labeled with 0.1 μCi/ml [³H]dThd for 6 hr 24 hr after mitogen stimulation. The vertical bars indicate standard error of the mean. At each time point the number of induced aberrations was calculated by subtracting the number of background breaks and the number of [³H]dThd-induced chromatid breaks from the total number of aberrations observed. For comparison, the number of X-ray-induced aberrations in cells not exposed to [³H]dThd is indicated by the dashed horizontal line. Reprinted with permission from Wiencke *et al.* (10).

Since DNA strand breaks serve as the inducing signal for poly(ADP-ribose) polymerase, the repair of these breaks could remove the signal and stop induction of the enzyme activity. Because poly(ADP-ribose) polymerase is active only during the time period between strand breakage and resealing, the role it plays in the adaptive response must occur sometime during this period. This time period has been identified by noting the effect of poly(ADP-ribose) polymerase inhibition at various times after the challenge dose exposure on the adaptive response (Fig. 3). Addition of 3AB up to two hr after the challenge dose could prevent the adaptive response (10). These studies demonstrate that whatever modulates the rejoining of chromatid breakage following the challenge dose of X-rays is dependent on the activity of poly(ADP-ribose) polymerase.

Sister Chromatid Exchange. Sister chromatid exchange (SCE) is a widely used cytogenetic assay for detecting exposure to DNA damaging agents (17). The mechanism by which SCE occurs and the biological significance of SCEs are, however, unknown. Poly(ADP-ribose) inhibitors increase SCE frequency in a dose dependent manner, the stronger the inhibitor the more SCEs are induced (18-20). Because inhibitors by themselves are only weakly cytotoxic, and do not induce either mutations or chromosome aberrations, much attention has focused on how the inhibitors increase SCEs. The majority of SCEs induced by 3AB appear to be the result of an interaction between 3AB and DNA that has incorporated the bromodeoxyuridine (BrdUrd) used for the visualization of SCEs. The majority of exchanges are induced during the second cell cycle when DNA replicates on a template containing BrdUrd (20, 21).

Nevertheless, 3AB also increased SCE levels when cells were cultured with [³H]thymidine instead of BrdUrd and SCEs analyzed after autoradiography (21). More recently, Morgan et al. (22) have shown by ring chromosome analysis that 3AB will increase SCE frequency in the absence of any thymidine analogue for the visualization of SCEs. The frequency of 3AB-induced SCEs determined by ring chromosome analysis is similar to that calculated by both Natarajan *et al.* (20) and Morgan and Wolff (21) when BrdUrd was not present in template DNA. Inhibitor-induced SCE frequencies can be reduced by co-culturing cells with the protease inhibitor antipain (23, 24) or by addition of exogenous βNAD⁺ (25).

A synergistic increase in SCEs has been observed when cells are exposed to 3AB and an alkylating agent (19, 26). The degree of synergism depends on the concentrations of both the 3AB and the alkylating agent, and varies with the alkylating agent used (26). The observed synergism does not appear to be related to potentiation of SCEs via an indirect effect of either proteases or topoisomerases (23). A synergistic increase was not observed when an inhibitor was combined with X-rays (27, 28) or ultraviolet light (19).

A Possible Role for ADP-Ribosylation in Damaged Cells. The role of poly(ADP-ribose) in normal and damaged cells remains unknown. The widely varying and often profound cytogenetic effects produced by the inhibition of ADP-ribosylation suggest that the polymer may well have a diverse role in response to DNA damage. In order to explain the plethora of responses when cells are subject to high levels of DNA breakage, the synthesis of poly(ADP-ribose) and concomitant decrease in NAD levels can be related to cellular lethality (1, 29-31). The polymer synthesis could be regarded as part of a balanced suicide response within the cell (32). Following DNA damage lethality results because of the rapid drop in cellular NAD and ATP levels (31). Binding of poly(ADP-ribose) polymers to numerous intracellular enzymes, e.g., proteases, topoisomerases, and endonucleases, could inhibit further enzymatic degradation, therefore potentially enhancing cellular viability. In each circumstance, inhibition of poly(ADP-ribose) synthesis could permit degradative enzyme action and result in increased DNA damage. This scheme would explain the cytogenetic effects related to an inhibition of polymerase. For example, after DNA damage, 3AB might permit activation of Ca^{2+}, Mg^{2+}-dependent endonucleases, which would otherwise have been inhibited by ADP-ribosylation (33, 34). A chaotic, nonspecific nuclease attack would follow, and DNA already containing strand breaks induced directly or indirectly as a consequence of damage would be subject to further strand breakage. This could result in enhanced toxicity, chromosome aberrations, and SCEs.

Acknowledgment. This work was supported by the Office of Health and Environmental Research, Department of Energy, contract no. DE-AC03-76-SF01012.

References

1. Cleaver, J. E., Morgan, W. F. (1985) Mutat Res 150: 69-76
2. Obe, G., Natarajan, A. T., Palitti, F. (1982) DNA Repair: Chromosome Alterations and Chromatin Structure, A. T. Natarajan, G. Obe, H. A. Altmann, (eds.), Amsterdam-Elsevier Biomedical Press, pp. 1-9
3. Wolff, S. (1978) DNA Repair Mechanisms, P. C. Hanawalt, E. C. Friedberg, C. F. Fox, (eds.), Academic Press New York, pp. 751-760
4. Heartlein, M. W., Preston, R. J. (1985) Mutat Res.148: 91-97
5. Kihlman, B. A., Anderson, H. C. (1986) Genetic Toxicology of Environmental Chemicals, Part A: Basic Principles and Mechanisms of Action, Alan R. Liss, Inc., pp. 395-402
6. Natarajan, A. T., Csukas, I., Degrassi, F., van Zeeland, A. A., Palitti, F., Tanzarella, C., DeSalvia, R., Fiore, M. (1982) Progress Mutat Res 4: 47-59
7. Wiencke, J. K., Morgan, W. F. (1987) Biochem Biophys Res Commu 143: 372-376
8. Pantelias, G. E., Politis, G., Sabani, C. D., Wiencke, J. K., Morgan, W. F. (1986) Mutat Res 174: 121-124
9. Wolff, S. (1972) Mutat Res 15: 435-444
10. Wiencke, J. W., Afzal, V., Oliveri, G., Wolff, S. (1986) Mutagenesis 1: 375-380
11. Wiencke, J. K. (1987) Exp Cell Res (1987) 171: 518-523
12. Jan, K. Y., Huang, R. Y., Lee, T. C. (1986) Cytogen Cell Genet 41: 202-208
13. Schwartz, J. L., Morgan, W. F., Brown-Lindquist, P., Afzal, V., Weichselbaum, R. R., Wolff, S. (1985) Cancer Res 45: 1556-1559
14. Hansson, K., Kihlman, B. A., Tanzarella, C., Palitti, F. (1984) Mutat Res 126: 251-258
15. Shadley, J. D., Wolff, S. (1987) Mutagenesis 2: 95-96
16. Olivieri, G., Bodycote, J., Wolff, S. (1984) Science 223: 594-597
17. Wolff, S. (1978) Mutagen-Induced Chromosome Damage in Man, H.J. Evans, D.C. Lloyd (eds.), Edinburgh University Press, Edinburgh, pp. 208-215
18. Hori, T. A. (1981) Biochem Biophys Res Commun 102: 38-45
19. Morgan, W. F., Cleaver, J. E. (1982) Mutat Res 104: 361-366
20. Natarajan, A. T., Csukas, I., van Zeeland, A. A. (1981) Mutat Res 84: 125-132
21. Morgan, W. F., Wolff, S. (1984) Cytogenet Cell Genet 38: 34-38
22. Morgan, W. F., Bodycote, J., Doida, Y., Fero, M. L., Hahn, P., Kapp, L. N. (1987) Mutagenesis 1: 453-459
23. Morgan, W. F., Doida, Y., Fero, M. L., Guo X.-C., Shadley, J. D. (1986) Environ Mutagen 8: 487-493
24. Schwartz, J. L., Weichselbaum, R. R. (1985) Environ Mutagen 7: 703-709

240

25. Schwartz, J. L. (1986) Carcinogenesis 7: 159-162

26. Schwartz, J. L., Morgan, W. F., Weichselbaum, R. R. (1985) Carcinogenesis 6: 699-704

27. Morgan, W. F., Schwartz, J. L., Murnane, J. P., Wolff, S. (1983) Radiat Res 93: 567-571

28. Morgan, W. F., Djordjevic, M. C., Jostes, R. F., Pantelias, G. E. (1985) Int J Radiat Biol 48: 711-721

29. Carson, D. A., Seto, S., Wasson, D. B., Carrera, C. J. (1986) Exp Cell Res 164: 273-281.

30. Cleaver, J. E. , Borek, C., Milam, K. M., Morgan, W. F. (1987) Pharmacol Therap 31: 269-293

31. Schraufstatter, I. U., Hyslop, P. A., Hinshaw, D. B., Spragg, R. G., Sklar, L. A., Cochrane, C. G. (1986) Proc Natl Acad Sci USA 83: 4908-4912

32. Berger, N. A. (1985) Radiat Res 101: 4-15

33. Nomura, H., Kitamura, A., Tanigawa, Y., Tsuchiya, M., Ueki, M., Sugimoto, O., Shimoyama, M. (1984) Biochem Biophys Acta 781: 112-120

34. Yoshihara, K., Tanigawa, Y., Burzio, L., Koide, S. S. (1975) Proc Natl Acad Sci USA 72: 289-293

241

Effect of 3-Aminobenzamide on the Induction of γ-Glutamyltranspeptidase Positive Foci in the Liver of Rats Treated with Various Chemical Carcinogens

Yoichi Konishi, Ayumi Denda, Masahiro Tsutsumi, Dai Nakae, and Seiichi Takahashi

Department of Oncological Pathology, Cancer Center, Nara Medical College, 840 Shijo-cho, Kashihara, Nara 634, Japan

Introduction

Biochemical evidence of the participation of poly(ADP-ribose) in DNA damage and repair (1, 2) suggests the involvement of poly(ADP-ribosyl)ation reactions in chemical carcinogenesis (3). We found that diethylnitrosamine (DEN) initiation of liver carcinogenesis was enhanced by inhibitors of ADP-ribosyl transferase (4, 5) and the foci initiated by DEN and 3-aminobenzamide (ABA) were found to be capable to develop into hepatocellular carcinomas without promotion by phenobarbitol (6). In the present experiment, we studied the effect of ABA on the induction of γ-glutamyltranspeptidase (GGT) positive foci in the liver of rats treated with various chemical carcinogens.

Results and Discussion

Male Wistar or Fischer 344 rats (Shizuoka Laboratory Animal Center, Shizuoka, Japan), weighing approximately 150 g each, were given a commercial stock diet (Oriental MF, Oriental Yeast Ind., Tokyo, Japan) or the same diet supplemented with 0.02% 2-acetylaminofluorene (AAF). Rats were weighed weekly and sacrificed under ether anesthesia. N-methyl-N-nitrosourea (MNU), N-nitrosobis(2-hydroxypropyl)amine (BHP), AAF and CCl_4 were purchased from Nakarai Chemical Ind., Kyoto Japan, DEN from Wako Pure Chemical Ind., Kyoto, Japan, benzo(a)pyrene (B(a)P) from Sigma Chemical Co., St. Louis, MO, and 1,2-dimethylhydrazine•2HCl (DMH) from Aldrich Chemical Co., Milwaukee, WI, USA. MNU (12 mg/ml in 0.1 M Na-citrate pH 6.0) was given intraperitoneally (i.p.) at a dose of 60 mg/kg body weight, B(a)P (30 mg/ml in corn oil) intragastrically (i.g.) at 200 mg/kg, DMH (20 mg/ml in 0.4 mM EDTA-saline, pH 6.6) i.p. at 100 mg/kg, BHP (125 mg/ml in saline) i.p. at 250 and 500 mg/kg, and DEN (10 mg/ml in saline) i.p. at 20 mg/kg. Those carcinogens were given 4 hr before i.p. administration of ABA (Tokyo Kasei Kogyo Co. Ltd., Tokyo, Japan, 300 mg/ml in DMSO) at a dose of 600 mg/kg. Rats were submitted to the experimental regimen for the assay of liver initiation as described previously

(7). Immediately after the rats had been killed, tissue slices taken from three lobes of the liver were histochemically stained for GGT by the method described previously (4). GGT positive foci were analyzed with Imagelyzer model HTB-c995 (Hammatsu Television Co., Ltd., Shizuoka, Japan) connected to Desktop Computer System-45 (Hewlett-Packard, Co., USA).

Effects of ABA on the induction of GGT-positive foci in the liver initiated by MNU, B(a)P, and DEN of Wistar and Fischer 344 rats are shown in Tables 1 and 2. ABA enhanced the initiation by B(a)P and DEN in both Wistar and Fischer 344 rats judging from the numbers of GGT-positive foci and percent liver area occupied with foci. In contrast, no enhancement of MNU was observed. The results of the comparison of the effects of ABA on the induction of GGT-positive foci in the liver initiated by 1, 2-DMH in Wistar and Fischer strains of rats are shown in Table 3. Enhancing effects of ABA on the initiation by 1, 2-DMH were detected in Wistar rats but not in Fischer 344 rats.

Table 1. Effects of 3-aminobenzamide on the induction of GGT positive foci by carcinogens in Wistar rat liver

| Carcinogens Dose (mg/kg) | ABA or DMSO | No. of rats | GGT Positive Foci | | |
			No. of foci/cm^2	Size of foci (mm^2)	% Area occupied with foci
MNU 60	ABA	6	7.1 ± 1.9	0.54 ± 0.20	4.1 ± 2.1
MNU 60	DMSO	9	6.3 ± 2.1	0.44 ± 0.22	2.6 ± 1.0
B(a)P 200	ABA	8	10.4 ± 2.6 ****	0.49 ± 0.20	4.8 ± 1.5*
B(a)P 200	DMSO	8	3.5 ± 1.4	0.59 ± 0.22	2.2 ± 1.1
DEN 20	ABA	7	11.1 ± 2.6 ***	0.54 ± 0.17	6.2 ± 2.7**
DEN 20	DMSO	6	6.7 ± 1.4	0.33 ± 0.23	2.2 ± 1.6

* $p < 0.05$, ** $p < 0.01$, *** $p < 0.01$, **** $p < 0.001$

The present results support our previous findings of enhancing effects ABA on liver carcinogenesis by DEN in rats. It has been reported that poly(ADP-ribose) synthetase inhibitors enhance streptozotocin-inducing killing of insulinoma cells by inhibiting the repair of DNA strand breaks (8) and an accumulation of DNA containing single-stranded regions occurred in the presence of ABA more predominantly in ataxia telangiectasia fibroblasts than in normal fibroblasts following damage by methylmethanesulphonate (9). In contrast, Cleaver and Morgan (10) reported that poly(ADP-ribose) synthesis is involved in the toxic effects of alkylating agents but does not regulate DNA repair. In the present study, an enhancing effect of ABA in the initiation by MNU and BHP (data not shown) were not observed. ABA effects may depend on differences of DNA damage by different chemicals and their repair.

Table 2. Effects of 3-aminobenzamide on the induction of GGT positive foci by carcinogenesis in Fischer F344 rat liver

Carcinogens Dose (mg/kg)	ABA or DMSO	No. of rats	GGT Positive Foci		
			No. of foci/cm^2	Size of foci (mm^2)	% Area occupied with foci
MNU 60	ABA	10	3.5 ± 1.1	0.23 ± 0.30	0.8 ± 0.6
MNU 60	DMSO	10	3.9 ± 2.3	0.28 ± 0.28	1.1 ± 1.0
B(a)P 200	ABA	9	8.8 ± 5.2 **	0.42 ± 0.42	3.7 ± 2.5 **
B(a)P 200	DMSO	10	2.0 ± 1.1	0.21 ± 0.26	0.4 ± 0.4
DEN 20	ABA	7	9.7 ± 2.1 **	0.26 ± 0.08 *	2.5 ± 1.0 **
DEN 20	DMSO	5	1.8 ± 0.6	0.13 ± 0.03	0.2 ± 0.1

* $p < 0.02$, ** $p < 0.01$

Table 3. Effects of 3-aminobenzamide on the induction of GGT positive foci by 1, 2-dimethylhydrazine 2HCl in Fischer F344 and Wistar rat liver

Strain of Rats	Treatments	No. of rats	GGT Positive Foci		
			No. of foci/cm^2	Size of foci (mm^2)	% Area occupied with foci
Fischer	DMH + ABA	9	8.1 ± 2.3	0.52 ± 0.55	4.2 ± 2.2
	DMH + DMSO	9	6.8 ± 1.4	0.39 ± 0.14	2.7 ± 1.2
Wistar	DMH + ABA	10	5.8 ± 2.9 *	0.13 ± 0.12	0.8 ± 0.7 **
	DMH + DMSO	10	1.3 ± 1.0	0.08 ± 0.05	0.1 ± 0.1

* $p < 0.001$, ** $p < 0.01$

Acknowledgements. Supported by a grant-in-Aid for Cancer Research from the Ministry of Education, Science and Culture (61010089) and that from the Ministry of Health and Welfare for the Comprehensive 10-year Strategy for Cancer Control, Japan.

References

1. Rankin, P.W., Jacobson, M.K., Mitchell, V.R., Busbee, D.L. (1980) Cancer Res 40: 1803-1807
2. Shall, S. (1982) ADP-Ribosylation Reactions, O. Hayaishi and K. Ueda (eds.), Academic Press, London, New York
3. Sugimura, T., Miwa, M. (1983) Carcinogenesis 12: 1503-1506
4. Takahashi, S., Ohnishi, T., Denda, A., Konishi, Y. (1982) Chem-Biol Interactions 39: 363-368
5. Takahashi, S., Nakae, D., Yokose, Y., Emi, Y., Denda, A., Mikami, S., Ohnishi, T., Konishi, Y. (1984) Carcinogenesis 7: 901-906
6. Konishi, Y., Takahashi, S., Nakae, D., Uchida, K., Tsutsumi, M., Shiraiwa, K., Denda, A. (1986) Tox Path 14: 483-488
7. Cayama, E., Tsuda, H., Sarma, D.S.R., Farber, E. (1978) Nature 275: 60-62
8. Yamamoto, H., Okamoto, H. (1982) FEBS Lett 145: 298-301
9. Ireland, C.M., Stewart, B.W. (1987) Carcinogenesis 8: 39-43
10. Cleaver, J.E., Morgan, W.F. (1985) Mutat Res 150: 69-76

A Role of Poly(ADP-Ribosyl)ation in Carcinogen-Inducible Transient DNA Amplification

Alexander Bürkle, Thomas Meyer[1], Helmuth Hilz[1], and Harald zur Hausen

Institut für Virusforschung, Deutches Krebsforschungszentrum, D-6900 Heidelberg 1, West Germany

Introduction

DNA amplification is viewed as an adaptive cellular response to environmental changes and genotoxic stress (1). It is an important mechanism in the development of drug resistance (1-4) and seems to play significant roles in the tumor initiation process induced by chemical or physical carcinogens (5) and tumor progression (6). Chemical or physical carcinogens induce transient amplification of viral DNA sequences in a number of papovavirus-transformed cell lines (7-9) which have been used as model systems. Peak amplification factors are usually reached a few days after induction, thereafter amplified sequences are gradually lost. Amplification of cellular sequences as a result of carcinogen treatment has also been reported (10-12).

Another response of most eukaryotic cells to carcinogen-mediated DNA strand breaks is poly(ADP-ribosyl)ation of a variety of nuclear proteins, catalyzed by poly(ADP-ribose) polymerase. We therefore studied a possible role of poly(ADP-ribose) synthesis in carcinogen-inducible DNA amplification in a SV40 transformed Chinese hamster cell line (CO 60) which shows amplification of integrated viral sequences after carcinogen treatment (5, 7). We found that inhibition of carcinogen-stimulated poly(ADP-ribose) synthesis by 3-aminobenzamide (3AB) is correlated with an enhancement of inducible DNA amplification in CO 60 cells (13).

Results and Discussion

Treatment of CO 60 cells with the alkylating agent N-methyl-N'-nitro-N-nitrosoguanidine (MNNG) (50 μM) induced about 20-fold amplification of SV40 DNA sequences within four days (Table 1). To find out whether poly(ADP-ribosyl)ation plays a role in the amplification process, 3AB (2 mM), a competitive inhibitor of poly(ADP-ribose) polymerase, was added one hour before MNNG treatment. In the presence of 3AB amplification induced by MNNG was enhanced to values about 50-fold above basal levels. Treatment of CO 60 cells with 2 mM 3AB alone did not induce any detectable amplification, nor was there any influence on cell proliferation.

---

[1]Institut für Physiologische Chemie der Universität Hamburg, D-2000 Hamburg 20, West Germany

Table 1. SV40 DNA amplification and poly(ADP-ribose) levels after treatment of CO 60 cells with MNNG in the presence or absence of 3AB

MNNG μM	3AB mM	AF[a]	poly(ADP-ribose)[b] (pmol/mg DNA)
0	0	1.0/1.0	12.1/10.3
0	2	0.9/0.9	1.9/7.2
50	0	20.6/20.1	64.5/52.7
50	2	53.7/47.0	24.9/8.7

[a]CO 60 monolayer cultures were treated as indicated. (Cultures were pretreated with 3AB for 1 hr.) Four days later, cells were harvested for DNA extraction and amplification factors (relative to solvent controls) were determined by quantative slot blot hybridization: Replica slot blots were hybridized with an SV40 probe and, as a control, with an albumin probe, as described recently (13). The results of two cultures treated in parallel are shown for each treatment.

[b]Determinations of poly(ADP-ribose) levels from cultures treated in parallel to those described above were performed as previously reported (14). TCA precipitation was done 45 min after MNNG treatment.

In order to see whether the action of 3AB may be mediated by an inhibition of nuclear poly(ADP-ribosyl)ation, two sets of experiments were performed. First, we analyzed the amounts of poly(ADP-ribosyl) groups accumulating in CO 60 cells after MNNG treatment in the presence or absence of 3AB. Cells were extracted with cold TCA while still attached to the dishes 45 min after addition of MNNG (14). This was the time of maximal cellular poly(ADP-ribose) content under these conditions (Bürkle, A., Meyer, T., zur Hausen, H., Hilz, H., unpublished). Table 1B shows the poly(ADP-ribose) levels in CO 60 cultures treated in parallel with those used for the amplification studies (Table 1A). MNNG treatment alone led to a nearly six-fold increase in poly(ADP-ribosyl) groups above controls. As expected, this increase was largely prevented by 2 mM 3AB.

Table 2. SV40 amplification factors after treatment of CO 60 cells with MNNG in the presence of varying concentrations of 3AB or 3ABA

MNNG (μM)	3AB (mM)	3ABA (mM)	AF[a]
0	0		1.0/1.0
50	0		26.2/23.2
50	0.1		25.1/24.4
50	0.3		25.7/30.8
50	1		40.3/35.9
50	2		44.5/48.6
50		2	27.2/25.8

[a]Procedures were the same as described in the footnote to Table 1, except for the presence of varying concentrations of 3AB or 3ABA as indicated.

In a second type of experiment we studied the dose-response relationship and compared the action of 3AB on MNNG-induced amplification with that of its noninhibitory analog 3-aminobenzoic acid (3ABA) (15). While an enhancement of amplification by 3AB was detectable even at a concentration of 0.3 mM, 3ABA was essentially without effect at 2 mM (Table 2). Taken together, these findings strongly favor the interpretation that inhibition of nuclear poly(ADP-ribosyl)ation is associated with an increase in the amplification response induced by an alkylating agent.

Table 3. SV40 amplification factors obtained with varying duration of 3AB exposure after MNNG treatment of CO 60 cells.

MNNG (25 μM, 30 min)	Duration of 3AB exposure (2 mM)	AF[a]
-	-	1.0/1.0
+	-	4.0/4.5
+	1 hr	4.4/4.3
+	4 hr	5.7/7.1
+	7 hr	9.2/7.1
+	24 hr	23.0/19.4
+	4 d	28.5/25.8

[a]CO 60 cells were treated with MNNG for 30 min as indicated. Then, medium containing 3AB (2 mM) was applied for the times indicated. Thereafter, medium was removed, cultures rinsed with PBS, and medium without 3AB was added for the rest of the incubation. All the cultures were harvested for slot blot hybridization on the fourth day.

Table 3 shows the influence of varying durations of 3AB exposure. In this experiment, SV40 sequences were amplified four-fold by a 30 min treatment with 25 μM MNNG alone. Subsequent presence of 3AB in the culture medium for only 1 hr, followed by incubation without 3AB, was not sufficient to enhance amplification. Suppression of poly(ADP-ribosyl)ation for at least 4 hr was required to increase MNNG-induced amplification. Presence of 3AB during the first 24 hr after the addition of MNNG (the period during which transient accumulation of poly(ADP-ribose) groups is measurable in these cells) was nearly as efficient as continuous 3AB exposure throughout the 4 days of incubation. The latter treatment led to amplification factors six-fold higher than those obtained in the absence of 3AB. These findings, as well as those shown in Table 2, point to a relation between the enhancing effect on amplification and that amount of poly(ADP-ribose) formation which is prevented by 3AB.

Enhancement of MNNG-induced amplification by 3AB was paralleled by a similar effect on MNNG-induced cytotoxicity, as expected. While 3AB (2 mM) alone did not inhibit cell growth, it greatly potentiated the cytotoxicity of MNNG, whereas 3ABA was without effect (data not shown). Hypothetically, the cytotoxic action of 3AB could be a non-random event, leading to a selective loss of non- or low level-amplifying cells in a heterogeneous cell population, thus artificially increasing the average amplification factor. We could, however, exclude this possibility: (i) Viable (adherent) and non-viable (floating) cells were routinely combined prior to amplification analysis. (ii) Separate analysis of adherent versus non-adherent cells as obtained after treatment with 50 µM MNNG showed comparable amplification factors (AF: 45.2 vs. 32.7, mean value of two parallel cultures, respectively). The enhanced amplification in response to 3AB, therefore, is obviously not a consequence of an altered cell population. Rather, the co-cytotoxic and the co-amplifying action of 3AB may point to a common pathway through which poly(ADP-ribosyl)ation exerts a protecting effect against the deleterious sequelae of exposure to genotoxic alkylating agents.

A rat pituitary cell line (F_1BGH_12C) was reported to transiently amplify prolactin DNA sequences in response to treatment with bromodeoxyuridine (12). We recently found that among other agents MNNG is an inducer of amplification in this cell line as well (Heilbronn, R. and Bürkle, A., unpublished). Our most recent and still preliminary data suggest that, quite analogous to the CO 60/SV40 system, prolactin DNA amplification induced by MNNG-treatment is enhanced in the presence of 3AB (1 mM). This suggests that poly(ADP-ribose) synthesis plays a role not only in the inducible amplification of integrated viral DNA, but also in that of cellular sequences.

Divergent results have been reported by Lambert *et al.* in a different model system for carcinogen-inducible DNA amplification (16). Benzo[a]pyrene-7, 8-diol-9, 10-oxide, the ultimate carcinogenic metabolite of benzo[a]pyrene, induces amplification of polyoma virus DNA in a polyoma-transformed rat cell line. In their experiment, the induction of amplification was inhibited by 3AB, though at a higher concentration (5 mM) than we used. In contrast to alkylating agents like MNNG, however, polycyclic aromatic hydrocarbons are poor inducers of DNA single-strand breaks (17) and therefore are likely to be poor activators of poly(ADP-ribose) synthesis. This difference as well as nonspecific side-effects of 3AB evident at a concentration of 5 mM (18) might account for the apparent discrepancy in the effects of 3AB on carcinogen-induced amplification.

DNA amplification is a biological phenomenon which appears to be of relevance for several aspects of carcinogenesis (1-6). Viewed together, our results strongly suggest a negative regulatory influence of poly(ADP-ribose) in the process of amplification, as it is shown that the inhibition of carcinogen-stimulated poly(ADP-ribose) synthesis by 3AB is correlated with an enhancement of the amplification response.

Acknowledgements. We wish to thank M. Müller and H. Lengyel for competent technical assistance, Dr. S. Lavi for CO 60 cells, Dr. G. Scherer for pSP6-c-abl, Dr. D.K. Biswas for F_1BGH_12C cells, Dr. R.A. Maurer for pPRL-1, and many of our colleagues for helpful discussions. This work was supported by the Deutsche Forschungsgemeinschaft. Data in Tables 1-3 have been reprinted with permission from ref. 13.

References

1. Stark, G.R., Wahl, G.M. (1984) Annu Rev Biochem 53: 447-491
2. Schimke, R.T. (1984) Cancer Res 44: 1735-1742
3. Roninson, I.B., Abelson, H.T., Housman, D.E., Howell, N., Varshavsky, A. (1984) Nature 309: 626-628
4. Riordan, J.R., Deuchars, K., Kartner, N., Alon, N., Trent, J., Ling, V. (1985) Nature 316: 817-819
5. Lavi, S. (1982) Gene Amplification, R.T. Schimke (ed.), Cold Spring Harbor Laboratory, Cold Spring Harbor, pp.225-230
6. Schwab, M., Ellison, J., Busch, M., Rosenau, W., Varmus, H.E., Bishop, J.M. (1984) Proc Natl Acad Sci USA 81: 4940-4944
7. Lavi, S. (1981) Proc Natl Acad Sci USA 78: 6144-6148
8. Baran, N., Neer, A., Manor, H. (1983) Proc Natl Acad Sci USA 80: 105-109
9. Lambert, M.E., Gattoni-Celli, S., Kirschmeier, P., Weinstein, I.B. (1983) Carcinogenesis 4: 587-593
10. Tlsty, T., Brown, D.C., Schimke, R.T. (1984) Mol Cell Biol 4: 1050-1056
11. Kleinberger, T., Etkin, S., Lavi, S. (1986) Mol Cell Biol 6: 1958-1964
12. Biswas, D.K., Hanes, S.D. (1982) Nucl Acids Res 10: 3995-4008
13. Bürkle, A., Meyer, T., Hilz, H., zur Hausen, H. (1987) Cancer Res 47: 3632-3636
14. Wielkens, K., Bredehorst, R., Hilz, H. (1984) Methods in Enzymology, vol. 106, part A, F. Wold, K. Moldave (eds.), Academic Press, Inc, Orlando, pp.472-482
15. Purnell, M.R., Whish, W.J.D. (1980) Biochem J 185: 775-777
16. Lambert, M.E., Pelligrini, S., Gattoni-Celli, S., Weinstein, I.B. (1986) Carcinogenesis 7: 1011-1017
17. Lubet, R.A., Kiss, E., Gallagher, N.M., Dively, C., Kouri, R.E., Schectman, L.M. (1983) J Natl Cancer Inst 71: 991-997
18. Milam, K.M., Cleaver, J.E. (1984) Science 223: 589-591

Abbreviations:
3AB - 3-aminobenzamide
3ABA - 3-aminobenzoic acid
AF - amplification factor
MNNG - N-methyl-N'-nitro-N-nitrosoguanidine
SV40 - Simian virus 40
TCA - trichloroacetic acid

Arrest of Tumor Growth by DNA-Site Inhibitors of Adenosine Diphosphoribose Transferase

Ernest Kun

Department of Pharmacology and the Cardiovascular Research Institute, The University of California - San Francisco, San Francisco, California 94143 USA

Introduction

The impressive experiments of Juarez-Salinas *et al.* (1) reinforced the prevailing view that DNA-damage induces an increased rate of ADP-ribosyl transferase activity in intact cells and this "alarm reaction" was assumed to represent an important, if not, the main cellular role of ADP-ribosyl transferase. Persuasive arguments of Shall (2) connected DNA-strand breaks and cytodifferentiation, notably the fusion of myoblasts, with an increased ADP-ribosyl transferase activity and implicated the involvement of ADP-ribosyl transferase in DNA repair or processes equivalent to repair as occurring in DNA rearrangements or rejoining, professing a physiological function of ADP-ribosyl transferase in complex processes of gene control of eukaryotes. Although uncertainty prevails regarding the exact involvement of ADP-ribosyl transferase in DNA repair (3-6), the importance of the above pioneering experiments (1, 2) consists of calling attention to the critical nature of DNA-ADP-ribosyl transferase interactions in cellular pathophysiology. More recent experiments with simpler reconstructed systems (7) revealed that not so much "DNA-breaks" but probably short activating (8) DNA-domains containing specific sequences (7) may be responsible for the activation of ADP-ribosyl transferase and such sequences, which may be unmasked by DNA damage, could be physiological components of DNA in chromatin. It follows that poly(ADP-ribose) polymerase may belong to a group of DNA-binding sequence seeking proteins. (A more direct experimental approach to this problem is presented by S. Sastry in this volume). These results tend to extradite the cellular function of ADP-ribosyl transferase from the realm of toxicology and place it in the highly intricate field of DNA-protein associations and physiological processes that depend on them, which in eukaryotes, represent gene regulation.

In highly developed eukaryotes gene regulation is expressed during differentiation (9), hormonally influenced processes (10) and the control of cellular phenotype, which are the most complex cell functions, and an involvement of ADP-ribosyl transferase is documented.

The regulation of cellular phenotype, notably of cell transformation and other expressions of malignancy is a highly elusive research field because at least two complex cell functions are obligatory components of the malignant phenotype: (a) the malignant transformation itself and (b) the unrestrained cell growth (or tumorigenicity). As will be discussed, ADP-ribosyl transferase appears to participate in an opposite manner in (a) and (b).

Results and Discussion

Detection of the malignant phenotype depends on both *in vitro* and *in vivo* experimental models. In cell cultures aberrant morphology and abnormal growth characteristics of morphologically transformed cells are the customary tests. Anchorage independent colony formation, and growth characteristics of soft agar colonies should go hand-in-hand with *in vivo* tumorgenesis tests of transformed cells to satisfy the complete spectrum of requirements outlined in (a) and (b). Since the experiments dealing with cell transformation are extremely laborious, most research workers, almost by necessity, confined their attention to one experimental approach. As it will be shown here, this leads to apparent controversy, which upon closer analysis does not exist.

When chemical carcinogens are used as cell transforming agents at toxic doses and transformation is assayed by cell morphology (Type II and III focus formation), simultaneous exposure to methane sulfonate and an inhibitor of ADP-ribosyl transferase (3-aminobenzamide, 1 to 3 mM) a significant increase in both cell toxicity and the appearance of transformed loci were observed (11). Notably, this enhancing effect of ADP-ribosyl transferase inhibition on transformation occurred only with the DNA-damaging carcinogen, not with 3-methylcholanthrene. In apparent agreement with these cell culture experiments, it was shown *in vivo* that toxic hepatocarcinogen-induced liver foci were enhanced by almost simultaneous treatment with 3-aminobenzamide (12). It is of further interest that inhibitors of ADP-ribosyl transferase enhance the transfection of 3T3 cells by SV40 DNA (13), using the Ca^{2+} procedure of transfection. The significant common denominator of these experiments (11-13) is that DNA fragmentation coincides with mutagen induced carcinogenesis, or the insertion of tumorigenic DNA into fragmented genomic DNA (13).

When *in vivo* carcinogenesis was induced by an apparently non-toxic concentration of carcinogens, the growth of soft-agar selected transformed colonies was severely depressed by treatment of cells by ADP-ribosyl transferase inhibitors (14, 15). Even though DNA damage could not be readily identified in these models, this cannot be rigorously excluded and the apparent discrepancy between results described in ref. 11, 12 and 14, 15 requires further analysis. Notably, when transformation is tested by the growth rate of colonies in soft agar (14, 15), not only the process of transformation, but growth rates of anchorage independent colonies are tested simultaneously; therefore, equating growth rates with transformation itself can be misleading. Confirmatory results to ref. 14, 15 were obtained by a morphologic transformation test which consisted of transformed colony

251

counts (16) but it cannot be rigorously excluded either that the depression of the growth rates of transformed cells did not obscure the actual tests for cell transformation itself.

Table 1. The effect of sustained inhibition of ADP-ribosyl transferase for 4 days by 1, 2-benzopyrone (BP) on DNA synthesis in intact oncogene transformed rat (14C) and mouse (3T3) cells.

Number	Cell Type and Treatment	[^{3}H]thymidine/µg DNA after 5 min labelling (CPM)
1	14C	6608
2	14C 0.1 mM BP	6315
3	14C 0.2 mM BP	4343
4	14C + 10^{-7} M DEX (48 hr)	6920
5	14C + 10^{-7} M DEX (48 hr) + 0.1 mM BP	4243
6	14C + 10^{-7} M DEX (48 hr) + 0.2 mM BP	2311
7	3T3	2172
8	3T3 + 0.2 mM BP	1648
9	3T3-SV40	4073
10	3T3-SV40 + 0.2 mM BP	1734

($n = 2, \pm 10\%$)

Resolution of this problem required the unambiguous separation of the process of transformation and tumorigenic growth characteristics of malignant cells (17). This was accomplished by the construction of hormone activatable oncogene containing cells (14C cells, containing MTV-Ejras gene construct). The specific DNA-site oriented inhibitors of ADP-ribosyl transferase had to be present for 3-4 cell cycles, i.e. cells or animals had to be pretreated for a prolonged time, before a significant inhibition of malignant growth was observed both in soft agar or *in vivo* (17). These results clearly define a separate action of ADP-ribosyl transferase inhibitors on malignant growth, which is an inhibitory action whereas previous experiments (11-13) strongly suggest that DNA damage involving initiation of transformation is enhanced by ADP-ribosyl transferase inhibition.

More recent results (Table 1) concerned with cellular DNA synthesis (18) demonstrate that prolonged inhibition of ADP-ribosyl transferase (for 4 days) by selective drugs depress DNA synthesis in malignant cells more than in non-malignant counterparts. This relative selectivity is sufficient to inhibit tumor growth *in vivo* and has been effective even in certain metastatic human tumors (17). More importantly, these results tend to indicate a possible cell growth regulatory function of ADP-ribosyl transferase in malignant cells, a problem that requires detailed studies at the macromolecular level.

Acknowledgements. This work was supported by grants HL-27317 (NIH) and AFO-SR-85-0377 and 86-0064 (AFOSR). Ernest Kun is the recipient of the Research Career Award of the USPH.

References

1. Juarez-Salinas, H., Sims, J.L., Jacobson, M.K. (1979) Nature 282: 740-741
2. Farzaneh, F., Zalin, R., Brill, D., Shall, S. (1982) Nature 300: 362-366
3. Strauss, B. (1984) Carcinogenesis 5: 577-582
4. James, R.L., Lehmann, A.R. (1982) Biochemistry 21: 4007-4013
5. Collins, A. (1985) Carcinogenesis 6: 1033-1036
6. Morgan, W.F., Cleaver, J.E. (1983) Cancer Res 43: 3104-3107
7. Hakam, A., McLick, J., Buki, K.G., Kun, E. (1987) Fed Eur Biochem Soc Lett 212: 73-78
8. Berger, N.A., Petzold, S.J. (1985) Biochemistry 24: 4352-4355
9. Caplan, A.I. (1985) ADP-Ribosylation of Proteins, F.R. Althaus, H. Hilz, S. Shall (eds.) Springer, New York pp. 388-396
10. Kun, E., Minaga, T., Kirsten, E., Hakam, A., Jackowski, G., Tseng, A, Jr., Brooks, M. (1986) Biochemical Actions of Hormones, vol. XIII, G. Litwack (ed.) Acad Press, New York pp. 34-53
11. Lubet, R.A., McCarvill, J.T., Putnam,D.L., Schwartz, J.L., Schechtman, L.M. (1984) Carcinogenesis 5: 459-462
12. Takahashi, S., Ohnishi, T., Dena, A., Konishi, T. (1982) Chem Biol Interactions 39: 363-368
13. Strain, A.J. (1985) Exper Cell Res 159: 531-535
14. Kun, E., Kirsten, E., Milo, G.E., Kurian, P., Kumari, H.L. (1983) Proc Natl Acad Sci USA 80: 7219-7223
15. Milo, G.E., Kurian, K., Kirsten, E., Kun, E. (1985) Fed Eur Biochem Soc Lett 179: 332-336
16. Borek, C., Morgan, W.F., Ong, A., Cleaver, J.E. (1984) Proc Natl Acad Sci USA 81: 243-247
17. Tseng, Jr., A., Lee, W.M.F., Kirsten, E., Hakam, A., McLick, J., Buki, K., Kun, E. (1987) Proc Natl Acad Sci USA 84: 1107-1111
18. Lönn, U., Lönn, S. (1985) Proc Natl Acad Sci USA 82: 104-108

Inhibition of SV40 Replication *In Vitro* by Poly(ADP-Ribosyl)ated Diadenosine Tetraphosphate

Jeffrey C. Baker, Stephen Smale, and Bruce N. Ames

Department of Biochemistry, University of California, Berkeley, California 94720 USA

Introduction

Although poly(ADP-ribosyl)ation is generally considered as a post-translational modification of protein, poly(ADP-ribose) polymerase may also catalyze the addition of poly(ADP-ribose) to the 5' dinucleotide diadenosine-5',5'''-P^1,P^4-tetraphosphate (Ap$_4$A) (1). 5'-Dinucleotides such as Ap$_4$A have been identified in a number of biological systems, including bacteria, plants, and mammalian cells, yet the physiological role of these compounds remain unclear (reviewed in 2 and 3). One proposed role for Ap$_4$A and related compounds is as "alarmones" or mediators of the cellular stress response (4, 5). This proposal is based on the observation that Ap$_4$A accumulates in cells following a number of physiological stresses, including heat shock and oxidative stress (4-8). Ap$_4$A has also been proposed to serve as a positive regulator of DNA replication (9, 10). Support for this model include reports of 50-1000 fold increases in the Ap$_4$A pools of cells in S phase (9, 10) and that Ap$_4$A initiates DNA synthesis in permeablized growth arrested cells (11). It should be noted, however, that these effects have not been observed by all investigators (5, 6, 12, 13). Ap$_4$A could regulate either the cellular stress response or the initiation of DNA replication through an Ap$_4$A binding protein that has been identified as part of the 640 kDa DNA polymerase alpha holoenzyme (EC 2.7.7.7) (14). Although neither the presence of this protein nor Ap$_4$A seem to alter the catalytic activity of DNA polymerase alpha *in vitro* (15), the presence of the Ap$_4$A binding protein allows Ap$_4$A to serve as a primer for DNA synthesis in the absence of an RNA primer (16).

We considered it possible that poly(ADP-ribosyl)ated Ap$_4$A could interact with the DNA polymerase alpha holoenzyme through the Ap$_4$A binding protein, and play a role in the repression of DNA synthesis following DNA damage. We have examined the effects of Ap$_4$A and poly(ADP-ribosyl)ated Ap$_4$A on the *in vitro* synthesis of DNA using a double stranded template which contains the SV40 origin of replication in an effort to evaluate the possible influence of the polymer on DNA replication.

Results

Synthesis and characterization of poly(ADP-ribosyl)ated Ap$_4$A.
Poly(ADP-ribosyl)ated Ap$_4$A differentially labeled in the Ap$_4$A and
poly(ADP-ribose) portions with ^{3}H and ^{32}P was synthesized by reacting 100
μM NAD (10 μCi ^{3}H-NAD), 4 mM Ap$_4$A (50-100 μCi α-^{32}P-Ap$_4$A), 5mM
MgCl$_2$, 250 μg calf thymus DNA, 5 units DNase I and 200 μl crude calf
thymus poly(ADP-ribose) polymerase in 50 mM HEPES, pH 7.5 at 37°C
(total volume of 400 μl). The reaction was terminated after 4 min by the
addition of 100 μl of 100% (w/v) trichloroacetic acid (TCA). Acid insoluble
material was washed once with 20% TCA and resuspended in 0.25 M
ammonium acetate, pH 6.6, 6 M guanidine hydrochloride. The sample was
adjusted to pH 8.8 and applied to a 0.5 ml column of dihydroxyboryl BioRex
(DHB-B, prepared as in ref. 17) which had been pre-equilibrated in 0.25 M
ammonium acetate, pH 8.8, 6 M guanidine hydrochloride. The column was
then washed successively with 0.25 M ammonium acetate, pH 8.8 and 1 M
ammonium bicarbonate, pH 9.0, and bound material was eluted in 5 ml
water and lyophilized. It should be noted that free Ap$_4$A is neither acid
precipitable nor does it bind to DHB-B under these conditions (18). The
recovery of acid insoluble radioactivity from the boronate column was
always greater than 90%. Lyophilized samples were re-dissolved in 0.25 M
ammonium acetate, pH 8.8, incubated with 5 units bacterial alkaline
phosphatase at 37°C for 30 min, and repurified over DHB-B as described
above. The phosphatase treatment served to remove ^{32}P from partially
degraded Ap$_4$A which could retain the label in the alpha-phosphate position.
The ratio of Ap$_4$A/ribosyl-adenosine in the poly(ADP-ribosyl)ated Ap$_4$A was
calculated from the specific activities of the ^{32}P- and ^{3}H-labeled moieties.

The product was characterized as poly(ADP-ribosyl)ated Ap$_4$A on the
basis of the following criteria: (i) Acid insoluble material containing both ^{3}H
and ^{32}P was retained by DHB-B in 0.25 M ammonium acetate, pH 8.8.
These conditions allow the retention of poly(ADP-ribose), but not free
Ap$_4$A. (ii) The control incubation shown in Fig. 1 demonstrates that the ^{32}P
label was resistant to the alkaline phosphatase treatment described as part of
our synthesis protocol and that the Ap$_4$A moiety was intact. (iii) ^{32}P-labeled
material eluted from DHB-B was sensitive to digestion with venom
phosphodiesterase (which cleaves phosphoanhydride linkages) and Ap$_4$A:
ADP phosphohydrolase from *Physarum polycephalum* (Fig. 1). (iv) ^{32}P-
labeled material comigrated with Ap$_4$A standard when analyzed by PEI-
cellulose thin layer chromotography following digestion with poly(ADP-
ribose) glycohydrolase (which degrades poly(ADP-ribose) to ADP-ribose
monomers) (Fig. 1). (v) Tritiated material was resistant to digestion with
alkaline phosphatase, sensitive to digestion with venom phosphodiesterase,
and comigrated with ADP-ribose when analyzed by PEI-cellulose thin layer

chromatography following digestion with poly(ADP-ribose) glycohydrolase (Fig. 1). This is consistent with the incorporation of tritiated material into poly(ADP-ribose). Control incubations performed in the absence of enzyme demonstrate that the poly(ADP-ribosyl)ated Ap$_4$A is stable to the incubation conditions employed.

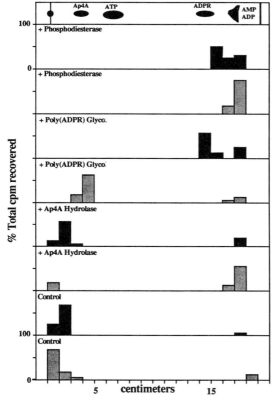

Fig. 1. Characterization of radiolabeled poly(ADP-ribosyl)ated Ap$_4$A by thin layer chromatography on PEI-cellulose. Poly(ADP-ribosyl)ated Ap$_4$A was digested with either snake venom phosphodiesterase (10 units/ml), Ap$_4$A:ADP phosphohydrolase from *Physarum polycephalum* (5 units/ml), or poly(ADP-ribose) glycohydrolase from calf thymus (0.5 units/ml)in 20 mM HEPES, pH 7.4, 10 mM MgCl$_2$, at 37°C for 60 min. Products of these digestions were subjected to thin layer chromatography on PEI-cellulose plates with a 0.5 M HCl solvent system. Following chromatography, the plates were briefly placed in a tank containing a beaker of NH$_4$OH, cut into 1 cm strips and assayed for radioactivity in a toluene based cocktail.

Tritiated counts (which represent the poly(ADP-ribose) portion of the polymer) are shown in black, while ^{32}P counts (which represent the Ap$_4$A portion of the polymer) are shown in grey. The elution profile of unlabeled carrier is shown at the top of the figure.

Venom phosphodiesterase digests both the ^{3}H- and ^{32}P-portions of the polymer to material which coelutes with AMP and a position consistent with the elution of phosphoribosyl adenosine. Poly(ADP-ribose) glycohydrolase digests the tritiated material to ADP-ribose and generates ^{32}P-labeled material which coelutes with Ap$_4$A. Ap$_4$A:ADP phosphohydrolase does not affect the elution of the tritiated counts but does cause more than 75% of the ^{32}P-labeled counts to be degraded. ^{32}P-labeled counts which remains at the origin probably represent poly(ADP-ribose) labeled in a terminal alpha phosphate. Control incubations performed in the absence of enzyme demonstrate that poly(ADP-ribosyl)ated Ap$_4$A is stable to the incubation conditions employed.

Although the length of the polymers was not determined, the ribosyl-adenosine/Ap$_4$A ratio in four different preparations was calculated from the specific activities of the ^{32}P and ^{3}H to be 7.5 ($\pm$ 0.5, n=4). This acid insoluble material is longer than reported by Yoshihara and Tanaka, who reported the synthesis of acid insoluble oligo(ADP-ribosyl)ated Ap$_4$A (1). Since it cannot be assumed that every poly(ADP-ribose) chain is this preparation is terminated by Ap$_4$A or that the polymers are homogeneous in length, the concentration of the poly(ADP-ribosyl)ated Ap$_4$A is given in terms of the Ap$_4$A moiety.

Effect of Poly(ADP-ribosyl)ated Ap$_4$A on DNA synthesis *in vitro*. The effect of poly(ADP-ribosyl)ated Ap$_4$A on DNA synthesis was addressed by examining the replication of supercoiled plasmid DNA (pSV08) containing an SV40 origin of replication inserted into a pBR322-derived vector by HeLa polymerase (19). As shown in Table 1, the replication of pSV08 was strongly inhibited by addition of poly(ADP-ribosyl)ated Ap$_4$A. The replication of pSV08 was dependent on the presence of both T-antigen and the SV40 origin of replication and was unaffected by addition of free Ap$_4$A, poly(ADP-ribose), or a mixture of Ap$_4$A and poly(ADP-ribose) (Table 1). Moreover, the inhibition of DNA replication was eliminated when the poly(ADP-ribosyl)ated Ap$_4$A was preincubated with venom phosphodiesterase, Ap$_4$A hydrolase, or poly(ADP-ribose) glycohydrolase (Table 1). The sensitivity of the poly(ADP-ribosyl)ated Ap$_4$A to these treatments is demonstrated in Fig. 1. These data indicate that *in vitro* replication of SV40 DNA is inhibited by poly(ADP-ribosyl)ated Ap$_4$A and that this inhibition requires both dinucleotide and poly(ADP-ribose) portions of the polymer.

Although the addition of equimolar amounts of free Ap$_4$A and poly(ADP-ribosyl)ated Ap$_4$A to the reaction mix had no detectable effect on the inhibition of DNA synthesis, the addition of a 1000 fold excess of free Ap$_4$A restored 60% of the DNA synthesis (Table 1). The addition of a 1000 fold excess of diguanosine-5',5'''-tetraphosphate (Gp$_4$G) had no effect on the inhibition of DNA synthesis by poly(ADP-ribosyl)ated Ap$_4$A. These data are consistent with poly(ADP-ribosyl)ated Ap$_4$A acting through an Ap$_4$A binding protein. If this were the case, however, the binding coefficients of free Ap$_4$A and poly(ADP-ribosylated) Ap$_4$A would be quite different.

Table 2 shows that neither Ap$_4$A, poly(ADP-ribose), or poly(ADP-ribosyl)ated Ap$_4$A had any effect on the catalyic activity of DNA polymerase as assessed by measuring DNA synthesis using activated salmon sperm DNA as a template. These data suggest that the inhibition of DNA synthesis by poly(ADP-ribosyl)ated Ap$_4$A occurs during the initiation or elongation of a double stranded super-coiled template.

Table 1. Effects of Ap_4A and poly(ADP-ribosyl)ated Ap_4A on T-antigen dependent replication of SV40 DNA *in vitro*.

	SAMPLE	DNA Synthesis (% of control)¶
1.	pSV08 (SV40 DNA template)	100.0
2.	pPM1 (SV40 template lacking an origin of replication)	2.0
3.	pSV08 (T antigen omitted)	1.7
4.	pSV08 + Ap_4A (200 nM)	117.9
5.	pSV08 + poly(ADP-ribose) ($2\mu M$ ribosyl-adenosine)	106.8
6.	pSV08 + poly(ADP-ribose) ($2\mu M$ rAdo) + Ap_4A (200 nM)	101.6
7.	pSV08 + Ap_4A -poly(ADP-ribose) (20 nM in Ap_4A)	4.2
8.	pSV08 + Ap_4A -poly(ADP-ribose) (20 nM in Ap_4A) + Ap_4A (2 μM)	60.1
9.	pSV08 + Ap_4A -poly(ADP-ribose) (20 nM in Ap_4A + Gp_4G(2 μM)	5.4
*10.	pSV08 + Ap_4A -poly(ADP-ribose) + SVPD	77.2
*11.	pSV08 + Ap_4A -poly(ADP-ribose) + Glycohydrolase	87.5
*12.	pSV08 + AP_4A -poly(ADP-ribose) + Ap_4A hydrolase	108.0
*13.	pSV08 + Ap_4A -poly(ADP-ribose) (No Addition)	3.6

*Denotes a preincubation with either SVPD, poly(ADP-ribose) glycohydrolase, Ap_4A hydrolase, or no addition, in 20 mM HEPES, pH 7.6, 10 mM $MgCl_2$, 37°C for one hr followed by boiling for 3 min to inactivate the enzyme. Except for these incubation conditions, samples 7 and 13 were identical.

¶DNA synthesis was assessed as described in Ref 19. Values given are the means of two separate experiments, each done in duplicate. While the absolute amounts of radioactivity varied somewhat from experiment to experiment, the values normalized to untreated control varied by less than 7%.

Table 2. Effects of Ap_4A and poly(ADP-ribosyl)ated Ap_4A on replication of activated salmon sperm DNA *in vitro*.

	Addition	DNA Synthesis (% of control)*
1.	None	100.0
2.	Ap_4A (200 nM)	114.2
3.	poly(ADP-ribose) (2 μM ribosyl-adenosine)	96.3
4.	poly(ADP-ribose) (2 μM rAdo) + Ap_4A (200 nM)	89.8
5.	poly(ADP-ribosyl)ated Ap_4A (20 nM in Ap_4A)	90.4

*DNA synthesis was assessed as described in Table 1 except that activated salmon sperm DNA was used as a template. Values given are the means of three separate experiments, each done in duplicate. While the absolute amounts of radioactivity varied somewhat from experiment to experiment, the values normalized to untreated control varied by less than 10%.

Discussion

Since the repair of DNA damage without loss of genetic information is predicated upon the presence of an undamaged template, it is important for the cell not to replicate damaged DNA or DNA in the process of being repaired. It has been suggested that poly(ADP-ribose) inhibits cell cycle progression following DNA damage (20, 21) and that Ap4A may serve as a

258

signal for stalled replication forks (22). Our data suggest a model consistent with these ideas in which an activated poly(ADP-ribose) polymerase inhibits the initiation of DNA replication by poly(ADP-ribosyl)ating Ap_4A. Poly(ADP-ribosyl)ated Ap_4A then transiently inhibits DNA polymerase alpha, via an interaction with the Ap_4A binding protein of the holoenzyme, with degradation by either poly(ADP-ribose) glycohydrolase or Ap4A phosphohydrolase releasing the inhibition.

Consistent with such a model is the observation that Ap_4A and poly(ADP-ribose) both accumulate following a number of conditions which cause an accumulation of DNA strand breaks(6, 23). While the activity of poly(ADP-ribose) polymerase is dependent upon the presence of DNA strand breaks for activity (24), it is unclear what mechanisms could couple Ap_4A accumulation to the presence of DNA strand breaks. Ap_4A is synthesized by some, but not all, aminoacyl-tRNA synthetases by the addition of ATP to the aminoacyl-adenylate form of the enzyme(25, 26). Although the aminoacylation of tRNA is generally considered to be a cytoplasmic event, a tryptophanyl-tRNA synthetase (EC 6.1.1.2) activity has been reported to copurify with the DNA polymerase alpha holoenzyme (27, 28). Tryptophanyl-tRNA synthetase is one of the aminoacyl-tRNA synthetases which are very efficient in catalyzing the synthesis of Ap_4A and related dinucleotides in vitro (25, 26). It is possible that either DNA strand breaks themselves, the production of poly(ADP-ribose) in response to DNA strand breaks, or alterations in chromatin structure caused by poly(ADP-ribosyl)ation of nuclear protein, stimulate the synthesis of Ap_4A by interacting with this nuclear aminoacyl-tRNA synthetase.

It should be noted that current assays for poly(ADP-ribose) in intact cells are unable to differentiate between free polymer and poly(ADP-ribosyl)ated Ap_4A. Preliminary data indicate that acid insoluble material accumulates in human fibroblasts in culture following treatment with N-methyl-N'-nitro-N-nitrosoguanidine (MNNG) which gives a positive response in a luciferin/luciferase bioluminescence assay for Ap_4A and is retained by DHB-B under conditions selective for binding of poly(ADP-ribose) and not Ap_4A (data not shown). Experiments are currently in progress to confirm the identity of this material as poly(ADP-ribosyl)ated Ap_4A.

In summary, we have found that poly(ADP-ribosyl)ated Ap_4A inhibits the in vitro replication of SV40 DNA. This inhibition was found to be dependent upon the presence of both an intact Ap_4A terminus and intact polymer and did not occur in a non-specific DNA synthesis assay which used activated salmon sperm DNA as a template. The ability of free Ap_4A to prevent this inhibition suggests that the interaction of the polymer could be an Ap_4A binding protein previously identified as being associated with DNA polymerase alpha. Since both Ap_4A and poly(ADP-ribose) accumulate in mammalian cells following treatments which are accompanied by an accumulation of DNA strand breaks, these data are consistent with a model

259

in which Ap_4A is poly(ADP-ribosyl)ated following DNA damage and transiently inhibits DNA replication.

Acknowledgements. This work was supported by NIEHS Center Grant ES01896. BNA was supported by NCI Outstanding Investigator Grant CA39910. JCB was supported by NIEHS Training Grant ES07075. We would like to thank Dr. Robert Tjian for his support during this research and Dr. M. K. Jacobson, Dr. L. D. Barnes, and Dr. G. G. Poirer for their generous assistance in providing us with poly(ADP-ribose) polymerase, AP_4A phosphohydrolase, and poly(ADP-ribose) glycohydrolase respectively.

References

1. Yoshihara, K., Tanaka, Y. (1981) J Biol Chem 256: 6756-6761
2. Baril, E.F., Coughlin, S.A., Zamecnik, P.C. (1985) Cancer Inv 3: 465-471
3. Bambara, R.A., Crute, J.J., Wahl, A.F. (1985) Cancer Inv 3: 473-479
4. Lee, P.C., Bochner, B.R., Ames, B.N. (1983) Proc Natl Acad Sci USA 80: 7496-7500
5. Bochner, B.R., Lee, P.C., Wilson, S.W., Cutler, C.W., Ames, B.N. (1984) Cell 37: 225-232
6. Baker, J.C., Jacobson, M.K. (1986) Proc Natl Acad Sci USA 83: 2350-2352.
7. Garrison, P.N., Mathis, S.A., Barnes, L.D. (1986) Mol Cell Biol 6: 1179-1186
8. Brevet, A., Plateau, P., Best-Belpomme, M., Blanquet, S. (1985) J Biol Chem 260: 5566-5570
9. Rapaport, E., Zamecnik, P.C. (1976) Proc Natl Acad Sci USA 73: 3984-3988
10. Weinman-Dorsch, C., Hedl, A., Grummt, I.,Albert, W., Ferdinand, F., Friis, R., Pierron, G., Moll, W. Grummt, F. (1984) Eur J Biochem 138: 179-185
11. Grummt, F. (1978) Proc Natl Acad Sci USA 75: 371-375
12. Segal, E., Le PecQ, J.-B. (1986) Exp Cell Res 167: 119-126
13. Edwards, M.J., Kaufman, W.K. (1982) Biochim Biophys Acta 721: 223-225
14. Baril, E.F., Bonin, P., Burstein, D., Mara, `K., Zamecnik, P.C. (1983) Proc Natl Acad Sci USA 80: 4931-4935
15. Ono, K., Iwata,Y., Nakamura, H., Matsukage, A. (1980) Biochem Biophys Res Commun 95: 34-40
16. Zamecnik, P.C., Rapaport, E., Baril, E.F. (1982) Proc Natl Acad Sci USA 79: 1791-1794
17. Baker, J.C., Jacobson, M.K. (1984) Analyt Biochem 141: 451-460
18. Alvarez-Gonzalez, R., Juarez-Salinas, H., Jacobson, E.L., Jacobson, M.K. (1983) Analyt Biochem 135: 69-77
19. Smale, S., Tjian, R. (1986) Mol Cell Biol 6: 4077-4087
20. Cleaver, J.E., Bodell, W.J., Gruenert, D.C., Kapp, L.N., Kaufman, W.K., Park, S.D., Zelle, B. (1982) Mechanisms in Chemical Carcinogenesis, C.C. Harris, P.C. Cerutti, (eds.), Liss, New York, pp. 409-418
21. Painter, R.B., Cramer, P., Howard, R., Young, B.R. (1982) Mechanisms in Chemical Carcinogenesis, C.C. Harris, P.C. Cerutti, (eds.), Liss, New York pp. 383-387
22. Varshavsky, A. (1983) J Theor Biochem 105: 707-714
23. Juarez-Salinas, H., Duran-Torres, G., Jacobson, M.K. (1984) Biochem Biophys Res Commun 122: 1381-1388
24. Benjamin, R.C., Gill, D.M. (1980) J Biol Chem 225: 10493-10501
25. Plateau, P., Mayaux, J.F., Blanquet, S. (1981) Biochemistry 20: 4654-4662
26. Blanquet, S., Plateau, P., Brevet, A. (1983) Molec Cell Biochem 52: 3-11
27. Rapaport, E., Zamecnik, P.C., Baril, E.F. (1981) Proc Natl Acad Sci USA 78: 838-842
28. Castriviejo, M., Fournier, M., Gatius, M., Gandar, J.C., Labouesse, B., Litvak, S. (1982) Biochem Biophys Res Commun 107: 294-301

Killing and Mutation of Density Inhibited V-79 Cells as Affected by Inhibition of Poly(ADP-Ribose) Synthesis

Nitaipada Bhattacharyya, Sharmila Das Gupta,
and Sukhendu B. Bhattacharjee

Crystallography and Molecular Biology Division, Saha Institute of Nuclear Physics,
I/AF, Salt Lake, Calcutta 700064, India

Introduction

It is now established that the synthesis of poly(ADP-ribose) is stimulated following DNA-damage (1-4). Inhibition of that enhanced synthesis has been found to increase cellular cytotoxicity depending upon the damaging agent; in some cases, inhibition enhances the effect, whereas, in other cases, it is of no apparent consequence. Amongst the first group are N-methyl-N'-nitro-N-nitrosoguanidine (5, 6) and H_2O_2 (7) and in the second, X-rays and UV light (6). It seemed that there was no general role for poly(ADP-ribose) in repair activity, although various suggestions regarding its involvement in repair had been made. It was clear that so far as normally growing mammalian cells were concerned, the role of poly(ADP-ribose) in repair remains yet to be clarified. Mammalian cells in a density-inhibited condition has been considered to be equivalent to cells in a tumor (8). It is of interest to see if cells in the density inhibited condition would behave in the same way as exponentially growing cells to inhibition of poly(ADP-ribose) synthesis following exposures to X-rays, UV light and H_2O_2, representing DNA damaging agents of group 1 and 2. It is expected that these results might lead to some clarification of the situation.

Results

Density inhibited cells were obtained by growing V-79 cells in 55mm plastic petri dishes. After 2 days of growth the medium was changed and the cells allowed to grow for another 36 hr. Cells, about 5×10^6 in 55mm petri dishes thus obtained were non-dividing and not synthesizing DNA and were called density inhibited cells (6). Fig. 1. shows the variation in survival ratios in UV irradiated density-inhibited V-79 cells, on delay in trypsinization after UV-irradiation. Here the effect after one dose of UV light has been shown, the delay in trypsinization being from 0-24 hr. It may be seen that with an increase in delay of up to 12 hr, there was a gradual increase in survival. The survival fraction at this dose without any delay was 1.6×10^{-2}. This went up to 1×10^{-1} in 12 hr. Thereafter, no change in the survival level was observed in the 24 hr period studied. When inhibitors of poly(ADP-ribose) such as 5 mM benzamide or 15 mM nicotinamide were used during this delay, the increase in survival was not observed. The survival level remained at 1.6×10^{-2} for benzamide treated cells. For nicotinamide treated cells, there was a slight increase probably because nicotinamide was a less efficient inhibitor for poly(ADP-ribose) synthesis. The figure also shows the kinetics of repair of potentially lethal damage in UV irradiated density inhibited cells. In this cell line, the repair is completed within 12 hr of holding.

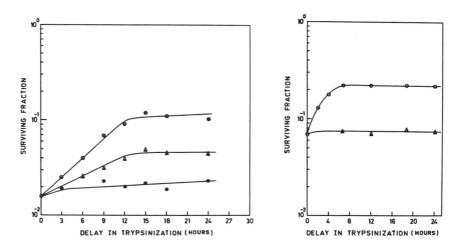

Fig. 1. (left) Survival of density inhibited UV irradiated (30 J/m2) V-79 cells as influenced by delay in trypsinization. The symbol (⊘) represents a delay in the absence of any inhibitors of poly(ADP-ribose) polymerase, (△) represents delay in presence of 15 mM nicotinamide and (●) represents delay in presence of 5 mM benzamide.

Fig. 2. (right) Survival of density inhibited X-ray irradiated (7 Gy) V-79 cells as influenced by delay in trypsinization. The symbol (⊘) represents a delay in the absence of any inhibitors of poly(ADP-ribose) polymerase and (△) represents a delay in presence of 5 mM benzamide.

Fig. 2 represents the effect of delay in trypsinization on the survival level after exposure of density inhibited cells to 7 Gy of X-rays. Here 6 hr were enough for attainment of the maximum level of survival. The presence of 5 mM BA inhibited this recovery.

Fig. 3. gives kinetics of recovery for density inhibited cells exposed to H_2O_2. Here, 4 hr of delay before trypsinization was sufficient to show the maximum recovery. The presence of benzamide clearly inhibited this recovery process. Fig. 4. shows the yield of mutants due to UV irradiation of density inhibited cells. The phenotypic end point selected for the study was resistance to 8-azaguanine. Immediately after UV exposure, cells were diluted for mutation expression or given a delay time before mutation

262

expression in the presence or absence of inhibitors. When there was no delay in trypsinization, 20 J/m² of UV light induced 12 mutants per 10⁵ viable cells. This value decreased to 6 mutants per 10⁵ viable cells when the cells were held in density inhibited condition for 24 hr in fresh medium after UV irradiation. A similar decrease was also evident at other doses of exposure as is seen from the figure. However, when benzamide was present during the period of incubation after UV irradiation, the yield of mutants again increased compared to the number obtained in the absence of benzamide. The yield of mutants from these density inhibited cells in the presence of benzamide was the same as that obtained on immediate trypsinization. The results obtained with X-ray exposed, density inhibited cells was of a similar nature to that reported earlier (6).

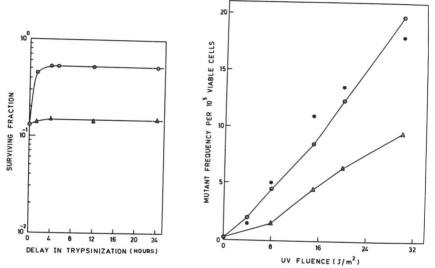

Fig. 3. (left) Survival of density inhibited cells on exposure to 30 μg/ml of H₂O₂ for 1 hr as influenced by a delay in trypsinization. The symbol (⊙) represents a delay in the absence of any inhibitors of poly(ADP-ribose) polymerase, and (△) represents a delay in the presence of 5 mM benzamide.

Fig. 4. (right) UV induced mutant frequency (resistance to 8-azaguanine) in density inhibited V-79 cells. The symbol (⊙) represents the yield without any delay in trypsinization; (△) represents 24 hr delay in trypsinization; (●) represents 24 hr delay in presence of 5mM benzamide.

Discussion

An improvement in survival due to delay in trypsinization after treatment with DNA damaging agents has been termed the repair of potential lethal damage (PLDR). It is seen that for density inhibited cells PLDR took place following exposures of V-79 cells to X-rays, ultraviolet light and H₂O₂, albeit with different rates. Normal human cells (9) or mouse cells (10) show such

263

repair after UV light exposures but Xeroderma pigmentosum cells do not show such repair activity (9, 11). Thus the observed cellular effects on holding the cells in a density inhibited condition after irradiation appear to reflect the repair potential of the irradiated cells. The effect activity of BA or NA shows a good correlation between the relative efficiency of inhibition of poly(ADP-ribose) activity and suppression of PLDR, as shown in Fig. 1. Thus even if the role of poly(ADP-ribose) in repair could be suspect in the case of exponentially growing cells, there is no doubt that this polymer played a part in PLDR.

Results of mutational studies given here and those reported earlier (6, 12, 13) show that PLDR was error-proof. It is this error-proof activity which is suppressed because of inhibition of poly(ADP-ribose) polymerase activity. It is of considerable interest that inhibitors of poly(ADP-ribose) synthesis suppress PLDR with respect to both lethality and mutagenesis.

References

1. Juarez-Salinas, H., Sims, J.L., Jacobson, M.K. (1979) Nature 282:740-741
2. Durkacz, B.N., Omidiji, O., Gray, D.A., Shall, S. (1980) Nature 283: 593-596
3. Benjamin, R.C., Gill, D.M. (1980) J Biol Chem 255: 10493-10510
4. Cleaver, J.E., Bodell, N.J., Morgan, W.F., Zelle, B. (1983) J Biol Chem 258: 9059-9068
5. Bhattacharyya, N., Bhattacharjee, S.B. (1983) Mutation Res 121: 287-292
6. Bhattacharjee, S.B., Bhattacharyya, N. (1986) Carcinogenesis 7: 1267-1271
7. Cantoni, O., Murray, D., Meyn, R.E. (1986) Biochim et Biophys Acta 867: 125-143
8. Hahn, G.M., Little, J.B. (1984) Curr Top in Radiat Res 8: 39-83
9. Maher, N.M., Dorney, D.J., Mendrala, A.L., Konze-Thomas, B., McCormick, J.J. (1979) Mutation Res 62: 311- 323
10. Nagasawa, H., Fornace Jr., A.J., Ritter, M.A., Little, J.B. (1982) Radiat Res 92: 483-496
11. Chan, G.L., Little, J.B. (1979) Mutation Res 63: 401-412
12. Rao, B.S., Hopwood, L.E. (1982) Int J Radiat Biol 42: 501-508
13. Iliakis, G. (1984) Mutation Res 126: 215-225

Influence of Benzamide and Nalidixic Acid on the Killing and Mutations Induced in V-79 Cells by N-Methyl-N'-Nitro-N-Nitrosoguanidine

Rita Ghosh (Datta) and Sukhendu B. Bhattacharjee

Crystallography and Molecular Biology Division, Saha Institute of Nuclear Physics, I/AF, Salt Lake, Calcutta-700064, India

Introduction

Relaxation in the chromatin superstructure favoring DNA accessibility could be an early event in the repair pathway in the eucaryotic genome (1). Such a condition can be achieved through poly(ADP- ribosyl)ation of histone H1 (2). Relaxation of DNA can also be achieved through the action of topoisomerases which catalyse the concerted nicking and closing of duplexes, relieving the topological constraints of super-twisting. Benzamide is a known inhibitor of poly(ADP-ribose) synthesis and nalidixic acid is an inhibitor of topoisomerase activity. The role of poly(ADP-ribose) in repair of the eucaryotic genome has been the subject of many investigations. Inhibitors of topoisomerases such as novobiocin and nalidixic acid have also been used in the elucidation of the role of topoisomerases in DNA repair activity (3-7). But, as in the case of poly(ADP-ribose), much remains to be explained. Furthermore, there are recent reports indicating that the topoisomerases are modified by poly(ADP-ribosyl)ation (8, 9). In both of these reports the enzymes were inactivated by ADP-ribosylation. In the present investigation we have used inhibitors of both enzymes, separately and in combination, to see if poly(ADP-ribose) and topoisomerase have anything common in their action. The end points chosen were killing and mutation induced by N-methyl-N'-nitro-N-nitrosoguanidine (MNNG). For mutational analysis, resistance to the drug 6-thioguanine has been used as the phenotypic expression of mutation.

Results

Fig. 1 shows the cytotoxicity of nalidixic acid in V-79 cells. Cells were exposed to nalidixic acid at different concentrations in complete growth medium (MEM) for 20 hr and then washed, diluted and replated in fresh medium without nalidixic acid. Viability was determined by counting colonies after about 7 days growth. It may be seen from the figure that nalidixic acid was not cytotoxic for this cell line in these experimental conditions up to a concentration of 200 µg/ml. For experimental purposes here, concentrations of nalidixic acid chosen were in the non-toxic range.

265

Fig. 2 shows the influence of different concentrations of nalidixic acid on the killing induced by MNNG in a 1 hr treatment; the MNNG containing medium was replaced by fresh medium containing nalidixic acid at different concentrations and cells were exposed for 20 hr. In this figure, two doses of MNNG, 0.5 µg/ml and 1µg/ml were used. Normally survival levels at these two doses were 0.55 and 0.30, respectively, when there was no other treatment. However, on exposure to nalidixic acid at different concentrations, the cell survival varied. It may be seen that maximal sensitization by nalidixic acid was attained at 100 µg/ml. No further sensitization could be obtained at higher concentrations of nalidixic acid.

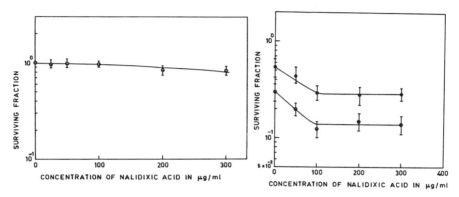

Fig. 1: (left) Cytotoxicity of nalidixic acid in V79 cells after a treatment time of 20 hr. The results are the mean of three experiments. The bars represent the standard deviations. Reprinted with permission from ref. 21.

Fig. 2: (right) Effect of varying concentrations of nalidixic acid on killing of V79 cells by MNNG in 1 hr. Two concentrations of MNNG have been shown: (●) represents 0.5 µg/ml and (◉) represents 1.0 µg/ml. Nalidixic acid treatment time was 20 hr after the MNNG treatment. The results are the average of 3 independent experiments. The bar represents the standard deviation.

Fig. 3 shows the survival of cells exposed to MNNG at various concentrations for 1 hr, followed by exposure to nalidixic acid at 100 µg/ml for 20 hr in fresh medium. It can be observed that nalidixic acid increased the killing by MNNG at all doses. The slope of the survival curve was also increased. Treatment of normal undamaged cells with 100 µg/ml of nalidixic acid did not affect their colony forming ability.

Fig. 4 compares the effects produced by benzamide and nalidixic acid upon killing of V-79 cells by MNNG. After a 1 hr treatment with MNNG, benzamide or nalidixic acid or both were added in fresh medium for 20 hr. As has been reported earlier from this laboratory, maximum potentiation of killing was observed at 5 mM (600 µg/ml) benzamide (10). Cells treated with 1 mM (120 µg/ml) benzamide and nalidixic acid (100 µg/ml), showed much more enhanced killing compared to cells which were treated with benzamide (1 mM) or nalidixic acid (100 µg/ml) alone. However, when MNNG exposed cells were treated with 5 mM benzamide, along with 100 µg/ml nalidixic acid, there was no further increase in killing compared to those treated cells exposed to 5 mM benzamide.

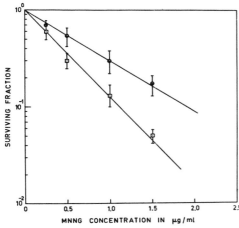

Fig. 3: Effect of treatment with 100 µg/ml of nalidixic acid for 20 hr on MNNG induced killing of V-79 cells. Symbol (⊙) represents the survival curve for 1 hr MNNG treatment alone and (□) represents that for cells exposed to nalidixic acid after MNNG treatment. The results are the average of 3 indepedent experiments. The bar represents the standard deviations. Reprinted with permission from ref. 21.

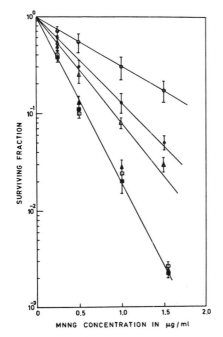

Fig. 4: Effect of nalidixic acid and benzamide, both separately and in conjunction with MNNG induced killing of V79 cells. Open symbols indicate treatment with benzamide for 20 hr after exposure to MNNG : $0 \, \mu g/ml$ benzamide (⊙), 120 µg/ml (1mM) benzamide (△), 600 µg/ml (5mM) benzamide (□), closed symbols indicate treatment with 100 µg/ml of nalidixic acid along with various concentrations of benzamide for 20 hr after MNNG treatment: 0 µg/ml benzamide + nalidixic acid (●), 120 µg/ml benzamide + nalidixic acid (▲), 600 µg/ml benzamide + nalidixic acid (■). Each point is the mean of three independent experiments and the bar represents the standard deviation.

Fig. 5 shows the effect of nalidixic acid on MNNG induced mutagenicity. Here the MNNG exposed cells were grown with or without 100 µg/ml of nalidixic acid in fresh medium for 20 hr. Thereafter, the cells were allowed

267

to grow for 9 days in normal growth medium for mutation expression with 3-4 subcultures during this period. After expression, cells were replated at 1 x 10^5 cells per 100 mm dish in selection medium containing 5µg/ml of 6-thioguanine. Clones resistant to 6-thioguanine were scored as mutants. Nalidixic acid at 100 µg/ml did not produce any 6-thioguanine resistant mutants above the background level (0.33 mutant/10^5 cells). It may be seen from the figure that the mutation frequency obtained at various doses of treatment with MNNG was less for cells exposed to nalidixic acid than to cells exposed to MNNG alone. With benzamide treatment following exposure to MNNG, there was also a decrease in mutation frequency which had been reported earlier (10). Simultaneous treatment with benzamide and nalidixic acid after MNNG exposure also suppressed the mutation induction (data not shown).

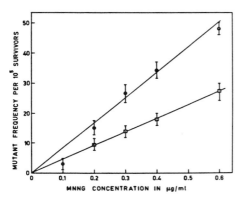

Fig. 5: Effect of nalidixic acid on MNNG induced mutations. The symbol (⊙) represents MNNG induced mutation frequency for 1 hr treatment at different concentrations and (□) represents the mutation frequency when MNNG exposure is followed by nalidixic acid treatment for 20 hr at a concentration of 100 µg/ml. The results are the mean of 3 independent experiments, the bars represent the standard deviations.

Discussion

The influence of the topoisomerase inhibitor nalidixic acid on MNNG induced killing and mutation is similar to the effects of inhibition of poly(ADP-ribose) synthesis following exposure to MNNG as reported earlier from our laboratory (10). It has also been observed in our laboratory that these two classes of inhibitors produced similar types of effects in cells exposed to ionizing radiation or UV light (unpublished observation). Furthermore, the two types of inhibitors have been found to complement each other.

It is now established that DNA damaging agents stimulate the activity of poly(ADP-ribose) polymerase (11-15). The nuclear polymer, poly(ADP-ribose) has diverse biochemical functions, all of which are not fully known. However, it is likely to be involved in DNA repair (12, 16-18). The roles of both poly(ADP-ribose) and topoisomerase in repair activity are not clear; our results indicate that the inhibitors could be acting through some common pathway. That topoisomerase inhibitors were affecting a pre-incisional step of DNA repair had been proposed by several authors (3, 4). Such proposals have also been put forward for poly(ADP-ribose) inhibitors (1). However, there is also evidence that the stage of repair that is affected is ligation for both topoisomerase inhibitors (19, 20) and for poly(ADP-ribose) inhibition (18). The effect of the two types of inhibitors could either be on a preincision step of repair or in the ligational stage. Our findings with mutational analysis give us some idea about the common pathway involved. If the effect was in the ligational stage, then because repair synthesis had already occurred, mutational induction should remain unaffected. However, we find in both cases mutation induction is suppressed. This indicates involvement of poly(ADP-ribose) or topoisomerase in the preincisional stage of repair.

References

1. de Murcia, G., Huletsky, A., Lamarres, D., Gadreau, A., Pouyet, J., Daune, M., Poirier, G.G. (1986) J Biol Chem 261:7011-7017
2. Poirier, G.G., de Murcia, G., Jongstra-Bilen, J., Niedergang, C., Mandel, P. (1982) Proc Natl Acad Sci USA 79: 3423-3427
3. Collins, A.R.S., Johnson, R.T. (1979) Nucleic Acid Res 7: 1311-1320
4. Mattern, M.R., Scudeiro, D.A. (1981) Biochem Biophys Acta 653: 248-254
5. Cleaver, J.E. (1982) Carcinogenesis 3: 1171-1174
6. Collins, A.R.S., Squires, S., Johnson, R.T. (1982) Nucleic Acid Res 10: 1203-1213
7. Mattern, M.R., Raone, R.F., Day, R.S. (1982) Biochem Biophys Acta 697: 6-13
8. Ferro, A.M., Olivera, B.M. (1984) J Biol Chem 259: 547-554
9. Darby, M.K., Schmitt, B., Jongstra-Belin, J., Vosberg, H. (1985) EMBO 4: 2129-2134
10. Bhattacharyya, N., Bhattacharjee, S.B. (1983) Mutat Res 121: 287-292
11. Berger, N.A., Sikorski, G.W., Petzold, S.J., Kurohara, K.K. (1980) Biochemistry 19: 289-293
12. Durkacz, B.W., Omidiji, O., Gray, D.A., Shall, S. (1980) Nature 283: 593-596
13. Nduka, N., Skidmore, C.J., Shall, S. (1980) Eur J Biochem 105: 525-530
14. Benjamin, R.C., Gill, D.M. (1980) J Biol Chem 255: 10493-10501
15. Benjamin, R.C., Gill, D.M. (1980) J Biol Chem 255: 10502-10508
16. Durkacz, B.W., Irwin, J., Shall, S. (1981) Biochem Biophys Res Commun 101: 1433-1441
17. Zwelling, L.A., Kerrigan, D., Pommier, Y. (1982) Biochem Biophys Res Commun 104: 897-902
18. Criessen, D., Shall, S. (1982) Nature 276: 271-272
19. Mattern, M.R., Zwelling, L.A., Kerrigan, D., Kohn, K.W. (1983) Biochem Biophys Res Comm 112: 1077-1084
20. Meechan, P.J., Killpack, S., Cleaver, J.E. (1984) Mutat Res 141: 69-73
21. Ghosh (Datta), R. and Bhattacharjee, B. (1988) Mutat. Res. 202: 71-75

Loss of Poly(ADP-Ribose) Transferase Activity in Liver of Rats Treated with 2-Acetylaminofluorene

C.F. Cesarone, A.I. Scovassi[1], L. Scarabelli, R. Izzo[1], M. Orunesu, and U. Bertazzoni[1]

Institute of General Physiology, Faculty of Sciences, University of Genoa, 16132-I Genoa, Italy

Introduction

The chronic exposure of rats to a feeding regimen containing sub-carcinogenic doses of N-2-acetylaminofluorene (2AAF) induces, in liver, multiple effects and the progression to hepatocellular carcinoma(s) (1). Among the effects induced at the hepatocellular level, which seem to play a crucial role in the initial steps of the carcinogenic process, are the alterations to DNA and the reduction in DNA repair capacity (1, 2). The presence of DNA damage is not sufficient *per se* to cause initiation of carcinogenesis since cell proliferation is also required for the induction of hepatic foci (3). Several enzymatic activities, involved in DNA excision repair primed by 2AAF exposure, such as UV- and AP-endonucleases, DNA polymerase β and DNA ligase II, were reported to be unmodified in the liver of treated animals (1, 4). We have investigated the possible involvement of poly(ADP-ribosyl) transferase in the carcinogenic process induced in rat liver by 2AAF exposure.

The *in vivo* experimental model used was essentially that developed by Teebor and Becker (5). It involves the treatment of male Wistar rats with a 4-cycle discontinuous feeding regimen containing 2AAF, each cycle being composed of a 3 week treatment and 1 week of recovery (Fig. 1). Moreover, we determined during the different cycles of treatment, the extent of DNA damage and repair by the alkaline elution technique and the activity and structure of poly(ADP-ribose) transferase by means of an activity gel procedure. In addition, we investigated the protective effect exerted on DNA and poly(ADP-ribose) transferase by N-acetylcysteine (NAC) during the 4-cycle treatment.

Results

The level of DNA damage produced in each treatment cycle after 3 weeks of 2AAF exposure, compared with the residual damage still present after the recovery week, is shown in Fig. 2. The decrease in DNA elution

---

[1]Institute of Biochemical and Evolutionary Genetics National Research Council (CNR), 27100-I Pavia, Italy

rate observed at the end of the first two cycles indicates that about 50% of the DNA damage is repaired during the recovery period. When the aminothiol, NAC, was supplemented in the carcinogen containing diet, the extent of DNA damage was significantly reduced and a complete repair was attained after one week of recovery. During the last two cycles of treatment the protective effect of NAC was drastically affected and the extent of DNA repair was similar to that observed for 2AAF-treated animals.

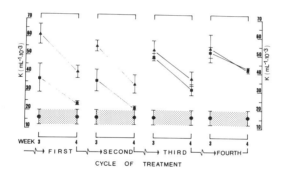

Fig. 1. Schematic representation of the experimental model for the discontinuous 4-cycle 2AAF exposure. Male Wistar rats (100 ± 10 g) were used. Each cycle was a 3 week treatment with 0.05% 2AAF supplied in the diet, followed by 1 week of recovery. NAC-supplemented animals received 0.1% N-acetylcysteine in the diet concomitantly with the carcinogen. Control rats received either a standard diet (24.5% protein) or a diet containing 0.1% NAC. After the third and the fourth week of each cycle, 5-6 animals, from each group, were sacrificed under slight ether anesthesia.

Fig. 2. Elution rate constants for liver DNA of rats exposed to a feeding regimen containing 0.05% 2AAF (▲) or 0.05% 2AAF + 0.1% NAC (■) determined after the third and the fourth week of each cycle of exposure, scheduled as described in Fig. 1. The alkaline elution of DNA was essentially techniques of Kohn et al. (6), combined with a fluorometric method for DNA determination (7). The elution rate constant ($K = - \ln Q_v/V$) was computed from the semilog plot of the DNA fraction retained on the filter (Q_v) versus the eluted volume (V). The values are proportional to the DNA damage and reflect the number of alkali labile lesions induced by 2AAF metabolites. Each value is the mean $\pm$ S.E. of at least 5 determinations, 8 for control animals (●), ($P = 0.05$).

The poly(ADP-ribose) transferase activity was determined at the end of each cycle of exposure using an activity gel technique. The active band of 116 kDa, evident in control rat liver extracts (Fig. 3A, lane 1), was undetectable after one cycle of treatment with 2AAF (lane 2). The active band reappeared progressively during additional 3 cycles (Fig. 3A, lanes 3 to 5). The protective effect exerted by NAC on the enzyme is clearly demonstrated in the autoradiogram shown in Fig. 3B. In this case the active band was still evident after the first cycle (lane 2), became barely visible after the second cycle, and returned to normal levels later.

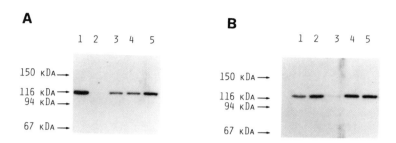

Fig. 3. Activity gel analysis of poly(ADP-ribose) transferase in liver extracts from rats exposed to a 4-cycle discontinuous feeding regimen containing 0.05% 2AAF (panel A) of 0.05% 2AAF + 0.1% NAC (panel B). Each autoradiogram shows: in lane 1 the poly(ADP-ribose) transferase activity band of control animals, and in lanes 2, 3, 4, and 5 the activity bands after the 1, 2, 3 and 4 cycle of treatment, respectively.

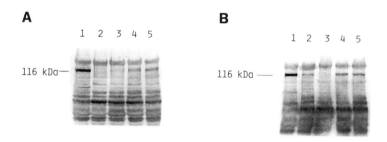

Fig. 4. Western blot analysis of poly(ADP-ribose) transferase in liver extracts from rats exposed to a 4-cycle discontinuous feeding regimen containing: 0.05% 2AAF (panel A) or 0.05% 2AAF + 0.1% NAC (panel B). Each section shows: in lane 1 the poly(ADP-ribose) transferase protein band of control animals, and in lanes 2, 3, 4, and 5 the poly(ADP-ribose) transferase immunoreactive peptide after the 1, 2, 3, and 4 cycle of treatment, respectively.

In order to ascertain that the loss in the poly(ADP-ribose) transferase band was due to a reduction in the amount of enzyme present in the hepatocytes of treated animals, rat liver extracts were analyzed by the western blot technique (9) using a polyclonal antibody against poly(ADP-ribose) transferase (donated by C. Niedergang). The protein band of 116 kDa, corresponding to poly(ADP-ribose) transferase, recognized in control liver (Fig. 4A, lane 1), became undetectable after the first and second cycle of exposure and reappeared progressively during the ongoing treatment. When 2AAF-exposed animals were supplemented with NAC, the immunoreactive band at 116 kDa was absent only after the second cycle of treatment (Fig. 4B). These results indicate that the loss of poly(ADP-ribose) transferase bands, observed in activity gels, results from a loss of enzyme protein.

Discussion

Several lines of evidence indicate that poly(ADP-ribose) transferase plays a central role in modulating the cellular response to DNA damage (10). As an experimental model to study DNA damage, DNA repair and poly(ADP-ribose) transferase activity, we used an *in vivo* system consisting of a discontinuous treatment of rats with 2AAF, known to induce liver carcinogenesis (1). The results obtained indicated that DNA was repaired with decreased efficiency throughout the treatment and that poly(ADP-ribose) transferase activity was markedly affected; it became undetectable on activity gels and on western blots after the first cycle of 2AAF exposure, and then reappeared progressively. A possible explanation of this phenomenon might be the observed loss of normal hepatocytes and the concomitant increase in 2AAF resistant cells in liver of treated rats (3).

It therefore can be suggested that the loss in poly(ADP-ribose) transferase activity could be related to a modified ratio between the two cell types. This hypothesis is also supported by the observation that liver protection by the aminothiol NAC prevents the loss in poly(ADP-ribose) transferase up to the second cycle of treatment.

In conclusion, our results indicate that:

1. The DNA damage induced in rat liver by a 4-cycle 2AAF exposure is only partially repaired. The concomitant treatment with NAC exerts a protective effect against DNA damage and allows a complete DNA repair during the first two cycles of treatment.
2. poly(ADP-ribose) transferase, visualized as an active band or as an immunoreactive peptide, was not detectable after the first cycle of 2AAF treatment. The presence of NAC during the exposure to the carcinogen prevents the loss of poly(ADP-ribose) transferase up to the second cycle.
3. The progressive reappearance of liver poly(ADP-ribose) transferase activity, during the treatment, seems to be related to the modified activity ratio between normal hepatocytes and resistant cells induced by the carcinogen (5).

Acknowledgements. This work was supported in part by the Italian National Research Council (CNR), Special Project "Oncology" contract n.86.00357.44 and by contract B16-158 of the Radiation Programme of the Commission of the European Communities. U. Bertazzoni is a scientific official of the CEC, Bruxelle, and R. Izzo is a recipient of a fellowship from the Fondazione Buzzati-Traverso, Roma. We wish to express our appreciation to Dr. V. Cicchetti, medical director of Zambon Farmaceutici, Bresso, Milano, for a grant to L. Scarabelli and for the gift of N-acetylcysteine. Figs. 2 and 4 were reprinted with permission from Cesarone *et al.*, Cancer Res. 48, 3581-3585, 1988.

References

1. Becker, F.F. (1981) Am J Pathol 105: 3-9
2. Cesarone, C.F., Scarabelli, L., Orunesu, M. (1986) Toxicol Path 14: 445-450
3. Farber, E. (1984) Cancer Res 44: 4217-4223
4. Chan, J.Y.H., Becker, F.F. (1985) Carcinogenesis 6: 1275-1277
5. Teebor,G.W., Becker, F.F. (1971) Cancer Res 31: 1-3
6. Kohn, K.W., Erickson, L.C., Grimek-Ewig, R.A., and Friedman, C.A. (1976) Biochemistry 15: 46294637
7. Cesarone, C.F., Bolognesi, C., Santi, L. (1979) Anal Biochem 100: 188-197
8. Scovassi, A.I., Stefanini, M., Bertazzoni, U. (1984) J Biol Chem 259: 10973-10977
9. Towbin, H., Staehelin, T., Gordon, J. (1979) Proc Natl Acad Sci USA 76: 4350-4354
10. Shall, S. (1984) Adv Radiat Biol 11: 1-69

V-79 Cell Variants Deficient in Poly(ADP-Ribose) Polymerase

Satadal Chatterjee, Shirley J. Petzold, Sosamma J. Berger, and Nathan A. Berger

Hematology/Oncology Division, Departments of Medicine and Biochemistry, R.L. Ireland Cancer Center, Case Western Reserve University, Cleveland, Ohio 44106 USA

Introduction

We devised a strategy to select cells with reduced levels of poly(ADP-ribose) polymerase based on our demonstration of the involvement of this enzyme in the suicide response after high levels of DNA damage (1, 2). We reasoned that high levels of DNA damage would activate the enzyme to deplete NAD and ATP pools in normal cells resulting in the cessation of all energy dependent processes and rapid cell death, whereas, cells deficient in the enzyme activity would avoid these metabolic perturbations and therefore be more likely to survive. Employing the strategy we isolated variants from V-79 Chinese hamster cells having 5-11% of the normal enzyme activity.

Results

V-79 cells were mutagenized with 1 µg/ml MNNG (N-methyl-N'-nitro-N-nitrosoguanidine) for 1 hr and allowed 7-10 days expression. Then the cells were exposed to very high concentrations of MNNG (10-20 µg/ml for 24 hr) with the expectation that the suicide response would be induced only in cells having the enzyme activity. The surviving colonies were isolated, expanded and subjected to repeated mutagenesis and selection procedures. Variants ADPRT 54 and ADPRT 351 containing 5-11% of normal enzyme activity were isolated from V-79 cells following 5 rounds of selection. Results of the first 3 rounds of selection were described previously (3).

To measure total poly(ADP-ribose) synthesis under conditions of maximum DNA damage, we incubated permeabilized cells (1×10^6) along with 250 µM [^{3}H]NAD (specific activity 19×10^3 dpm/nmol), 0.04% Triton X-100 and 300 µg/ml DNase for different periods of time at 37°C and measured the total incorporation of [^{3}H]ADP-ribose from [^{3}H]NAD into trichloroacetic acid insoluble products (3). Fig. 1 shows the time courses of poly(ADP-ribose) synthesis in V-79 and variant cell lines. For the parental V-79 cells polymer synthesis increased for 30 min followed by a plateau. The variants reached saturation within 2 min. Most strikingly, there was a marked difference in the maximum amount of polymer synthesized by V-79 cells compared to the variants. Also the initial rates of polymer synthesis were much lower in the variants than in V-79 as detailed in the inset of

275

Fig. 1. Measurements of NAD glycohydrolase, poly(ADP-ribose) glycohydrolase and phosphodiesterase showed that all of these enzyme activities were lower in the variants than in the V-79 cells. Thus the apparent decrease in poly(ADP-ribose) synthesis in the variant cells could not be explained by rapid degradation of substrate or polymer.

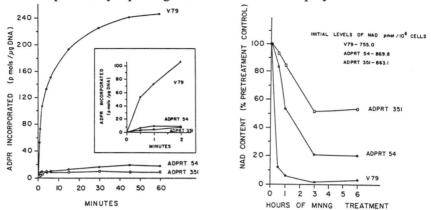

Fig. 1. Time course of poly(ADP-ribose) synthesis in V-79 and variant Chinese hamster cell lines measured as incorporation of [³H]ADP-ribose from [³H]NAD in permeabilized cells treated with DNase as described in text. Initial part of the reaction is shown in the insert. V-79 cells, (●); variant cell lines ADPRT 54, (○); ADPRT 351, (□).

Fig. 2. NAD⁺ lowering in V-79 and variant cell lines in response to treatment of cells with 10 μg/ml MNNG. MNNG was added at 0 time and then samples collected in duplicate at the indicated time for enzymatic cycling measurements of NAD⁺ (8). V-79 cells, (●); variant cell lines ADPRT 54, (○); ADPRT 351, (□).

Fig. 2 shows the effect on NAD levels of treating intact cells with high dose MNNG (10 μg/ml) for up to 6 hr. The initial rates of NAD lowering in the variants were much slower than those in parental V-79 cells. In addition, V-79 cells almost completely depleted their NAD within 3 hr, whereas, the variants ADPRT 54 and ADPRT 351 retained more than 20% and 50% of their NAD levels, respectively.

To examine whether the decreased synthesis of poly(ADP-ribose) in the variants was associated with a change in the size of the enzyme or due to any alteration in the pattern of ADP-ribosylation of acceptor proteins, we incubated permeabilized cells with [³²P]NAD and DNase for 15 min at 37°C and subjected the product to SDS polyacrylamide gel electrophoresis and autoradiography. Fig. 3 shows that each of the cell lines had radioactive bands of molecular weight 116,000, corresponding to the enzyme. The autoradiograph also shows that automodified poly(ADP-ribose) polymerase from ADPRT 351 did not make sufficiently long chain polymers to produce a band which migrated at the top of the gel. For ADPRT 54, the automodified enzyme showed a relative decrease of molecules with polymers of intermediate chain length (≈200,000 molecular weight). Other

ADP-ribosylated protein bands appeared to be qualitatively similar in variants and V-79 cells although the extent of ADP-ribosylation of these bands was much less in the variants than in the parental V-79 cells.

Table 1. Characteristics of cell lines with reduced poly(ADP-ribose) polymerase activity

Cell Line	DNA Content $\mu g/10^6$ cells	Protein Content $\mu g/10^6$ cells	Doubling Time (hr)	Sister Chromatid Exchange per Chromosome
V-79	6.1	118.21	10	0.29 + 0.12
ADPRT 54	13.65	217.83	68	1.78 + 0.19
ADPRT 351	15.37	298.16	50	3.21 + 0.51

For measurement of sister chromatid exchanges, cells were grown for two cycles in medium containing 10 μM bromodeoxyuridine. Differentiation of sister chromatids was achieved by the method of Goto *et al.* (5). At least 20 metaphases were analyzed to determine sister chromatid exchanges for each clone. Assays for protein and DNA were performed as previously described (6, 7).

As a preliminary characterization of the possible effects of decreased poly(ADP-ribose) polymerase activity in the variants, we compared their doubling time and spontaneous sister chromatid exchange frequency in addition to measuring their DNA and protein contents. Column 3 of Table 1 shows that the doubling time of V-79 cells was 10 hr, whereas, for the variants it ranged from 50-68 hr. The variants also showed a significant increase in baseline sister chromatid exchange rate compared to their parental line. In fact the sister chromatid exchange per chromosome was an order of magnitude higher in one of the variants than in the parental line. Their DNA and protein contents were also approximately 2 fold higher than that in V-79 cells.

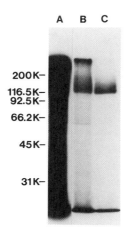

Fig. 3. Autoradiograph of [^{32}P]ADP-ribosylated proteins in V-79 and variant cells. Cells were permeabilized and 1×10^6 cells incubated at 37°C for 15 min with 1 μM [^{32}P] NAD (specific activity 7800 cpm/pmol) and 300 μg/ml DNase to induce maximal DNA damage. Reactions were stopped by addition of sample solution and boiled 2 min. Solubilized samples were electrophoresed on a SDS, 9% polyacrylamide slab gel with a 3% polyacrylamide stacking gel and the autoradiographed. Lanes A, B, and C represent V-79, variant ADPRT 54 and variant ADPRT 351, respectively. Autoradiographs were exposed for 24 hr. Molecular weights indicated on the left of autoradiograph were determined by Coomassie Brilliant Blue staining of standards included in the slab gel.

277

Discussion

Isolation and characterization of variants deficient in known enzyme activities play an important part in elucidation of the role of those enzymes in cellular processes. We took advantage of the involvement of poly(ADP-ribose) polymerase in the suicide response to develop a selection procedure for cells deficient in this enzyme activity. We have shown that a 10-20 fold decrease in poly(ADP-ribose) synthesis in the variants under maximum DNA damaging conditions was not due to an increase in either poly(ADP-ribose) glycohydrolase or NAD glycohydrolase activities. We have also shown that induction of high levels of DNA damage by MNNG in intact cells did not result in drastic depletion of NAD in the variants as it did in the parental cell line. Moreover, the variants showed qualitative as well as quantitative differences in the auto ADP-ribosylation pattern of the enzyme. All these observations point to the fact that the variants are truly defective in poly(ADP-ribose) polymerase activity.

Our preliminary characterization showed that the variants had 5-7 times longer doubling times than their parental lines. They also showed nearly a 10 fold increase in spontaneous sister chromatid exchange rates. Previous studies (4) showed that treatment of the cells with 3-aminobenzamide, a potent inhibitor of the enzyme activity, significantly increased that baseline sister chromatid exchange frequency. Thus the possibility of involvement of poly(ADP-ribose) polymerase in regulation of sister chromatid exchanges seems very likely. Further studies will be required to determine exactly which phenotypic characteristics are associated with poly(ADP-ribose) polymerase deficiency. Perhaps the availability of the cloned gene will make it possible to reconstitute these cells and identify which functions are directly associated with the activity of poly(ADP-ribose) polymerase.

Acknowledgements. These studies were supported by NIH grants GM32647, CA35983 and CA43703.

References

1. Berger, N.A. (1985) Rad Res 101: 4-15
2. Berger, S.J., Sudar, D.C., Berger, N.A. (1986) Biochem Biophys Res Commun 134: 227-232
3. Chatterjee, S., Petzold, S.J., Berger, S.J., Berger, N.A. (1987) Exp Cell Res, In press
4. Oikawa, A., Tohda, H., Kanai, M., Miwa, M., Sugimura, T. (1980) Biochem Biophys Res Commun 97: 1311-1316
5. Goto, K., Maeda, S., Kano, Y., Sugiyama, T. (1978) Chromosoma 66: 351-359
6. Lowry, O.H., Rosebrough, N.J., Farr, A.L., Randall, R.J. (1951) J Biol Chem 193: 265-275
7. Downs, T.R., Wilfinger, W.W. (1983) Analyt Biochem 131: 538-547
8. Berger, N.A., Berger, S.A., Sikorski, G.W., Catino, D.M. (1982) Exp Cell Res 137: 79-88

Cytotoxicity of TPA-Stimulated Granulocytes and Inhibitors of ADP-Ribosylation

Thomas Meyer, Andreas Ludwig, Helgard Lengyel, Gity Schäfer, and Helmuth Hilz

Institut für Physiologische Chemie, Universität Hamburg, 2 Hamburg 20, West Germany

Introduction

Reactive oxygen species as derived from stimulated phagocytes are able to induce DNA fragmentation (1, 2). Since fragmented DNA is known to activate poly(ADP-ribose) polymerase (3), it could be expected that target cells respond with an increased poly(ADP-ribosyl)ation of their nuclear proteins. Oxygen species thought to be derived from stimulated phagocytes like H_2O_2 or O_2^- radicals indeed induced an activation of poly(ADP-ribose) polymerase (4, 5). Also, self-activation in monocytes stimulated to produce O_2^- has been reported (6). Because of the possible role of reactive oxygen species produced by phagocytic cells in cytotoxic and carcinogenic side effects of inflammation (7), we have analyzed the action of TPA-stimulated human granulocytes on DNA fragmentation and ADP-ribosylation patterns in co-cultivated fibroblasts. We also studied the effect of ADP-ribosylation inhibitors and compared the results with those produced by an alkylating agent.

Results and Discussion

3-aminobenzamide affords partial protection against the cytotoxic effects of stimulated granulocytes. Human granulocytes incubated with TPA produce O_2^- radicals that can be measured by their reductive activity on cytochrome c (8). These radicals or their breakdown products are cytotoxic to cells (7). However, when 3T3 fibroblasts were co-cultivated with TPA-stimulated granulocytes in the presence of 3-aminobenzamide, the reduction of proliferative capacity was partially prevented. Protection was found with concentrations usually considered to be specific for inhibition of poly(ADP-ribose) polymerase (<5mM); (Fig. 1). 3-Aminobenzoate, which is not an inhibitor of the polymerase, had no significant inhibitory potential. The effect of 3-aminobenzamide in the granulocyte system is opposite to its co-cytotoxic action seen in combination with an alkylating agent (9, 10) (Fig. 1).

Similar to alkylating agents, the reactive oxygen species released from stimulated phagocytes are known to have clastogenic and genotoxic effects (1, 2). However, significant differences in detail emerged when these two treatments were compared. When exposed to TPA-stimulated granulocytes, DNA fragmentation in 3T3 cells exhibited an unusual alkaline elution profile. The elution pattern was biphasic, indicating the existence of two populations of fragmented DNA. The treatment also induced poly(ADP-ribosyl)ation of nuclear proteins. The time course was somewhat retarded and the extent of modification reached only two thirds of that of treatment with equitoxic doses of DMS (Fig. 2). Interestingly, treatment of 3T3 cells in the presence of 3-aminobenzamide partially suppressed fragmentation, especially that of the less fragmented fraction. This again is unlike that observed following alkylation, where the nicotinamide analog increased DNA fragmentation.

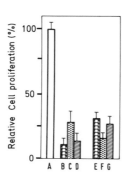 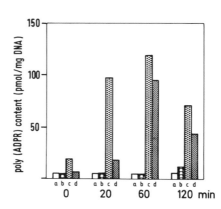

Fig. 1. (left) Protection and enhancement of cytotoxicity by 3-aminobenzamide in 3T3 cells. Monolayer cultures were treated for 60 min with human granulocytes + 25 ng/ml TPA in Krebs-Ringer phosphate (KRP) or with 100 μM dimethylsulfate (DMS) in KRP ± 2.5 mM 3-aminobenzamide (ABA) or 3-aminobenzoate (AB). After incubation for 3 days in DME medium ± ABA or AB, the increase in cell number was determined. Mean values from triplicate. A: control; B-D: TPA + granulocytes; E-G: + DMS; C and F: ABA; D and G: + AB.

Fig. 2. (right) Polymeric ADP-ribose residues in 3T3 fibroblast cultures after treatment with DMS or stimulated granulocytes. Cultures were treated with DMS or stimulated granulocytes as described in Fig. 1. After the times indicated, cells were precipitated *in situ* with cold TCA and analyzed for poly(ADP-ribose) as described previously (11). a: controls (+ granulocytes); b: + TPA (25 ng/ml); c: + 200 μM DMS; d: + TPA + granulocytes.

Protection by 3-aminobenzamide is not mediated by an inhibition of ADP-ribosylation.

At high concentrations (20 mM), benzamide and analogs inhibited O_2^- production from TPA-stimulated granulocytes, while the corresponding acids did not (Table 1). However, when analyzed in the cytotoxicity test, it was found that benzamide and nicotinamide at 2.5 mM concentrations did not prevent but enhanced cytotoxicity. This pointed to a special scavenger function of ABA in the prevention of cell damage as induced by reactive oxygen species. Protection by ABA was also observed in benzoylperoxide induced damage, and in two stage carcinogenesis in mice. In the latter system, TPA induces local inflammation with the accumulation of granulocytes and the development of multiple papillomas. It seems likely the reactive oxygen species formed under these conditions are scavenged by 3-aminobenzamide.

Table 1. Inhibition of O_2^- production from TPA-stimulated human granulocytes by inhibitors of poly(ADP-ribose) polymerase.

Additions	O_2^- Producing Activity (Relative increase in A_{550}/min)
none	100
benzamide	37 ± 12
3-aminobenzamide	65 ± 16
nicotinamide	68 ± 15
benzoate	91 ± 6
3-aminobenzoate	119 ± 17
nicotinate	110 ± 10

Human granulocytes were incubated at 25°C in one ml Krebs-Ringer phosphate containing 25 ng TPA, 25µM cytochrome c and the additions at 20 mM final concentration. The initial increase in A550 was registered. Mean values ± s.d. from three separate experiments.

From our data, the following conclusions can be drawn: Treatment of 3T3 fibroblasts with human neutrophils in the presence of the tumor promoter TPA induced DNA fragmentation, poly(ADP-ribosyl)ation of nuclear proteins, and cytotoxicity that was partially prevented by 2.5 mM 3-aminobenzamide. This effect of ABA was not mediated by an inhibition of poly(ADP-ribosyl)ation since other inhibitors did not prevent the cytotoxic action. ABA also inhibited cytotoxicity of the tumor promoter benzoylperoxide and it proved to be a potent anti-tumor promoter in mouse skin. This dualistic action of 3-aminobenzamide, inhibition of poly(ADP-ribose) polymerase and scavenging of reactive oxygen species, might explain divergent reports on the function of ADP-ribosylation in cell damage and carcinogenesis.

Acknowledgements. This work was supported by the Deutsche Forschungsgemeinschaft, SFB 232.

References

1 Birnboim, H. (1982) Science 215: 1247-1249
2. Emerit, I., Cerutti, P. (1981) Nature 293: 144-146
3. Benjamin, R.C. and Gill, D.M. (1980) J Biol Chem 255: 10502-10508
4. Schraufstätter, I.V., Hyslop, P.A., Hinshaw, D.B., Sprugg, R.G., Sklar, L.A., Cochram, C.W. (1986) Proc Nat Acad Sci USA 83: 4908-4912
5. Carson, D.A., Seto, S., Wasson, D.B. (1986) J Exp Med 163: 746-751
6. Singh, N., Poirier, G., Cerutti, P. (1985) Biochem Biophys Res Commun 126: 1208-1214

7. Del Maestro, R.F. (1984) Free Radicals in Molecular Biology, Aging and Disease A. Armstrong *et al*. (eds.) Raven Press, pp.87-102
8. Cohen, H.J., Chavanice, M.E. (1978) J Clin Invest 61: 1081-1087
9. Nduka, N., Skidmore, C.J., Shall, S. (1980) Eur J Biochem 105: 525-530
10. Hilz, H. (1981) Hoppe Seyl Z Physiol Chem 362: 1415-1425
11. Kreimeyer, A., Wielckens, K., Adamietz, P., Hilz, H. (1984) J Biol Chem 259: 890-896

3-Aminobenzamide Enhances X-Ray Induction of Chromosome Aberrations in Down Syndrome Lymphocytes

Renate A. MacLaren, William W. Au, and Marvin S. Legator

Division of Environmental Toxicity, Department of Preventive Medicine and Community Health, University of Texas Medical Branch, Galveston, Texas, 77550 USA

Introduction

The presence of strand breaks in cellular DNA induces the activity of poly(ADP-ribose) polymerase, which catalyzes the formation of poly(ADP-ribose) from units of NAD+ (1). Overall, the function of poly(ADP-ribose) is not well understood. However, studies with inhibitors of the polymerase have been useful in determining some aspects of the polymer's function. For example, 3-aminobenzamide (3AB) is a potent inhibitor of the enzyme. When used in combination with alkylating agents or X-rays it has been shown to delay DNA strand rejoining (2) and to increase the incidence of (a) cell transformation (3), (b) cytotoxicity (4), (c) sister chromatid exchanges (5), and (d) chromosome abberations (5). These findings suggest that poly(ADP-ribose) may play a role in DNA repair (5, 6).

Another approach that has been used to probe the functions of the polymer involves the study of repair deficient cell lines. Cell lines derived from patients with ataxia telangiectasia (7), xeroderma pigmentosum (8), Cockayne's Syndrome (8), and Fanconi's anemia (9) have been tested for polymerase activity. Only the cells from Fanconi's anemia patients exhibited a decreased ability to form poly(ADP)ribose. Apparently, this results from their inherent lower basal levels of the precursor, NAD+ (9).

Another genetic disorder that may involve a repair deficiency is Down Syndrome (DS). Cells from individuals with DS contain an extra copy of chromosome 21. Some investigators have shown that DS lymphocytes exhibit a higher incidence of X-ray induced chromosome abberations than normal lymphocytes (10-13). This suggests that DS may be repair deficient (10-13). However, the nature of this repair defect remains to be determined. To clarify whether DS is repair deficient, the ability of X-irradiation to cause chromosome aberrations in DS and normal lymphocytes was tested in the presence or absence of 3AB.

Results and Discussion

When normal lymphocytes were irradiated with either 150 or 300 rads and incubated in the presence or absence of 3AB for 1, 3, and 6 hr prior to mitogenic stimulation, no significant difference in the frequencies of dicentrics, or deletions was detected (Fig. 1-4). In normal lymphocytes treated with 150 rads and 3AB (Fig. 2) there was a time dependent increase in the deletion frequency as compared with those cells treated with 150 rads alone. However, in normal lymphocytes treated with 3AB and 300 rads (Fig. 3, 4), the aberration frequency decreased after 1 hour in 3AB and then returned to the same level as those cells treated with 300 rads alone. Heartlein and Preston (14), Morgan *et al.* (15), Wiencke and Morgan (16), and Pantelias *et al.* (17), using similar experimental conditions, also found no increase in the frequency of chromosomal aberrations in normal Go lymphocytes treated with 3AB and x-rays. However, Natarajan *et al.* (5) found a 2-fold greater frequency of chromosomal aberrations following treatment of normal Go lymphocytes with X-rays and 3AB.

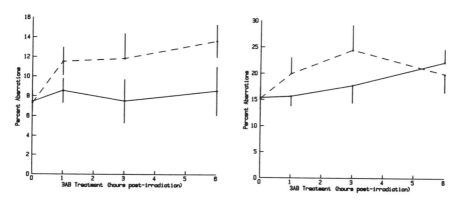

Fig. 1. (left) Dicentric frequency of normal (–) and Down Syndrome (- -) lymphocytes after X-ray with 150 rads in the presence of 3 mM 3AB. Whole blood was treated with 3 mM 3AB for 30 min prior to X-irradiation. After X-ray the cells were incubated in the presence of 3mM 3AB for an additional 0, 1, 3, or 6 hr before PHA stimulation. At the end of the incubation period, the cells were washed twice with culture medium and then returned to RPMI 1640 medium containing 2% PHA. Six hr after PHA stimulation 1 mM bromodeoxyuridine was added to each culture. Forty-eight hr after the addition of PHA, 2 μg of colcemid was added to the cultures to induce mitotic arrest, and 2 hr later the lymphocytes were harvested. The lymphocytes were stained using the fluorescence plus Geimsa method. One hundred metaphases from first division cells were analyzed for chromosome aberrations from each culture.

Fig. 2. (right) Deletion frequency of normal (–) and Down Syndrome(- -) lymphocytes after X-ray with 150 rads in the presence of 3 mM 3AB.

Poly(ADP-ribose) polymerase is known to be activated by the presence of DNA strand breaks. Inhibition of the enzyme by 3AB should theoretically cause an increase in the frequency of chromosomal aberrations after X-ray exposure. Our results and those of other investigators (14-17) indicate that poly(ADP-ribose) polymerase may not play a significant role in the repair of DNA damage that is important for the formation of chromosome aberrations in normal Go lymphocytes.

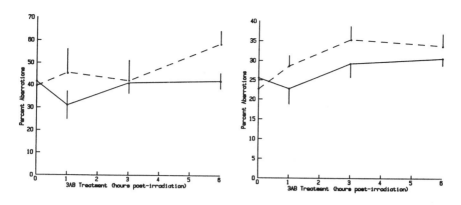

Fig. 3. (left) Dicentric frequency of normal (–) and Down Syndrome (- -) lymphocytes after X-ray with 300 rads in the presence of 3 mM 3AB.

Fig. 4. (right) Deletion frequency of normal (–) and Down Syndrome (- -) lymphocytes after X-ray with 300 rads in the presence of 3 mM 3AB.

In contrast, we have found that DS lymphocytes are more sensitive to treatment with 3AB and X-rays than cells treated with X-rays alone (Fig. 1-4). The differences are more evident at 150 rads (Fig. 1, 2) than at 300 rads (Fig. 3, 4). After 150 rads the dicentric frequency increases with the length of time of 3AB incubation. Although a similar trend is present after 300 rads (Fig. 3) the frequency of dicentrics decreases after 3 hr. The deletion frequency of cells treated with 150 rads and 3AB (Fig. 2) also show a similar trend. However, when the cells are treated with 300 rads, the deletion frequency (Fig. 4) increases after 6 hr in 3AB. Although there is not a definite time dependent dose response, it is clear that when DS lymphocytes are treated with 3AB they are consistently more sensitive to X-irradiation than when they are treated with X-rays alone.

There is some controversy in the literature in regard to the sensitivity of DS lymphocytes to the induction of chromosome aberrations by X-rays. Several investigators have reported X-ray induced aberration frequencies more than 50% greater than those induced in normal lymphocytes (12, 18). Others have found that the increase is less than 40% (10, 11). These variations could be explained by differences in the mean age of the DS subjects and differences in experimental conditions. Under our experimental conditions (incubation 1-6 hr prior to PHA stimulation), DS lymphocytes were more radiosensitive than normal lymphocytes only in the presence of 3AB. Additional studies are in progress to determine if 3AB causes an accumulation of DNA strand breaks in DS lymphocytes.

References

1. Ueda, K., Hayaishi, O. (1985) Ann Rev Biochem 54: 73-100
2. Zwelling, L.A., Kerrigan, D., Pommier, Y. (1982) Biochem Biophys Res Commun 104: 897-902

3. Kasid, U.N., Stefanik, D.F., Lubet, R.A., Dritschilo, A., Smulson, M.E. (1986) Carcinogenesis 7: 327-330
4. Oleinick, N.L., Evans, H.H. (1985) Radiat Res 101: 29-46
5. Natarajan, A.T., Csukas, I., Degrassi, F., van Zeeland, A.A. (1982) Prog Mutat Res 4: 47-59
6. Morgan, W.F., Cleaver, J.E. (1983) Cancer Res 43: 3104-3107
7. Zwelling, L.A., Kerrigan, D., Mattern, M.R. (1983) Mutat Res 120: 69-78
8. Fujiwara, Y.K., Goto, K., Yamamoto, K., Ichihashi, I. (1983) ADP-Ribosylation, DNA Repair and Cancer, M. Miwa *et al.* (eds.), Japan Sci Soc Press, Tokyo, pp. 209-218
9. Berger, N.A., Berger, S.J., Catino, D.M. (1982) Nature 299: 271-273
10. Lambert, B., Hansson, K., Bui, T.H., Funes-Cravioto, F., Lindstein, J., Holmberg, M., Strausmanis, R. (1976) Ann Hum Genet 39: 293
11. Preston, R.J. (1981) Environ Mutagen 3: 85-89
12. Morimoto, K., Kaneko, T., Lijima, L., Koizumi, A. (1984) Cancer Res 44: 1499-1504
13. Countryman, P.I., Heddle, J.A., Crawford, E. (1977) Cancer Res 37: 52-58
14. Heartlein, M.W., Preston, R.J. (1985) Mutat Res 148: 91-97
15. Morgan, W.F., Djordjevic, M.C., Jostes, R.F., Pantelias, G.E. (1985) Int J Radiat Biol 5: 711-721
16. Weincke, J.K., Morgan, W.F. (1986) Environ Mutagen 8: (Suppl. 6): 91
17. Pantelias, G.E., Politis, G., Sabani, C.A., Weincke, J.K., Morgan, W.F. (1986) Mutat Res 174: 121-124
18. Sasaki, M.S., Tonomura, A. (1968) Jpn J Hum Genet 14: 81-92

Abbreviations:
3AB - 3-aminobenzamide
DS - Down Syndrome
PHA - Phytohemagglutinin

3-Methoxybenzamide, A Possible Initiator for DMBA-Induced Carcinogenesis

Edward G. Miller, Francisco Rivera-Hidalgo, and William H. Binnie

Department of Biochemistry, Baylor College of Dentistry, Dallas, Texas, USA

Introduction

During the last few years this laboratory has formulated a hypothesis (Fig. 1) linking nicotinamide, NAD, and poly(ADP-ribose) polymerase to carcinogenesis. The factors that have led to the development of this hypothesis are as follows: (i) The fact that poly(ADP-ribose) polymerase utilizes NAD as a substrate to form poly(ADP-ribose) (1). (ii) The ability of this nuclear enzyme to respond to damage in DNA (2-4). (iii) Evidence (5, 6) indicating that poly(ADP-ribose) polymerase has a role in DNA repair. (iv) The fact that carcinogens damage DNA and the possibility that this damage is the driving force in carcinogenesis (7, 8).

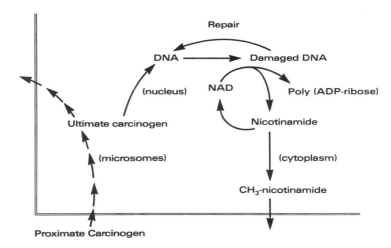

Fig. 1. A hypothesis linking nicotinamide, NAD and poly(ADP-ribose) polymerase to carcinogenesis.

When combined, these factors suggest that there should be an inverse relationship between the activity of this enzyme and the rate of tumorigenesis. Support for this idea can be found in studies with nicotinamide (9-11) and benzamides (12, 13). In the nutritional studies (9-11) it was found that an increase in dietary nicotinamide could inhibit the action of three different carcinogens; bracken fern (9), N-nitrosodimethylamine (10) and N-nitrosomethylbenzylamine (11). The data from the experiments (12, 13) with benzamides, which are inhibitors of poly(ADP-ribose) polymerase activity (14), have shown that these compounds can accelerate the formation of precancerous lesions in the liver of rats treated with N-nitrosodimethylamine (12) and enhance the development of hepatomas in fish treated with methylazoxymethanol acetate (13). This study represents another attempt to test this hypothesis. For the experiment the hamster cheek pouch model for oral carcinogenesis was utilized.

Results

Forty female Syrian Golden hamsters were obtained from the Charles River Breeding Laboratories, Wilmington, MA. The animals were given 10 days to adjust to their surroundings. Water and food were given ad libitum. After this period of adjustment, the animals were randomly divided into two equal groups. From each group 16 animals were selected. The left buccal pouch of the hamsters in Group 1 was painted three times weekly, first with a light application of dimethylsulfoxide (DMSO) and then with a heavier application of a 0.5% solution of the carcinogen, 7,12-dimethylbenz(a)anthracene (DMBA). For group 2, the basic treatment was the same. The only difference was that the DMSO contained 2.5% 3-methoxybenzamide. The remaining hamsters, four per group, served as controls. The left buccal pouches were treated three times per week, either with DMSO and mineral oil (controls for Group 1) or with DMSO containing 2.5% 3-methoxybenzamide and mineral oil (controls for Group 2).

Table 1. Tumor incidence, number, size and mass.

Group	No. of Animals	No. of Tumor Bearing Animals	No.of Tumors	Avg. No. of Tumors	Avg. Tumor Size (mm)	Avg.Tumor Mass(mm)
1	13	9 (69%)	25	1.9 ± 0.5[a]	2.3 ± 0.3[a]	4.7 ± 1.1[a,b]
2	12	10 (83%)	45	3.8 ± 1.0[a]	2.6 ± 0.2[a]	9.9 ± 2.9[a,b]

[a]Mean ± S.E.

[b]Statistically significant (Student's t-test for unpaired data) at $p < 0.05$.

After 50 treatments all of the hamsters were sacrificed. Tumors, when present grossly, were counted and measured, and the tissues were processed by routine histological techniques. The data from the macroscopic and microscopic observations was used to determine tumor incidence, number, size and type. During the experiment 7 animals from the two experimental groups died. Most of the deaths were due to respiratory infections. All of the animals that died prematurely were excluded from the study. At the end of the experiment there were 13 hamsters in Group 1 and 12 in Group 2.

The data for tumor incidence, number, size and mass is given in Table 1. As indicated 9 of the 13 animals in Group 1 and 10 of the 12 animals in Group 2 presented gross tumors. In most of these animals (17/19), multiple tumors were seen. The total number of tumors found in the hamsters in Groups 1 and 2 were 25 and 45, respectively. The average number of tumors per animal was 1.9 (Group 1) and 3.8 (Group 2). The average size of the tumors in the two groups was essentially the same. The calculated values for tumor mass, which is the number of tumors times the average diameter of the tumors in mm, indicated that the total tumor burden for an average animal in Group 2 was approximately twice as great as the tumor burden for an animal in Group 1. The difference in tumor mass was significant, $p < 0.05$. The microscopic observations showed that the left buccal pouches of all of the hamsters treated with DMBA exhibited pathologic changes. Even in the animals without apparent tumors there were signs of dysplasia. All of the tumors (Groups 1 and 2) demonstrated similar pathology. The malignancies ranged from small to large papillary carcinomas, all of which were well differentiated. Invasion into the underlying corium was minimal. Sections from the pouches of the control animals were normal.

Fig. 2. A two-phase mechanism for chemical carcinogenesis in the hamster buccal pouch.

From these findings several points can be made. One, the 3-methoxybenzamide affected tumor numbers but did not affect tumor type or size. The second point relates to a prevalent theory (15) for carcinogenesis. As illustrated (Fig. 2), this theory divides oral carcinogenesis into a two step-process. The first step, called initiation, ends with the production of latent tumor cells. The second step, called promotion, yields cancer cells. It is known that a number of agents magnify the effectiveness of carcinogens. Most of these agents are called promoters because their effects are found in the second step of the process. In other words, promoters reduce the time that it takes for latent tumor cells to become cancer cells. For this reason, the tumors will appear earlier and will attain a larger size in animals that are treated with a promoter plus a carcinogen as compared to the same animals that are treated with just the carcinogen. As indicated 3-methoxybenzamide did not affect the size of the DMBA-induced tumors, only the number.

From our observations, it was also apparent that the tumors appeared at essentially the same time in the hamsters in the two experimental groups. These results indicate that 3-methoxybenzamide is not a promoter, but an initiator. If this proves to be true, then it is possible that the activity of poly(ADP-ribose) polymerase may be critical during the initiation phase of carcinogenesis.

References

1. Chambon, P., Weill, J.D., Doly, J., Strosser, M.T., Mandel, P. (1966) Biochem Biophys Res Commun 25: 638-643
2. Miller, E.G. (1975) Biochim Biophys Acta 395: 191-200
3. Jacobson, E.L., Smith, J.Y., Mingmuang, M., Meadows, R., Sims, J.L., Jacobson, M.K. (1984) Cancer Res 44: 2485- 2492
4. Miller, E.G. (1986) Pro Soc Exp Biol Med 183: 221-226
5. Althaus, F.R., Lawrence, S.D., Sattler, G.L., Pitot, H.C. (1982) J Biol Chem 257: 5528-5535
6. Creissen, D., Shall, S (1982) Nature 296: 271-272
7. Farber, E. (1972) Current Research in Oncology. C.B. Anfinsen, M. Potter, A.N. Schechter, (eds.) Academic Press, New York, p.95.
8. Cleaver, J.E., Bootsma, D.A. (1975) Ann Rev. Genet 9: 19-38
9. Pamukeu, A.M., Milli, U., Bryan, G.T. (1981) Nutr Cancer 3: 86-93
10. Miller, E.G., Burns, H. (1984) Cancer Res 44: 1478-1482
11. van Rensburg, S.J., Hall, J.M., Gathercole, P.S. (1986) Nutr Cancer 8: 163-170
12. Takahashi, S., Nakae, D., Yokose, Y., Emi, Y.,Denda, A., Mikami, S., Ohnishi, T., Konishi, Y.(1984) Carcinogenesis 5: 901-906
13. Miwa, M., Kondo, T., Isikawa, T., Takayama, S., Sugimura, T. (1985) ADP-ribosylation of Proteins F.R. Althaus, H. Hilz, S. Shall, (eds.) , Springer-Verlag, Berlin, Heidelberg, New York, Tokyo, p.480
14. Purnell, M.R., Whish, W.J.D. (1980) Biochem J 185: 775-777
15. Odukoya, O., Shklar, G. (1982) Oral Surg 54: 547-552

Abbreviations:
DMBA - 7, 12-dimethylbenz(a)anthracene

Synergistic Induction of Sister Chromatid Exchanges and Cell Division Delays in Ehrlich Ascites Tumor Cells Treated with Cytoxan and Caffeine *In Vivo*

D. Mourelatos, J. Dozi-Vassiliades, A. Kotsis, and C. Gourtsas

Department of General Biology, Faculty of Medicine, Aristotelian University, Thessaloniki 54006, Greece

Introduction

Rodent tumor cells, like human mutant cell lines, are sensitive to effects of methylxanthines on lethality, cell cycle delays and chromosome aberrations after DNA damage by anti-tumor drugs. Enhanced cytogenetic damage by cytoxan was observed when Ehrlich Ascites tumor (EAT) cells *in vivo* were exposed to non-toxic concentrations of caffeine. One hr before intraperitoneal (ip) injection of 5-bromodeoxyurididine (BdUrd) absorbed to activated charcoal, EAT-bearing mice treated with cytoxan ip appear to have a dose dependent increase in SCE rates and cell division delays. The potential application of caffeine to clinical cancer therapy is poorly understood and few studies have been done relating to animal models or in human tumor cells in culture. Results from experiments with caffeine applying the SCE-assay are contradictory. Enhancement and attenuation of cytogenetic damage by anti-tumor agents has been demonstrated by post-treatment with caffeine *in vitro* and *in vivo* (1-2). We report here experiments that were designed to study the effects of caffeine on cytogenetic damage caused by cytoxan in EAT-cells *in vivo*. This is, to our knowledge, the first investigation testing the SCE-assay *in vivo* for its ability to identify the response of tumor cells to anti-tumor schemes.

Results

EAT-cells were maintained by serial transfer through peritoneal cavities of 7-9 week-old Balb-C mice with an average weight of 25 g. Four days after transplantation of the tumor, animals received an ip injection of cytoxan (Asta Werke), a metabolically activated chemotherapeutic, at a dose of 8 to 16 μg/g body wt. Immediately after, when required, caffeine at 130 μg/g body wt was also injected ip. One hr later animals were injected ip with 1 ml BdUrd-activated charcoal (Sigma and Merck, respectively) suspension with a BdUrd dose of 1 mg/g body wt. Prior to harvest animals were treated for 2 hr with 0.1 ml of 1 mg/ml colchicine (Sigma). Animals were sacrificed by ether narcosis after 26 or 48 hr. Cells were washed once with 0.075 M KCl and treated with the same hypotonic solution for 25 min. They were fixed with methanol glacial:acetic acid 3:1 and air dried. The

291

differential staining of sister chromatids was carried out by a modification of the FPG technique (3). Briefly, air dried slides were stained for 12 min in 10 μg/ml 33258 Hoechst dissolved in H_2O and mounted with cover slips in MacLlvaines buffer (pH 8.0) on them. Slides were exposed for 30 min to long wave UV light emitted from a lamp (Osram 300 W) located at a distance of 10 cm from the slides. Then the cover slips were removed and the slides were stained with 3% Giemsa solution (Merck) for 8 min.

Table 1. Synergistic induction of sister chromatid exchanges and cell division delays in Ehrlich ascites tumor cells treated with cytoxan and caffeine *in vivo*.

Treatment	No. of cells	SCEs/cell ± S.E.	(Range)	Proliferation Rate Index
1st Experiment (26 h)				
Control	30	7.70 ± 0.2	(4-12)	2.17
CAF (130 μg/g)	30	7.93 ± 0.27	(5-13)	2.01
CYT (8 μg/g)	45	14.82 ± 0.34	(3-27)	2.13
CYT(8 μg/g)	45	17.03 ± 0.40	(8-34)	1.93
+ CAF(130 μg/g)		(15.05)*		(1.97)*
2nd Experiment (48 h)				
Control	25	7.21 ± 0.26	(4-14)	2.56
CAF (130 μg/g)	25	8.49 ± 0.44	(4-15)	2.48
CYT (12 μg/g)	25	22.32 ± 0.41	(10-39)	2.36
CYT(12 μg/g)	25	30.22 ± 0.58	(14-49)	2.12
+ CAF(130 g/)		(23.60)*		(2.28)*
3rd Experiment (48 h)				
Control	25	8.01 ± 0.19	(5-11)	2.51
CAF (130 μg/g)	25	9.85 ± 0.40	(4-20)	2.46
CYT (12 μg/g)	25	20.40 ± 0.51	(13-31)	2.24
CYT(12μg/g)	25	29.47 ± 0.64	(17-45)	1.98
+ CAF(130 μg/g)		(22.24)*		(2.19)*
4th Experiment (26 h)				
Control	30	12.67 ± 0.25	(5-28)	1.27
CAF (130 μg/g)	30	12.71 ± 0.27	(6-25)	1.25
CYT (12μg/g)	50	20.50 ± 0.22	(12-40)	1.18
CYT(12μg/g)	50	30.75 ± 0.31	(16-51)	1.06
+ CAF(130 μg/g)		(20.54)*		(1.16)*
5th Experiment (26 h)				
Control	30	11.65 ± 0.24	(7-22)	1.37
CAF (130 μg/g)	30	12.73 ± 0.30	(7-23)	1.34
CYT (12 μg/g)	27	22.68 ± 0.31	(11-34)	1.28
CYT(12μg/g)	28	32.39 ± 0.36	(18-48)	1.09
+ CAF(130μg/g)		(23.76)*		(1.25)*
6th Experiment (26 h)				
Control	25	8.80 ± 0.16	(6-17)	2.10
CAF (130 μg/g)	25	10.00 ± 0.26	(4-20)	2.05
CYT (16 μg/g)	25	40.40 ± 0.88	(17-60)	1.42
CYT(16μg/g)	25	56.80 ± 0.79	(42-75)	1.22
+ CAF(130μg/g)		(41.60)*		(1.37)*

*The expected value if the increases above background for cytoxan and caffeine were independent and additive.

Preparation of BdUrd absorbed on activated charcoal was by the method of Russev and Tsanev modified by Kanda, Kato and Morales-Ramirez (4). Activated charcoal was washed several times with 2.5% NaOH, distilled H_2O, 2.5% HCl and finally distilled H_2O until the original water pH was attained. The charcoal was then dried for 2 hrs at 105°C. An aqueous solution of BdUrd was prepared (25 mg/ml) and sterilized by filtration. Sterilized activated charcoal was added to the BdUrd solution at a concentration of 100 mg/ml and stirred for 2 hr at room temperature before use.

Caffeine induced a small although significant increase in SCEs and cell division delays ($P < 0.05$) at a concentration of 130 μg/g body wt (Table 1). The results in Table 1 indicate that when caffeine was injected ip in mice previously injected ip with cytoxan at doses sufficiently high to induce SCEs and cell division delays, the effect of caffeine was synergistic, the SCE level achieved being consistently much higher than that expected by the simple addition of the effects of cytoxan and caffeine. The SCE frequencies in the presence of caffeine were significantly higher ($p < 0.01$) than those in the absence of caffeine (Table 1). The results in Table 1 also reveal that caffeine in combination with cytoxan acts synergistically on cell division delay, the cell division delay achieved being higher than that expected by the simple additive effects of caffeine and cytoxan for cell division delays. The increases in cell division delays by cytoxan in the presence of caffeine were significantly higher ($P < 0.01$) than those in the absence of caffeine.

The preparations were scored for cells in their first mitosis (both chromatids dark staining), second mitosis (one chromatid of each chromosome dark staining), and third and subsequent divisions (a portion of chromosomes with both chromatids light staining), and suitably spread second-division cells were scored blindly for SCEs. Proliferation Rate Indices (PRIs) were also scored blindly. The PRI was calculated as $(1M_1 + 2M_2 + 3M_3+)/100$, where M_1 is the percent value of cells in the first, M_2 in the second, and M_3+ in the third and higher divisions. For PRIs at least 100 cells were scored. Since cell cycle delays were observed during these treatments, we had to insure that the SCE frequency did not vary as a result of altered growth. Hence at the cytoxan concentration of 12 μg/g body wt, multiple periods of treatment were applied to rule out this possibility (Table 1). We observed no difference in the synergistic effects by cytoxan and caffeine on SCEs and PRIs when animals were sacrificed after 26 or 48 hr. (Experiments 2 - 5, Table 1). In each experiment in the group of six, the increases in SCE rates and in cell division delays by cytoxan in the presence and in the absence of caffeine were established (Table 1) and the paired t test was used to prove whether or not the SCE frequencies and PRIs in the presence of caffeine are significantly different than those in the absence of caffeine. The correlation between SCEs and PRI values was also calculated.

Discussion

It has been considered desirable to develop systems for measuring *in vivo* the genetic damage caused by potential chemotherapeutic schemes. Prediction of *in vivo* tumor response to chemotherapeutics and of human tumor cell chemosensitivity by the *in vitro* SCE-assay has been proposed (5, 6). Caffeine exhibits a variety of modifying effects in mammalian cells which have been treated with DNA damaging agents. One common hypothesis is that caffeine inhibits repair processes of DNA by inhibiting poly(ADP-ribose) polymerase (7). Poly(ADP-ribose) participates in DNA repair and has been postulated to function in the ligational step of DNA repair (7,8). Alkylating agents lower intracellular NAD levels by stimulating the activity of poly(ADP-ribose) polymerase (7). Cells bearing DNA lesions are reported to possess both poly(ADP-ribose)-dependent and poly(ADP-ribose)-independent DNA repair pathways. Thus the use of poly(ADP-ribose) polymerase inhibitors could give negative responses. However, apart from inhibiting poly(ADP-ribose) synthesis, caffeine exhibits a variety of modifying effects in mammalian cells that have been treated with DNA damaging agents. One common hypothesis is that caffeine inhibits repair processes (9). An alternative mechanism is that caffeine acts in G_2 to induce cells to undergo mitosis before the completion of DNA repair (9). Understanding the necessary conditions for enhanced cytogenetic damage by caffeine in tumor cells may provide a rational basis for the future design of cancer treatment protocols in animal models and in humans. For example, the data presented here may facilitate investigation of the hypothesis that SCE induction by anti-tumor agents *in vitro* can be positively correlated with *in vivo* tumor response and that the SCE assay could be used to predict both the sensitivity of human tumor cells to chemotherapeutic agents and the heterogeneity of drug sensitivity within individual tumors (5, 6). Furthermore, the SCE assay has been proposed as having predictive value as a clinical assay for drugs for which a strong correlation between cell killing and induction of SCEs has been established (6). In the present study, for the six experiments performed (Table 1), a sound correlation between SCE enhancement and cell division delay ($P<0.02$) was observed. Chemically induced cytotoxicity, in that it delays cell turnover times, is clearly manifested as a change in the relative proportions of cells in their 1st, 2nd and subsequent divisions (8). Studies designed to search for a relationship between SCE induction and other expressions of genotoxicity have shown a positive relationship between SCEs and reduced cell survival and alteration in cell cycle kinetics (8). These findings suggest that a common element, possibly a particular type of DNA damage produced by certain agents, was responsible for inducing SCEs and reducing cell survival and cell growth (8).

It has been proposed that successful DNA repair prior to S phase removes damage that might otherwise give rise to SCEs. In EAT-cells, interference by caffeine with DNA repair of cytoxan-induced DNA damage would lead to an increase in the number of incompletely repaired lesions at the time the cells reach S phase *in vivo*. These lesions may subsequently give rise to SCEs and cell division delays (2). Selective toxicity in tumor versus normal cells treated with alkylating agents has been indicated in some studies which showed less caffeine effects in normal human cells and in other studies with human or rodent tumors transplanted into mice and subsequently treated with anti-tumor agents (9). We have encouraging results for prediction of *in vivo* anti-tumor effects of chemotherapeutics by the *in vivo* SCE-assay.

References

1. Kato, H. (1974) Exp Cell Res 85: 239-247
2. Mourelatos, D., Dozi-Vassiliades, J., Tsigalidou-Balla, V., Granitsas, A. (1983) Mutat Res 121: 147-152
3. Goto, K., Maeda, S., Kano, Y., Sugiyama, T. (1978) Chromosoma 66: 351-359
4. Morales-Ramirez, P., Vallarino-Kelly, T., Rodriguez-Reyes, R. (1984) SCEs 25 Years of Experimental Research., N.Y. Tice, A. Hollaender (eds) Plenum Press New York, pp 591-611
5. Tofilon, P.J., Basic, I., Milas, L. (1985) Cancer Res 45: 2025-2030
6. Deen, D.F., Kendall, L.A., Marton, L.J., Tofilon, P.J. (1986) Cancer Res 46: 1599-1602
7. Durkacz, B.W., Omidiji, O., Douglas, A., Shall, S. (1980) Nature 283: 593-596
8. Morris, S.M., Heflich, R.H. (1984) Mutat Res 126: 63-71
9. Fingert, H.J., Chang, J.D., Pardee, A.B. (1986) Cancer Res 46: 2463-2467

Inhibition of Poly(ADP-Ribosyl)ation and Sister Chromatid Exchange Induction

Jeffrey L. Schwartz and Paul Song

Department of Radiation Oncology, University of Chicago, Chicago, Illinois 60637 USA

Introduction

Sister chromatid exchanges (SCEs) represent reciprocal exchanges between homologous daughter chromosomes. Although their mechanism of formation and their significance are unknown, they are thought of as cytological markers of some process important in carcinogenesis, possibly recombination (1). Exposure of cells to inhibitors of poly(ADP-ribose) polymerase, such as 3-aminobenzamide (3AB), will result in an increase in the baseline frequency of SCEs (2-4). At concentrations that dramatically increase SCE frequency, 3AB is nontoxic, nonmutagenic and noncarcinogenic (5, 6). Furthermore, while 3AB will interact with certain DNA damaging agents to affect cytotoxicity, mutagenicity and carcinogenicity (5, 6), it does not interact (directly) with these agents to affect SCE frequency (7, 8). Thus 3AB appears to be an ideal agent with which to probe SCE induction in the absence of other possibly confounding lesions.

SCE induction requires DNA strand breakage. Potent DNA strand breaking agents, such as ionizing radiation, however, are poor SCE inducers (9). It seems reasonable, therefore, to assume that the breaks involved in SCE formation are induced at specific sites in the genome and/or at specific times. 3AB therefore could be acting to increase the baseline frequency of SCEs by: (i) Increasing the frequency of the site-specific DNA strand breakage events that are required for SCE formation, and/or (ii) Increasing the probability of a crossover (SCE) event, by delaying the ligation of these site-specific breaks, altering chromatin structure, or affecting the activity of enzymes such as topoisomerases which facilitate exchange events. 3AB exposure is thought to affect many different phenomena including DNA synthesis, DNA strand break ligation, *de novo* nucleotide biosynthesis, DNA topoisomerase and endonuclease activity, and chromatin structure (10, 11). Alteration in any one of these processes could possibly result in an increased SCE frequency. While many of these potential targets have been examined for their role in 3AB-mediated SCE formation, none has been conclusively shown to be involved in SCE induction. In the present study, we have determined the time of action of 3AB on SCE formation. SCEs are S phase phenomena. Cells must pass through S phase in order for SCEs to be induced (12, 13). It is thought that SCEs are induced near the replication fork. Determining whether 3AB acts at the replication fork, ahead of the replication fork or behind the replication fork could provide a clue to its mechanism of action.

Results and Discussion

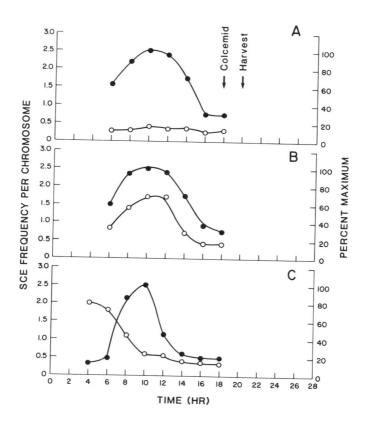

Fig. 1. SCE frequency per cell (0------0) and relative rate of DNA synthesis (X-----X) as a function of time after a 2 hr exposure to 0.2 μCi/ml [³H]thymidine (panel A), 20 mM 3AB plus [³H]thymidine (panel B), or 10⁻⁵ M MNNG plus [³H]thymidine (panel C). Cells were cultured continuously with 10⁻⁵ M bromodeoxyuridine for a total of 24 hr. Rate of DNA synthesis is expressed as a percent of the maximum [³H]thymidine incorporated per cell. Results are the mean of 2-3 determinations.

To determine when 3AB exerts its effect on SCE induction, asynchronous, exponentially-growing Chinese hamster ovary (CHO) cells were exposed to 20 mM 3AB along with 0.2 μCi/ml [³H]thymidine (20 Ci/mmol) for 2 hr and then mitotic cells were shaken off every 2 hr for 12-14 hr and the frequency of SCEs as well as the uptake of [³H]thymidine per cell determined. By analyzing cells from sequential mitotic shake-offs, the effect of 3AB exposure at different parts of the cell cycle could be examined. As shown in Fig. 1A, baseline SCE frequencies remained constant throughout S phase. The peak of [³H]thymidine incorporation occurred 8-10 hr prior to harvest. The 3AB treatment induced SCEs throughout S phase (Fig. 1B). The peak effect, however, was in mid-S phase, and the time course for SCE induction by 3AB essentially mirrored [³H]thymidine incorporation. Therefore, 3AB induced SCEs primarily when exposure occurred close to the replication fork.

In contrast to the results with 3AB, when cells were exposed to the monofunctional alkylating agent, N-methyl-N'-nitro-N-nitrosoguanidine (MNNG), SCE frequency remained low in those cells exposed in mid to late S phase and was highest in cells exposed in G1 or early S phase. Thus, as has been shown by others with ultraviolet light or 8-methoxypsoralen-induced DNA lesions (12, 13), MNNG exposure resulted in lesions that lead to SCE formation only when the lesions were produced prior to the replication of the chromosome, i.e., ahead of the replication fork.

As noted previously (4), 3AB treatment tends to prolong S phase. These earlier studies (4) also ruled out any effects of 3AB on DNA synthesis initiation or chain elongation. Taken together, these findings would suggest that the target for 3AB action is close to but just behind the replication fork. It would appear therefore, that there exists a sensitive period during DNA replication, very close to the replication fork, where SCEs can be induced by inhibitors of poly(ADP-ribose) synthesis. It is likely that the target is some protein associated with replication such as a DNA topoisomerase. Interestingly, in a recent publication by Dillehay *et al.* (14), it was noted that DNA topoisomerase II inhibitors induce SCEs in early to mid-S phase, like 3AB. It is unlikely that the effect of 3AB is due to a direct effect on DNA topoisomerase II, because there is apparently no interaction between 3AB and topoisomerase II inhibitors in SCE induction (15). The similarity between inhibitors of topoisomerase II and inhibitors of poly(ADP-ribose) synthesis in their action on SCE frequency, however, suggests that they might share a common component.

In conclusion, 3AB exposure will transiently produce changes just behind the replication fork that lead to SCE induction. The ability to identify where in the genome 3AB acts could ultimately help in identifying the processes involved in SCE induction.

Acknowledgements. This work was supported by Grant CA 14599 from the NIH.

References

1. Wolff, S. (1977) Annu Rev Genet 11: 183-201
2. Oikawa, A., Tohda, H., Kanai, M., Miwa, M., Sugimura, T. (1980) Biochem Biophys Res Commun 97: 1311-1316
3. Morgan, W.F., Cleaver, J.E. (1982) Mutat Res 104: 361-366
4. Schwartz, J.L., Morgan, W.F., Kapp, L.N., Wolff, S. (1982) Exp Cell Res 143: 377-832

5. Schwartz, J.L., Morgan, W.F., Brown-Lindquist, P., Afzal, V., Weichselbaum, R.R., Wolff, S. (1985) Cancer Res 45: 1556-1559
6. Lubet, R.A., McCarvill, J.T., Putnam, D.L., Schwartz, J.L., Schectman, L.M. (1984) Carcinogenesis 5: 459-462
7. Schwartz, J.L., Morgan, W.F., Weichselbaum, R.R. (1985) Carcinogenesis 6: 699-704
8. Schwartz, J.L. (1986) Carcinogenesis 7: 159-162
9. Morgan, W.F., Schwartz, J.L., Murnane, J.P., Wolff, S. (1983) Radiat Res 93: 567-571
10. Althaus, F.R., Hilz, H., Shall, S. (eds.) (1985) ADP-Ribosylation of Proteins. Springer-Verlag, Berlin, Heidelberg, New York
11. Shall, S. (1984) Adv Radiat Biol 11: 2-64
12. Wolff, S., Bodycote, J., Painter, R.B. (1974) Mutat Res 25: 73-81
13. Latt, S.A., Loveday, K.S. (1978) Cytogenet Cell Genet 21: 184-200
14. Dillehay, L.E., Denstman, S.C., Williams, J.R. (1987) Cancer Res 47: 206-209
15. Morgan, W.F., Doida, Y., Fero, M.L., Xi-Cang, G., Shadley, J.D. (1986) Environ Mutagenesis 8: 487-493

Abbreviations:
3AB - 3-aminobenzamide
CHO - Chinese hamster ovary
MNNG - N-methyl-N'-nitro-N-nitrosoguanidine
SCE - sister chromatid exchange

Differential Effect of Benzamide on Repair-Deficient and Proficient Strains of L5178Y Lymphoblasts Exposed to UVC or X-Radiations

Irena Szumiel, Danuta Wlodek, and Karl Johan Johanson[1]

Department of Radiobiology and Health Protection, Institute of Nuclear Chemistry and Technology, PL-03-195 Warsaw, Poland

Introduction

Much of the evidence concerning poly(ADP-ribose) polymerase (adenosine diphosphate ribosyl transferase; EC.2.4.99) involvement in DNA repair relies on the cellular response to treatment with DNA damaging agents in combination with poly(ADP-ribose) polymerase inhibitors. One possible experimental approach is to examine the effects of combined treatment, using cell variants differing either in DNA repair ability or in poly(ADP-ribose) metabolism. A convenient model, used in this work, consists of two closely related strains of L5178Y murine lymphoma cells L5178Y-R (LY-R) and L5178-S (LY-S) (1,2). The two strains are closely related in that LY-R cells (transplantable in DBA/2 mice, stable *in vivo*) undergo conversion into LY-S cells (non-transplantable, stable *in vitro*) upon *in vitro* cultivation. They also display inverse cross-sensitivity to DNA damaging agents (2). LY-R cells are more sensitive than LY-S cells to UVC and UVA radiations, Pt complexes and 9-aminoacridine derivatives. Conversely, LY-S cells are more sensitive than LY-R cells to X and γ-rays and several alkylating agents.

Sensitivities to X, γ and UVC-radiations have been related to impaired DNA repair processes: LY-S cells have been found to repair DNA dsb (double strand breaks) more slowly than the more radiation-resistant LY-R cells (3); LY-R cells are nucleotide excision-deficient, in contrast with the more UVC-resistant LY-S cells (4, 5). In this work we review experimental data (5, 6-8) concerning LY cell response to combined UVC + Bz (benzamide) and Bz in combination with X or γ rays. These results are best explained by assuming an impaired DNA ligase function in LY-S cells.

Poly(ADP-ribose) polymerase Activity. Both LY strains were similarly able to synthesize poly(ADP-ribose) in response to DNA degradation

---

[1]Department of Radioecology, The Swedish University of Agricultural Sciences, S-750 07 Uppsala, Sweden

induced by hypotonic cold shock. The procedure, necessary to achieve ^{3}H-incorporation from [^{3}H]NAD+ during a 30 min incubation induced DNA damage corresponding to that inflicted by about 17 Gy of γ-rays. Additional γ-irradiation (10 Gy) had only minor effects on ^{3}H-incorporation (6). Sensitivity to continuous Bz treatment assessed by its influence on growth rate was the same in both LY strains, when related to the number of cell divisions instead of time. A 3 hr treatment changed the cloning efficiency to the same extent (7).

Response to X or γ-Rays and Bz. The response to ionizing radiation and 2 mM Bz was examined in 3 laboratories with the use of different radiation sources. Since no significant differences have been reported between the effects of X and γ-rays on LY cells we discuss the X and γ-ray data jointly. The effects of combined X (γ) + Bz treatment were evaluated on survival, chromosomal damage, NAD+ content and DNA repair. As seen from Table 1 LY-S cells were sensitized by Bz treatment, in contrast with LY-R cells. The post-irradiation decrease in NAD+ content was more pronounced in LY-S than LY-R cells. The effect of 2 mM Bz on rejoining of DNA sb induced by 5 Gy of γ-rays was examined by a double-labelling modification of the DNA unwinding method (7). There was a delay in sb rejoining that was more marked in S than in R cells. In LY-S cells Bz treatment increased DNA sb frequency by 40% at 7.5 min after irradiation and by 28% at 15 min. The highest increase for LY-R cells was 20% at 15 min; two hr after irradiation there was no difference between Bz-treated and untreated cells of both strains.

Table 1. Response of LY cells to X or γ + 2 mM Bz treatment.

Cells	Survival Curve Parameters		Chromatid Aberr-ations/100 cells[1]		NAD+ content, % Control[1]	
	X	X + Bz	X	X + Bz	23 min[2]	60 min[2]
LY-R	n=1.4 D_o=0.97 Gy	n=1.4 D_o=0.97 Gy	38	43	90	118
LY-S	n=1 D_o=0.53 Gy	n=1 D_o=0.43 Gy	84	180	74	97

[1]Maximal values, scored 18 h after X-irradiation (2 Gy for Ly-R cells, 1 Gy for LY-S cells).
[2]Time after irradiation with 5 Gy γ-rays.

Thus far, we have no direct proof for a casual relationship between this observation and the effects at the chromosomal and cellular levels. However, the differently delayed sb rejoining in the presence of Bz and the concomitant decrease in NAD+ content allow us to conclude that in LY-R and LY-S cells ligation depends on ADP-ribosylation to a different extent.

We assume a low DNA ligase I activity in LY-S cells and an ADP-ribosylation-mediated activation of ligase II (9-11). Correlation of ligase activity and radiation sensitivity was reported and a ligase defect indicated in 46BR cells (12) and Bloom's syndrome cells (13, 14). So, it seems plausible that the high radiation sensitivity of LY-S cells is related to an impaired ligase function.

Response to UVC Radiation and Bz. Similar features of the cellular response were examined after UVC + Bz treatment as in the case of X or γ rays + Bz (5). The results are summarized in Table 2. In contrast with LY-R cells and numerous other cell lines, LY-S cells were sensitized to UVC radiation by Bz. As in the case of X or γ + Bz treatment, the distinct features of LY-S cell response to UVC + Bz treatment can be explained by low ligase I activity and hence, the necessity to activate ligase II. In LY-R cells, as in xeroderma pigmentosum cells (15), the rate of DNA incisions seems too low to activate poly(ADP-ribose) polymerase. On the other hand, LY-R cells may synthesize NAD^+ more efficiently than LY-S cells (Table 1), and be able to maintain a stable NAD^+ level in spite of poly(ADP-ribose) polymerase activation. In fact, when a double-labelling modification of the DNA unwinding method was used to examine the difference in sb frequency between Bz-treated and untreated UVC irradiated cells, the results were identical for LY-R and LY-S cells. We interpreted this as an indication of a Bz-sensitive base excision repair system (16,17) operating in both cell strains (in contrast with a nucleotide excision functional only in LY-S strain). Lack of nucleotide excision presumably makes LY-R cells sufficiently sensitive to UVC radiation that the base excision repair is not limiting for survival.

Table 2. Response of LY cells to UVC + 2 mM Bz treatment.

Cells	Survival Curve Parameters		Chromatid Aberr-ations/100 cells[1]		NAD+ content, % control	
	UVC	UVC + Bz	UVC	UVC + Bz	60 min[2]	90 min[2]
LY-R	n=1.3 D_o=2.8 J/m²	n=1.5 D_o=2.5 J/m²	46	40	100	101
LY-S	n=1.8 D_o=9.0 J/m²	n=0.9 D_o=7.1 J/m²	66	99	72	75

[1]Maximal values (20 h after irradiation) induced by exposure in Fischer's medium to 10 J/m² (LY-R) or 30 J/m² (LY-S).
[2]Time after exposure to UVC (15 J/m²) in PBS with 5% bovine serum.

Conclusion

The results reviewed above point to an impaired UV-endonuclease function in the LY-R strain and low basal activity of DNA ligase in LY-S cells. Since LY-R cells are reproducibly converted into LY-S cells with concerted change in all phenotypic features, concomitant mutations in the genes that code the 2 enzymes seem implausible, especially without a clear selective advantage connected with such mutations. However, since the differences in R and S karyotypes are due to numerous chromosomal rearrangements (private communications, O. Rosiek, J. Hozier and C. Sanchez) there might be an effect of position on the transcriptional expression of the genes coding the respective repair enzymes.

It is also possible to explain the difference in X and UVC sensitivities of LY strains by assuming just one difference in chromatin structure. Both UV-endonuclease and ligase activities depend on chromatin organization. The difference in these activities in LY strains may be the reflection of a difference in chromatin architecture rather than of molecular defects in specific enzymes. This assumption would also be compatible with the altered requirements for ADP-ribosylation of chromatin proteins in DNA repair processes.

Acknowledgements. Supported by the Polish Government Cancer Research Program (CPBR 11.5/101) and the Swedish Natural Sciences Research Council.

References

1. Alexander, P., Mikulski, Z.B. (1961) Nature 192: 572-573
2. Beer, J.Z., Budzicka, E., Niepokojczycka, E., Rosiek, O., Szumiel, I., Walicka, M. (1983) Cancer Res 43: 4736-4742
3. Wlodek, D., Hittleman, W.N. (1987) Radiat. Res. 112; 146-155
4. Szumiel, I., Wlodek, D., Johanson, K-J. (1988) Photochem Photobiol, 48: 201-204
5. Szumiel, I., Wlodek, D., Niepokojczycka, E., Johanson, K-J. (1989) J Photochem Photobiol, in press
6. Szumiel, I., Wlodek, D., Johanson, K-J, Sundell-Bergman, S. (1984) Br J Cancer 49 Suppl. VI: 33-38
7. Johanson, K-J., Sundell-Bergman, S., Szumiel, I., Wlodek, D. (1985) Acta Radiol Oncol 24: 451-457
8. Szumiel, I., Wlodek, D., Johanson, K-J. (1989) Acta Oncol, In press
9. Lehmann, A.R., Broughton, B.C. (1984) Carcinogenesis 5: 117-119
10. Shall, S. (1984) Adv Radiat Biol 11: 1-69
11. Murray, B., Irwin, J., Creissen, D., Tarassoli, M., Durkacz, B.W. (1986) Mutat Res 165: 191-198
12. Henderson, L.M., Arlett, C.F., Harcourt, S.A., Lehmann, A.R., Broughton, B.C. (1985) Proc Natl Acad Sci USA 82: 2044-2048
13. Chan, J.Y.H., Becker, F.F., German, J., Ray, J.H. (1987) Nature 325: 357-359
14. Willis, A.E., Lindahl, T. (1987) Nature 325: 355-357
15. Berger, N.A., Cohen, J.J. (1982) ADP-ribosylation Reactions, Biology and Medicine, O. Hayaishi, K. Ueda (eds.) Academic Press, New York, pp. 547-560
16. Ahnström, J.M. (1986) Genetic Toxicology of Environmental Chemicals. C. Ramel, B. Lambert, J. Magnuson (eds.) Alan R. Liss Inc., New York, pp. 275-282
17. Erixon, K. (1986) Mechanisms of DNA Damage and Repair, M.G. Simic, L. Grossman, A. D. Upton (eds.) Plenum Press, New York, pp.159-170

Nicotinamide Deficiency and Sister Chromatid Exchanges

Manoochehr Tavassoli, K. Lindahl-Kiessling[1], and S. Shall

Cell and Molecular Biology Laboratory, University of Sussex, Brighton, BN1 9QG, United Kingdom

Introduction

The analysis of sister chromatid exchanges (SCE) is frequently used as an end-point for the detection of genotoxic effects of both chemicals and radiation. A good correlation between SCE induction and the Ames' test has been repeatedly demonstrated, although certain compounds fail this correlation (1). Usually, agents that induce elevated levels of SCE have also been demonstrated to be mutagenic (2). However, in all these correlations there is one notable set of exceptions: benzamide and its derivatives. These compounds are excellent at inducing sister chromatid exchanges, but do not apparently themselves damage DNA. The mode of action of these aromatic carboxamides are consequently of considerable interest.

Benzamide and its analogues, such as 3-aminobenzamide are efficient inhibitors of nuclear ADP-ribose transferase, (poly(ADP-ribose) polymerase), (E.C.:2.4.2.30) (3, 4). This enzyme activity is required for efficient DNA excision repair (5), probably because it regulates DNA ligase activity (6). Consequently, the SCE-inducing capacity of poly(ADP-ribose) polymerase inhibitors has been interpreted as a consequence of delayed rejoining of DNA strand-breaks which are considered to occur spontaneously in DNA which contains bromodeoxyuridine (9). The source of these hypothetical DNA breaks is a matter of dispute (8), but Natarajan and his co-workers (9, 10) have suggested that replication on template DNA which contains either [^{3}H]thymidine or bromodeoxyuridine (both of which are used to visualize SCE) results in DNA strand-breaks which constitute the foundation for the induction of SCE by benzamides. The major evidence which they adduce for this inference, is their observation that benzamides induce only a few SCE when present during the first cell cycle with bromodeoxyuridine, but that they produce 7 to 10 times more SCE when present during the second cell cycle when bromodeoxyuridine is already incorporated into the template DNA.

There are, however, certain puzzling features; for example, why does 3-aminobenzamide induce any SCE during the first cell cycle, presumably on a normal template. In addition, recent experiments with 3-aminobenzamide-resistant variant cells are difficult to reconcile with this hypothesis (11).

---

[1]Institute of Zoophysiology, Uppsala University, S-751-22, Uppsala, Sweden

Consequently, we have made a further investigation into this hypothesis utilizing nicotinamide starvation to deplete the cell of NAD+, and thus to inhibit formation of poly(ADP-ribose). We propose additional considerations to explain the induction of SCE by benzamides.

Results

L1210 cells proliferate normally in medium with a lowered nicotinamide level (5, 12, 13). In the present experiments, we also observed that L1210 cells proliferated at the normal rate for at least 24 hr; that is, for two generations, in nicotinamide-free RPMI 1640 medium, supplemented with undialyzed 5% (v/v) human serum. Consequently, it was possible to investigate and to compare the frequency of SCE-induction in cells grown for two cycles in both complete and in nicotinamide-deficient medium. The interaction of 10mM benzamide, 3-aminobenzamide or 3-methoxybenzamide in cells with lowered cellular NAD levels was examined. The frequencies of SCE were determined with a concentration of bromodeoxyuridine (BrdUrd) of 20 μM. We found a pronounced synergistic interaction between the poly(ADP-ribose) polymerase inhibitors and the nicotinamide deficiency; the combination of poly(ADP-ribose) polymerase inhibitors and lowered cellular NAD+ levels gave a synergistic increase in the level of SCE of up to eight-fold. For example, the median frequency of SCE induced by 1.0 mM benzamide increased from 22.0 in normal complete medium to 52.0 in nicotinamide-deficient medium; this represents a synergistic increase of 29.0. The synergistic increments observed with 1.0 mM 3-aminobenzamide and 1.0 mM 3-methoxybenzamide were somewhat less; 13.5 and 10.0 SCE, respectively. BrdUrd by itself induces some SCE; but this frequency increased significantly in nicotinamide-free medium (Table 1).

Table 1. Interaction of nicotinamide deprivation and inhibitors of poly(ADP-ribose) polymerase in the induction of SCE.

| | Sister Chromatid Exchange Frequency (SCE/cell) | | | |
	Normal Medium	Nicotinamide Deficient Medium	Δ	P
BrdUrd (20 μM)	7.0	8.0	1.0	>0.10
BrdUrd (20 μM) + Benzamide (1.0 mM)	22.0	52.0	29.0	<0.001
BrdUrd (20 μM) + 3-aminobenzamide (1.0 mM)	18.0	32.5	13.5	<0.001
BrdUrd (20 μM) + 3-methoxybenzamide (1.0 mM)	28.5	39.5	10.0	0.002>>0.001
BrdUrd (80 μM)	13.5	14.0	0.5	>0.10

The increased frequency of SCE induced by benzamide in nicotinamide-deficient medium was dose dependent at low concentrations but leveled off at around 1 mM (Fig. 1), similar to the effect of benzamide in normal medium. BrdUrd alone was also dose dependent; at 20 µM there were 7.0 - 8.0 SCE per cell and at 80 µM BrdUrd there were 13.5 - 14.0 SCEs per cell (Table 1). Consequently, we have investigated further the effect of different concentrations of BrdUrd on the induction of SCE by poly(ADP-ribose) polymerase inhibitors (Tables 2-4). Again, the synergistic increment of the poly(ADP-ribose) polymerase inhibitors was observed at all four BrdUrd concentrations. However, it was noted that at the three lowest concentrations there was no concentration dependence. Only at the very highest concentration of 50.0 µM, when the degree of BrdUrd incorporation is levelling off (Fig. 2), is there any increase beyond an additive effect. The incorporation of BrdUrd into DNA increases with increasing external BrdUrd concentration to a maximum at 40 µM, and then declines at higher concentrations (Fig. 2). Moreover, nicotinamide deficiency itself has no effect on the induction of SCE in these experiments (Fig. 2).

Table 2. Effect of the concentration of bromodeoxyuridine on the induction of SCE by 300 µM benzamide.

| BrdUrd (µM) | Sister Chromatid Exchange Frequency (SCE/cell) | | |
	No Inhibitor	+ Benzamide (300 µM)	Increment
5.0	5.0	14.5	+ 9.5
10.0	7.0	13.0	+ 6.0
20.0	7.0	15.5	+ 8.5
50.0	10.5	23.0	+ 12.5

Table 3. Effect of concentration of bromodeoxyuridine on the induction of SCEs by 3.0 mM 3-aminobenzamide.

| BrdUrd (µM) | Sister Chromatid Exchange Frequency (SCE/cell) | | |
	No Inhibitor	+ 3-aminobenzamide (3.0 mM)	Increment
5.0	6.0	28.0	22.0
10.0	7.5	28.0	20.5
20.0	8.0	27.5	19.5
50.0	8.5	34.5	26.0

Table 4. Effect of concentration of bromodeoxyuridine on the induction of SCE by 300 μM methoxybenzamide.

	Sister Chromatid Exchange Frequency (SCE/cell)		
BrdUrd	No Inhibitor	+ 3-methoxybenzamide (300 μM)	Increment
5.0	4.5	12.0	+ 7.5
10.0	6.0	11.0	+ 5.0
20.0	7.5	14.0	+ 6.5
50.0	11.5	25.5	+ 14.0

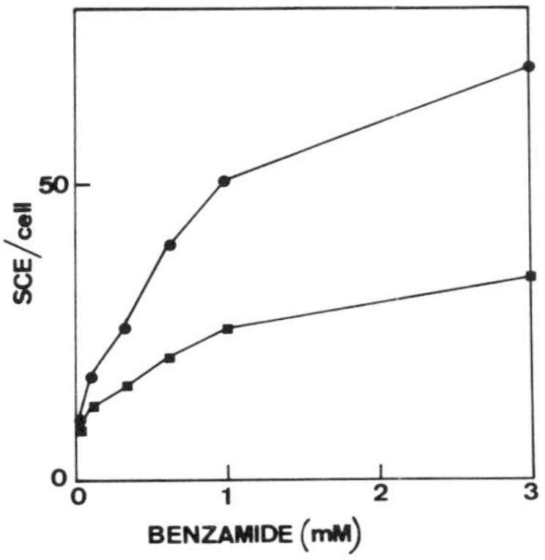

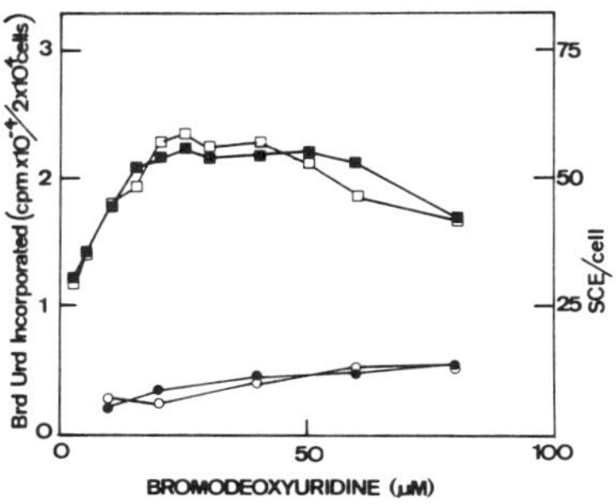

Fig. 1. (left) Induction of SCE by benzamide in normal and nicotinamide deficient medium. Complete medium consists of nicotinamide-free RPMI-1640 plus 1 μg/ml nicotinamide and 5% undialysed human serum. (●), nicotinamide-free medium; (■)nicotinamide added back to the medium.

Fig. 2. (right) Incorporation of bromodeoxyuridine and the frequency of sister chromatid exchanges in L1210 cells in normal and nicotinamide-deprived medium at varying concentrations of external bromodeoxyuridine. Complete medium consists of nicotinamide-free RPMI-1640 plus 1μg/ml nicotinamide and 5% undialysed human serum. (●), 0 SCE frequency; (■, □) incorporation of radioactive BrdUrd. Open symbols, nicotinamide-free medium; closed symbols, complete medium.

Discussion

The molecular mechanism by which the benzamides induce SCE is of interest because they seem to be unique in being neither DNA-damaging agents nor mutagens. In addition, there is a need to determine whether the induction of SCE by the benzamides are mediated by inhibition of nuclear poly(ADP-ribose) polymerase activity or by some other means. The present report confirms that 3-aminobenzamide can induce SCE (14) and demonstrates clearly that there is a synergistic interaction between nicotinamide deprivation and benzamide, 3-aminobenzamide and 3-methoxybenzamide. In our experiments deprivation of nicotinamide alone does not enhance the spontaneous level of SCE. In contrast, Hori (15) found a two-fold increase in SCE per cell after 72 hr of nicotinamide deprivation. This latter observation has been taken as evidence for the involvement of poly(ADP-ribose) polymerase activity in the induction of SCE. However, the consequences of nicotinamide deprivation for such a long period as 3 days (15) might include perturbations of nucleotide pools.

We confirm and extend the observation of (15) that there is a synergistic interaction between nicotinamide deprivation and the benzamides. Naively, one might argue that if they are both acting on the poly(ADP-ribose) polymerase activity, then they should be only additive. However, this approach takes no note of the complex dynamic aspects of this system (13, 16). Our main interest has been to consider the mechanism of induction of SCE, particularly in comparison to other agents which induce SCE. The apparent difference to be expected of the benzamides may be more apparent than real if the underlying mechanism is in fact dependent on BrdUrd, which is itself a mutagen. Natarajan and co-workers (9, 10) have demonstrated a correlation between the amount of BrdUrd incorporated into the DNA during the previous cell cycle and the frequency of SCE by poly(ADP-ribose) polymerase inhibitors. The implication is that somehow DNA synthesis on a BrdUrd substituted template causes DNA damage or DNA strand-breaks whose repair is hindered by the presence of poly(ADP-ribose) polymerase inhibitors. However, nicotinamide deprivation might be expected also to lower the effective poly(ADP-ribose) polymerase activity. If this were so, and if poly(ADP-ribose) polymerase inhibition increased SCEs, then we expect that nicotinamide deprivation alone would increase SCE frequency. We did not observe such an increase as (15) has reported under the conditions of these experiments. Nonetheless, we do observe under the same combinations of both nicotinamide and poly(ADP-ribose) polymerase inhibitors a substantial synergism between nicotinamide deprivation and the benzamides (Table 1, Fig. 1). In our experiments, the nicotinamide deprivation (which had no effect on SCE frequency by itself) showed a very dramatic effect in the presence of the benzamides. Thus, we may wonder whether the benzamides are indeed working via poly(ADP-ribose) polymerase activity.

If the benzamides are preventing the repair of damage or breaks induced by the BrdUrd then we may predict that, at higher concentrations of BrdUrd, they would show higher frequencies of SCE. However, we do not find a dose-dependence on BrdUrd as anticipated (Tables 2-4). We only observe an increase in SCE at 50.0 μM BrdUrd, which is a level of BrdUrd which is toxic to the cells (Fig. 2) and may seriously disturb the deoxynucleotide pool. We note that even 300 μM benzamide or 3-methoxybenzamide has a synergistic effect on the induction of SCE. This excludes the involvement of some cytoplasmic mono(ADP-ribosyl) transferase activities because Rankin and Jacobson (personal communication) have estimated the K_i values of benzamide for cytoplasmic enzymes to be about 3 mM, whereas the apparent half maximum effect based on the inhibition of SCE frequency is about 500 µM. We are therefore not able to reconcile easily our results with the hypothesis that benzamides inhibit poly(ADP-ribose) polymerase which act on lesions or breaks induced by the incorporation of BrdUrd. At this stage the precise mechanism of induction of SCE by benzamides seems unclear.

References

1. Lindahl-Kiessling, K., Bhatt, T.S., Karlberg, I., Coombs, M.M. (1984) Carcinogenesis 5: 11-14
2. Carrano, A.V., Thompson, L.H., Lindl, P.A., Minkler, J.L. (1978) Nature 271: 551-553
3. Shall, S. (1975) Biochemistry (Tokyo) 77: 2
4. Purnell, M.R., Whish, W.J.D. (1980) Biochem J. 185: 757-777
5. Durkacz, B.W., Omidiji, O., Gray, D.A., Shall, S. (1980) Nature 283: 593-596
6. Creissen, D., Shall, S. (1982) Nature 296: 271-272
7. Natarajan, A.T., Csukas, I., van Zeeland, A.A. (1981) Mutat Res 84: 125-132
8. Morgan, W.F., Wolff, S. (1984) Cytogenet Cell Genet 38: 34-38
9. Natarajan, A.T., Zoukas, I., van Zeeland, A.A. (1981) Mutat Res 84: 125-132
10. Zwanenburg, T.S.B., Natarajan, A.T. (1984) Cytogenet Cell Genet 38: 278-281
11. Lindahl-Kiessling, K., Shall, S. (1984) Sister Chromatid Exchanges. R.R. Tice, A. Hollaender (eds.), Plenum Press, New York, Part A, pp. 305-311
12. Jacobson, E.L., Smith, J.Y., Mingmuang, M., Meadows, R., Sims, J.L., Jacobson, M.K. (1984) Cancer Res 44: 2485-2492
13. Shall, S. (1984) Nucl Acids Res Symp Ser 13 pp. 143-191
14. Oikawa, A., Tohda, H., Kaina, M., Miwa, M., Sugimura, T. (1980) Biochem Biophys Res Commun 97: 1311-1316
15. Hori, T.A. (1981) Biochem Biophys Res Commun 102: 38-45
16. Shall, S. (1984) Adv Rad Biol 11: 1-69

ADP-Ribosyl Transferase in the Differentiation of Protozoa and Embryonic Mouse Thymus

Gwyn T. Williams, Dale R. Taylor, Eric J. Jenkinson, Emma Griffin, and Marguerite E. Hill

Department of Anatomy, University of Birmingham Medical School, Birmingham B15 2TJ, United Kingdom

Introduction

Studies using a number of experimental systems have suggested that ADP-ribosyl transferase activity may be required for the differentiation of several vertebrate cell types (1-6). Although they are also eukaryotes, the kinetoplastid protozoa, several of which cause serious human and animal diseases, are only very distantly related to the vertebrates. The most important of these protozoa belong to genus *Trypanosoma* or genus *Leishmania*. The ancestors of the protozoa diverged from those of the vertebrates well before the ancestors of yeasts diverged from those of the higher eukaryotes. Comparison of the roles of the ADP-ribosyl transferase in protozoan differentiation with that found in the differentiation of mammalian cells should therefore indicate which functions of ADP-ribosyl transferase are restricted to particular groups of eukaryotes, and which have been strongly conserved throughout evolution. Clearly, roles played by ADP-ribosyl transferase which have been conserved over such a long period must be important components of cell differentiation.

Results and Discussion

ADP-ribosyl transferase in the differentiation of *Leishmania*. Inhibitors of ADP-ribosyl transferase have been shown to reduce the rate of differentiation of *Trypanosoma cruzi* amastigotes at concentrations which do not significantly affect proliferation (7-9). The differentiation of the related species *Leishmania mexicana* can also be studied *in vitro*, and preliminary experiments have indicated that ADP-ribosyl transferase inhibitors also affected *L. mexicana mexicana* (10, 11). However, using this system it was not possible to distinguish unequivocally between inhibition of differentiation and selective toxicity to undifferentiated cells. We have now used a *L. mexicana amazonenis* differentiation system to provide a clear demonstration of specific inhibition of differentiation produced by ADP-ribosyl transferase inhibitors.

Leishmania species exist in several morphological forms: the non-motile amastigote which normally grows inside mammalian macrophages, and the promastigote which proliferates inside the insect vector and is motile

310

by virtue of a singular flagellum. The promastigote form can be grown axenically at 27°C, and the effect of ADP-ribosyl transferase inhibitors on its proliferation can be determined. Dose response curves were obtained for several inhibitors of ADP-ribosyl transferase, and those for 3-methoxybenzamide (12) and theophylline (13), which is less specific for ADP-ribosyl transferase, are shown in Fig. 1. These studies allowed us to select optimal inhibitor concentrations for investigation of specific effects on differentiation, i.e. the maximum concentration which had no significant effect on the promastigote proliferation rate. This was 2.5 mM for 3-methoxybenzamide and 0.25 mM for theophylline.

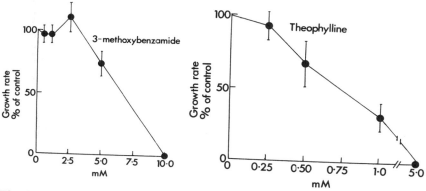

Fig. 1. Inhibition of *Leishmania mexicana amazonensis* promastigote proliferation by inhibitors of ADP-ribosyl transferase. Promastigotes of strain LV78 were grown at 27°C in LIT medium (21) with 10% fetal calf serum in the presence of a range of concentrations of inhibitors. Cell densities were determined after 48 hrs incubation and growth rates were calculated for control and experimental cultures. The growth rate at each concentration of inhibitor was expressed as a percentage of the growth rate in control cultures. The starting cell density was 1.3×10^6 promastigotes/ml. The mean and standard error (except where encompassed by the symbol) of four cultures are shown for each concentration of inhibitor.

The non-motile amastigote form can be grown at 34°C in the mouse macrophage cell line J774 (14) and induced to differentiate to the promastigote form by incubation in the appropriate medium at 27°C. The promastigotes which form in these cultures are easily detected by their motility and characteristic morphology. Under standard conditions, amastigote numbers decrease only marginally as differentiation proceeds, since some proliferation of these cells does occur at 27°C. In the presence of 2.5 mM 3-methoxybenzamide, the production of promastigotes was inhibited (Fig. 2a). In addition, the number of amastigotes in these same cultures did not decline in parallel with the numbers in control cultures, but increased significantly over the 65 hr incubation period (Fig. 2b). These observations cannot be explained by toxicity or inhibition of proliferation, whether these effects were general or were restricted to either one of the two morphological forms. Theophylline and other ADP-ribosyl transferase

311

inhibitors also produced this specific inhibition of differentiation, together with the increase in the numbers of undifferentiated cells. These simultaneous measurements of amastigote and promastigote numbers therefore suggest that 3-methoxybenzamide is preventing amastigotes from differentiating to promastigotes, and that the amastigotes subsequently proliferate and accumulate in the culture. This phenomenon is similar to that observed with differentiating chick myoblasts, where the addition of 3-aminobenzamide also resulted in an increase in the cell number when compared to untreated cultures (15). Such effects suggest that ADP-ribosyl transferase activity can be required very early in differentiation, before an irrevocable decision to differentiate immediately has been made. This phenomenon is not universal, however. Studies on the human promyelocytic cell line HL-60 have suggested that inhibition of ADP-ribosyl transferase activity in some cases results in the death of differentiating cells at concentrations of inhibitor which do not affect cells which are only proliferating (4). This indicated that the HL-60 cells did not revert to simple proliferation when the differentiation pathway was blocked.

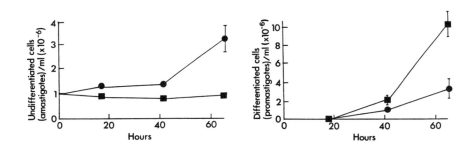

Fig. 2. Effect of 3-methoxybenzamide on the differentiation of *Leishmania mexicana amazonensis* amastigotes. Infected J774 cell cultures were homogenized and the amastigotes released were puified using Percoll (14). The amastigotes were washed with LIT medium (21) and resuspended in LIT medium with 10% fetal calf serum. Cultures were pre-incubated at 34°C for 4 hrs at 9.5 x 10^5 cells/ml in the presence of 2.5 mM 3-methoxybenzamide and in untreated control cultures. Differentiation was then induced by transferring the cultures to 27°C. Differentiation was monitored microscopically and the concentrations of promastigotes (Fig. 2a) and amastigotes (Fig. 2b) were determined. (●) Cultures containing 2.5 mM 3-methoxybenzamide. (■) Controls. Mean and standard errors (unless encompassed by the symbols) of at least four cultures are shown for each time point.

ADP-ribosyl transferase in the differentiation of embryonic mouse thymus. The precursors of T-lymphocytes will only develop normally when growing within the thymus. However, such lymphocyte precursors do produce mature T cells in thymic lobes surgically removed from 14-day mouse embryos and kept in organ culture (16). Thymocytes will still rearrange their T cell receptor ß chain genes under such conditions (17). These organ cultures can be treated with chemical inhibitors to examine effects on the differentiation process. We have previously shown that 3-methoxybenzamide inhibits the expression of the lymphocyte differentiation marker Lyt-2 in this system (18). We have now examined the effects of low concentrations of 3-methoxybenzamide (0.32-1.0 mM) on the appearance of the T cell receptor on the surface of these thymocytes in organ culture.

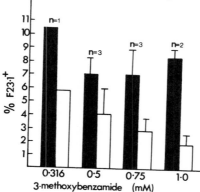

Fig. 3. Effect of 3-methoxybenzamide on the appearance of the T-cell receptor during the development of mouse thymic lobes in organ culture. Thymuses were removed from 14-day mouse embryos and were maintained *in vitro* in the presence and absence of 3-methoxybenzamide (18). After 12-14 days, thymocytes were harvested and labelled with monoclonal antibody F-23.1 which recognizes the products of a V ß gene family present on about 25% of peripheral T cells (22). The percentage of cells which were surface-labelled with the antibody was determined by fluorescence microscopy. The clear bars indicate the percentage of cells labelled with F-23.1 in cultures maintained continuously in the indicated concentrations of 3-methoxybenzamide. Filled bars show the corresponding percentage in control cultures run in parallel. The standard errors of the indicated number of separate experiments (n) is given in each case. Each experimental and control result was obtained using a minimum of 5 thymic lobes.

Fig. 3 shows that T cell receptor expression was partially inhibited at all concentrations tested. Interestingly, thymocyte proliferation was also reduced with the inhibitor at concentrations of 0.5 mM and above. It is possible that this inhibition of proliferation was itself the result of inhibition of differentiation, as thymocyte growth appears to be stimulated by interleukin-2 and the most likely sources of this growth factor within the thymus are the thymocytes at a more advanced stage at differentiation. Inhibition of thymocyte maturation is therefore likely to reduce proliferation through the reduction in interleukin-2 production.

Thymus organ cultures provide the only *in vitro* system yet available for studying generation of diversity in normal cells of the immune system. Central to this process is the rearrangement of T cell receptor gene segments to produce functional genes. Since this gene rearrangement must at some stage involve joining DNA molecules, and the involvement of nuclear ADP-ribosyl transferase in DNA repair is now widely accepted, it is possible that ADP-ribosyl transferase activity might be required for such rearrangement.

We have previously used Southern hybridization analysis of DNA from organ cultured thymocytes to monitor ß-chain gene rearrangement (17, 19). However, we could detect no inhibition of ß-chain gene rearrangement at concentrations of 3-methoxybenzamide which did not inhibit proliferation. In future experiments, the effects of higher concentrations of inhibitors will be examined by compensating for the possible reduction in the amount of interleukin-2 produced by the organ cultures.

Inhibitors of ADP-ribosyl transferase affect the cell differentiation of widely differing cell types and species. For the *Leishmania*, in particular, the effects of the inhibitors have been clearly demonstrated to be differentiation-specific. There are several possible explanations for such striking effects. Amongst these, one attractive hypothesis is that inhibition of nuclear ADP-ribosyl transferase results in the disruption of structural changes in chromatin required for differentiation (1, 2, 4, 6, 20). However, possibly for technical reasons, we have not yet seen any clear indications of this using thymus organ cultures. A second possibility is that inhibition of membrane ADP-ribosyl transferase blocks transduction of the extracellular signals involved in initiating differentiation. Similarly, there is as yet little independent evidence to support this explanation. Further studies of the systems described here, together with detailed comparisons between them, should prove very useful in investigating these hypotheses.

Acknowledgements. We thank the Wellcome Trust and the U.K. Medical Research Council for financial support, and Dr. R. Neal for supplying *Leishmania mexicana amazonensis* strain LV78.

References

1. Farzaneh, F., Shall, S., Brill, D., Zalin, R. (1982) Nature 300: 362-366
2. Johnstone, A.P., Williams, G.T. (1982) Nature 300: 368-370
3. Francis, G.E., Gray, D.A., Berney, J.J., Wing, M.A., Guimares, J.E.T., Hoffbrand, A.V. (1983) Blood 62: 1055-1062
4. Farzaneh, F., Meldrum, R., Shall, S. (1987) Nucl Acids Res 15: 3493-3502
5. Brac, T. and Ebisuzaki, K. (1985) ADP-Ribosylation of Proteins, F.R. Althaus, H.Hilz, S. Shall (eds.), Springer-Verlag, Berlin, pp. 446-452
6. Williams, G.T., Johnstone, A.P. (1983) Biosci Rep 3: 815-830
7. Williams, G.T. (1983) Exp Parasitol 56: 409-415
8. Williams, G.T. (1984) J Cell Biol 99: 79-82
9. Williams, G.T. (1985) Curr Top Micro Immunol 117: 1-22
10. Capaldo, J., Coombs, G.H. (1983) Parasitol 87: xli
11. Williams, G.T. (1985) ADP-Ribosylation of Proteins, F.R. Althaus, H.Hilz, S. Shall (eds.), Springer-Verlag, Berlin, pp. 358-366
12. Purnell, M.R., Whish, W.J.D. (1980) Biochem J 185: 775-777
13. Levi, V., Jacobson, E.L., Jacobson, M.K. (1978) FEBS Lett 88: 144-146
14. Chang, K.-P. (1980) Science 209: 1240-1242
15. Farzaneh, F. (1982) D Phil Thesis, Unversity of Sussex
16. Van Ewijk, W., Jenkinson, E.J., Owen, J.J.T. (1982) Eur J Immunol 12: 262-271
17. Williams, G.T., Kingston, R., Owen, M.J., Jenkinson, E.J., Owen, J.J.T. (1986) Nature 324: 63-64

18. Mirza, I.H., Jenkinson, E.J., Williams, G.T. (1985) ADP-Ribosylation of Proteins, F.R. Althaus, H. Hilz, S. Shall (eds.), Springer-Verlag, Berlin, pp. 429-432
19. Williams, G.T. (1987) Gene 53: 121-126
20. Khan, Z., Francis, G.E. (1987) Blood 69: 1114-1119
21. Camargo, E.P. (1964) Rev Inst Med Trop Sao Paulo 6: 93-100
22. Staertz, U.D., Rammensee, H.G., Benedetto, J., Bevan, M.J. (1985) J Immunol 134: 3994-4000

ADP-Ribosylation of Leukemia Cell Membrane Proteins Correlates with Granulocyte Colony Stimulating Factor Dependent Differentiation

Ming-Chi Wu, M. Rebecca Zaun, Nasreen Aboul-Ela, and Myron K. Jacobson

Department of Biochemistry, Texas College of Osteopathic Medicine, University of North Texas, Denton, Texas 76203 USA

Introduction

Modulation of myelopoiesis is a receptor mediated process which requires colony-stimulating factors (CSFs) for differentiation and maturation (1). The genes for several of these factors have been cloned and their products have been characterized. Thus, much is known regarding the interaction of these factors with their respective target cells (2). However little is known concerning the metabolic reactions that follow receptor binding such as subsequent signal transduction and other events which lead to the initiation of differentiation.

Several laboratories have reported that benzamide or benzamide derivatives, which are effective inhibitors of ADP-ribose transfer reactions, can inhibit differentiation of myeloid cells (3-5). The mechanism of this inhibition has not been investigated at the molecular level. We report here studies of the inhibition of granulocyte colony-stimulating factor (G-CSF)-induced differentiation by benzamide on a murine myelomonocytic leukemia cell line, WEHI-3BD$^+$. We also report that ADP-ribosylation of membrane proteins in these cells is correlated with the inhibition of differentiation by benzamide.

Results and Discussion

We have previously reported that inhibitors of ADP-ribose transfer reactions inhibited G-CSF-induced myeloid differentiation of WEHI-3BD$^+$ cells at concentrations which had little effect on proliferation (6). The data of Fig. 1 show the effect of benzamide on a recombinant G-CSF-induced proliferation and differentiation. These data confirm our earlier observation showing that 2 mM benzamide inhibits by 50% the G-CSF dependent differentiation and has little effect on proliferation. This result suggests that an ADP-ribose transfer reaction may be involved in G-CSF dependent differentiation of these cells. In an attempt to determine whether this is the case, membrane preparations from WEHI cells were incubated with

[^{32}P]NAD during stimulation with G-CSF. Following incubation, the membranes were washed, and proteins were resolved by polyacrylamide gel electrophoresis and the dried gel was subjected to autoradiography. Prominent bands were radiolabeled with approximate molecular weights of 67, 45, and 20 kDa.

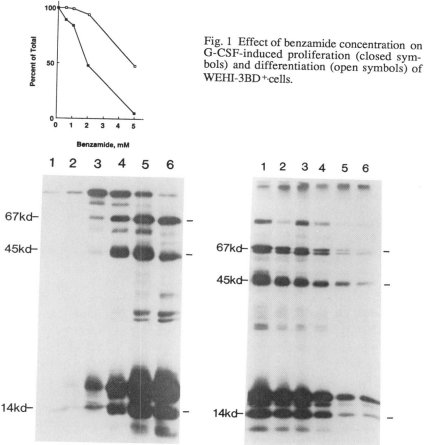

Fig. 1 Effect of benzamide concentration on G-CSF-induced proliferation (closed symbols) and differentiation (open symbols) of WEHI-3BD$^+$cells.

Fig. 2. (Left) Time course of labeling of WEHI membranes with [^{32}P]NAD in the presence of G-CSF. Incubation at 37°C for 0 min, (1); 5 min, (2); 10 min, (3); 30 min, (4); 60 min, (5); and 120 min, (6).

Fig. 3. (Right) Effect of varying concentrations of benzamide on the labeling of WEHI membranes with [^{32}P]NAD in the presence of G-CSF. Incubation was at 37°C for 60 min in in benzamide at 0 mM, (1); 0.5 mM, (2); 1.0 mM, (3); 2.0 mM, (4); 5.0 mM, (5); or 10 mM, (6).

The data of Fig. 2 show a time course of incorporation of [^{32}P]NAD into WEHI membrane proteins in the presence of G-CSF. At early times, the most prominent labeling was observed in a band at approximately 100 kDa.

317

With longer incubation times, this labeling progressively decreased concomitant with labeling of bands of lower molecular weight. Benzamide inhibited the labeling in a dose dependent manner as shown in Fig. 3. The concentration range where inhibition was observed was similar to the range where inhibition of differentiation was observed (Fig. 1).

Membrane preparations frequently contain active NAD glycohydrolases, which can rapidly convert NAD to ADP-ribose and readily form covalent non-enzymatic adducts with proteins (7). To determine whether the membrane labeling observed above utilized NAD or ADP-ribose as substrate, [^{32}P]NAD was converted to [^{32}P]ADP-ribose with NAD glycohydrolase. Radiolabeled ADP-ribose and radiolabeled NAD were purified by paper chromatography followed by chromatography on DHB-Sepharose. Fig. 4 shows the results of an experiment in which [^{32}P]NAD or [^{32}P]ADP-ribose was incubated with membranes in the presence of G-CSF. Also shown is a control using [^{32}P]NAD that was not subjected to the chromatographic purification steps. The autoradiogram was prepared as described for Fig. 2. The data clearly indicate that both NAD preparations resulted in membrane labeling while the ADP-ribose was utilized very poorly. These data suggest that NAD rather than ADP-ribose was the substrate for the membrane labeling.

For further analysis of protein labeling, molecular sieve chromatography was utilized. Membranes incubated with [^{32}P]NAD were subjected to centrifugation and dissolved in buffer containing guanidinium chloride and subjected to molecular sieve chromatography in the presence of guanidinium chloride. Radiolabel was detected in the high molecular weight region of the molecular sieve column eluate and a second peak, presumably from trapped low molecular weight material, was also observed (data not shown). The material in the high molecular peak was examined by SDS polyacrylamide gel electrophoresis and revealed the same labeling pattern shown in Figs. 2 through 4.

The above studies utilized NAD labeled in the adenosine proximal phosphate. To determine if other portions of the NAD molecule were incorporated into the membrane proteins, we utilized a preparation of NAD in which ^{3}H was present in the adenine ring and ^{14}C was present in the nicotinamide proximal ribose. Reversed phase chromatography was utilized to show that the high molecular weight material following incorporation contained a ^{3}H to ^{14}C ratio which was indistinguishable from that of the NAD utilized. This result, along with the observation that the radiolabel remains with high molecular weight material, demonstrates that NAD is utilized to label proteins and that the labeled proteins contain the adenine ring, adenosine proximal phosphate and the nicotinamide proximal ribose of NAD. These observations are consistent with the hypothesis that the modified material represents proteins modified by ADP-ribose.

The stability of the putative ADP-ribose protein linkages was also studied. The known protein:mono-ADP-ribosyltransferases modify three different amino acid residues in target proteins. These include a glycosylic linkage to a guanidinium nitrogen of arginine, an imidazole nitrogen of a hypermodifed histidine (diphthamide), and the formation of a thioglycosidic linkage to a cysteine residue. We have compared the stability of the adducts formed in WEHI membranes with model conjugates containing ADP-ribose bound to arginine, hypermodifed histidine or cysteine residues (Table 1).

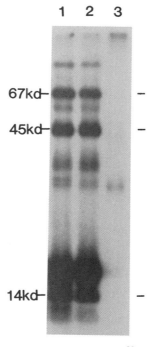

Fig 4. Labeling of WEHI membrane proteins with [^{32}P]NAD or [^{32}P]ADP-ribose. Lane 1 [^{32}P]NAD was not subjected to purification; lane 2, purified [^{32}P]NAD; lane 3, purified [^{32}P]ADP-ribose. Equal amounts of radiolabel were used in each incubation.

For these studies radiolabeled WEHI membrane proteins were mixed with model conjugates and, after incubation, the stability was monitored by molecular sieve chromatography. Release of arginine:ADP-ribose linkages can be selectively obtained by incubation in the presence of neutral hydroxylamine. Under conditions where ADP-ribose was quantitatively released from arginine the adducts from WEHI cell membranes were stable. Cysteine can be selectively released by treatment with Hg^{2+}. Conditions which resulted in the quantitative release of ADP-ribose from cysteine residues did not result in significant release of the WEHI cell adducts. The N-glycosylic linkage to the imadazole ring of diphthamide is stable even in the presence of M NaOH. When WEHI cell proteins were incubated with M NaOH, the N-glycosylic linkage of diphthamide was completely stable but

319

quantitative release of radiolabel from the WEHI proteins was observed. Thus these data indicate that the nature of the protein:ADP-ribose linkage in the WEHI membranes differs from any of the known mono-ADP-ribosyl-protein adducts.

Table 1

Comparison of WEHI Mono-ADP-Ribosylated Proteins to ADP-Ribose:Protein Conjugates of Known Linkage

Conditions	Acceptor Amino Acid	% Released	
		Standard	WEHI
NH_2OH	Arginine	93	17
$HgCl_2$	Cysteine	96	9
NaOH	Histidine	15	93

Protein from incubation of WEHI membranes with [^{32}P]NAD was purified by molecular sieve chromatography and mixed with model conjugates.

Thus it can be concluded that the inhibition by benzamide of WEHI cell differentiation can be segregated from effects on cell proliferation and can be correlated with radiolabeling of membrane proteins with [^{32}P]NAD. Evidence presented here shows that these radiolabeled membrane proteins contain the adenine ring, adenosine proximal phosphate and the nicotinamide proximal ribose of NAD, suggesting that the protein is modified with ADP-ribose. Furthermore, the linkage of ADP-ribose to membrane protein does not appear to be any of the ADP-ribose:protein linkages previously reported. Whether this ADP-ribosylation of membrane proteins serves any function in signal transduction in G-CSF-induced differentiation is an interesting question which remains to be investigated. Since the linkage of ADP-ribose to protein is different from previously characterized ADP-ribose modifications of protein, it is interesting to speculate that these observations may represent a new class of signal transducing proteins. Further characterization of this metabolism is currently in progress.

Acknowledgements This work was supported in part by NIH grants AM31624 and CA43894 and The Robert A. Welch Foundation (B1058).

References
1. Metcalf, D. (1986) Blood 76: 257-267
2. Clark, S.C. and Kaman, R. (1987) Science 236: 1229-1237
3. Colon-Otero, G., Sando, J.J., Sims, J.L., McGrath, E., Jensen, D.E. and Quesenberry, D.J. (1987) Blood 70: 686-693
4. Exley, R., Gordon, J. and Clemens, M.J. (1987) Proc. Natl. Acad. Sci. USA 84: 6467-6470
5. Francis, G.E., Gray, D.A., Berney, J. Jr., Wing, M.A., Guimaraes, J.E.T. and Hoffbrand, A.V. (1983) Blood 62: 1055-1062
6. Wu, M.-C., Zahn, M.R. and Wu, F.-M. (1989) FEBS Letts 244:338-342
7. Hilz, H. Koch, R., Fanick, W., Klapproth, K and Adamietz, P. (1984) Proc Natl Acad Sci USA 81:3929-3933

Poly(ADP-Ribose) Transferase in Lizard Oviduct

Gaetano Ciarcia, Hisanori Suzuki[1], and Massimo Lancieri[2]

Dipartmento di Biologia Evolutiva e Comparata, University di Napoli, Napoli, Italy

Introduction

Poly(ADP-ribose) transferase often undergoes a drastic change in its activity following the action of various biological effectors in a variety of cell systems or tissues. An increase in poly(ADP-ribose) transferase activity was usually observed in those systems in which the biological effectors were followed with cell proliferation, for instance, in immature quail oviduct after estrogen treatment (1). On the contrary, a drastic decrease in poly(ADP-ribose) transferase activity was often detected in a number of cells after treatment with effectors used to induce an arrest of cell proliferation such as interferons (2) and retinoic acid (3). Virus gene expression of mouse mammary tumor induced after glucocorticoid treatment seems to be strictly correlated with a marked decrease in poly(ADP-ribosyl)ation of HMG 14 and 17 in pituitary gland cells (4).

In the present work, we have analyzed poly(ADP-ribose) transferase activity and characterized histone-rich fractions in the oviduct of the mediterranean lizard, *Podarcis s.sicula* Raf., during the reproductive cycle, for studying possible involvement of poly(ADP-ribose) transferase in tissue proliferation and gene expression. The lizard oviduct shows a well defined seasonal growth cycle regulated by gonadal hormones (winter stasis, preovulatory and ovulatory phase). In spayed females, estradiol can induce morphological and biochemical changes characterizing mature oviduct (5), including increases in RNA and protein synthesis and in some hydrolytic enzyme activities (6).

Results and Discussion

As shown in Fig. 1, poly(ADP-ribose) transferase activity in lizard oviduct nuclei undergoes a marked variation during the reproductive cycle. An increase in poly(ADP-ribose) transferase activity was observed at the beginning of the recovery phase (the middle of March-April) with the peak (12 times the activity detected during the reproductive stasis and secretory period) about the first week of May after the oviduct had reached its maximum growth and was ready to be functional. Oviduct growth followed the increase in poly(ADP-ribose) transferase activity. Enzyme activity in

[1]Instituto di Chimica Biologica, University di Verona, Verona, Italy
[2]Dipartimento di Biologia Generale e Molecolare, University di Napoli, Napoli, Italy

321

cytoplasmic fractions of the organ was very low and no significant variation was observed during the reproductive cycle (data not shown). A similar relationship between poly(ADP-ribose) transferase activity and oviduct proliferation was also observed in the spayed lizard organ after estradiol valerate treatment. No appreciable variation in poly(ADP-ribose) transferase activity was observed in the nuclei of the non estrogen-target organs such as muscle and small intestine (data not shown). These data suggest a possible involvement of poly(ADP-ribose) transferase activity in estrogen-induced cell proliferation, confirming previous observations in quail oviduct (1).

Table 1. Amino acid composition of the lizard specific nuclear protein (SNP).

	native[a]		estrogen induced[a]	
Asx	2.64	± 0.30	2.34	± 0.30
Thr	3.51[b]	± 0.28	4.12[b]	± 0.49
Ser	3.58[b]	± 0.35	3.69[b]	± 0.29
Glx	3.93	± 0.19	3.94	± 0.21
Pro	20.46	± 1.38	17.01	± 1.56
Gly	12.61	± 1.28	12.74	± 1.40
Ala	6.87	± 0.29	9.33	± 0.35
1/2Cys	n.d.[c]		n.d.[c]	
Val	3.77	± 0.20	3.83	± 0.18
Met	0.16	± 0.05	traces	
Ileu	0.21	± 0.08	0.41	± 0.10
Leu	3.33	± 0.13	2.96	± 0.18
Tyr	12.61[b]	± 0.43	12.51[b]	± 0.54
Phe	traces		traces	
His	11.83	± 0.38	11.47	± 0.45
Lys	4.76	± 0.09	5.07	± 0.11
Arg	7.26	± 0.21	7.93	± 0.39
Trp	2.39	± 0.12	2.58	± 0.14
Asx + Glx	6.57		6.28	
Lys + Arg	12.02		13.00	

Proteins, obtained from gels by electrodialysis (14) and dialyzed against water, were hydrolyzed and analyzed on a Beckman 119 automatic amino acid analyzer. For tryptophan determination, proteins were hydrolyzed with 3 M mercaptoethanolamine sulfonic acid and analyzed according to Penke et al. (15).Amino acid values are expressed as mole %. a: mean values of three separate measurements; b: extrapolated to zero time hydrolysis; c: not determined. Reprinted with permission from ref. 7.

A peculiar protein with high mobility on SDS-PAGE was detected in histone-rich fractions of the lizard oviduct nuclei at the recovery phase (Fig. 2). A similar protein appeared transiently in the oviduct nuclei of spayed females slightly before the organ maturity, 15 days after estradiol administration (7). These proteins shared very similar amino acid composition (Table 1) and were designated "specific nuclear protein (SNP)". Amino acid composition of SNP was very peculiar, being rich in proline, glycine, tyrosine and histidine and poor in methionine and isoleucine. The quantity of this protein was higher than the histone H-1. On electrophoretic analysis SNP showed a molecular weight of about 9.9 kDa and was very basic. These data indicated strongly that SNP was not a proteolytic product of known proteins such as histones and high mobility protein groups (8). A higher level of SNP was detected in the uterine tract (Fig. 2). SNP was neither present in oviduct cytoplasm nor in small intestine and liver (Fig. 3).

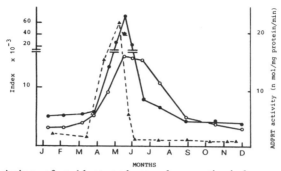

Fig. 1. Variation of oviduct and gonado-somatic indexes and poly(ADP-ribose)transferase activity in the oviduct during the reproductive cycle of the lizard *Podarcis s.sicula*Raf. The poly(ADP-ribose) transferase assay was performed as described in Ciarcia *et al.* (7). (●) gonado/somatic index; (○) oviduct/somatic index; (▲) poly(ADP-ribose) transferase activity in nuclear extracts. Arrow indicates when SNP appeared.

The transient appearance of massive amounts of SNP during a well defined period in the reproductive cycle raised questions about its possible physiological functions in oviduct nuclei. At present no data is available in this regard. A transient appearance of SNP occurred soon after poly(ADP-ribose) transferase achieved maximum activity in the oviduct of both native lizards and estrogen-treated spayed females. *De novo* synthesis of a specific oviductal protein, avidin, was found to occur in quail oviduct with a decrease in poly(ADP-ribose) transferase activity (1). A prompt decrease in poly(ADP-ribosyl)ation of HMG 14 and 17 with parallel induction of the mouse mammalian tumor virus gene in pituitary gland cells after glucocorticoid administration has been reported (4). Our present data add another example on the appearance of a specific protein (SNP) which coincides with a prompt decrease in poly(ADP-ribose) transferase activity, although it remains to be determined if its appearance in lizard oviduct nuclei is due to an estrogen-regulated gene induction. A possible involvement of poly(ADP-ribose) transferase in the estrogen-related appearance of SNP was further suggested by experiments with 3-aminobenzamide, a potent inhibitor of the enzyme (9). Administration of this inhibitor (0.8 g/Kg body weight) to intact and spayed estrogen-treated lizards prevented the appearance of SNP. The same treatment did not cause any apparent change in oviduct maturation (data not shown). An estrogen-induced increase in poly(ADP-ribose) transferase activity could be caused by a variety of factors: an increase in the level of putative activator(s) of poly(ADP-ribose) transferase, a decrease in endogenous poly(ADP-ribose) transferase inhibitor concentrations, or inhibition of enzymes involved in the removal of poly(ADP-ribose) from nuclear proteins such as poly(ADP-ribose) glycohydrolase (10) and ADP-ribosyl protein lyase (11).

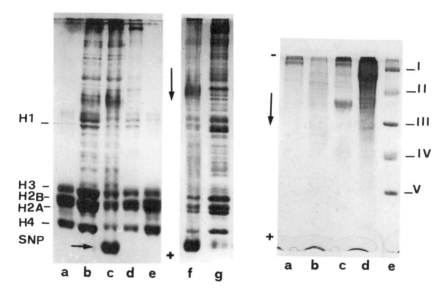

Fig. 2. (left) SDS-PAGE of histone-rich fractions in oviduct of intact lizards during the reproductive cycle. SDS-PAGE was carried out according to Laemmli (16). Extraction of the histone-rich fractions from nuclei was performed as described in Ciarcia *et al.* (7). a and e: calf thymus; b: oviduct *in toto* during the winter stasis; c: oviduct *in toto* during the recovery phase; d: oviduct *in toto* during the post-ovulatory period; f: uterine tract during the recovery phase; g: tubal tract during the recovery phase.

Fig. 3. (right) SDS-PAGE of cytoplasmic fractions from different tissues in spayed and estradiol treated lizards. Cytoplasmic fractions were prepared as in Ciarcia *et al.* (7). a: liver of spayed animals treated with four doses (2.4 µg); b: liver of spayed animals, control; c: oviduct of spayed animals treated with four doses (2.4 µg); d: oviduct of spayed animals, control; e: low molecular weight markers I: phosphorylase B 92.5 kDa; II: bovine serum albumin 66.2 kDa; III: ovalbumin 45.0 kDa; IV: Carbonic anhydrase 31.0 kDa; V: soybean trypsin 21.5 kDa.

Our previous data demonstrated a strict relationship between the endogenous level of the core of 2', 5'-oligoadenylates, especially $A_{2p}A_{2p}A$, and poly(ADP-ribose) transferase activity. $A_{2p}A_{2p}A$ was shown to be a potent inhibitor of poly(ADP-ribose) transferase activity *in vitro* (12). A rapid decrease in the $A_{2p}A_{2p}A$ level and a concomitant increase in poly(ADP-ribose) transferase activity with a parallel increase in proliferation were observed in the uterus of an immature rat after estrogen administration (13). These data suggest the possibility that an increase in poly(ADP-ribose) transferase activity in lizard oviduct could be also due to a rapid decrease in the $A_{2p}A_{2p}A$ level, although any data regarding this possibility is not

324

available in the lizard organ. The possibility that an estrogen-induced increase in poly(ADP-ribose) transferase activity could be due to an enhancement of the expression of the gene for the enzyme cannot be ruled out.

References

1. Müller, W.E.G., Zahn, R.K. (1976) Mol Cell Biochem 12: 147-159
2. Suhadolnik, R.J., Sawada, Y., Gabriel, J., Reichnlach, N.L., Henderson, E.E. (1984) J Biol Chem 259: 4764-4769
3. Ohashi, Y., Ueda, K., Hayaishi, O., Ikai, K., Niwa, O. (1984) Proc Natl Acad Sci USA 81: 7132-7136
4. Tanuma, S., Johnson, L.D., Johnson, G.S. (1983) J Biol Chem 258: 15371-15375
5. Ciarcia, G., Angelini, F., Picariello, O., D'Alterio, E. (1982) Boll Zool 49: 40
6. Botte, V., Granata, G. (1977) J Endocrinol 73: 535-536
7. Ciarcia, G., Lancieri, M., Suzuki, H., Manzo, C., Vitale, L., Tornese-Buonamassa, D., Botte, V. (1986) Mol Cell Endocrinol 47: 235-241
8. Mayes, E.L.V., Johns, E.W. (1982) The HMG Chromosomal Proteins, E.W. Johns (ed.), Academic Press, London, pp. 223-247
9. Purnell, M.R., Whish, W.J.D. (1980) Biochem J 185: 775-777
10. Miwa, M., Tanaka, M., Matsushima, T., Sugimura, T. (1974) J Biol Chem 249: 3475-3482
11. Okayama, H., Honda, M., Hayaishi, O. (1978) Proc Natl Acad Sci USA 75: 2254-2257
12. Pivazian, A.D., Suzuki, H., Vartanian, A.A., Zhelkovsky, A.M., Farina, B., Leone, E., Karpeisky, M.Y. (1984) Biochem Int 9: 143-152
13. Suzuki, H., Tornese-Buonamassa, D., Weisz, A. (1986) Biol Chem Hoppe Seylor 367 Suppl.: 275
14. Stephens, R.E. (1975) Anal Biochem 65: 369-379
15. Penke, B., Ferenczi, R., Kovacs, K. (1974) Anal Biochem 60: 45-50
16. Laemmli, U.K. (1970) Nature 227: 680-685

Abbreviations:
HMG - High Mobility Group
SDS-PAGE - Polyacrylamide Gel Electrophoresis in Sodium Dodecyl Sulfate
SNP - Specific Nuclear Protein

DNA Damage Does Not Induce Lethal Depletion of NAD During Chicken Spermatogenesis

Montserrat Corominas and Cristobal Mezquita

Molecular Genetics Research Group, Department of Physiological Sciences, Faculty of Medicine, University of Barcelona, 08028 Barcelona, Spain

Introduction

The nuclear enzyme poly(ADP-ribose) polymerase transfers ADP-ribose from NAD to chromosomal proteins forming a nucleic acid-like polymer with a branched structure (1). Poly(ADP-ribose) polymerase activity is involved in processes where DNA breaking and rejoining take place (2). We have previously shown that the content of poly(ADP-ribose), poly(ADP-ribose) polymerase activity and the turnover of ADP-ribosyl residues were maximal in a fraction containing meiotic and premeiotic cells and decrease during the differentiation of the germinal cell line, especially at the end of spermatogenesis (3). It has been postulated that genotoxic actions in a variety of cells produce, through a stimulation of poly(ADP-ribose) biosynthesis, suicidal NAD depletion (4, 5). In order to examine if the germinal cell line might prevent, through such a suicide mechanism, the transmission of genotoxic damage to the embryo, we have determined the levels of NAD at successive stages of chicken spermatogenesis after treatment of cells with bleomycin, an antitumor antibiotic that cleaves double stranded DNA (6).

Results

Poly(ADP-ribose) polymerase activity after bleomycin treatment. Chicken testis cells at different stages of spermatogenesis, separated by centrifugal elutriation as described in (7), and spermatozoa obtained from the vas deferens, were treated with bleomycin for 1 hr and the poly(ADP-ribose) polymerase activity was assayed in nuclei isolated from the different cell types. Incubation with bleomycin resulted in stimulation of the enzymatic activity in premeiotic and meiotic cells, round spermatids and in elongated spermatids and was ineffective in spermatozoa (Fig. 1).

Effect of bleomycin treatment on poly(ADP-ribose) levels. Chicken testis cells at successive stages of spermatogenesis were incubated with bleomycin for 1 hr. Trichloroacetic acid precipitates from the different cell types were used for analysis of polymeric ADP-ribose by fluorescence methods as described by Jacobson et al. (8). The poly(ADP-ribose) content increased, relative to DNA, after bleomycin treatment in premeiotic and meiotic cells, round spermatids and in elongated spermatids. No changes were observed in mature spermatozoa (Fig. 2).

326

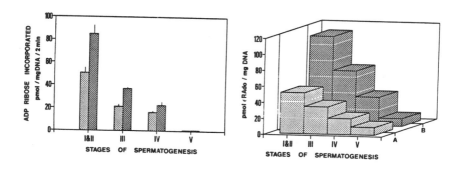

Fig. 1. (left) ADP-ribose incorporation by nuclei after treatment of chicken testis cells with bleomycin. Intact chicken testis cells, separated by centrifugal elutriation, were incubated for 1 hr in a medium with (hatched bars) or without bleomycin (dotted bars). Nuclei were isolated from treated and untreated cells and incubated at 37°C for 2 min in a medium containing [^{14}C]-NAD. The incubation was terminated by adding 20% trichloroacetic acid and the precipitates were collected on Whatman GF/A filters. I & II, premeiotic and meiotic cells; III, round spermatids; IV, elongated spermatids; V, spermatozoa from the vas deferens.

Fig. 2. (right) Levels of polymeric ADP-ribose after bleomycin treatment in cells at successive stages of spermatogenesis. The content of polymeric ADP-ribose was determined as described. I & II, premeiotic and meiotic cells; III, round spermatids; IV, elongated spermatids; V, mature spermatozoa. Control cells (dotted bars). Cells treated with bleomycin for 1 hr (hatched bars).

Effect of bleomycin treatment on NAD content. The NAD content was measured in acid-soluble extracts following treatment with bleomycin and precipitation of cells with 20% trichloroacetic acid as described by Alvarez-Gonzalez *et al.* (9). Incubation with the antibiotic resulted in a slight decrease of the NAD content, relative to DNA, in premeiotic and meiotic cells, round spermatids and in elongated spermatids and was ineffective in mature spermatozoa (Fig. 3).

DNA integrity after bleomycin treatment. The DNA integrity was studied by nucleoid sedimentation analysis (10, 11) and by denaturing agarose gel electrophoresis (12-14). The effect of bleomycin treatment on the sedimentation of nucleoids derived from premeiotic and meiotic cells and round spermatids is shown in Fig. 4. The relative sedimentation rate decreased after the treatment with the antibiotic, due to release of supercoiling within nucleoids, leading to a more relaxed structure. This technique cannot be applied to determine the integrity of the DNA in nuclei of spermatids containing the nucleoprotamine complex. The induction of DNA strand scissions in these nuclei was detected by electrophoresis in denaturing agarose gels. Bleomycin induced a 20% increase in DNA strand scissions in elongated spermatids but no damage could be detected in mature spermatozoa.

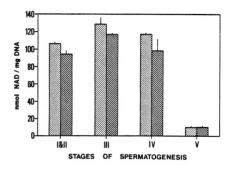 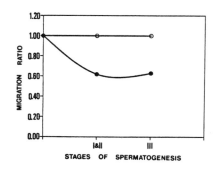

Fig. 3. (left) Effect of bleomycin treatment on NAD content in cells at successive stages of spermatogenesis. The NAD content was assayed in the acid-soluble extracts. I & II, premeiotic and meiotic cells; III, round spermatids; IV, elongated spermatids; V, spermatozoa from the vas deferens. Control cells (dotted bars), cells treated with bleomycin for 1 hr (hatched bars).

Fig. 4. (right) Effect of bleomycin on the sedimentation of nucleoids. Cells were exposed to bleomycin for 1 hr and prepared for centrifugation. The migration ratio (vertical axis) is expressed as the ratio of the distance migrated by damaged nucleoids relative to the distance migrated by undamaged nucleoids centrifuged at the same time in one of the tubes. I & II, premeiotic and meiotic cells; III, round spermatids. Control cells (O), cells treated with bleomycin for 1 hr (●).

Discussion

The results of this study, in accordance with previous results (3), demonstrate that the induction of DNA strand breaks increases the content of poly(ADP-ribose) polymer and the poly(ADP-ribose) polymerase activity in meiotic and premeiotic cells and also in round and elongated spermatids but not in mature spermatozoa. Treatment with bleomycin induces DNA strand break formation in chicken testis cells at successive stages of spermatogenesis with the exception of mature spermatozoa. The invulnerability of the DNA of chicken spermatozoa to bleomycin treatment has been previously reported (15).

Numerous reports have demonstrated the NAD lowering effects of DNA damaging agents (4). The reduction is caused by an increase in poly(ADP-ribose) polymerase activity by the presence of strand breaks in DNA. In certain cells, an excessive activation of the poly(ADP-ribose) polymerase with drastic reduction in the intracellular NAD pools may induce lethal ATP depletion, a situation known as a "suicide response" (4). However other cells are remarkably resistant to suicidal NAD depletion (16). Chicken testis cells at successive stages of spermatogenesis showed minimal changes in NAD levels after bleomycin treatment. The same results were obtained when the alkylating agent N-methyl-N'-nitro-N-nitrosoguanidine was used

328

instead of bleomycin (results not shown). Germinal cells therefore possess the capacity to avoid the typical "suicide response" observed in other cells when treated with these damaging agents. The ability to maintain constant levels of NAD may be related to the presence of effective mechanisms of replenishment in chicken testis cells with high levels of poly(ADP-ribose) polymerase activity and high turnover of ADP-ribosyl residues. In germinal cells with low enzymatic activity and lack of turnover, particularly in spermatozoa, where no poly(ADP-ribose) polymerase activity can be induced, the NAD level could be maintained in the absence of excessive consumption by poly(ADP-ribose) polymerase reactions. The relevance of the germ line to hereditary stability suggests that special mechanisms may exist to avoid the transmission of genotoxic damage to the embryo. The stability observed in NAD levels in chicken testis cells after bleomycin treatment indicates that suicidal depletion of NAD is not likely to be involved in germ cell and gamete selection and other mechanisms should be responsible for the accurate transmission of genetic information.

Acknowledgements. We would like to express our appreciation to Dr. F.R. Althaus. In his laboratory, at the institute of Pharmacology and Biochemistry, University of Zurich, M. Corominas performed NAD analyses. We would like to thank also Dr. R. Alvarez-Gonzalez for helpful and stimulating discussions. This work was supported by a grant from CAICYT (3182/83).

References

1. Hayaishi, O., Ueda, K. (1982) ADP-Ribosylation Reactions: Biology and Medicine, O. Hayaishi, K. Ueda, (eds.) Academic Press, New York
2. Benjamin, R.C., Gill, D.M. (1980) J Biol Chem 255: 10493
3. Corominas, M., Mezquita, C. (1985) J Biol Chem 260: 16269
4. Berger, N.A. (1985) Radiation Res 101: 4
5. Carson, D.A., Seto, S., Wasson, D.B., Carrera, C.J. (1986) Exp Cell Res 164: 273
6. Hecht, S.M. (1986) Fed Proc 45: 2784
7. Meistrich, M.L. (1977) Methods in Cell Biol XV: 15
8. Jacobson, M.K., Payne, D.M., Alvarez-Gonzalez, R., Juarez-Salinas, H., Sims, J.L., Jacobson, E.L. (1984) Methods Enzymol 106: 483
9. Alvarez-Gonzalez, R., Eichenberger, R., Loetscher, P., Althaus, F. (1986) Anal Biochem 156: 473
10. Cook, P.R., Brazell, I.A., Jost, E. (1976) J Cell Sci 22: 303
11. Weniger, P. (1979) Int J Rad Biol 36: 197
12. Maniatis, T., Fritsch, E.F., Sambrook, J. (1982) Molecular Cloning Cold Spring Harbor Laboratory), Cold Spring Harbor, New York, p. 280
13. Maniatis, T., Fritsch, E.F., Sambrook, J. (1982) Ibid, p. 171
14. Kohen, R., Szyf, M., Chevion, M. (1986) Anal Biochem 154: 485
15. Young, R.J., Sweeney, K. (1979) Gamete Res 2: 265
16. Alvarez-Gonzalez, R., Eichenberger, R., Althaus, F. (1986) Biochem Biophys Res Commun 138: 1051

Divergence of Poly(ADP-ribose) Polymerase Activity, Poly(ADP-ribose) Levels and the Effect of Polymerase Inhibitors in Differentiating 3T3-L1 Preadipocytes

Onno Janssen and Helmuth Hilz

Institut für Physiologie Chemie, Universität Hamburg, West Germany

Introduction

Modification of nuclear proteins by poly(ADP-ribose) has been implicated as an essential step in the differentiation of a number of different systems. However, in some systems a transient rise of poly(ADP-ribose) as a prerequisite of differentiation was deduced from experiments with inhibitors [e.g. mesodermal chick limb cells (1), myoblasts (2) HL 60 cells to macrophages (3) and lymphocytes (4)]. In others, a transient decrease of poly(ADP-ribose) polymerase activity was thought to be required for the induction of differentiation, as inhibitors of poly(ADP-ribosyl)ation were able to replace the inducers [e.g. erythroid differentiation of Friend leukemia cells (5), HL 60 cells to granulocytes (6), murine embryo carcinoma cells (7) and mouse mammary gland differentiation (8)].

Even contradictory results for one and the same differentiation system were reported: In the conversion of 3T3-L1 preadipocytes to adipocytes reports of an early, transient decrease of poly(ADP-ribose) polymerase activity as determined in isolated nuclei (9) and a stimulation of differentiation by nicotinamide (10) contrast with a paper reporting the prevention of preadipocyte differentiation by nicotinamide and benzamide (11). In none of these reports, however, was protein-bound poly(ADP-ribose) determined. Here, we describe an analysis of mono(ADP-ribose) and poly(ADP-ribose) status in 3T3-L1 preadipocytes during differentiation to adipocytes and the effect of various poly(ADP-ribose) polymerase inhibitors.

Results and Discussion

Poly(ADP-ribose) polymerase activity versus amounts of poly(ADP-ribose) in differentiating 3T3-L1 cells. When confluent 3T3-L1 cells were treated with insulin, dexamethasone and methylisobutylxanthine and then with insulin alone, they differentiated within about 6 days to adipocytes (13). Beginning with day 4, enzymes of the adipocyte phenotype like glycerophosphate dehydrogenase were induced from low basal levels to high activities and the cells started to accumulate large amounts of fat which are deposited in the form of droplets. When we analyzed intrinsic poly(ADP-ribose) polymerase activity in permeabilized cells, a rise of activity was

induction. However, these changes in poly(ADP-ribose) polymerase activity were not accompanied by corresponding changes of the reaction product: protein-bound poly(ADP-ribose) residues remained unchanged throughout the entire period of differentiation (Fig. 1).

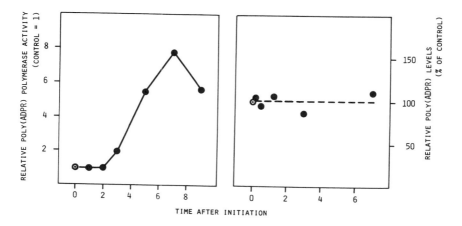

Fig. 1. Poly(ADP-ribose) polymerase activity and poly(ADP-ribose) levels during differentiation. Cells were grown to confluence and initiated with insulin/dexamethasone/methylisobutylxanthine as described in (12). Poly(ADP-ribose) polymerase activity was determined in permeabilized cells, and poly(ADP-ribose) levels were quantiated (15) after TCA precipitation of cell cultures *in situ*.

Influence of benzamide on the expression of the differentiation markers. Since it was possible that only a small fraction of poly(ADP-ribosyl) proteins were involved in differentiation, we analyzed the influence of benzamide as an inhibitor of poly(ADP-ribose) polymerase (13) and of ADP-ribosyl transferases (14) on the expression of glycerophosphate dehydrogenase. As shown in Table 1, benzamide at 3 mM, strongly increased induction of the enzyme. This result pointed to an involvement of some sort of ADP-ribosylation in the differentiation process.

Table 1. Benzamide-enhanced induction of differentiation in 3T3-L1 cells.

Benzamide Concentration	Glycerophosphate Dehydrogenase Activity $(mU \times cell^{-1} \times 10^{-6})$	(%)
0 mM	67.5 ± 10.9	100
0.3 mM	64.6 ± 10.2	96
1 mM	92.1 ± 19.4	135
3 mM	134.1 ± 17.0	199
10 mM	6.1 ± 1.0	9

Determination of glycerophosphate dehydrogenase activity was performed in triplicate 10 days after initiation with insulin/dexamethasone/methylisobutyl-xanthine and addition of benzamide.

Modification of histones and of HMG proteins. Histones are the major acceptors of mono(ADP-ribosyl) residues formed in response to alkylation of cells (15, 16). When we determined modification of histones present in the H_2SO_4-soluble fraction, no significant differences were seen between control and differentiating cells. We also analyzed the modification of HMG proteins as present in the perchloric acid extracts of cells previously labeled *in vivo* with [³H]adenosine. There was a general decrease of total protein-bound [³H]adenine, to about one third in differentiating cells, which pointed to a corresponding decrease in the specific radioactivity of the precursor pool. However, none of the HMG proteins showed a specific change under these conditions, which makes participation in differentiation of one of the HMG proteins in its ADP-ribosylated form very unlikely.

Effect of benzoate and 3-aminobenzoate on differentiation. The failure to find differentiation-associated changes in the ADP-ribosylation status of 3T3-L1 cells indicated that the differentiation accelerating action of benzamide was a side effect. We therefore studied the influence of an expanded number of nicotinamide analogs. It was found that not only benzamide, 3-aminobenzamide and nicotinamide, but also the corresponding acids were able to induce insulin primed differentiation (induction of glycerophosphate dehydrogenase and fat deposition) (Table 2). Since the acids are not inhibitors of ADP-ribosylation, it is clear that their action on chromatin function involves a process that is not linked to covalent modification by ADP-ribose.

Table 2. Induction of differentiation by benzoate and analogs.

Additions	Induction of G3PDH ($mU \times cell^{-1} \times 10^{-6}$)	(%)
None	0.42± 0.14	69
Insulin	0.61± 0.09	100
Insulin + benzamide	10.93± 0.24	1792
Insulin + 3-aminobenzamide	4.65± 1.32	762
Insulin + nicotinamide	5.30± 1.37	869
Insulin + benzoate	9.43± 1.40	1546
Insulin + 3-aminobenzoate	3.12± 2.06	511
Insulin + nicotinate	3.18± 0.45	521

Glycerophosphate dehydrogenase activity was determined 10 days after initiation with insulin. Benzamide, 3-aminobenzamide, benzoate and 3-aminobenzoate: 3mM. Nicotinamide and nicotinate: 5 mM. Triplicate values ± s.d.

From these data, several conclusions can be drawn. First, poly(ADP-ribose) polymerase activity as determined in isolated nuclei or in permeabilized cells does not necessarily reflect the endogenous ADP-ribosylation status. Second, differentiation of 3T3-L1 preadipocytes does not appear to require changes in the ADP-ribosylation status since neither poly(ADP-ribose) levels, nor mono(ADP-ribosyl)ated histones, nor the modification of HMG proteins showed specific changes during differentiation. Third, not only were inhibitors of poly(ADP-ribose) polymerase like benzamide, 3-aminobenzamide and nicotinamide co-inducers of 3T3-L1 cell differentiation, but conversion was also strongly enhanced by the corresponding acids (benzoate, 3-aminobenzoate, nicotinate), which do not inhibit the polymerase. Finally, high concentrations of benzamide (>5 mM) interfered with differentiation. Since poly(ADP-ribose) polymerase *in situ* is inhibited completely (>99%) at 2-5 mM benzamide, the inhibitory action at 10 mM appears to be a non-specific effect which does not relate to ADP-ribosylation.

Acknowledgements. This work was supported by the Deutsche Forschungsgemeinschaft.

References

1. Caplan, A.I., Rosenberg, M.J. (1975) Proc Natl Acad Sci USA 72: 1852-1857
2. Farzaneh, F., Zalin, R., Brill, D., Shall, S. (1982) Nature 300: 362-366
3. Francis, G.E., Gray, D.A., Berney, J.J., Wing, M.A., Guimaraes, J.E.T., Hoffbrand, A.V. (1983) Blood 62: 1055-1062
4. Johnstone, A.P., Williams, G.T. (1982) Nature 300: 368-370
5. Morika, E., Tanaka, K., Nokuo, T., Ishizawa, M., Ono, T. (1979) Gann 70: 37-46
6. Damji, N., Khoo, K.E., Booker, L., Browman, G.P. (1986) Am J Hemat 21: 67-78
7. Ohashi, Y., Ueda, K., Hayaishi, O., Ikai, K., Niwa, O. (1984) Proc Natl Acad Sci USA 81: 7132-7136
8. Bolander, F.F. (1986) Biochem Biophys Res Commun 137: 359-363
9. Pekala, P.H., Lane, M.D. (1981) J Biol Chem 256: 4871-4687
10. Pekala, P.H., Moss, J. (1983) Mol Cell Biochem 53/54: 221-232
11. Lewis, J.E., Shimizu, Y., Shimizu, N. (1982) FEBS Lett 146: 37-41
12. Green, H., Meuth, M. (1974) Cell 3: 127-133
13. Rubin, C.S., Hirsch, A., Fung, C., Rosen, O.M. (1978) J Biol Chem 253: 7570-7578
14. Moss, J., Stanley, S.J., Watkins, P.A. (1980) J Biol Chem 255: 5838-5840
15. Kreimeyer, A., Wielckens, K., Adamietz, P., Hilz, H. (1984) J Biol Chem 259: 890-896
16. Kreimeyer, A., Adamietz, P., Hilz, H. (1985) Biol Chem Hoppe-Seyl 366: 537-544

Differences in DNA Supercoiling and ADP-Ribosylation During Granulocytic and Monocytic Differentiation

Z. Khan and G.E. Francis

Molecular Cell Pathology Unit, The Royal Free Hospital School of Medicine, London NW3 2PF, United Kingdom

Introduction

Neutrophil-granulocytes and monocytes arise from a common precursor cell. We have previously shown that monocytic differentiation is selectively inhibited by compounds which inhibit ADP-ribosyl transferase and that stimulation of these precursor cells by their combined growth and differentiation stimulus is accompanied by rapid activation of the enzyme (1). The object of the current investigation was to use inducers of monocytic and granulocytic differentiation to compare the changes in DNA supercoiling and in ADP-ribosylation which precede differentiation to these two lineages.

Results and Discussion

Changes in DNA supercoiling were analyzed using nucleoid sedimentation as previously described and are shown in Fig. 1 (2). Human granulocyte-macrophage precursor cells were prepared from bone marrow samples (obtained with consent from healthy volunteers) by density fractionation and adherence (2). Cells were exposed to two inducers of granulocytic differentiation (retinoic acid and G-CSF) and two inducers of monocytic differentiation (phorbol myristate acetate and vitamin D3). Treatment with differentiation inducers and culture conditions are described elsewhere (2).

Since monocytic differentiation is inhibited by 3-MB and other inhibitors of ADP-ribosyl transferase (1, 2), it seems likely that these patterns reflect differences in changes in chromatin structure during the induction of differentiation to the two lineages and that only the changes in monocytic differentiation require ADP-ribosyl transferase. With the monocytic inducers in the presence of 3-MB, it is apparent that the transient relaxation of supercoiling which is associated with granulocytic differentiation, is also occurring during monocytic differentiation. Further investigation of this second type of change suggests it may be mediated by topoisomerase II (3) since during retinoic acid induced granulocytic differentiation the transient relaxation of supercoiling, the transient appearance of protein-associated DNA breaks and granulocytic differentiation are all inhibited by VP16-213.

Ribosylation of nuclear and cytoplasmic proteins during the induction of differentiation by RET and PMA was assessed by slab gel electrophoresis.

Non-adherent cells (5 x 10⁶) were exposed to the differentiation inducers for 15-120 min and permeabilized on ice for 5-10 min in 50 µl hypotonic buffer consisting of 10 mM Tris-HCl (pH 7.8), 4 mM MgCl, 1 mM EDTA and 5 mM DTT. After permeabilization 25 µl of reaction mix was added to 50 µl of cell suspension to produce a final concentration of 33 mM Tris-HCl (pH 7.8), 0.67 EDTA, 2.67 mM MgCl, 1.67 mM DTT and 10 µM (5 µCi [³²P]NAD, 0.3-10 x 10⁵ mCi/mmol). Tubes were transferred to 26°C and incubated for 10 min. The reaction was stopped by adding 675 µl of hypotonic buffer as above containing 5 mM PMSF, left on ice for 10 min, homogenized with 5 strokes and centrifuged at 200 xg for 10 min. The supernatant was removed and proteins precipitated with 10% TCA. Pellet proteins were also precipitated by resuspension in hypotonic buffer containing 10% TCA and left on ice for 10 min. Cytoplasmic and nuclear proteins were centrifuged and the pellet washed twice in 1:1 ether:ethanol. Samples were resuspended in 50 µl loading buffer, consisting of 0.01 M sodium phosphate (pH 6.8), 2% 2-mercaptoethanol, 10% glycerol, 0.002% bromophenol blue and boiled for three min.

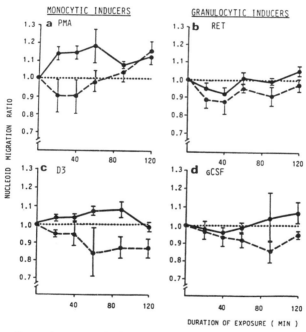

Fig. 1. Migration ratios on nucleoid sedimentation gradients following treatment with differentiation inducers, in the presence (dashed line) and absence of 5 mM 3-methoxybenzamide: A) 10⁻⁹ M PMA; B) 10⁻⁶ M RET; C) 10⁻⁹ M Vit D3; D) G-CSA (50% umbilical cord conditioned medium). Results are means of five, six, five and two independent experiments for A-D respectively. With the monocytic inducers (PMA and Vit D3) there was an increase in sedimentation. In contrast, with the granulocytic inducers (RET, G-CSA) there was a transient retardation of sedimentation. The increased sedimentation rate induced by the monocytic inducers was abolished by 3-MB.

Acrylamide gel electrophoresis was carried out using a 7.5% continuous SDS slab gel system. Both the separating gel buffer and the electrode buffer were 25 mM sodium phosphate pH 6.8. Gels were run at a constant current of 50 mA/gel for 3-4 hr. They were stained for 60 min at room temperature in 65% acetic acid, 30% methanol and 5% water containing 1% Coomasie blue and destained in 25% methanol, 10% acetic acid for 2 hr. Dried gels were autoradiographed for 2 weeks at -70°C.

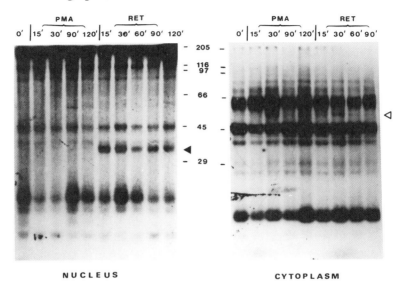

Fig. 2. Gel electrophoresis of nuclear and cytoplasmic ADP-ribosylated proteins. Cells were treated with PMA or RET for 15-120 min.

Fig. 2 shows gels of nuclear and cytoplasmic proteins. Numerous ribosylated proteins were found both in the nuclear and cytoplasmic fractions. Most proteins were stably ribosylated. Two proteins in this experiment (arrows) showed induced ribosylation, a nuclear protein with RET only and a cytoplasmic protein with both RET and PMA. Other proteins in this and additional experiments varied in intensity with the two treatments, but it is difficult to exclude slight variations in loading in an individual experiment. In order to overcome this and since the distribution of ribosylated proteins varied somewhat, we performed repeated experiments in an attempt to identify proteins both on the basis of their apparent molecular weight and on their behaviour during induction (Fig. 3). Ribosylated proteins were classified into three groups: 1) those more ribosylated during RET induction than PMA induction; 2) those more ribosylated during PMA induction than RET induction; 3) those ribosylated approximately equally or those where ribosylation was unchanged by induction.

336

Fig. 3 shows the results from 11 independent experiments. The majority of ribosylated proteins were of the third type. Variation in the distribution of the three groups of proteins with respect to molecular weight suggests heterogeneity of the proteins ribosylated and/or the extent of ribosylation under different biological conditions. Further experiments are in progress using these and additional inducers of monocytic and granulocytic differentiation to determine whether these differences reflect lineage specific events. Variation between individual experiments probably reflects heterogeneity of samples with respect to the distribution of progenitor and more mature cell types contained. Having established the technique using enriched progenitors we plan to confirm findings using highly purified granulocyte-macrophage precursor cell preparations.

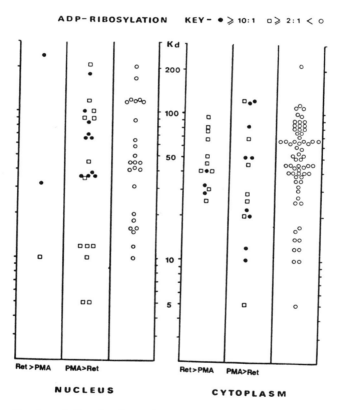

Fig. 3. Classification of ribosylated proteins by behavior with induction and molecular weight. Autoradiographs of 30 min RET and PMA treatment and untreated controls were analyzed by scanning densitometry and classified as showing unequal ribosylation if the ratio of absorption was >= 10:1 (filled circles) or >= 2:1 (squares) or approximately equal ribosylation (open circles). Left hand columns show proteins more heavily ribosylated with RET than PMA, central columns the converse and right hand columns those with approximately equal ribosylation.

337

The results presented here suggest a role for ADP-ribosyl transferase in regulation of lineage switching in haemopoietic cells. Monocytic but not granulocytic differentiation of granulocyte-macrophage precursor cells requires and is preceded by an ADP-ribosyl transferase dependent increase in supercoiling (which independent experiments have shown is accompanied by the ligation of pre-existing DNA breaks (4)). This change occurs within 20 min, well before overt signs of differentiation in this cell system and may thus represent an early regulatory event.

The analysis of ribosylated proteins and enzyme assays during induction (2) suggest that changes in ribosylation occur not only with monocytic but also with granulocytic differentiation and yet such differentiation is not readily inhibited by ADP-ribosyl transferase inhibitors. One can envisage a number of situations where ribosylation might accompany events in differentiation but not be essential for them. For example, Ferro and Olivera (5) have suggested that failure to ribosylate topoisomerases does not inhibit their action but renders the enzyme more error-prone and increases the rate of induction of sister chromatid exchanges. In addition ADP-ribosylation might be involved in epiphenomena of granulocytic differentiation not detected by the criteria of our differentiation assays (we evaluated production of chloroacetate esterase, capacity to reduce nitroblue tetrazolium and loss of proliferative capacity). Further analysis of the proteins ribosylated with a range of inducers of both granulocytic and monocytic differentiation and the relation of ribosylation to other events like the activation of topoisomerases will allow us to dissect further the differentiation process.

Acknowledgements: Some of the data shown in this manuscript have been reprinted with permission from ref. 2.

References

1. Francis, G.E., Gray, D.A., Berney, J.J., Wing, M.A., Guimaraes, J.E.T., Hoffbrand, A.V. (1983) Blood 62: 1055-1062
2. Khan, Z., Francis, G.E. (1987) Blood 69: 1114-1119
3. Francis, G.E., Berney, J.J., North, P.S., Khan, Z., Jacobs, P., Wilson, E.L. (1986) Proc of the First Conf on Diff Therapy
4. Francis, G.E., Ho, A.D., Gray, D.A., Berney, J.J., Wing, M.A., Yaxley, J.J., Ma, D.D.F., Hoffbrand, A.V. (1984) Leukemia Research 8: 407-416
5. Ferro, A.M., Olivera, B.M. (1984) J Biol Chem 259: 547-554

Abbreviations:
PMA - phorbol myristate acetate
RET - all-trans retinoic acid
Vit D3 - 1-25, dihydroxy vitamin D3
G-CSA - granulocytic colony stimulating activity
3-MB - 3-methoxybenzamide
DTT - dithiothreitol
PMSF - phenylmethyl-sulphonyl fluoride

Niacin Analogs That Induce Differentiation of Friend Erythroleukemia Cells are Able to Cause DNA Hypomethylation

Jim R. Kuykendall and Ray Cox

Cancer Research Laboratory, Veterans Administration Medical Center, and Department of Biochemistry, University of Tennessee-Memphis, Memphis, Tennessee 38163 USA

Introduction

Friend erythroleukemia cells (FELCs) are retrovirus-transformed murine leukemia cells of the erythroid lineage, blocked in a relatively early stage of differentiation (1). These cells can be induced to differentiate into erythroid-like cells able to synthesize hemoglobin mRNA, heme, and hemoglobin (Hb) (2, 3). Nicotinamide (NAm), a potent inhibitor of poly(ADP-ribose) synthetase, has been shown to be a moderate inducer of FELCs differentiation (4). Other inhibitors of poly(ADP-ribose) synthetase were also found to be capable of inducing differentiation (5). These investigations hypothesized that the inhibition of poly(ADP-ribose) turnover was the primary biochemical mechanism leading to differentiation of FELCs by NAm and related compounds.

A number of investigators have proposed that cytosine methylation by DNA methylase may play a role in regulation of gene expression (6-9). Christman (10-12) found that DNA isolated from induced cultures of FELCs was hypomethylated when compared to control cultures, as judged by the ability of the DNA to serve as a better methyl-acceptor in an *in vitro* methylation reaction. Several studies have shown a transient decrease in the methylation of cytosine residues of the globin genes and their flanking sequences, during expression, in erythroid tissues from a number of species (13-15). The purpose of this study was to determine if DNA methylation was altered during induction of differentiation by NAm, and if so, by what mechanism. Additionally, we examined the properties of NAm analogs to see if this class of compounds was able to induce differentiation by some common pathway.

Results and Discussion

Mechanism of NAm inhibition of DNA methylation. FELCs were cultured in the presence of increasing concentrations (2.5-40 mM) of NAm for four days. Differentiation status was determined using the hemoglobin detection assay of Kaiho and Mizuno (16). Cell growth was inhibited by all concentrations of NAm tested, while cell viability was not affected up to 10 mM. All concentrations were able to cause induction of differentiation, with

20 mM as the optimum dose, giving 65% hemoglobin-positive cells by 96 hr.

FELCs were cultured for 36 hr at these increasing NAm concentrations, DNA and crude nuclear extracts were isolated, and DNA methylase was assayed as previously described (17). DNA from these cells exhibited a dose-dependent hypomethylation, measured by increased methyl accepting ability, which peaked with 20 mM NAm treatment (Fig. 1). DNA isolated from cells treated with 40 mM NAm was not hypomethylated and was presumed to be due to lack of DNA synthesis and cell division, caused by cytotoxic effects of this compound. DNA methylase activity (Fig. 1) was found to be normal at non-cytotoxic doses of NAm. NAm added to the cell-free assay system was unable to significantly inhibit the DNA methylase activity. These results suggest that the DNA from FELCs grown in NAm is hypomethylated and that the activity of the DNA methylase enzyme is not directly altered by NAm.

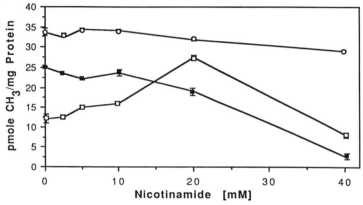

Fig. 1. Dose response of NAm on DNA methylation of FELCs. DNA methyl-accepting ability and DNA methylase activity were assayed as previously described in reference (17). The cells were cultured for 36 hr with varying concentrations of NAm, the effect of NAm on DNA methylase activity (O) with control enzyme and FELCs DNA from 20 mM NAm treated culture; (□), DNA methyl-accepting ability; (■), DNA methylase activity. All data points represent the mean of triplicate assays with standard error.

Crude nuclear extracts and DNA were isolated from cells which were cultured for four days in the presence of 20 mM NAm. The methyl-accepting ability of the DNA was double that of control DNA at 24 hr, but gradually declined to that of control values over the next 72 hr (Fig. 2). The methylation of newly synthesized DNA was examined using the ^{32}P-labeling procedure of Christman and the percent 5-methylcytosine (5 mC) was likewise found to undergo a transient decrease, which peaked at 24-48 hr. The DNA methylation returned to normal by 72 hr and was above control levels at 96 hr, indicating that a stimulation of methylation was occurring at 72-96 hr (Fig. 2). Changes in methylation of newly synthesized DNA seemed to parallel those found in total DNA, showing that alterations in methyl-accepting ability of the DNA were due to generation of new hemi-methylated sites. These data further suggest that DNA methylation was indirectly altered by NAm.

Excess NAm in cell cultures is metabolized primarily by enzymatic methylation, catalyzed by nicotinamide methyltransferase, forming 1-methylnicotinamide (1-MNA), the primary excretory form (19, 20). This leads to depletion of cellular S-adenosylmethionine (AdoMet), the methyl donor substrate, and an accumulation of S-adenosylhomocysteine (AdoHcy), a potent feedback inhibitor of enzymatic methylation reactions (21). Cellular levels of AdoMet and AdoHcy were measured as previously described (22). When FELCs were incubated with 20 mM NAm for four days, the AdoMet levels declined to almost half that of control cells by 72 hr, with a concommitant increase in AdoHcy concentrations, which peaked at 48 hr (Fig. 3). The decline in AdoHcy levels after this time seemed to follow the return of DNA methylation status by 96 hr, as determined earlier. Thus, excess cellular NAm levels result in a depletion of AdoMet and an increase in AdoHcy, indirectly causing hypomethylation of the DNA.

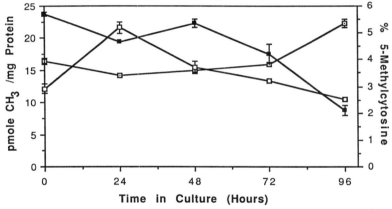

Fig. 2. Time curve for 20 mM NAm on DNA methylation of FELCs. DNA methylase was assayed as previously described (17). (■), DNA methylase activity; (□), DNA methyl-accepting ability; (□), 5-mC content of newly synthesized DNA (18). All data points represent the mean of triplicate assays with standard error.

Nicotinamide analogs as inducers of differentiation. Methylated analogs of NAm, a group of compounds differing only in the presence and/or position of a single methyl group, provide a model system for the study of erythroleukemic differentiation. These compounds show diverse effects on cell growth and viability, DNA methylation, ADP-ribosylation and differentiation. This allows a dissection of the differentiation process into a set of cause and effect relationships, rather than mere correlations. FELCs were incubated with each compound at 2.5, 5, 10 and 20 mM concentrations. The optimum inducer concentrations were determined by

341

percent Hb-positive staining cells, as well as the concentration of Hb from whole cell extracts. N'-methylnicotinamide (N'-MNA) was found to be the best inducer, with an optimal concentration of 5 mM, followed by 6-methylnicotinamide (6-MNA) at 20 mM, NAm at 20 mM, and 5-methylnicotinamide (5-MNA) at 10 mM. 1-Methylnicotinamide did not serve as an inducer.

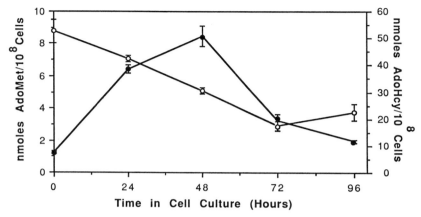

Fig. 3. Effect of NAm on cellular levels of AdoMet and AdoHcy. FELCs were cultured in 20 mM NAm and the concentrations of AdoMet and AdoHcy were determined at various times (22). (□), AdoMet; (■), AdoHcy. All data points represent the mean of triplicate assays with standard error.

Niacin analogs were compared at 10 mM for effects on induction of differentiation, Hb accumulation, poly(ADP-ribose) synthetase *in vitro,* and on DNA methylase activity in cell-free assays and in growing cells (Table 1). None of these compounds were active as inhibitors of DNA methylase activity in cell-free assays. In contrast, these compounds were found to have differential effects on poly(ADP-ribose) synthetase activity in isolated nuclei, as measured by a modified procedure of Hayaishi (23). Inhibition of poly(ADP-ribose) synthetase by these compounds did not, as previously hypothesized, correlate with their ability to induce Hb synthesis and differentiation. N'-MNA and 6-MNA were not able to inhibit poly(ADP-ribose) synthetase activity in isolated nuclei when present at 5 mM, whereas NAm and 5-MNA both possess inhibitory activity *in vitro*. When these compounds were incubated at 10 mM in cell culture for 24 hr, all those able to induce differentiation were also able to cause DNA hypomethylation, as judged by their effect on methyl-accepting abilities of DNA isolated from these cells.

These data are in complete agreement with previously published reports that N'-MNA could not serve as an inhibitor of poly(ADP-ribose) synthetase activity *in vitro* (5), and that 5-MNA was a good inhibitor (24). N'-MNA is the best inducer of the methylated pyridines, while 5-MNA is the poorest, indicating that inhibition of poly(ADP-ribose) synthetase activity may not be the primary pharmacological effect leading to the induction of differentiation by this group of compounds. Rather, these data suggest that the inhibition of DNA methylation is specific for inducer-capable analogs,such as N'-MNA, NAm and 6-MNA. The effect on DNA methylation status of DNA from cells cultured with each analog closely correlates to the inductive abilities of the drug. This work supports the hypothesis that DNA methylation is an important part of the early processes leading to chemical induced maturation of murine erythroleukemia cells.

Table 1. Comparison of the effects of niacin analogs on nuclear events relating to the induction of differentiation of FELCs. Compounds were present at 10 mM in all assays, except ADP-ribose transferase activity, where they were 5 mM. All data represent the mean of triplicate assays with standard error, except % ADP-ribose transferase activity, where only the mean was included.

			Niacin Analogs			
	Control	NAm	1-MNA	N'-MNA	5-MNA	6-MNA
% ADP-ribose transferase activity *in vitro*	100	26	74	85	22	83
DNA methylase[a] activity *in vitro*	42.3 ± 0.8	45.1 ± 0.8	49.7 ± 0.6	43.5 ± 1.7	43.9 ± 0.7	43.7 ± 0.9
μg Hb/10^6 Cells	0.9 ± 0.2	2.9 ± 0.7	ND	9.4 ± 1.9	0.9 ± 0.2	3.0 ± 0.7
% Hb-positive staining cells	6.3 ± 0.3	50.6 ± 1.2	4.0 ± 0.5	76.6 ± 2.7	17.0 ± 2.0	62.3 ± 2.5
Methyl-accepting[a] ability of DNA	17.1 ± 1.2	31.9 ± 0.6	14.5 ± 0.7	57.5 ± 2.2	10.9 ± 0.1	35.5 ± 0.5

[a]Values are expressed as pmol CH_3/mg protein

Acknowledgements. These studies were supported by the United States Veterans Administration (4323-01) and USPHS Research Grant CA-15189 from the National Cancer Institute.

References

1. Friend, C., Scher, W., Holland, J.G., Sato, T. (1971) Proc Natl Acad Sci USA 68: 378-382
2. Marks, P.A., Rifkind, M. (1978) Ann Rev Biochem 47: 419-448
3. Rifkind, R.A., Sheffery, M., Marks, P.A. (1984) Adv Cancer Res 42: 149-163
4. Morioka, K., Tanaka, K., Nokus, T., Ishizawa, M., Ono, T. (1979) Gann 70: 37-46
5. Terada, M., Fujiki, H., Marks, P., Sugimura, T. (1979) Proc Natl Acad Sci USA 76: 6411-6414
6. Burdon, R.H., Adams, R.P.L. (1980) Trends Biochem Sci 5: 294-297
7. Razin, A, Riggs, A.D. (1980) Biochem Biophys Acta 782: 331-342
8. Ehrlich, M., Wang, R.Y.H. (1981) Science 212: 1350-1357
9. Schapiro, F. (1983) Biomedicine and Pharmacol Therapy 37: 173-175
10. Christman, J.K., Price, P., Pedriman, L., Acs, G. (1977) Eur J Biochem 81: 53-61
11. Christman, J.K., Weich, N., Schonenbrum, B., Schneiderman, N., Acs, G. (1980) J Cell Biol 83: 366-370

12. Christman, J.K. (1984) Current Topics in Microbiology and Immunology, Vol. 108: 49. Springer-Verlag, New York and Berlin
13. Waalwijk, C., Flavell, R.A. (1978) Nucl Acids Res 5: 4631-4641
14. van der Ploeg, L.H.T., Flavell, R.A. (1980) Cell 19: 947-958
15. Busslinger, M., Hurst, J., Flavell, R.A. (1983) Cell 34: 197-206
16. Kaiho, S., Mizuno, K. (1985) Anal Biochem 149: 117-120
17. Cox, R., Goorha, S. (1986) Carcinogenesis 7: 2015-2018
18. Christman, J.K. (1982) Anal Biochem 119: 38-48
19. Stanulovic, M., Chaykin, S. (1971) Arch Biochem Biophys 145: 35-42
20. Chang, M.L.W., Johnson, B.C. (1959) J Biol Chem 234: 1817-1821
21. Cox, R., Prescott, C., Irving C.C. (1977) Biochem Biophys Acta 474: 493-499
22. Cox, R. (1983) Cancer Letters 17: 295-300
23. Kawachi, M., Ueda, K., Hayaishi, O. (1980) J Biol Chem 242: 816-819
24. Clark, J.B., Ferris, G.M., Pinder, S. (1970) Biochem Biophys Acta 238: 82-85

Abbreviations:
FELCs - Friend erythroleukemia cells
Hb - hemoglobin
NAm - nicotinamide
1-MNA - 1-methylnicotinamide
AdoMet - S-adenosylmethionine
AdoHcy - S-adenosylhomocysteine
5-MNA - 5-methylnicotinamide
6-MNA - 6-methylnicotinamide
N'-MNA - N'-methylnicotinamide
5-mC - 5-methylcytosine

In Vivo Poly(ADP-ribose) Levels in Carcinogen Treated Human Promyelocytic Leukemia Cells

Bhartiben N. Patel, Melvyn B.J. Dover, and Christopher J. Skidmore

Department of Physiology and Biochemistry, University of Reading, Reading RG6 2AJ, United Kingdom

Introduction

The cellular role of nuclear ADP-ribosylation has been analyzed traditionally by the investigation of the activity of ADP-ribosyl transferase (NAD$^+$:protein ADP-ribosyl transferase, EC 2.4.2.30) and by the use of enzyme inhibitors (1). ADP-ribosyl transferase is activated when strand breaks are induced in DNA by alkylating agents and x-rays (2, 3) or during the process of cell differentiation (4, 5). Inhibitors of the enzyme inhibit most measures of DNA repair (3) although nicotinamide can stimulate unscheduled DNA synthesis (6). 3-Aminobenzamide in particular inhibits differentiation of myoblasts (7) and of macrophage precursor cells. It is only with the recent introduction of reliable methods for measuring the steady-state levels of poly(ADP-ribose) that it has been possible to look at ADP-ribosylation directly *in vivo*. We have initiated a study comparing the behavior of ADP-ribosylation following DNA damage and during differentiation. We use the human cell line HL60, promyelocytic cells which have the capacity to differentiate into either granulocyte or macrophages in culture following treatment with phorbol-12-myristate 13-acetate or retinoic acid, respectively (5). We present here initial studies in which the levels of poly(ADP-ribose) *in vivo* in undifferentiated HL60 cells are measured following treatment with the powerful mutagen N-methyl-N'-nitro-N'-nitrosoguanidine (MNNG). We also investigate the effect of ADP-ribosyl transferase inhibitors on the steady state level of ADP-ribosylation.

Results and Discussion

Trichloroacetic acid precipitates from HL60 cells treated with MNNG were analyzed for the presence of ADP-ribosyl transferase. Poly(ADP-ribose) was isolated by affinity chromatography and, after enzymatic digestion of the polymer, the fluorescent derivatives, ethenoribosyladenosine (εRAdo) and ethenodiribosyladenosine (εR$_2$Ado), generated from linear and branched residues respectively, were identified and quantified by HPLC (8). The presence of εR$_2$Ado from branched residues is not routinely detected in our samples. The relationship between fluorescence peak height and amount of εRAdo applied is linear from 0-6 pmoles.

345

A time course of the effect of 500 μM MNNG at 10 min showed a large peak (Fig. 1). This transient increase is similar to that in other studies (6). After 30 min the polymer level has stabilized at approximately 15 pmol εRAdo/10⁸ cells, still some 4-fold higher than the basal level in these cells. HL60 cells behave in a similar manner to other human and animal cells in responding to MNNG (6). The short half-life of this polymer, also observed in other systems, indicates that degradation is a major factor in determining the levels of poly(ADP-ribose) in cells.

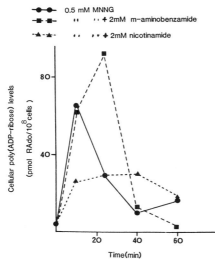

Fig. 1. Effect of inhibitors on cellular poly(ADP-ribose) levels in MNNG-treated HL60 cells. HL60 cells were cultured in RPM1 1640 medium buffered with 20 mM Hepes, pH 7.8 and supplemented with 10% horse serum, penicillin (50 units/ml), streptomycin (50 μg/ml) and 7.5% sodium bicarbonate at 37°C. HL60 cells were treated with 0.5 mM MNNG dissolved in DMSO. Treatments were stopped at different times of incubation by treating the cells with 10 ml 20% TCA. The poly(ADP-ribose) content was determined in the TCA pellet according to Jacobson *et al.* (8). Control cells contained 4.1 pmol of εRAdo/10⁸ cells. These values remained constant in cells incubated at 30°C without the carcinogen. Reprinted with permission from ref. 9.

The effect of inhibitors of ADP-ribosyl transferase on the increase in steady-state polymer levels is also shown in Fig. 1. Both 3-aminobenzamide and nicotinamide at 2 mM delayed the transient rise in polymer levels so that the peak occurred at around 25 min. 3-Aminobenzamide also inhibited the accumulation of polymers by 50% indicating that the increase in polymer levels is mediated by ADP-ribosyl transferase. Nicotinamide (2 mM) however, stimulated the accumulation of polymer.

Since the medium in which HL60 cells are grown contains only 8 μM nicotinamide, these cells may not be saturated with NAD⁺ and nicotinamide may be stimulating NAD⁺ production, and hence polymer synthesis. Fig. 2 shows the effect of supplementing the medium with 2 mM nicotinamide 24 hr before MNNG treatment. While this gave only a modest increase in the control level of poly(ADP-ribose), a 100% increase in the level attained following 500 μM MNNG was observed even when no added nicotinamide was present during the incubation. Incubation with 2 mM nicotinamide during MNNG treatment gave a nearly 200% increase in polymer levels.

346

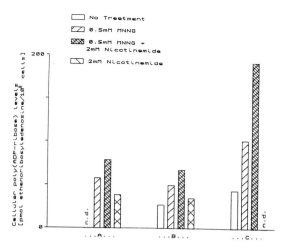

Fig. 2. Cellular poly(ADP-ribose) levels of HL60 cells grown in the absence and presence of nicotinamide prior to MNNG treatment for 10 min at 30°C. Cells were cultured, treated and extracted as in the legend to Fig. 1. A: Cells in 3 day culture medium. B: Fresh medium 24 hr prior to experiment. C: Fresh medium supplemented with 2mM nicotinamide 24 hr prior to experiment.

So in this cell line 2 mM nicotinamide stimulated poly(ADP-ribose) levels presumably via its effect on NAD⁺ synthesis. This is in contrast to the inhibition of polymer accumulation seen in fresh human lymphocytes (6). This cell-type specific variation in the effect of nicotinamide underlines its inadequacy as an inhibitor of ADP-ribosylation in whole cell studies.

References

1. Ueda, K., Hayaishi, O. (1985) Ann Rev Biochem 54: 73-100
2. Berger, N.A., Sikorski, G.W., Petzold, S.J., Kurohara, K.K. (1979) J Clin Invest 63: 1164-1171
3. Shall, S. (1984) Adv Rad Biol 11: 1-69
4. Francis, G.E., Gray, D.A., Berney, J.J., Wing, M.A., Guimares, J.E.T., Hoffbrand, A.V. (1983) Blood 62: 1055-1062
5. Damji, N., Khoo, K.E., Booker, L., Browman, G.P. (1986) Am J Hematol 21: 67-78
6. Sims, J.L., Berger, S.J., Berger, N.A. (1981) J Supramol Struct Cell Biochem 16: 281-288
7. Farzaneh, F., Zalin, R., Brill, D., Shall, S. (1982) Nature 300: 362-366
8. Jacobson, M.K., Payne, D.M., Alvarez-Gonzalez, R., Juarez-Salinas, H., Sims, J.L., Jacobson, E.L. (1984) Methods Enzymol 106: 483-494
9. Patel, B.N., Dover, M.B.J., Skidmore, C.J. (1988) Biochem Soc Trans 16: 869-870

Activation-Inactivation of Poly(ADP-Ribose) Transferase of Mammalian Cells Exposed to DNA Damaging Agents

A.I. Scovassi, M. Stefanini, R. Izzo, P. Lagomarsini, and U. Bertazzoni

Istituto di Genetica Biochimica ed Evoluzionistica del C.N.R., Pavia, Italy

Introduction

The molecular mechanism of poly(ADP-ribose) transferase activation following the treatment of mammalian cells with DNA damaging agents is still unknown. Several lines of evidence indicate that the activation process does not require *de novo* synthesis of the enzyme and that the extent of increase in activity is correlated to the type of DNA damage (1). In particular, the response of poly(ADP-ribose) transferase seems to be proportional to the number of single strand breaks introduced in the DNA by different agents. From various data it appears that in conditions of extreme DNA damage the activation of the enzyme may induce the depletion of cellular NAD and cellular death (2). We have studied the response of poly(ADP-ribose) transferase of different mammalian cells treated with monofunctional and bifunctional alkylating agents including DMS, MMS, and EMS. An activity gel and western blot were used to analyze poly(ADP-ribose) transferase. The results we have obtained indicate that the poly(ADP-ribose) transferase response to alkylation damage to mammalian cells is showing two distinct phases: an activation step at lower mutagen concentrations followed by a decrease in enzyme activity at higher doses.

Results

We have analyzed the variations and the structural properties of poly(ADP-ribose) transferase in human quiescent lymphocytes and CHO cells treated with DNA-damaging agents. The preparation of samples and the activity gel technique have been performed essentially as previously described (3-5). The dose-response relationship of different mutagens to the poly(ADP-ribose) transferase of Go human lymphocytes treated with mutagens is reported in Fig. 1. We have analyzed the effect of a 1 hr treatment with increasing concentrations of DMS, MMS and EMS on the appearance of poly(ADP-ribose) transferase on activity gel autoradiograms. Compared to untreated lymphocytes (panel A, lane 1), the incubation with 0.01 mM DMS (lane 2) caused a significant increase in the intensity of the active band of the enzyme, which became much greater after treatment with 0.1 mM DMS (lane 3). At a higher dose of mutagen (1 mM, lane 4), no

348

activation of the enzyme was evident. On the contrary, the activity band became barely visible, indicating that high doses of alkylating agents may induce a loss in the activity of the enzyme. Also in the case of MMS treatment a similar result was obtained (Fig. 1, panel B); in particular, the activation was evident with 0.1 mM MMS (lane 2) and was magnified when the lymphocytes were treated with 1 mM MMS (lane 3), remained at the same level with 10 mM MMS. When the lymphocytes were treated with 50 mM MMS, a notable decrease in the active band of the enzyme was observed (lane 5). The effect of the alkylating agent EMS on poly(ADP-ribose) transferase activity band is presented in panel C. In this case no enhancing effect was evident at 1 mM EMS (lane 2) whereas a notable decrease was observed at 10 and 100 mM EMS (lanes 3 and 4).

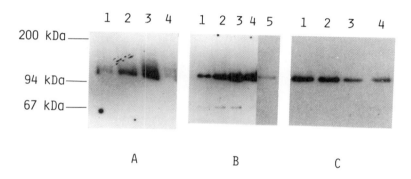

Fig. 1. Detection by activity gel of poly(ADP-ribose) transferase catalytic peptides in human lymphocytes treated with different DNA damaging agents. Human lymphocytes (4-5 x 10⁶) freshly separated from peripheral blood were incubated for 1 hr with 0, 0.01, 0.1, 1 mM DMS (panel A, lanes 1-4), 0, 0.1, 1, 10, 50 mM MMS (panel B, Lanes 1-5) or 0, 1, 10, 100 mM EMS (panel C, lanes 1-4). After treatment, lymphocytes were collected and washed twice with PBS and the pellet was kept in liquid nitrogen until use. Total extract and activity gel procedure were performed as described (4). Protein mass markers loaded onto the same gel were: myosin (200 kDa), phosphorylase b (94 kDa) and BSA (67 kDa).

In order to confirm the activation-inactivation pattern observed in human lymphocytes for poly(ADP-ribose) transferase, we have treated CHO cells in culture with different concentrations of DMS. A much stronger signal was obtained with 1 mM DMS (Fig. 2, lane 3) than in untreated cells (lane 1) while the activity band was barely visible after treatment with 10 mM DMS (lane 4). To study in more precise terms the kinetics of disappearance of the active band of the enzyme, we have treated CHO cells with 10 mM DMS for different times. It appears that the intensity of the catalytic band decreased proportionally with time and completely disappeared after 60 min of cell treatment (Fig. 2B). The CHO samples treated with 0.1, 1 and 10 mM DMS were also analyzed by the western blot technique (Fig. 3). No significant

349

variations could be seen in the intensity of the protein band of 116 kDa at different concentrations of DMS. This suggests that the increase in enzyme activity, as observed in the activity gel, was not due to *de novo* synthesis and that the disappearance of the activity band at high doses of alkylating agent (obtained at 10 mM DMS) was not the result of a loss of the protein from the gel. Since the transferase has an absolute requirement for DNA and the activation process implies that the enzyme binds very tightly to damaged DNA, a legitimate question is whether the activation in the gel requires the presence of a stable nicked DNA-poly(ADP-ribose) transferase complex. From preliminary experiments, it appears that the active band of the enzyme is not dependent on the presence of DNA added to the gel. It may be found that fragments of DNA with ends are strictly bound to the enzyme and are carried through the activity gel procedure.

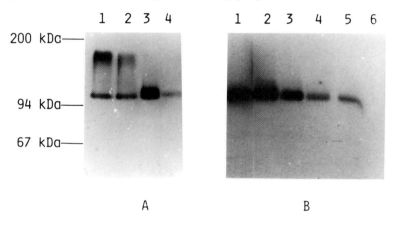

Fig. 2. Effect of DMS treatment on poly(ADP-ribose) transferase of CHO cells. Panel A: CHO cells (4-5 x 10⁶) were treated with 0, 0.1, 1, 10 mM DMS for 1 hr (lanes 1-4). Panel B: cells were submitted to 10 mM DMS treatment for 0, 5, 10, 20, 40, 60 min (lanes 1-6). Treatments were performed as described in the legend of Fig. 1.

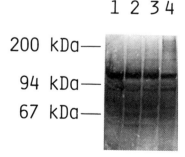

Fig. 3. Western blot analysis of poly(ADP-ribose) transferase from CHO cells. Total extracts from CHO cells were submitted to the immunoblot analysis following the usual procedure (4, 7). The polyclonal antibody against calf thymus purified poly(ADP-ribose) transferase was donated by Dr. Niedergang (Strasbourg). Lanes: 1, control cells; 2-4, cells treated with 0.1, 1, 10 mM DMS, respectively.

Discussion

The activation of poly(ADP-ribose) transferase in cells treated with DNA-damaging agents is considered to be strictly dependent on the appearance of newly formed single strand breaks in DNA. We have further explored this mechanism utilizing the human lymphocyte system and the activity gel technique. In quiescent human lymphocytes, although poly(ADP-ribose) transferase activity was barely detectable before treatment, activation by DMS and MMS was promptly obtained. From the dose-response relationship, it appeared that DMS was 10-fold more effective than MMS in poly(ADP-ribose) transferase activation. That is expected from the number of single-stranded breaks produced by these methylating agents. When the lymphocytes were treated with very high doses of mutagens, a tendency of the active band of the enzyme to disappear was observed. Several hypotheses can be formulated for this result: i) the massive dose of mutagen causes a rapid depletion of NAD and ATP pools and, therefore, a rapid cellular death (2, 6); ii) poly(ADP-ribose) transferase, with its autoribosylation sites fully saturated, has reduced or no capacity to further bind ADP-ribose; iii) the enzyme molecule is damaged or has lost its ability to renature *in situ* during the activity gel procedure.

In CHO cells treated with DMS, the same overall observation as in human lymphocytes was obtained, although the concentrations of the mutagen needed for the activation of poly(ADP-ribose) transferase were about 10-fold higher. Also the disappearance of the active band at very high doses of DMS was readily observable in CHO cells. From the western blot analysis of CHO samples it appeared that no variations in the intensity of the 116 kDa protein band could be observed after treatment with low and high concentrations of DMS, suggesting that the activation of poly(ADP-ribose) transferase was reflecting a change in the intrinsic activity of the enzyme and that its subsequent inactivation is not causing a loss of the enzyme from the gel. Finally, preliminary results on the importance of DNA for poly(ADP-ribose) transferase activity suggest that the active band is not dependent on the presence of exogenous DNA and that an enzyme-bound DNA is needed for activity.

Acknowledgements. This work was supported in part by contract BI6-158-I from the Radiation Protection Programme of the Commission of the European Communities. U.B. is a scientist official of CEC; R.I. is recipient of a fellowship from Fondazione Buzzati-Traverso and P.L. from Associazione Italiana Ricerca sul Cancro.

References

1. Shall, S. (1984) Advances Radiation Biol 11: 1-69
2. Berger, N.A. (1985) Radiation Res 101: 4-15
3. Scovassi, A.I., Stefanini, M. and Bertazzoni, U. (1984) J Biol Chem 259: 10973-10977

4. Scovassi, A.I., Izzo, R., Franchi, E. and Bertazzoni, U. (1986) Eur J Biochem 159: 77-84
5. Bertazzoni, U., Scovassi, A.I., Mezzina, M., Sarasin, A., Franchi, E. and Izzo, R. (1986) Trends Genet 2: 67-72
6. Gaal, J.C., Smith, K.R., Pearson, C.K. (1987) TIBS 12: 129-130
7. Towbin, H., Stahelin, T., Gordon, J. (1979) Proc Natl Acad Sci USA 76: 4350-4354

Abbreviations:
BSA - bovine serum albumin
DMS - dimethyl sulfate
EMS - ethyl methane sulfonate
MMS - methyl methane sulfonate
PBS - phosphate buffer saline

Aspects of NAD Metabolism in Prokaryotes and Eukaryotes

Baldomero M. Olivera, Kelly T. Hughes, Pauline Cordray, and John R. Roth

Department of Biology, University of Utah, Salt Lake City, Utah 84112 USA

Introduction

Our laboratories have investigated NAD metabolism in both prokaryotic and eukaryotic systems. Background information in this area has previously been reviewed (1-3). The work we describe here on eukaryotic systems is initial experiments to examine the possible role of ADP-ribosylation in modulating synaptic transmission in the nervous system. In this article, we also summarize our present understanding of NAD metabolism and its regulation in the bacterium *Salmonella typhimurium,* and present preliminary data for a link between NAD turnover cycles and DNA repair in this bacterium.

Results and Discussion

Regulation of NAD metabolism in *Salmonella typhimurium*. Despite the central role of NAD metabolism in all cells, there is little information regarding the molecular mechanisms for regulating NAD levels in any cell. Our initial approach to this problem in the bacterium *Salmonella typhimurium* was to systematically define all the genes involved in NAD biosynthesis. Previous to our work, three structural genes involved in *de novo* NAD biosynthesis had been identified, *nad*A, *nad*B and *nad*C. These genes are involved in the biosynthesis of the key metabolic intermediate nicotinic acid mononucleotide (NaMN) from aspartate and dihydroxyacetone phosphate. Mutations in any of these genes cause the cell to become auxotrophic, requiring an exogenous source of the pyridine ring (see Fig. 1). Nicotinic acid mononucleotide can be synthesized either *de novo,* or from exogenous nicotinamide or nicotinic acid. Two genes involved in the salvage of the pyridine ring, *pnc*A and *pnc*B were previously identified.

The conversion of nicotinic acid mononucleotide to NAD proceeds in two metabolic steps, with nicotinic acid dinucleotide (NaAD) as an intermediate (4); there are no alternative metabolic steps for the synthesis of NAD from NaMN. Thus, the genes that control these two metabolic steps should be essential for viability, and conditional-lethal mutations should be recovered.

We recently identified and characterized both of the genetic elements which control the two essential terminal metabolic steps of the NAD biosynthetic pathway. The structural gene encoding the NaAD pyrophosphorylase enzyme has been designated the *nad*D locus; it is located at 14 min on the *Salmonella typhimurium* chromosome (5). We also identified and mapped mutations affecting the final synthetic step; this gene has been designated *nad*E (6, K. Hughes, B. Olivera and J.R. Roth, unpublished results). As expected, neither of these genes can be completely deleted. Temperature-sensitive mutations have been obtained for both *nad*D and *nad*E; the mutant cells die at the high temperature, and temperature sensitivity of either NaAD pyrophosphorylase (*nad*D) and NAD synthetase (*nad*E) can be demonstrated *in vitro*. Thus, all of the structural genes for the biosynthesis of NAD either *de novo* (from aspartate and glycerol) or by a salvage pathway from nicotinamide have apparently been defined.

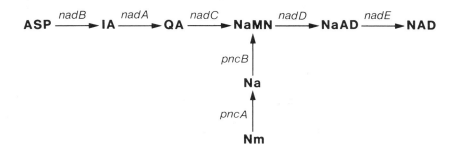

Fig. 1. NAD biosynthetic pathways in enteric bacteria. The genes controlling each metabolic step are indicated. Abbreviations used are: ASP, aspartate; IA, iminoaspartate; QA, quinolinic acid; Nm, nicotinamide; Na, nicotinic acid; NaMN, nicotinic acid mononucleotide; NaAD, nicotinic acid adenine dinucleotide.

How is the activity of this biosynthetic pathway regulated so that the cell has the desired level of NAD? It has previously been shown that if an exogenous source of the pyridine ring is available, the *de novo* synthesis is suppressed (7). One mechanism for regulating the *de novo* biosynthesis of NAD is through feedback inhibition. *In vitro,* the *nad*B enzyme is subject to feedback inhibition by NAD. We recently obtained mutants in the *nad*B locus which were feedback resistant (5,8). These mutants were recovered by screening for resistance to an analog of nicotinamide (6-aminonicotinamide); a significant fraction of such mutants map at the *nad*B locus, within the *nad*B structural gene. The continued production of NAD by the *de novo* pathway, even in the presence of exogenous precursors leads to a higher ratio of authentic NAD to 6-aminonicotinamide-substituted NAD; this allows the bacterium to escape killing in the presence of the analog. The frequency of *nad*B mutants that are resistant to the analog suggest that the *nad*B enzyme is the primary locus of feedback inhibition. This might have been anticipated from the fact that the *nad*B enzyme, L-aspartate oxidase, catalyzes the first metabolic step in the NAD biosynthetic pathway.

In addition, the *de novo* biosynthetic pathway is regulated by repression. Recently, definitive genetic evidence was obtained for the presence of a repressor acting on the transcription of the first two biosynthetic genes (*nad*A & B) (9). This repressor is encoded by the *nad*I locus (9,10); both null mutations (which are derepressed for *de novo* NAD biosynthesis) as well as super-repressor mutations (which have the same phenotype as nicotinamide auxotrophs) have been obtained at the *nad*I locus.

In summary, the levels of NAD in *Salmonella typhimurium* appear to be regulated both by feedback inhibition as well as by repression. The structural genes for the proteins that are involved in these regulatory mechanisms have been identified, although a detailed biochemical characterization of the *nad*I repressor protein and a definition of the regulatory site on the *nad*B enzyme remain to be carried out.

A possible regulatory locus affecting both NAD and DNA metabolism in *Salmonella typhimurium*. The mechanisms described above are conventional for regulating biosynthetic pathways. However, there are a number of aspects of NAD metabolism which are unusual. NAD metabolism provides not only a cofactor for oxidation and reduction reactions, but in bacteria, also provides a substrate for ADP-ribosylation reactions, and a source of ribose for vitamin B12 biosynthesis. Cleavage of the NAD pyrophosphate linkage provides a source of energy for DNA ligation, thereby linking NAD metabolism to DNA metabolism (1,3). We recently defined a genetic locus, here referred to as *xth*R which suggests yet another potential link between NAD and DNA metabolism, specifically DNA repair. The locus was discovered by the isolation of an unusual mutation which was resistant to the nicotinamide analog, 6-aminonicotinamide. In addition to mutants mapping at the *pnc*A, *pnc*B or *nad*B genes, a class of mutations generically called "*pnc*X" map close to, but not at the *pnc*A structural gene (5). The "*pnc*X" mutants have significantly lower (but clearly detectable) levels of nicotinamide deamidase; this confers the 6-aminonicotinamide resistant phenotype. The *xth*R mutant, a unique "*pnc*X" mutation, differed from all such other mutants in that it was found to be *ts*-lethal. Since lac operon fusions are available for the *pnc*A locus, double mutants combining *xth*R and the lac fusion at *pnc*A reveal that the lower enzyme levels in *xth*R are due to decreased transcription of the *pnc*A gene.

When the *ts* lethality was investigated, it was found that these cells were filamentous. In enteric bacteria filamentous growth is characteristically observed upon induction of the SOS system in response to DNA damage. It was determined that the *xth*R cells were in fact constitutively SOS-induced. Examination of genetic loci that mapped in the "*pnc*X" region revealed that the *xth*A locus in *E. coli* which encodes exonuclease III mapped in this general region (11). We therefore assayed exonuclease III levels in *xth*R mutants and found that compared to wild type strains, the levels of *xth*A gene product, exonuclease III, was approximately 40-fold higher.

The *ts* lethality can be reversed by second site mutations, which map at loci different from the original *xth*R locus. Preliminary results suggest that one of these second site mutations may be at the structural gene for exonuclease III, i.e., at the *xth*A locus. The levels of exonuclease III assayed in extracts of the revertant are below detectable limits and the revertant is sensitive to hydrogen peroxide.

Why would a 40-fold increase in the level of exonuclease III lead to an SOS constitutive phenotype, which in turn, is presumably responsible for the *ts* lethality? Exonuclease III digests DNA $3' \Rightarrow 5'$ starting at a nick or a gap structure. The 3' OH terminus of a DNA strand where hydrolysis occurs is the same site for priming DNA chain growth by DNA polymerases. Wherever a gap occurs, there would therefore be competition between exonuclease III and DNA polymerase I or III for the 3' OH primer structure. If exonuclease III were overproduced 40-fold, this should lead to the persistence of gaps in the DNA structure: polymerases would be unable to efficiently convert a gap to a nick which can be sealed by ligase. Because of the presence of excess exonuclease III, gap structures would persist thereby stimulating the *rec*A *lex*A complex to cleave the *lex*A repressor, inducing an SOS response (for a general review of the relevant DNA enzymology, see 12).

A central question is why should there be a regulatory locus which, when mutated, lowers an enzyme of NAD metabolism 8-fold (nicotinamide deamidase) and increases the level of exonuclease III 40-fold? One intriguing possibility is that this is part of the response of *Salmonella typhimurium* to superoxide and oxygen radical production by white blood cells. *Salmonella typhimurium* is one of the few bacteria which can survive being phagocytized once it invades an avian or mammalian host. Phagocytes produce high levels of superoxide and oxygen radicals, damaging bases in the bacterial genome. *Salmonella typhimurium* must repair such oxygen damage to survive phagocytosis. Exonuclease III is believed to be the major enzyme which causes an incision event in DNA after an oxidized base has been removed by the glycosylases which recognize the damaged base. Thus, under conditions of oxygen radical attack, it may be desirable to raise the levels of exonuclease III.

Why might it be desirable to coordinately lower the levels of nicotinamide deamidase? The 8-fold depression in nicotinamide deamidase activity causes excretion of nicotinamide; *xth*R mutants have shown to be "feeders" for nicotinamide auxotrophs, indicating that these strains continuously excrete nicotinamide into the medium. This is presumably a consequence of the pyridine nucleotide cycle, shown in Fig. 2. Nicotinamide deamidase is not only an enzyme for the salvage of exogenous pyridine, but it is part of a NAD recycling pathway, i.e., a "pyridine nucleotide cycle"

356

(13). Any intracellular nicotinamide which is produced from NAD, either directly or indirectly (from NMN), must be deamidated to nicotinic acid before it can be resynthesized into NAD. When nicotinamide deamidase levels fall, intracellular nicotinamide is not efficiently recycled, and a significant fraction of nicotinamide produced by pyridine nucleotide cycles is excreted into the medium. Thus, lowering nicotinamide deamidase levels shunts the normal pyridine nucleotide cycle to cause excretion of nicotinamide.

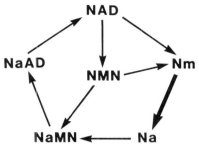

Fig. 2. Intracellular pyridine nucleotide cycles in enteric bacteria. The breakdown and resynthesis of NAD occurs in bacteria by the metabolic steps shown above. All abbreviations are as in Fig. 1, with the addition of NMN, nicotinamide mononucleotide. The metabolic step catalyzed by nicotinamide deamidase, the levels of which are reduced by an *xth*R mutation (see text) is shown by the bold arrow.

Nicotinamide would be endogenously released within the eukaryotic nucleus as a by-product of the synthesis of poly(ADP-ribose). We raise the possibility that the presence of nicotinamide signals the phagocytizing cell that its own poly(ADP-ribose) synthetase is turned on (presumably in response to DNA breaks in its nuclear DNA). Such a signal may lead the phagocyte to reduce production of superoxide and oxygen radicals. We further speculate that *Salmonella* may have exploited the system to protect itself from the host phagocyte. It may produce nicotinamide to induce the host to stop free radical generation. Although this hypothesis is a very specific one, the discovery of a mutant that seems to affect two apparently unrelated enzymes, nicotinamide deamidase and exonuclease III, suggests a biological link between NAD and DNA metabolism in the bacterium *Salmonella typhimurium* which needs to be further defined.

Pyridine nucleotide cycles in eukaryotes and prokaryotes. We previously demonstrated that pyridine nucleotide cycles are a significant fraction of total NAD biosynthesis in both eukaryotic and prokaryotic cells. In enteric bacteria such as *Escherichia coli*, NAD turnover accounts for approximately 30% of total NAD biosynthesis (14). Surprisingly, this fraction is even higher in eukaryotic cells in culture. Our measurements on HeLa D98AH2 in culture for example indicate that greater than 90% of NAD biosynthesis compensates for NAD breakdown (15).

Since pyridine nucleotide turnover can be such a dominant feature of the total biosynthesis of NAD in a eukaryotic cell, animal tissues where intracellular NAD breakdown occurs at the greatest rate may be the tissues with the greatest demand for the pyridine ring, since recycling is never perfect. We have suggested elsewhere that unexplained symptoms of pellagra, (i.e., the 3-D's of pellagra: dermatitis, diarrhea and dementia) are more rationally explained by considering NAD turnover than by oxidation/reduction effects (1). Dermatitis and diarrhea could very well be due to increased NAD turnover in skin cells (particularly as a result of sunlight induced DNA damage) and in the cells of the intestinal lining (which are exposed to environmental damaging agents). Why dementia is a prominent symptom of a niacin deficiency remains unexplained. For these reasons, we have begun to investigate the biochemical events which consume NAD in neuronal tissue.

Our laboratory previously investigated nuclear poly(ADP-ribosyl)ation and the ADP-ribosylation of DNA topoisomerase; in this report, however, we would like to explore the possibility that ADP-ribosylation reactions are important in membrane phenomena, particularly signal transduction in neurons. Our initial experiments are discussed in the following section.

ADP-ribosylation of membrane proteins; studies on synaptosomes.
As a model system for neuronal membranes, our laboratories have examined the synapse in the Torpedo electric organ. This preparation has the advantage of being predominantly of one synaptic type, quite analogous to the vertebrate neuromuscular junction (from which it is believed to have been derived).

When synaptic membrane fractions are incubated with radioactive NAD, and the proteins are displayed on an SDS gel, a complex labelling pattern is seen, suggesting endogenous ADP-ribosylation of multiple membrane proteins (16). When cholera toxin is added, a protein band with a molecular weight identical to G_s is labelled in addition to the background labeling seen without cholera toxin; this result is shown in Fig. 3. Auto-ADP-ribosylation of cholera toxin can also be detected under these conditions. The labeling of the endogenous "G_s-like" band is intensified in the presence of GTP or GTPβS and is suppressed if GDPβS is added. If exogenous G_s is added, it is found that using the appropriate incubation mixture, G_s is preferentially ADP-ribosylated over endogenous membrane proteins (Fig. 4.). Under the incubation conditions promoting preferential labeling of exogenous G_s, an endogenous band ($M_r \approx 70$ K) is also preferentially labeled, such that it and exogenous G_s become the most prominently labeled species (see Fig. 4, lane D). The labeling of the 70 K band and G_s are competitive to some degree; as more G_s is added, 70 K labeling is suppressed. This suggests that exogenous G_s and the 70 K band are both modified by the same transferase.

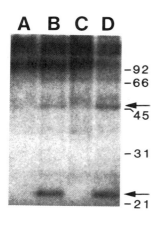

A B C D

−92
−66
45
−31
−21

Fig. 3. Preferential labeling of "Gs-like" band in *Torpedo* membrane by choleratoxin. Reaction mixtures contain 0.05 M K phosphate buffer, pH 7.5, 1 mm DTT, 4 mM MgCl$_2$, 1 mM EDTA, 120 nM [^{32}P]NAD (131 C$_i$/mmole) and 9.8 µg membrane protein from the fraction of *Torpedo* electric organ membranes which banded at the 0.8 M sucrose layer (in a discontinuous sucrose gradient comprised of 0.32 M, 0.8 M and 1.3 M sucrose in 5 mM HEPES buffer, pH 7.4). Lanes B and D also contained 1.7 µg of the freshly activated choleratoxin. The incubation mixtures analyzed in Lanes C and D had 0.1 mM GTPγS. Apart from auto ADP-ribosylation of the choleratoxin, the major new labeled band in lanes with choleratoxin is a band with Mr ≈ 48 K, with an electrophoretic mobility indistinguishable from exogenous bovine Gs under these conditions (the arrows indicate the bands specifically labeled in the presence of choleratoxin).

A B C D E F G H I

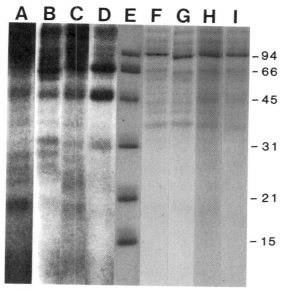

−94
−66
−45
−31
−21
−15

Fig. 4. Labeling of bovine Gs by an endogenous ADP-ribosyl transferase in *Torpedo* membranes. Lanes A-D show auto radiograms of stained gels in lanes F-I; lane E is a MW standard. The incubation mixtures analyzed in lanes A and B (gels F, G) have no exogenous Gs added; lanes C and D (gels H, I) have exogenous purified bovine brain Gs (1.5 µg). In lanes B and D, reaction mixtures contained 0.05 M K phosphate buffer, pH 7.5, 1 mM DTT, 0.1 mM CaCl$_2$, 120 nM [^{32}P]NAD (specific activity 200 Ci/mmol), and *Torpedo* electric organ membranes (9.8 µg protein). Reaction mixtures analyzed in A and C contained the same buffer and membrane protein, but enough EDTA was present to chelate all of the Ca^{++} in the reaction mixture (in addition to the buffer constituents in B and D, 0.28 mM EDTA, 14 mM Tris, 0.84 mM NaN$_3$ and 56 mM NaCl were present). The *Torpedo* membrane fraction used for these experiments was the same fraction as in the legend to Fig. 3.

359

Taken together, these results suggest that synaptic membranes have an active ADP-ribosyl transferase, and several receptor targets for mono-ADP-ribosylation. In addition, there is a cholera toxin target with the same molecular weight as authentic G_s which appears to be preferentially labeled. Finally, there is an ADP-ribosyl transferase which can preferentially label exogenous G_s (and an endogenous 70 K band) under appropriate reaction conditions. These results generally support the notion that synaptic membranes may have a high demand for mono-ADP-ribosyl transferase reactions. Thus, the shortage of niacin at a systematic level would affect such synaptic protein modification reactions, and perhaps inhibition of mono-ADP-ribosylation reactions at synapses could contribute to the dementia so often observed in pellagra. Such a hypothesis remains to be explored as more experimental data about NAD breakdown reactions in nervous tissue begin to emerge.

Acknowledgements. This work was supported by Grants GM25654 and GM 26107.

References

1. Ferro, A.M., Olivera, B.M. (1987) Intracellular Pyridine Nucleotide Degradation and Turnover in Pyridine Nucleotide Coenzymes: Chemical, Biochemical and Medical Aspects, Vol. 2B D. Dolphin, R. Poulson, O. Avramovic (eds.) pp. 25-77
2. White, III. H.B., (1982) J. Everse, B.M. Anderson,, K.S. You (eds.), Pyridine Nucleotide Coenzymes. Academic Press, Inc., New York
3. Foster, J.W., Moat, A.G. (1980) Microbiol Rev 44: 83-105
4. Preiss, J., Handler, P. (1958) J Biol Chem 233: 483-492
5. Hughes, Kelly T., Cookson, Brad T., Ladika, Dubravka, Olivera, B.M., Roth, John R. (1983) J Bacteriol 154: 1126-1136
6. Hughes, K.T, Ladika, D., Roth, J.R., Olivera, B.M. (1983) J Bacteriol 155: 213-221
7. McLaren, J., Ngo, T.C., Olivera, B.M. (1973) J Biol Chem 248: 5144-5149
8. Cookson, B.T., Olivera, B., Roth, J.R. (1987) J Bacteriol 169: 4285-4293
9. Zhu, N., Olivera, B.M., Roth, J.R. (1988) J Bact 170: 117-125
10. Holley, E.A., Spector, M.P., Foster, J.W. (1985) J Gen Microbiol 131: 2759-2770
11. White, B.J., Hochhauser, S.J., Cintro, N.M., Weiss, B. (1976) J Bacteriol 126: 1082-1088
12. Kornberg, A. (1980) DNA Replication. W.H. Freeman & Co., San Francisco
13. Gholson, R.K. (1966) Nature 212: 933-935
14. Manlapaz-Fernandez, P., Olivera, B.M. (1973) J Biol Chem 248: 5067-5073
15. Hillyard, D., Rechsteiner, M., Manlapaz-Ramos, P., Imperial, J.S., Cruz, L.J., Olivera, B.M. (1981) J Biol Chem 256: 8491-8497
16. Lester, H.A., Steer, M.L., Michaelson, D.M. (1982) J Neurochem 38: 1080-1086

An Approach to the Selective Targeting of NAD Consuming Reactions in Cultured Mammalian Cells

Myron K. Jacobson, Patrick W. Rankin, and Elaine L. Jacobson

Departments of Biochemistry and Medicine, Texas College of Osteopathic Medicine, University of North Texas, Fort Worth, Texas 76107 USA

Introduction

Many studies concerned with elucidating the biological functions of ADP-ribose transfer reactions have utilized compounds originally identified as inhibitors of poly(ADP-ribose) polymerase (1, 2). Studies using these inhibitors have shown that they alter many processes including cellular recovery from DNA damage (3, 4), sister chromatid exchange (5), malignant transformation (6-8), DNA replication (9) and cellular differentiation (10, 11). While the effects of these inhibitors have been generally attributed to inhibition of poly(ADP-ribose) polymerase, the lack of absolute selectivity with regard to ADP-ribose transferring or other enzymes makes the link to poly(ADP-ribose) polymerase a tenuous one. In this study, we have examined compounds originally identified as inhibitors of poly(ADP-ribose) polymerase for their relative inhibition of several classes of ADP-ribose transferring enzymes and for their efficacy of inhibition of poly(ADP-ribose) polymerase *in vivo*.

Results

Comparison of effect of inhibitors on poly(ADP-ribose) polymerase and mono(ADP-ribosyl) transferase A *in vitro*. To examine inhibition patterns *in vitro*, a partially purified preparation of poly(ADP-ribose) polymerase from calf thymus (12) was compared with homogeneous turkey erythrocyte NAD^+:arginine mono(ADP-ribosyl) transferase A (13). They are referred to here as polymerase and transferase, respectively. The NAD^+ concentration of both assays was 300 μM which was selected because it approximates the estimated intracellular concentration of NAD^+ in cultured C3H10T1/2 mouse cells (14). To evaluate each compound, varying concentrations were competed against 300 μM NAD^+. Table 1 lists IC_{50} values (concentration of inhibitor necessary for 50% inhibition) for 23 compounds examined.

The data show that IC_{50} values of all compounds tested were much higher for transferase than for polymerase. The largest difference was seen for benzamide in which the IC$_{50}$ value for the transferase was higher by more than three orders of magnitude; this suggests that the benzamides may prove very useful in discriminating between effects on polymerase and transferases in intact cells. However, since multiple endogenous transferases are present in cells, the utility of this approach would depend upon whether the inhibition pattern of transferase A is representative of other endogenous transferases. Additionally, it was also of interest to determine the effects of these inhibitors on of a third class of NAD$^+$ consuming enzymes, NAD$^+$ glycohydrolase. Thus, inhibition curves were generated for 4 additional transferases and for NAD$^+$ glycohydrolase. Fig. 1 shows *in vitro* inhibition by benzamide of each of the 5 endogenous transferases and NAD$^+$ glycohydrolase in relation to the polymerase. The data show that all of the transferases had similar inhibition patterns while the NAD$^+$ glycohydrolase was less sensitive to inhibition than the transferases.

Table 1. Effects of selected compounds on poly(ADP-ribose) polymerase and mono(ADP-ribosyl) transferase A *in vitro*

Compound	Poly(ADP-ribose) Polymerase IC$_{50}$ (μM)	Mono(ADP-ribosyl) Transferase A IC$_{50}$ (μM)
Benzamide	3.3 ± 0.28	4,100 ± 220
3-Methoxybenzamide	3.4 ± 0.31	2,700 ± 250
3-Aminobenzamide	5.4 ± 0.40	3,000 ± 970
5-Bromodeoxyuridine	15 ± 1.3	590 ± 94
Nicotinamide	31 ± 5.9	3,400 ± 410
Thymidine	43 ± 5.2	1,900 ± 300
Theophylline	46 ± 15	2,800 ± 460
5-Methylnicotinamide	70 ± 3.3	NI*
ZnCl$_2$	77 ± 21	11,000 ± 1,600
2-Aminobenzamide	100 ± 26	3,000 ± 970
Theobromine	110 ± 19	NI
Pyrazinamide	130 ± 18	NI
4-Aminobenzamide	400 ± 33	NI
Caffeine	1,400 ± 240	NI
Hypoxanthine	1,700 ± 190	NI
1-Methylnicotinamide	1,700 ± 250	NI
3-Isobutyl-1-methylxanthine	3,100 ± 310	NI
Isonicotinate hydrazide	4,800 ± 570	NI
8-Methylnicotinamide	7,800 ± 770	23,000 ± 8,500
Benzoate	NI*	NI*
3-Aminobenzoate	NI	NI
4-Aminobenzoate	NI	NI
Nicotinoate	NI	NI

*IC$_{50}$ greater than 30,000 μM

Inhibition patterns of poly(ADP-ribose) polymerase *in vivo*. The effect of benzamide on the accumulation of ADP-ribose polymers *in vivo* was examined following treatment of C3H1OT1/2 cells with the alkylating agent N-methyl-N'-nitro-N-nitrosoguanidine (MNNG) (Fig. 2). The inhibitor concentrations shown represent the concentrations in the culture medium for the *in vivo* experiments. The *in vivo* and *in vitro* inhibition patterns derived were very similar.

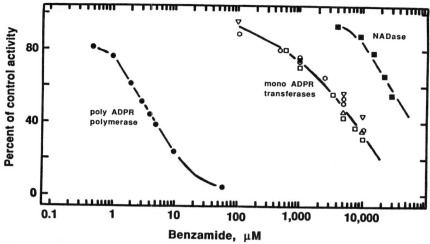

Fig. 1. Comparison of *in vitro* inhibition curves for poly(ADP-ribose) polymerase (●), mono(ADP-ribosyl) transferases A (△), A' (□), B (▽), C (◇), C' (○) and NAD⁺ glycohydrolase (■).

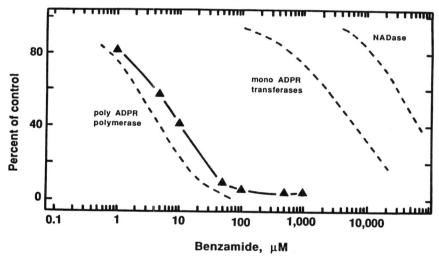

Fig. 2. Effect of a varying concentration of benzamide in the culture medium on the accumulation of ADP-ribose polymers following MNNG treatment. The dashed lines represent the *in vitro* inhibition curves shown in Fig. 1.

Discussion

A number of studies have quantitatively evaluated the inhibition of poly(ADP-ribose) polymerase *in vitro* (12, 17-19) but very little quantitative information concerning the inhibition of other classes of ADP-ribose transferring enzymes or on the efficacy of inhibitors in intact cells has been reported. The existence of multiple endogenous mono(ADP-ribosyl) transferases (20) makes information concerning the selectivity of these inhibitors especially important with regard to assessing the scope of mono(ADP-ribose) metabolism in total cellular ADP-ribosylation and in assigning biological effects of these inhibitors. While the use of inhibitors in intact cells has many limitations, the lack of genetic approaches to ADP-ribosylation in mammalian cells makes their use important in understanding the physiological functions of ADP-ribosylation. While this study has focused on the selectivity of these inhibitors between different classes of ADP-ribosyl transferring enzymes, the possible action of the inhibitors on unrelated enzymes must also be a concern when evaluating biological effects of these compounds.

The results presented here are in general agreement with previous studies that have shown that benzamide and its derivatives are very potent inhibitors of poly(ADP-ribose) polymerase *in vitro* (1, 2). The results shown here have demonstrated that the benzamides are also effective inhibitors of poly(ADP-ribose) polymerase *in vivo* at concentrations in the medium which are much lower than those generally used in studies designed to assess the effects of ADP-ribosylation inhibitors on biological responses (1-11). The large quantitative differences between the *in vitro* inhibition curves of poly(ADP-ribose) polymerase and the mono(ADP-ribosyl) transferases examined suggests that it may be possible to inhibit poly(ADP-ribose) polymerase with minimal effects on mono(ADP-ribosyl) transferases.

The large quantitative differences between IC_{50} values observed for poly(ADP-ribose) polymerase and mono(ADP-ribosyl) transferases may prove useful in assessing the mechanism of the biological effects of the benzamides. For example, benzamide and its derivatives have been shown to be relatively non-toxic to cells alone but they enhance the cyotoxicity of DNA alkylating agents (9, 10). Previous studies have utilized concentrations in the culture medium of 1 mM and higher to achieve an enhancement of cytotoxicity. In view of the results described here, we have re-examined the co-cytotoxic property of benzamide at lower concentrations and have observed maximal enhancement of the cytotoxicity of MNNG at micromolar concentrations (E.L. Jacobson *et al.*, unpublished results). When quantitatively comparing the effects of inhibitors on biological end-points such as cell survival, an exact correspondence between inhibition curves of the target enzyme and the biological effect is not necessarily expected since the process must be inhibited to the extent that it becomes rate-limiting for the end-point measured. Nevertheless, the fact that benzamide was co-cytotoxic at similar concentrations to those that inhibited poly(ADP-ribose) synthesis *in vivo*, which is far below that where effects on other ADP-ribosyl transfer enzymes are expected, argues that poly(ADP-ribose) polymerase is the likely target for the co-cytotoxic effects of benzamide. Further, it suggests that even relatively low levels in tissues may be effective in enhancing the cytoxicity of chemotherapeutic DNA damaging agents.

Acknowledgements. This work was supported by Grant No. CA43894 from the National Institutes of Health and the Texas VFW Ladies Auxiliary Cancer Research Fund. We thank Dr. Paul F. Cook for assistance with the determination of IC_{50} values, Dr. Joel Moss for the mono ADP-ribosyl transferases and Kay Hartman for help in preparation of the manuscript.

References

1. Purnell, M.R., Whish, W. (1979) Biochem J 185: 775-777
2. Sims, J.L., Sikorski, G.W., Catino, D.M., Berger, S.J., Berger, N.A. (1982) Biochemistry 21: 1813.
3. Nudka, N., Skidmore, C.J., Shall, S. (1980) Eur J Biochem 105: 525-530
4. Jacobson, E.J., Smith, J.Y., Wielckens, K., Hilz, H., Jacobson, M.K. (1985) Carcinogenesis, 6: 715-718
5. Oikawa, A., Tohda, H., Kanai, M., Miwa, M., Sugimura, T. (1980) Biochem Biophys Res Commun 97: 1311-1316
6. Jacobson, E.L., Smith, J.Y., Nunbhadki, V., Smith, D.G., (1985) *In* ADP-Ribosylation of Proteins, F.R. Althaus, H. Hilz and S. Shall, Eds.Springer-Verlag, Berlin, Heidelberg, pp. 277-283
7. Kasid, U.N., Stefanik, D.F., Lubet, R.A., Dritschilo, A., Smulson, M.E., (1986) Carcinogenesis 7: 327-330
8. Borek, C., Morgan, W.F., Ong, A. and Cleaver, J.E. (1984) Proc Natl Acad Sci 81: 243-247
9. Lönn, U., Lönn, S. (1986) Proc Natl Acad Sci USA 82: 104-108
10. Farzaneh, F., Zalin, R., Brill, D., Shall, S. (1982) Nature 300: 362-366
11. Althaus, F.R., Lawrence, S.D., He, Y.Z., Sattler, G.L., Tsukada, Y., Pitot, H.C. (1982) Nature 300: 366-368
12. Oka, J., Ueda, K., Hayaishi, O., Komura, H., Nakanishi, K. (1984) J Biol Chem 259: 986-993
13. Moss, J., Stanley, S.J., Watkins, P.A. (1980) J Biol Chem 255: 5838-5840
14. Jacobson, E.L., Jacobson, M.K. (1976) Arch Biochem Biophys 175: 627-634
15. Jacobson, E.L., Smith, J.Y., Mingmuang, M., Meadows, R., Sims, J.L., Jacobson, M.K. (1984) Cancer Res 44: 2485-2492
16. Cleland, W.W. (1979) Methods Enzymol 63: 103-139

Pyridine Nucleotide Metabolism as a Target for Cancer Chemotherapy

Nathan A. Berger, Sosamma J. Berger, and Nora V. Hirschler

Departments of Medicine and Biochemistry, R.L. Ireland Cancer Center, Case Western Reserve University, Cleveland, Ohio 44106 USA

Introduction

Pyridine nucleotide metabolism has an essential role in regulating multiple and diverse aspects of cellular metabolism ranging from carbohydrate utilization to DNA repair processes. As shown in Fig. 1, most cells convert nicotinamide to pyridine nucleotides via NMN pyrophosphorylase followed by NMN:ATP adenylyltransferase. The resultant product, NAD, is used in oxidation-reduction reactions leading ultimately to the synthesis of ATP. It is also important as a co-factor for dehydrogenase enzymes that provide essential components for cell growth such as IMP dehydrogenase, whose activity is required to provide guanine nucleotides.

NAD is also phosphorylated by NAD kinase to yield NADP. This is used in oxidation-reduction reactions, for example as a co-factor for the dehydrogenase enzymes in the pentose phosphate shunt. The energy derived from NADPH is not transferred directly into synthesis, rather it is usually used for bioreductive synthetic processes. NADP-dependent pathways are also important for maintaining glutathione in reduced form, producing pentose sugars for nucleotide synthesis and formation of some of the lipids involved in maintaining membrane composition. Thus, alterations in cellular NAD levels can interfere with ATP generation, as well as the products of many NAD dependent dehydrogenase reactions and energy dependent processes in general. Alterations in cellular NADP levels can interfere with the cell's ability to maintain reduced glutathione to detoxify chemotherapeutic agents, block the synthesis of nucleotides required for replication and interfere with lipid synthesis required for maintenance of cell membrane integrity.

NAD also serves as a substrate for ADP-ribosyl transferases. Mono-ADP-ribosyl transferases appear to function mainly in the cytoplasm where they may be involved in regulatory functions such as allosteric modifiers. Poly(ADP-ribose) synthesis is important in regulation of nuclear events, in particular, the DNA repair process; interference with poly(ADP-ribose) synthesis has an inhibitory effect on DNA repair synthesis. Thus, NAD metabolism, ADP-ribosylation and the interrelation of these processes serve as important connections between regulation of nuclear events and control of a wide variety of cellular processes. The central regulatory role of pyridine

366

nucleotide metabolism, in such a wide range of cellular processes, makes it an excellent target for biochemical modulation of cancer chemotherapy.

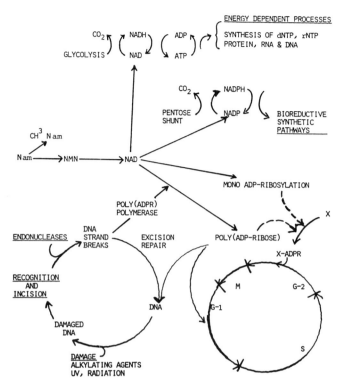

Fig. 1. Interrelation of NAD and poly(ADP-ribose) metabolism with other cellular processes.

Results and Discussion

The interrelationships described above provide a basis for evaluating the possibility that chemotherapeutic or metabolic manipulation can be used to produce drastic alterations in cell functions. We have shown that one mechanism of cell death, which we have called Type 1 or the Suicide mechanism, is mediated by activation of poly(ADP-ribose) polymerase and the consequent depletion of cellular energy metabolites (1, 2). This enzyme is activated by DNA strand breaks to cleave NAD at the glycosylic bond between the nicotinamide and adenosine diphosphoribose moieties (3). The latter are joined by the same enzyme into linear or branched chain polymers of ADP-ribose (3). Poly(ADP-ribose) polymerase activity and utilization of its substrate, NAD, are proportional to the number and duration of DNA strand breaks (4, 5). As shown in Fig. 2, DNA damage can sufficiently

367

activate this enzyme to deplete cellular NAD pools (1, 2). Consumption of NAD pools results in loss of ability to synthesize ATP, with consequent depletion of ATP and loss of all energy-dependent functions (1, 2). The result is an inability to phosphorylate and use glucose, loss of the ability to conduct DNA, RNA, or protein synthesis, and an inability to maintain membrane integrity (1, 2, 6, 7). The consequence is cell death.

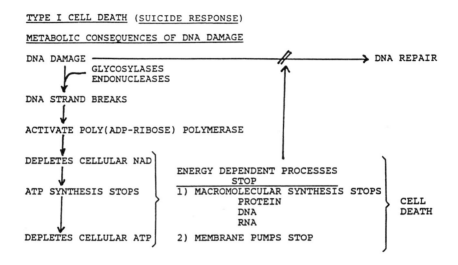

Fig. 2. Schematic pathway for the involvement of poly(ADP-ribose) polymerase in the suicide response to DNA damage.

Fig. 3 shows NAD, ATP and glucose-6-phosphate levels following DNA damage induced by N-methyl-N'-nitro-N-nitrosoguanidine (MNNG). The glucose-6-phosphate level showed a biphasic change: first, the pool size decreased, then increased as NAD depletion led to a blockade in the glycolytic pathway. Ultimately, glucose-6-phosphate disappeared from the cells as their NAD was depleted and they could no longer phosphorylate glucose to start it through the glycolytic pathway. In all cases the association with activation of poly(ADP-ribose) polymerase was shown by the use of nicotinamide or 3-aminobenzamide which partially inhibited the enzyme and concomitantly retarded the metabolic consequences (6, 7).

Based on this pathway of cell death, it appears likely that agents which can alter NAD or ATP metabolism should be useful in sensitizing cells to the cytotoxic effects of some chemotherapeutic agents whose mechanism of action involves the production of DNA damage. Both 6-aminonicotinamide (6-AN) and Tiazofurin (Taz) have already been shown to interfere at multiple steps of pyridine nucleotide metabolism. Taz has been shown to produce the Taz analog of NAD (8, 9), to interfere with NAD synthesis and

NAD dependent dehydrogenases (10), reduce guanine nucleotide synthesis (8, 9) and inhibit poly(ADP-ribose) polymerase (10). 6-AN has also been shown to produce the 6-AN analogs of NAD and NADP (11), interfere with glycolysis, inhibit ATP production, inhibit pyridine nucleotide dependent dehydrogenases (12, 13), and inhibit poly(ADP-ribose) polymerase (14). The simultaneous use of these two agents offers the possibility of potentiating the effects of other cytotoxic agents by establishing a more effective blockade of pyridine nucleotide synthesis and utilization pathways that can be attained with either agent alone.

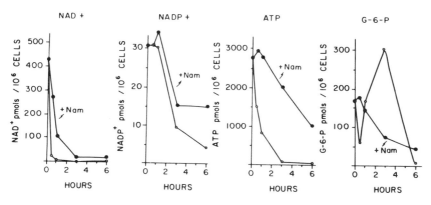

Fig. 3. L1210 cells in log phase growth were incubated in regular growth medium or in medium containing 2 mM nicotinamide. MNNG was added to a final concentration of 5 µg/ml. At the indicated times, cells were removed for counting and perchloric acid extracts were made, processed and metabolites analyzed utilizing enzymatic cyclic techniques described above. Cells for 0 time values were removed just before addition of MNNG. From left to right graphs are NAD+, NADP+, ATP and G-6-P. (O) Cultures treated with MNNG. (●) Cultures pretreated with 2 mM nicotinamide before treatment with MNNG.

The above proposal was evaluated in mice inoculated with L1210 leukemia using BCNU as a model for a clinically effective chemotherapeutic agent in combination with 6-AN and Taz to suppress NAD levels. In these experiments, mice were injected with L1210 leukemia and treatment started 24 hr later. The increase in life span for each agent given alone, either in single or multiple doses, is shown by the open bars in Fig. 4. Survival determinations were performed daily for the first 30 days of the experiment. Survivals were also examined at 30 and 60 days after treatment and are reported for each treatment regimen at the right of the illustration. Each of the individual agents produced small increases in life span, however, none produced a significant increase in long term survival.

The solid dark areas in Fig. 4 illustrate the anticipated additive effects for each combination based on the sums of the effects of each of their independent constituents. The striped area in each bar represents the synergistic effect above and beyond the anticipated additive effect. Additive effects were produced when Taz was administered in combination with 6-AN. There was no synergism produced by these two agents in combination. However, combinations of BCNU with 6-AN produced synergistic increases in life span. In addition to the synergistic increase in life span produced by 6-AN and BCNU, this combination also produced an increase in the long term survivors at 30 and 60 days.

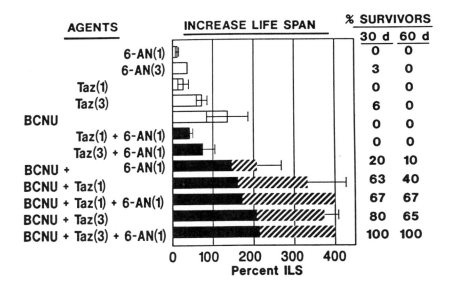

Fig. 4. Mice were injected with L1210 leukemia and treatment started 24 hr later. The number in parentheses indicate the number of days each drug was given. When drugs were given as single agents they were all administered on the first day of therapy. Doses used were Taz, 12 mg/mouse; 6-AN, 0.4 mg/mouse and BCNU, 0.4 mg/mouse. The bar graph represents the percent increase in life span calculated as indicated in the text. Values are expressed as the mean ±S.D. of at least 3 separate experiments with 5 mice in each group. The experiments using 3 doses of 6-AN were conducted with 5 animals only. Open bars indicate increased life span (ILS) for single agents, black bars indicate additive effect of ILSs expected from each combination determined by adding ILS produced by each individual agent. Striped bars represent synergistic ILS above that expected from the additive effects. The total bars (black and striped) represent experimentally derived values. Long term survivors of 30 and 60 days are indicated at right for each drug regimen.

The combination of BCNU with 3 days of Taz resulted in a marked increase in long term survivors. Addition of one course of 6-AN to this regimen resulted in long term survival of all animals. Thus, the addition of 6-AN to the already effective regimen of BCNU and Taz resulted in apparent cures for L1210 leukemia.

370

In summary, we have shown that pyridine nucleotide metabolism can be altered to cause cell death. This pathway involves the activation of poly(ADP-ribose) polymerase by DNA strand breaks, the subsequent consumption of NAD and depletion of ATP leading to cessation of all energy dependent functions, loss of cell volume and finally cell death. We have also shown that agents such as Taz and 6-AN can be used to potentiate the metabolic effects of chemotherapeutic agents that activate poly(ADP-ribose) polymerase. These observations provide new insights into the biochemical mechanism of cell death and a new focus for combination chemotherapy, biochemical modulation and investigations of drug sensitivity and resistance.

Acknowledgements. These studies were supported by NIH grants GM32647, CA35983, and CA43703.

References

1. Sims, J.L., Berger, S.J., Berger, N.A. (1983) Biochemistry 22: 5188-5194
2. Berger, N.A. (1985) Rad Res 101: 4-15
3. Hayaishi, O., Ueda, K. (1977) Annu Rev Biochem 46: 95-116
4. Berger, N.A., Sikorski, G.W., Petzold, S.J., Kurohara, K.K. (1979) J Clin Invest 63: 1164-1171
5. Cohen, J.J., Catino, D.M., Petzold, S.J., Berger, N.A. (1982) Biochemistry 21: 4931-4940
6. Berger, S.J., Sudar, D.C., Berger, N.A. (1986) Biochem Biophys Res Commun 134: 227-232
7. Das, S.K., Berger, N.A. (1986) Biochem Biophys Res Commun 137: 1153-1158
8. Streeter, D.G., Robins, R.K. (1983) Biochem Biophys Res Commun 115: 544-550
9. Carney, D.N., Ahluwalia, G.S., Jayaram, H.N., Cooney, D.A., Johns, D.G. (1985) J Clin Invest 75: 175-182
10. Berger, N.A., Berger, S.J., Catino, D.M., Petzold, S.J., Robins, R.K. (1985) J Clin Invest 75: 702-709
11. Dietrich, L.S., Friedland, I.M., Kaplan, L.A. (1958) J Biol Chem 233: 964-968
12. Dietrich, L.S., Kaplan, L.A., Friedland, I.M., Martin, D.S. (1958) Cancer Res 18:1272-1280
13. Kolbe, H., Keller, K., Lange, K., Herken, H. (1976) Biochem Biophys Res Commun 73: 378-382
14. Berger, S.J., Manory, I., Sudar, D.C., Krothapalli, D., Berger, N.A. (1987) Exp Cell Res 169: 149-157

Pivotal Role of Poly(ADP-Ribose) Polymerase Activation in the Pathogenesis of Immunodeficiency and in the Therapy of Chronic Lymphocytic Leukemia

Carlos J. Carrera, Shiro Seto, D. Bruce Wasson, Lawrence D. Piro, Ernest Beutler, and Dennis A. Carson

Department of Basic and Clinical Research, Research Institute of Scripps Clinic, La Jolla, California 92037 USA

Introduction

Activation of poly(ADP-ribose) polymerase plays an essential role in mediating the toxic effects of DNA damage in non-dividing lymphocytes. In cells with extensive DNA damage, excessive poly(ADP-ribose) polymerase activity can exhaust the available NAD pool, leading to a depletion of cellular ATP and to toxic perturbations of intermediary metabolism (1). Various agents known to damage DNA have been shown *in vitro* to elicit the biochemical sequelae of poly(ADP-ribose) polymerase activity in resting human lymphocytes (2). For example, exposure of lymphocytes to xanthine oxidase plus hypoxanthine causes massive DNA damage and a lethal fall in cellular NAD and ATP levels, attributable to excessive poly(ADP-ribose) polymerase activity (3). This model system suggests that poly(ADP-ribose) polymerase activation may contribute to immune dysfunction in certain chronic inflammatory states, in which stimulated neutrophils release toxic oxygen species.

Exposure of lymphocytes to deoxyadenosine (dAdo) plus deoxycoformycin (dCF) also results in the accumulation of DNA strand breaks and the activation of poly(ADP-ribose) polymerase (4, 5). dCF is an inhibitor of adenosine deaminase, and when administered to cancer patients, dCF causes dAdo to increase in the plasma. Elevated plasma dAdo is also the most striking biochemical change in children with inherited adenosine deaminase deficiency, a condition that causes severe combined immunodeficiency disease.

Certain lymphoid cells are especially sensitive to high dAdo levels when the deaminase is inhibited because they efficiently phosphorylate and retain the compound in the form of deoxyATP. Small amounts of deoxyATP are needed for DNA synthesis by all eukaryotic cells. However, deoxyATP in great excess inhibits ribonucleotide reductase, blocking the synthesis of the other vital deoxynucleotides in rapidly dividing cells (8). Increased cellular deoxyATP is also toxic to resting or slowly proliferating lymphocytes in which the activity of ribonucleotide reductase is barely detectable. In these cells deoxyATP blocks the ongoing maintenance repair of DNA, resulting in the accumulation of unrepaired single-strand DNA breaks (6, 7). The subsequent decline in NAD and ATP, and the loss of cell viability can be partially blocked by inhibitors of poly(ADP-ribose) polymerase. These observations implicate excessive ADP-ribosylation in the lymphocyte-specific toxicity of dAdo both in congenital adenosine deaminase deficiency and in dCF chemotherapy.

The exquisite sensitivity of resting lymphocytes to DNA damage and poly(ADP-ribose) polymerase activation may be due, in part, to a decreased capacity of the cells to synthesize NAD. Human lymphocytes possess an intact pyridine nucleotide cycle (10). The cells synthesize NAD from either nicotinamide or nicotinic acid, and they release nicotinamide as a by-product of ADP-ribosylation reactions. NMN pyrophosphorylase is the rate-limiting enzyme in NAD biosynthesis from nicotinamide, and the content of this enzyme in unstimulated human lymphocytes is quite low (9). We have examined the rate of NAD turnover in resting lymphocytes in order to quantitate the contribution of ADP-ribosylation to the overall consumption of NAD (10). In addition, we present here preliminary results of *in vitro* biochemical studies performed on malignant lymphocytes from patients with chronic lymphocytic leukemia (CLL), who were treated with 2-chlorodeoxyadenosine.

Results and Discussion

NAD turnover studies. Pulse-chase experiments with [^{14}C]-nicotinamide show that half of the labelled NAD in mature lymphocytes disappears during 8-12 hr of *in vitro* culture (Fig. 1). If 3-aminobenzamide is added, more than 50% of the radioactivity persists beyond 24 hr. These results indicate that most of the NAD turnover in non-dividing human lymphocytes is due to ADP-ribosylation reactions. Exposure of the cells to 10 μM dAdo causes a marked increase in the turnover of NAD (Fig. 1). The enhanced NAD consumption is associated with a 50% fall in cellular NAD pools after 24 hr (10). Thus, an increase in NAD utilization by lymphocytes for ADP-ribosylation, in response to accumulating DNA strand breaks, cannot be matched by a compensatory increase in NAD synthesis, and the NAD pool becomes depleted.

Effects of CdA on CLL cells *in vitro*: Clinical correlation. To exploit the mechanism of dAdo toxicity in chronic lymphoid malignancies, our laboratory synthesized the congener drug CdA (11). Unlike dCF, CdA is not an inhibitor of adenosine deaminase, but is rather a substrate analogue of dAdo that completely resists degradation by the enzyme. Pre-clinical studies showed that CdA was toxic in the nanomolar range both to resting lymphocytes and to T lymphoblast lines. As we observed with dAdo plus dCF, peripheral blood lymphocytes exposed to CdA *in vitro* accumulate DNA damage and show a drastic fall in cellular NAD (4). Over the next 24 hr, ATP declines in the CdA-treated cells, and they ultimately lyse.

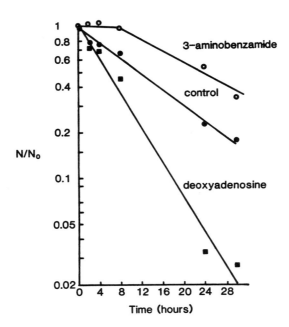

Fig. 1. NAD turnover in non-dividing human lymphocytes. Resting lymphocytes at a density of 2 x 10⁶/ml were preincubated for 20 hr in nicotinamide-free medium supplemented with 20% autologous plasma and 1.8 μM [^{14}C]nicotinamide. Cells were then washed and resuspended in medium containing 1.8 μM unlabelled nicotinamide, either without (closed circles) or with 3-aminobenzamide, 5 mM (open circles) or 10 μM dAdo plus 1 μM dCF (closed squares). At the indicated times, cells were extracted and the NAD content was determined by an enzyme cycling assay. Extracts were also fractionated by thin-layer chromatography to determine the radioactivity remaining in the NAD pool. N/No represents the fraction of the initial [^{14}C]NAD remaining after the elapsed times.

Table 1. *In Vitro* Effect of CdA on the NAD and ATP Content of CLL Cells

Assay	Culture	Responders (n=11)	Non-Responders (n=5)
		(pmol/10⁶ cells)	
NAD	No Drug	57.2 ± 16.5	50.0 ± 19.9
	CdA 1 μM	45.0 ± 16.7	49.5 ± 21.3
	Mean % Decrease	21%	1.5%
ATP	No Drug	563 ± 175	405 ± 136
	CdA 1 μM	511 ± 183	436 ± 136
	Mean % Decrease	9.6%	-7.9%

CLL Cells isolated from peripheral blood by density gradient centrifugation were cultured for 12 hr in the absence or presence of 1 μM CdA in nicotinamide-free RPMI medium, supplemented with L-glutamine 2 mM and 20% autologous plasma. Neutralized perchloric acid extracts were analyzed for NAD content by an enzyme cycling assay, and for ATP content by HPLC.

Several patients with CLL have received intravenous CdA on protocol study at our institution with an encouraging response rate of approximately 60%. *In vitro* studies were performed on pre-treatment leukemic cell samples from 16 patients who were considered evaluable for treatment response. All patients had previously received chemotherapy, and all were classified Rai Stage III or IV with one exception (immune hemolytic anemia). CdA was given by continuous intravenous infusion at 0.1 mg/kg per day for 7 days. For this analysis, a response was defined simply as a reduction in the circulating lymphocyte count by greater than 50%. Treatment cycles were repeated at intervals of 3-6 weeks, and the CdA was discontinued if no response occurred after two treatments.

Eleven patients responded to therapy with a significant decrease in leucocytosis, whereas the remaining 5 patients did not. There was no difference between the CLL cells from responders and those from non-responders in the amount of DNA damage induced by a 12-hr incubation with 1 μM CdA. Table 1 presents the results of NAD and ATP determination on CLL cells cultured in the presence of 1 μM CdA . There was little difference between the mean NAD content of control cells (no CdA) from responders vs. non-responders. However, exposure to CdA caused NAD to fall significantly only in cells from patients that subsequently responded to treatment with CdA ($p=0.004$ by Wilcoxon's one-sample test). The mean fall in NAD in responders' cells was 21% following the short incubation with CdA. This value is significantly different from the 1.5% mean NAD change occurring in cells from patients who proved to be resistant to CdA therapy ($p=0.017$ by Wilcoxon's two-sample test). Although ATP content appears to decline more in cells from responders than from non-responders, the change is not yet significant by the 12 hr incubation time point.

Fig. 2 illustrates the effect of CdA *in vitro* on NAD levels in CLL cells from the responders compared to non-responders. It is evident that the NAD fall is not consistent among patients within the responding group. Fig. 3 illustrates the time-course of response in a patient treated because of refractory anemia. Although he was initially dependent upon frequent transfusions, this patient's hemoglobin level was well maintained following three cycles of CdA, without further transfusion support.

Most of the NAD turnover in non-dividing human lymphocytes is attributable to ADP-ribosylation reactions. A rise in ADPRT activity, in response to increasing DNA strand breaks, causes NAD to be consumed without a compensatory rise in NAD synthesis. Thus, excessive poly(ADP-ribose) polymerase activity, relative to NAD synthetic capacity, provides a common mechanism by which quiescent lymphocytes suffer lethal NAD depletion following exposure to agents that damage DNA. The deaminase-resistant congener CdA produces a prompt reduction of the leucocytosis in

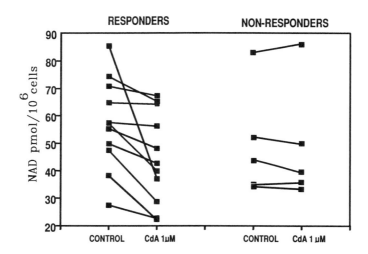

Fig. 2. Effect of CdA *in vitro* on cellular NAD in CLL. Cells from CLL patients were incubated for 12 hr without (control) or with 1 μM CdA in nicotinamide-free medium supplemented with 20% autologous plasma. Cells were then extracted, and NAD content was determined by an enzymatic cycling assay. The change in NAD content in paired control and treated samples from each patient is shown. Values from patients who subsequently responded to CdA therapy are compared to the results from patients who were resistant to the drug.

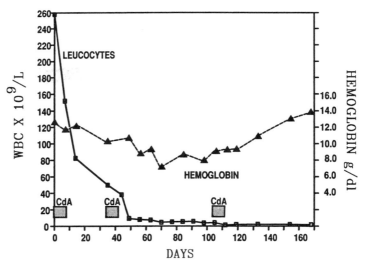

Fig. 3. Hematologic response to CdA therapy in a patient with CLL. CdA 0.1 (mg/kg per day) was administered by continuous intravenous infusion for 7 days at the times indicated (shaded). Prior to the third cycle of CdA, the patient required frequent transfusions for refractory anemia.

376

two-thirds of CLL patients who receive the drug by intravenous infusion. Based on *in vitro* evidence, the mechanism of CdA action similarly involves the formation of DNA strand breaks and NAD depletion, consistent with the activation of poly(ADP-ribose) polymerase in the malignant lymphocytes. Further studies will determine whether or not measurements of NAD consumption *in vitro* may predict which CLL patients are most likely to benefit clinically from CdA therapy.

Acknowledgements. This work was supported in part by grants CA01100, AR25443, GM23200, and RR00833 from the National Institutes of Health. Dr. Carrera is a Special Fellow of the Leukemia Society of America. This is publication number 4879BCR from the Research Institute of Scripps Clinic, La Jolla, California.

References

1. Berger, N.A. (1985) Radiat Res 101: 4-15
2. Berger, N.A., Sikorski, G.W., Petzold, S.J., Kurohara, K.K. (1979) J Clin Invest 63: 1164-1171
3. Carson, D.A., Seto, S., Wasson, D.B. (1986) J Exp Med 163: 746-751
4. Seto, S., Carrera, C.J., Kubota, M., Wasson, D.B., Carson, D.A. (1985) J Clin Invest 75: 377-383
5. Brox, L., Ng, A., Pollack, E., Belch, A. (1984) Cancer Res 44: 934-937
6. Farzaneh, F., Shall, S., Johnstone, A.P. (1985) FEBS Lett 189: 62-66
7. Seto, S., Carrera, C.J., Wasson, D.B., Carson, D.A. (1986) J Immunol 136: 2839-2843
8. Carrera, C.J., Carson, D.A. (1987) The Molecular Basis of Blood Diseases. A.W. Nienhuis, P.Leder, P.W. Majerus (eds.) W.B. Saunders Co., Philadelphia, pp. 407-449
9. Berger, S.J., Manory, I., Sudar, D.C., Berger, N.A. (1987) Exp Cell Res 169: 149-157
10. Carson, D.A., Seto, S., Wasson, D.B. (1987) J Immunol 138: 1904-1907
11. Carson, D.A., Wasson, D.B., Beutler, E. (1984) Proc Natl Acad Sci USA 81: 2232-2236
12. Birnboim, H.C., Jevcak, J.J. (1981) Cancer Res 41: 1889-1892

Abbreviations:

dAdo - 2'-deoxyadenosine
dCF - deoxycoformycin
NMN - nicotinamide mononucleotide
CdA - 2-chloro-2'-deoxyadenosine
CLL - chronic lymphocytic leukemia

Active Transport of Nicotinamide and the Adenine Moiety of NAD by Metabolically Controlled NAD Catabolism

Ronald W. Pero[1], Anders Olsson[1], Tor Olofsson[2], and Elisabeth Kjellén[1,3]

Division of Biochemical Epidemiology, Preventive Medicine Institute of the Strang Clinic, New York, New York 10016 USA

Introduction

Levels of coenzymes, NAD(H) and NADP(H), are highly regulated to assure a catalytic balance to meet the various metabolic needs of tissues. There are two major enzymes which catabolize these coenzymes, and thus, they have a regulatory potential on pool sizes. They are NAD glycohydrolase and (ADP-ribosyl) transferase. The function of both enzymes still remains unclear but it has been postulated that NAD glycohydrolase is involved in the catabolic control of NAD(H)/NADP(H) levels in cells (1, 2), and that (ADP-ribosyl) transferase activity is associated with the cellular processes of DNA repair (3-6), differentiation (7, 8) and gene expression (9).

Our laboratory has been concerned with using human blood cells as a model system to study the role of NAD metabolism in the mechanism of sensitization of hyperthermia and gamma radiation (10-12). In an earlier study of patients with acute myeloid leukemia (13) it was shown that the *in vitro* growth patterns of isolated marrow cells in agar and the *in vitro* drug sensitivity of the clonogenic cells to cytosine arabinoside correlated to the constitutive levels of (ADP-ribosyl) transferase. In addition, nicotinamide (NAM), an inhibitor of (ADP-ribosyl) transferase, was an effective radiosensitizer of C3H mammary adenocarcinomas even at relatively low doses (14-16). Both of these findings have inspired us to investigate NAD catabolism in a well established human leukemic cell line (K-562) for the purposes of understanding if NAD catabolism regulates either NAM uptake and/or (ADP-ribosyl) transferase activity. Our data indicate an active transport of NAM and ADP-ribose by NAD glycohydrolase which is, in turn, coupled to the level of nuclear (ADP-ribosyl) transferase activity.

---

[1]Department of Molecular Ecogenetics, Wallenberg Laboratory, University of Lund, S-220 07 Lund, Sweden

[2]Department of Internal Medicine, Division of Hematology, University of Lund, S-221 85 Lund, Sweden

[3]Department of Oncology, University of Lund, S-221 85 Lund, Sweden

378

Results and Discussion

Utilization of exogenously supplied NAD. The K-562 cells used in our experiments were grown and subcultured every fifth day in RPMI 1640 medium with 10% fetal bovine serum. Cell death varied considerably with the number of days in culture and the dead cells were removed by centrifugation through an Isopaque-Ficoll cushion (1.077 gm/ml) at 400 x g for 20 min. Cells collected at the interphase zone were used for all of our experiments and they were always >95% viable by trypan blue exclusion. K-562 cells prepared in this way were incubated with [³H]adenine labeled NAD dissolved in physiologic saline. The uptake and (ADP-ribosyl) transferase activity was assessed as described as in Fig. 1. Although nucleotides are not considered to permeate viable cells, K-562 cells readily took up and utilized exogenously supplied NAD to support (ADP-ribosyl) transferase activity (Fig. 1). The internalization of NAD was supported by the fact that the cells utilized NAD for (ADP-ribosyl) transferase activity, and yet, the cells were still >95% viable after the exposure to NAD in physiologic saline. In addition, [³H]adenine label was readily recoverable in the cytosol that was produced after the cells were washed and permeabilized as described elsewhere (17).

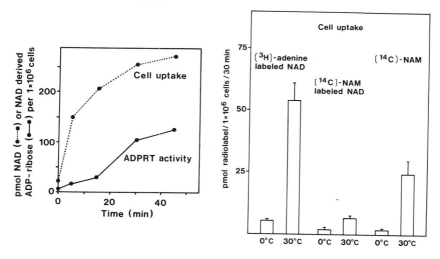

Fig. 1. (left) Uptake and utilization of exogenously supplied NAD by K-562 cells. About 1 x 10⁶ cells were incubated in physiologic saline supplemented with 250 μM [³H]adenine labeled NAD at 30°C. Uptake was estimated after washing twice with physiologic saline and then counting the radioactivity present in aliquots of cells lyzed in distilled water. (ADP-ribosyl) transferase was quantified by TCA precipitation as described elsewhere (13).

Fig. 2. (right) Evidence for active transport of NAD catabolic products in K-562 cells. Cell uptake was measured after exposure to 250 μM doses of the indicated compounds in physiologic saline as described in Fig. 1.

379

Active transport of NAD catabolic products. In order to determine whether NAD or some catabolic product(s), was being transported intracellularly, K-562 cells were cultured in saline for 30 min at 0° C and 30° C in the presence of [³H]adenine labeled NAD, [¹⁴C]NAM labeled NAD or [¹⁴C]NAM and the uptake of radioactivity was quantified. There was no uptake of the [¹⁴C]NAM labeled NAD clearly indicating the lack of transport of the NAM moiety of NAD (Fig. 2). However, [¹⁴C]NAM was taken up readily by K-562 cells in the absence of phosphorylated analogues of NAD. The NAM intracellular transport was about one-half as active as the transport for the [³H]adenine labeled moiety of NAD. Essentially no cellular uptake was observed for either NAM or the adenine moiety of NAD at 0 °C. These data were taken as evidence for an active transport mechanism of NAM and ADP-ribose in K-562 cells which involves, in part, NAD catabolism.

Effectors of active transport. NAD glycohydrolase is well known to be located on the outer surface of plasma membranes (18, 19) and it has a phosphate binding site (20). Both these characteristics have suggested to us the possibility that this enzyme may be involved in the active transport of the NAD catabolic product, ADP-ribose. Further support for this hypothesis is presented in Fig. 3. Phosphorylated analogues of NAD at 8-20 fold excess concentrations were effective inhibitors of the uptake and utilization of the adenine moiety of NAD when estimated by (ADP-ribosyl) transferase activity in viable K-562 cells. Even NAM and BAM were effective inhibitors of the active transport indicating an affinity for these compounds at the NAD binding site. Although NAM and BAM are direct inhibitors of (ADP-ribosyl) transferase, and as a consequence they could have affected uptake, this would be unlikely for NAM since it is not actively transported in the presence of NAD (Fig. 2). We have clarified this point by also incubating viable K-562 cells in the presence of [¹⁴C]NAM with and without various NAD analogues and then assessing their effects on cell uptake (Fig. 4). As expected, NAM uptake was blocked by NAD and its phosphorylated analogues which were also shown to block the transport of the adenine moiety of NAD in Fig. 3. BAM was only a weak inhibitor of NAM cell uptake, presumably because it was not dependent on active transport. Thus, it could cause an inhibition indirectly by inhibiting (ADP-ribosyl) transferase if the transport and utilization of NAM for NAD synthesis were coupled. Nevertheless, these experiments do demonstrate that both NAM and the adenine moiety of NAD are actively taken up by K-562 cells via binding at the same receptor site.

Characterization of the NAD catabolism. The fact that the NAM moiety of NAD is not transported intracellularly in viable K-562 cells has allowed us to assess the NAD catabolic activity on the outer surface of K-562 cells by measuring the generation of NAM in the extracellular medium.

380

This catabolic activity (Panel A, Fig. 5) could then be compared with the total catabolic activity of NAD by measuring NAM generation in permeabilized cells (Panel C, Fig. 5).

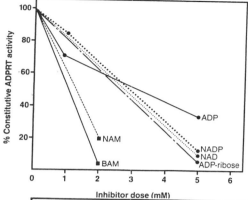

Fig. 3. (left) Inhibition of the utilization of exogenously supplied NAD by co-incubation with NAD analogues. NAD utilization was assessed by estimating (ADP-ribosyl) transferase in viable K-562 cells as described in Fig. 1.

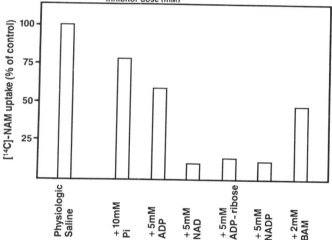

Fig. 4. (right) Inhibition of nicotinamide uptake by phosphorylated analogues of NAD and benzamide in cultured K-562 cells. The uptake of 250 μM [14C]NAM by cells cultured in physiologic saline for 30 min at 30°C was determined as described in Fig. 1.

In addition, the portion of (ADP-ribosyl) transferase activity surviving removal of radioactive poly(ADP-ribose) by poly(ADP-ribose) glycohydrolase activity could be estimated and compared for its contribution to NAD catabolism by measuring the TCA precipitable adenine labeled NAD in permeabilized cells (Panel B, Fig. 5). When these assay procedures for NAD catabolism were carried out at saturating concentrations of NAD, there was about 300 pmol of NAM generated extracellularly and about 300 pmol of ADP-ribose consumed by (ADP-ribosyl) transferase whereas > 2000 pmol of NAM were generated from the overall cell catabolism of NAD. Since about 1400 pmol of the NAM generated from NAD catabolism could not be accounted for, we have sought an explanation via poly(ADP-ribose) glycohydrolase activity. We have estimated this enzymatic activity by first incubating permeabilized cells for 15 min in the presence of adenine-labeled NAD, then adding 2 mM BAM and continuing the incubation for an additional 7.5 min (Panel D, Fig. 5). Under these conditions, removal of the prelabeled poly(ADP-ribose) in the absence of (ADP-ribosyl) transferase activity (i.e. BAM inhibition >90%) is a rough estimation of poly(ADP-ribose) glycohydrolase activity. There was considerable activity of this enzyme present in permeabilized K-562 cells (Panel D, Fig. 5). It could, in turn, account for the increased generation of NAM in permeabilized cells via a severe underestimation of total (ADP-ribosyl) transferase activity due to high turnover of the poly(ADP-ribose) polymer because of the glycohydrolase activity. This interpretation is consistent with the kinetic data. NAM generation in permeabilized cells had a $K_m = 136\,\mu M$ which is very close to the reported values for nuclear (ADP-ribosyl) transferase (21). However, the (ADP-ribosyl) transferase activity estimated minus poly(ADP-ribose) glycohydrolase activity had a much higher K_m of $750\,\mu M$ (Panel B, Fig. 5), indicating a slower accumulation of protein bound ADP-ribose product.

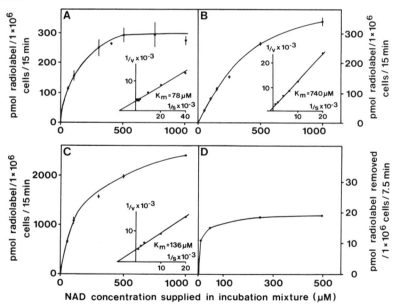

Fig. 5. The catabolism of NAD in K-562 cells analyzed by 4 different techniques. (A) NAM produced extracellularly from [^{14}C]NAM labeled NAD by viable K-562 cells (i.e. NAD glycohydrolase activity); (B) TCA precipitable [^{3}H]adenine labeled NAD in permeabilized K-562 cells (i.e. (ADP-ribosyl) transferase-panel D activity); (C) NAM produced from [^{14}C]NAM labeled by permeabilized K-562 cells (i.e. sum of panels A, B and D activities); (D) K-562 cells (i.e. poly(ADP-ribose) glycohydrolase activity).

Coupling of plasma membrane NAD catabolism and (ADP-ribosyl) transferase activity. We have studied the coupling of NAD glycohydrolase activity (i.e. plasma membrane NAD catabolism) and (ADP-ribosyl) transferase activity in viable K-562 cells in two different ways: (i) by providing a demand for intracellular NAD by depleting the internal pool with a treatment of 400 μM H_2O_2 and (ii) by examining the activities of these NAD catabolic pathways in relation to the cellular growth state. In the presence of an increased NAD demand, both the activities of NAD glycohydrolase and (ADP-ribosyl) transferase are increased to about the same extent (Fig. 6). Since H_2O_2 induces extensive DNA damage, then the increased (ADP-ribosyl) transferase activity is principally nuclear in origin. Furthermore, nuclear (ADP-ribosyl) transferase activity is also believed to relate to cellular differentiation (7, 8, 13). In this regard it is interesting to observe in Fig. 7 that as K-562 cells achieve growth arrest, NAD glycohydrolase, (ADP-ribosyl) transferase and cell uptake of the adenine moiety of NAD all increase in proportion. Taken together these data support a strong coupling of these two catabolic activities of NAD.

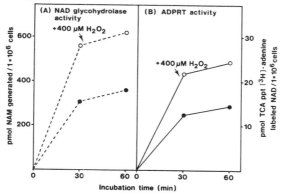

Fig. 6. The effect of increasing cellular demand for NAD by H_2O_2 exposure on NAD catabolism in cultured K-562 cells. The cells were exposed to $\pm$ 400 μM H_2O_2 for 30 min at 37°C in saline and then extracellular generation of NAM (i.e. NAD glycohydrolase) and (ADP-ribosyl) transferase estimated as described elsewhere (Fig. 1, ref. 13, 20).

Identification of the active transport mechanism. We believe the mechanism for the uptake of NAM and the adenine moiety of NAD by K-562 cells is via binding and internalization on NAD glycohydrolase. Our evidence, some of which we have not illustrated here, can be summarized as follows: (i) The location is known to be on the outer surface of the plasma membrane (18, 19) where we observe NAM generation from NAD (Fig. 5). (ii) We have detected both NAM and ADP-ribose in the extracellular medium following incubation with NAD, the products of hydrolase activity. (iii) The K_m of 78 μM (Fig. 5) is close to reported values for membrane bound NAD glycohydrolase (50 μM, ref. 19). (iv) There is a known

383

phosphate binding site on NAD glycohydrolase (20) which is a necessary prerequisite for the selective transport of the ADP-ribose moiety of NAD (Fig. 2, 3). (v) K-562 plasma membrane generation of NAM from NAD is relatively insensitive to inhibition by 5 mM thymidine compared to the (ADP-ribosyl) transferase activity. (vi) Crude microsomal and plasma membrane fractions from K-562 cells are enriched for NAM generation from NAD compared to (ADP-ribosyl) transferase activity. However, we should caution that we have not yet been able to completely separate NAD glycohydrolase activity from (ADP-ribosyl) transferase activity. It has been technically difficult to prepare purified plasma membranes from K-562 cells. Therefore, we cannot exclude a plasma membrane (ADP-ribosyl) transferase activity associated with the NAD glycohydrolase activity, which we have concluded is present.

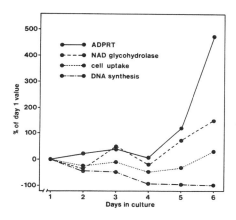

Fig. 7. The effect of the cell cycle on the coupling of NAD catabolic events in K-562 cells. 15 x 10^6 cells were placed in 75 ml fresh medium and cultured without any further medium change for 1-6 days. Dead cells were removed by gradient centrifugation and the indicated parameters estimated (for details see Results and Discussion). Day 1 values were: (ADP-ribosyl) transferase in permeabilized cells (11.8 pmol TCA precipitable NAD/1 x 10^6 cells/30 min) (13); NAD glycohydrolase as NAM generated extracellularly (185 pmol/1 x 10^6 cells/30 min), uptake of [^{3}H]adenine labeled NAD (64 pmol/1 x 10^6 cells/30 min) (Figure 1); DNA synthesis (140,000 cpm [^{3}H]dThd/μg DNA/3 hr, 22 Ci/mmol).

Possible importance for coupled activities between NAD glycohydrolase and (ADP-ribosyl) transferase. Neoplastic cells have elevated levels of (ADP-ribosyl) transferase compared to normal cell levels (13). Using K-562 cells as an example, elevated (ADP-ribosyl) transferase levels may also indicate high poly(ADP-ribose) glycohydrolase activities which would place a tremendous demand on neoplastic cells to maintain NAD pools. This, in turn, may result in a high level of NAD glycohydrolase in order to increase a salvage pathway for uptake of NAM and other NAD catabolic products for the maintenance of NAD pools. Under these conditions, there would be a preferential uptake of NAM by neoplastic cells which may have special relevance in the design of drugs for sensitizing radio- and chemotherapies by (ADP-ribosyl) transferase inhibition.

References

1. Editorial. (1984) Nutritional Reviews 42 (2): 62-64
2. McCreanor, G.M., Bender, D.A. (1983) Biochim Biophys Acta 759: 222-228
3. Durkacz, B.W., Imidijii, O., Gray, D.A., Shall, S, (1980) Nature 283: 593-596
4. Zwelling, L.A., Kerrigan, D., Pommier, F. (1982) Biochem Biophys Res Commun 104: 897-902
5. Althaus, F.R., Lawerance, S.D., Sattler, G.L., Pitot, H.C. J Biol Chem 257: 5528-5535
6. Creissen, D., Shall, S. (1982) Nature 296: 271-272
7. Farzaneh, F., Zalin, R., Brill, D., Shall, S. Nature 300: 262-266
8. Johnstone, A.P., Williams, G.T. (1982) Nature 300: 368-370
9. Althaus, F.R., Lawerance, S.C., He, Y.-Z., Sattler, G.L., Tsukada, Y., Pitot, H.C. (1982) Nature 300: 366-368
10. Jonsson, G.G., Eriksson, G., Pero, R.W. (1984) Radiation Res 97: 97-107
11. Jonsson, G.G., Eriksson, G., Pero, R.W. (1985) Radiation Res 102: 241-253
12. Kjellén, E., Jonsson, G.G., Pero, R.W., Christensson, P.-I. (1986) Int J Radiat Res. 49: 151-162
13. Pero, R.W., Olofsson, T., Gustavsson, A., Kjellén, E. (1981) Carcinogenesis 6: 1055-1058
14. Jonsson, G.G., Kjellén, E., Pero, R.W. (1984) Radiother Oncol 1: 349-353
15. Jonsson, G.G., Kjellén, E., Pero, R.W., Cameron, R. (1985) Cancer Res 45: 3609-3614
16. Kjellén, E., Pero, R.W., Cameron, R., Ranstam, J. (1986) Acta Radiologica Oncol 25: 281-284
17. Berger, N.A. (1978) Methods in Cell Biology, D.M. Prescott (ed.) Academic Press, New York, pp. 325-340
18. Artman, M., Seeley, R.J. (1978) Arch Biochem Biophys 195: 121-127
19. Amar-Costesec, A., Prado-Figueroa, M., Beaufay, H., Nagelkerke, J.F., Van Berkel, T.J.C. (1985) J Cell Biol 100: 189-197
20. Lory, S., Carrol, S.F., Bernard, P.D., Collier, R.J. (1980) J Biol Chem 255: 12011-12015
21. Benjamin, R.C., Cook, P.F., Jacobson, M.K. (1985) ADP-ribosylation of Proteins, F.R. Althaus, H. Hilz, S. Shall (eds.), Springer-Verlag, Berlin-Heidelberg, pp. 93-97

Abbreviations:
NAM - nicotinamide
BAM - benzamide
TCA - trichloroacetic acid

The Effect of Extracellular Calcium on NAD Metabolism in Human Diploid Cells

Michael R. Duncan, Patrick W. Rankin[1], Robert L. King, Myron K. Jacobson[1], and Robert T. Dell'Orco

The Samuel Roberts Noble Foundation, Inc., Ardmore, Oklahoma 73402 USA

Introduction

The growth of human diploid fibroblast-like cells (HDF) in culture can be inhibited by reducing the level of extracellular Ca^{2+} in the incubation medium (1). The mechanism(s) by which lowered extracellular Ca^{2+} inhibits the growth of HDF is unclear. Certain evidence, however, indicates that it is the result of complex alterations in the Ca^{2+} pools, involving both the mobilization of intracellular stores and plasma membrane transport (2, 3). Mono(ADP-ribosyl) transfer reactions have been implicated in both of these processes (4, 5). Additionally, certain studies have provided evidence for the involvement of GTP-binding proteins (G-proteins) in the regulation of Ca^{2+} mobilization (6); and others have linked the regulation of G-protein function to mono(ADP-ribosyl)ation catalyzed by bacteria toxins (5, 7, 8). The present studies were undertaken to determine if extracellular Ca^{2+} depletion of HDF affects cellular NAD metabolism. A lowering of NAD pools in the absence of altered NAD biosynthesis would suggest a relationship between the effects of Ca^{2+} depletion and ADP-ribosylation reactions since they are NAD consuming processes.

Results

Fig. 1 shows the total NAD content of CF-3 HDF following replacement of the complete incubation medium with Ca^{2+} depleted medium. In Ca^{2+} depleted medium, a progressive depletion of up to 40% of the total NAD pool occurred. During this period there were no detectable changes in cell morphology or protein content. Further, when normal levels of Ca^{2+} were restored to the medium, the NAD content returned to normal. The steady state levels of NAD reflect a balance between the rate of NAD biosynthesis and the rate of NAD consuming reactions (9). The decrease in NAD observed in Ca^{2+} depleted medium could be the result of alterations in either or both of these rates; therefore the effect of Ca^{2+} depletion on NAD biosynthesis was examined. Cultures were placed in control or Ca^{2+} depleted medium for 2 hr followed by the addition of $[^{14}C]$-nicotinamide. The total

---

[1]The Department of Biochemistry, Texas College of Osteopathic Medicine, University of North Texas, Fort Worth, TX 76107 USA

NAD pool and the radiolabel in NAD were determined 1 and 2 hr later. No appreciable effect of Ca^{2+} depletion on the absolute amount of nicotinamide converted to NAD was observed; however, the specific radioactivity of the NAD pool was higher at both time points in Ca^{2+} depleted medium (10.1 and 17.6 for control and Ca^{2+} depleted, respectively, at the 2 hr time point). These results indicate that Ca^{2+} deprivation does not significantly reduce the rate of NAD biosynthesis but appears to be associated with a depletion of the NAD pool.

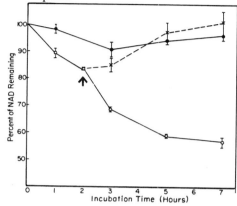

Fig. 1. NAD content of CF-3 cells incubated in Ca^{2+} depleted medium O--- O), Ca^{2+} supplemented medium (●-●), and Ca^{2+} depleted medium supplemented with normal levels of Ca^{2+} (1.8 mM) after 2 hr of incubation (x---x). Arrow indicates time of Ca^{2+} addition. Values are Mean ± SEM for three experiments. Ca^{2+} depleted medium was McCoy's 5a prepared without nicotinamide, nicotinic acid, and $CaCl_2$. It was supplemented with 10% FBS treated as described (14). NAD determinations were performed as described (14). Reprinted with permission from ref. 14.

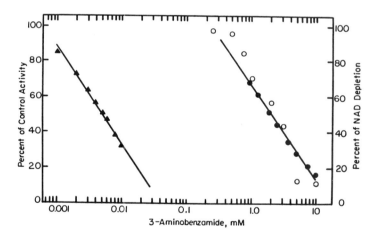

Fig. 2. Inhibition of NAD depletion in Ca^{2+} depleted medium by 3-aminobenzamide. CF-3 cells were placed in Ca^{2+} depleted medium containing the indicated concentrations of 3-aminobenzamide and 3 hr later were extracted and assayed for NAD (16). Control cultures without 3-aminobenzamide had a 40% reduction in NAD content. Values shown (O) are the percentage of NAD depleted in the absence of 3-aminobenzamide and represent the mean of duplicate determinations. The relative activities of poly(ADP-ribose) polymerase (▲) and mono(ADP-ribosyl) transferase (●) are also shown as a function of 3-aminobenzamide concentration. Partially purified poly(ADP-ribose) polymerase was obtained from beef thymus (12); and purified mono(ADP-ribosyl)transferase was obtained from turkey erythrocytes (13) and was provided by Dr. Joel Moss (NIH). Reprinted with permission from ref. 14.

387

The results of the biosynthesis experiments suggest that NAD depletion is due to an increased rate of NAD consuming reactions. NAD can be consumed by ADP-ribosyl transfer reactions which can be inhibited by 3-aminobenzamide (10). When 5 mM 3-aminobenzamide was added to the culture medium, total inhibition of NAD depletion was observed (data not shown). Since ADP-ribosyl transfer reactions can be catalyzed by nuclear poly(ADP-ribose) polymerase and by cytoplasmic mono(ADP-ribosyl) transferases (11), the inhibition of these enzymes by 3-aminobenzamide was quantitatively compared. Fig. 2 shows dose response curves for the inhibition of purified poly(ADP-ribose) polymerase (12) and mono(ADP-ribosyl) transferase (13) by this compound. The activity of poly(ADP-ribose) polymerase was much more sensitive to inhibition with a 50% inhibitory concentration (IC_{50}) of approximately 5.5 μM under the assay conditions used compared to an IC_{50} of 2000 μM for the mono(ADP-ribosyl) transferase. The effect of different concentrations of 3-aminobenzamide on NAD depletion in CF-3 cells incubated with Ca^{2+} depleted medium was also determined (Fig. 2). These cell experiments indicated an IC_{50} similar to that of mono(ADP-ribosyl) transferases and argue that NAD depletion in CF-3 cells was due to a stimulation of cellular mono(ADP-ribosyl)ation.

Discussion

The results presented here show that one of the earliest effects of the lowered extracellular Ca^{2+} in cultures of HDF is a rapid loss of NAD which can be reversed by restoring the Ca^{2+} to normal levels (Fig. 1). The observed loss in intracellular NAD is the result of NAD consuming reaction(s) since the biosynthesis of NAD was unaffected in Ca^{2+} depleted medium. The inhibition of the NAD loss by 3-aminobenzamide indicated that NAD is being consumed as a substrate for ADP-ribosylation reactions; the concentrations required for inhibition closely correlated with those needed to inhibit mono(ADP-ribosyl)ation reactions (Fig. 2). Thus, we conclude that extracellular Ca^{2+} depletion stimulates mono (ADP-ribosyl)ation reactions in cultured HDF.

The mono(ADP-ribosyl) transfer reactions detected here may have a G-protein as an acceptor of the ADP-ribose moiety. G-proteins have been shown to be involved in Ca^{2+} mobilization, and mono(ADP-ribosyl)ation reactions have been linked to their regulation (5-8). Therefore, the stimulation of the mono(ADP-ribosyl)ation of a G-protein, an NAD consuming reaction, could explain the loss of NAD in response to low extracellular Ca^{2+} levels in HDF.

Acknowledgements Supported in part by NIH Grant CA43894.

References

1. Boynton, A.L., Whitfield, J.F., Isaacs, R.J., Tremblay, R. (1977) J Cell Physiol 92: 241-248
2. Tupper, J.T., Ryals, W.T., Bodine, P.V. (1982) J Cell Physiol 110: 29-34
3. Owen, N.E., Villereal, M.L. (1983) J Cell Physiol 117: 23-29
4. Richter, C., Winterhalter, K.H., Baumhuter, S., Lotscher, H-R., Moser, B. (1983) Proc. Natl Acad Sci USA 80: 3188-3192
5. Knoop, F.C., Thomas, D.C. (1984) Inter J Biochem 16: 275-280
6. Gomperts, B.D. (1983) Nature 306: 64-66
7. Holz, G.G., Rane, S.G., Dunlap, K. (1986) Nature 319: 670-672
8. Heschler, J., Rosenthal, W., Trautwein, W., Schultz, G. (1987) Nature 325: 445-447
9. Jacobson, M.K., Levi, V., Juarez-Salinas, H., Barton, R.A., Jacobson, E.L. (1980) Cancer Res. 40: 1797-1802
10. Hayaishi, O., Ueda, K. (1982) ADP-ribosylation Reactions, Biology and Medicine, O. Hayaishi, K. Ueda (eds) Academic Press, New York
11. Althaus, F.R., Hilz, H., Shall, S. (eds.) (1985) ADP-ribosylation of Protein, Springer-Verlag, New York
12. Benjamin, R.C., Gill, D.M. (1980) J Biol Chem 255: 10502-10508
13. Moss, J., Stanley, S.J., Watkins, P.A. (1980) J Biol Chem 255: 5838-5840
14. Swierenga, S.H.H., MacManus, J.P. (1982) J Tissue Cult Meth 7: 1-3
15. Duncan, M.R., Rankin, P.R., King, R.L., Jacobson, M.K., Dell'Orco, R.T. (1988) J Cell Physiol 134: 161-165

Reduced Tumor Progression *In Vivo* by an Inhibitor of Poly(ADP-Ribose) Synthetase (3-Aminobenzamide) in Combination with X-Rays or the Cytostatic Drug DTIC

F. Darroudi, T.S.B. Zwanenburg, A.T. Natarajan[1], O. Driessen[2], and A. van Langevelde[2]

Department of Radiation Genetics and Chemical Mutagenesis, State University of Leiden, Leiden, The Netherlands

Introduction

Poly(ADP-ribose) synthetase activity is strongly stimulated by DNA strand breaks (1-3). Both ionizing radiations and monofunctional alkylating agents increase the activity of poly(ADP-ribose) synthetase with a concomitant (transient) decrease in cellular NAD content (4). The involvement of poly(ADP-ribose) synthetase in DNA damage and/or repair is supported by the observation that inhibitors of this enzyme have been found to retard DNA strand rejoining and potentiate the biological effects (e.g., cell killing, chromosomal alterations) of X-rays and chemical mutagenic carcinogens *in vitro* (5-11) and *in vivo* (10,12). In view of the ability of inhibitors of poly(ADP-ribose) synthetase to increase cell killing without enhancing the mutagenic effect of alkylating agents (9,13), the utility of these inhibitors for improvement of chemotherapy of tumor cells has been investigated *in vitro* and *in vivo*. They were found to increase the antitumor activity of N-methyl-N-nitrosourea (14) and bleomycin (10,15). Differential radiosensitization of human tumor cells by 3-aminobenzamide (3AB) *in vitro* correlates well with the clinical radiocurability of tumors *in vivo* (16).

In the present study two animal models, DBA2 mice and Syrian golden hamsters were employed. The modulating effect of 3AB on *in vivo* tumor progression in combination with X-rays or the cytostatic drug dacarbazine (DTIC) was studied.

Results

Syrian golden hamsters bearing Greene melanoma: The modulative effect of 3AB on the progression of Greene melanomas in Syrian golden hamsters after treatment with X-rays or DTIC was studied. This tumor was successfully retransplanted more than 20 times before our

[1]J.A. Cohen Institute, Interuniversity Institute of Radiation Protection, Leiden, Netherlands
[2]Department of Pharmacology, State University of Leiden, Leiden, Netherlands

experiment. Greene melanoma tissue was implanted subcutaneously into the abdomen of male Syrian golden hamsters (80-120 g). After 10-11 days, when a palpable tumor of 2 to 5 g had grown at the site of implantation, treatments were started. A). X-ray-treatment: X-rays were generated by an ENRAF apparatus at 150 kV, 6 mA at a dose rate of 5 rad/sec. During the irradiation the animals were kept anaestesized. The animals were exposed locally at the site of tumor implantation to fractionated doses of X-rays of 16, 32 or 48 Gy (4 times 4, 8 or 12 Gy respectively with 24 hr intervals between fractions). B). DTIC treatment: Dacarbazine (DTIC-dome, Miles Pharmaceutical, Connecticut, USA) was dissolved in Hank's BSS (with 20 mM Hepes buffer). The animals received two intraperitoneal (i.p.) injections of DTIC (200 mg/kg) with a 72 hr interval. In both types of experiments 6-10 hamsters were used per point. When 3AB was used, animals were injected i.p., 1 hr before X-irradiation or immediately after DTIC injection at a dose range of 0-449 mg/kg.

The first day of the treatment tumors were measured in three dimensions using a vernier. The product of these parameters was fixed as 100% and each time a tumor was measured, its size was expressed relative to the value on the first day. From these experiments the delay in growth induced by the treatment was calculated. Fig. 1 shows that there was a good linear correlation between the product of the three parameters and the volume of the tumor (fixed at 95% of its weight) based on 59 tumors which were measured, dissected and weighed.

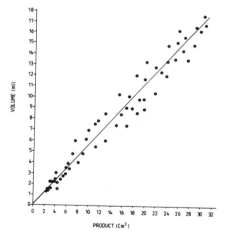

Fig. 1. Relation between tumor size and its volume in Syrian golden hamsters bearing Greene melanoma. 59 tumors were measured, dissected and weighed.

Data presented in Fig. 2a show that the untreated tumors doubled their size in 2-3 days. 3AB (449 mg/kg) alone did not influence the growth kinetics of the non-irradiated tumors. When a fractionated dose of 16 Gy was given (4 times 4 Gy), a delay in growth of the tumors was induced and this effect was increased significantly by 3AB (Fig. 2a). In combined treatment, delay in the growth of the tumors was also dependent on the 3AB concentration (Fig. 2b).

From three separate experiments, it was calculated that the specific growth delay (the time interval relative to control tumors which was required to double the volume of tumors measured at the start of treatment) induced by X-rays and the combination of X-rays and 3AB (449 mg/kg) was about 1.5 and 4 days respectively. At fractionated dose levels of 4 times 8 Gy these figures were about 4 and 8.4 days respectively (Fig. 3). When the dose was increased to 4 times 12 Gy, the average size of the tumors fell to below its value on day zero, but the initial difference between the treatment with and without 3AB was absent (Fig. 4). Due to severe damage of the skin of the X-irradiated animals examination beyond day 14 was not possible.

Treatment of hamsters bearing Greene melanoma with DTIC effectively delayed the growth of tumors and in combined treatment with 3AB (449 mg/kg) this effect was more pronounced. The specific growth delay for DTIC and DTIC in combination with 3AB was about 2.5 and 5 days respectively (Fig. 5). Due to cytotoxicity of DTIC (2 x 200 mg/kg), 50% of the hamsters died within 7 days after the first injection, but when 3AB was combined with DTIC, this frequency was reduced to 25% (data evaluated from 40 hamsters).

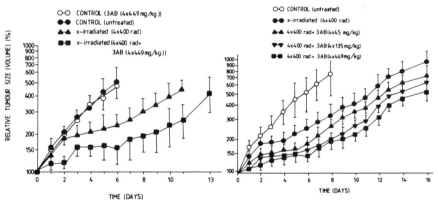

Fig. 2a. Reduced tumor progression by fractionated dose of X-rays of 16 Gy (4 x 4 Gy with 24 hr intervals between fractions) in the presence or absence of 3AB in Syrian golden hamsters bearing Greene melanoma.

Fig. 2b. Relation between increased therapeutic effect of X-rays (4 x 4 Gy) and concentration of 3AB in Syrian golden hamsters bearing Greene melanoma.

DBA2 mice infected with lymphoma cells (L1210):

Lymphocytic leukemia L1210 cells were kept in cultures in modified Fischer's medium. DTIC was dissolved in Hank's BSS (with 20 mM Hepes buffer). For treatment L1210 cells (10^5/mouse) were injected i.p. A group of mice received, in addition, an i.p. injection with DTIC (100-500 mg kg) or 3AB (0-675 mg/kg) or combined treatment (simultaneous injection of both chemicals into the mice 24 hr after injection with leukemic cells). Five male DBA2 mice were used for each dose. The end-point studied was the median survival time.

Results obtained from three separate experiments indicated that 3AB alone had no therapeutic effect and it did not increase the median survival of infected mice (Fig. 6). DTIC at 500 mg/kg was toxic to the mice. At 300 mg/kg it increased the median survival about 35% when compared to the controls, and in combination with 3AB (449 mg/kg) the median survival increased to 80% (Fig. 6). In separate experiments we have tested the influence of different doses of 3AB (10 doses at a range of 0-675 mg/kg) in combined treatment with DTIC (300 mg/kg). There was a positive trend for an increase in median survival with increasing 3AB concentration (Fig. 7).

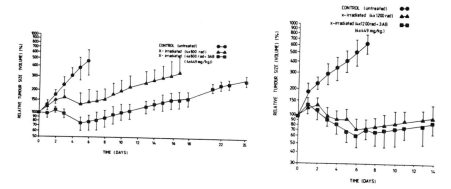

Fig. 3. (left) Reduced tumor progression by fractionated dose of X-rays of 32 Gy (4 x 8 Gy with 24 hr intervals between fractions) in the presence or absence of 3AB in Syrian golden hamsters bearing Greene melanoma.

Fig. 4. (right) Reduced tumor progression by fractionated dose of X-rays of 48 Gy (4 x 12 Gy with 24 hr between fractions) in the presence or absence of 3AB in Syrian golden hamsters bearing Greene melanoma.

Discussion

In the present study we found that 3AB alone has no therapeutic effect and no influence on the growth kinetics of untreated Greene melanoma or survival of untreated leukemic mice. X-rays and DTIC treatment alone could induce growth delay of the melanotic tumors in Syrian golden hamsters. Additional treatment with 3AB resulted not only in an increase in the therapeutic effect of X-rays or DTIC, but also in the survival of DTIC treated hamsters and leukemic DBA2 mice.

DTIC was selected on the basis of the observation that this chemical is an effective drug against malignant melanoma (17). It is a bifunctional alkylating agent needing metabolic transformation to its monofunctional derivatives to be effective (18). It has been shown earlier that inhibitors of poly(ADP-ribose) synthetase specifically potentiate the effects of monofunctional and not polyfunctional alkylating chemicals (9,11). Damage to normal tissue is usually the major limiting factor in any approach to the radiation or

393

chemotherapeutic treatment of malignant tumors. However, NAD$^+$ pools have been found to be lower in a variety of tumors than in normal and adult tissues (19). It is reasonable to assume that the low NAD$^+$ pool together with inhibition of poly(ADP-ribose) synthetase, following DNA damage (e.g. induced by radiation or cytotoxic agents) should result in accumulation of DNA strand breaks. The combined treatment can thus be more toxic for tumor than for normal tissue, which has a higher level of NAD$^+$ available. In conclusion, the data presented here indicate the possibility of using poly(ADP-ribose) synthetase inhibitors together with X-radiation or cytostatic drugs in the treatment of malignant tumors for improving conventional tumor therapy.

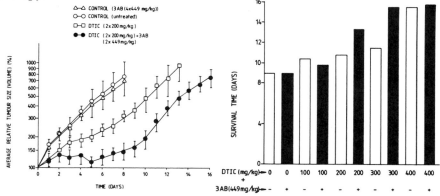

Fig. 5. (left) Reduced tumor progression by DTIC (2 x 200 mg/kg with 72 hr intervals) and in combination with 3AB in Syrian golden hamsters bearing Greene melanoma.

Fig. 6. (right) Median survival time of leukemic DBA2 mice treated with DTIC in the presence or absence of 3AB.

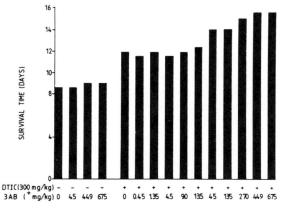

Fig. 7. Relation between increased therapeutic effect of DTIC and concentration of 3AB in leukemic DBA2 mice.

Acknowledgements. This investigation was financially supported by the Queen Wilhelmina Fund (The Netherlands) (IKW 84-39).

394

References

1. Halldersson, H., Gray, D.A., Shall, S. (1978) FEBS Letters 85: 349-352
2. Benjamin, R.C., Gill, D.M. (1980) J Biol Chem 255: 65-69
3. Cohen, J.J., Berger, N.A. (1981) Biochem Biophys Res Commun 98: 268-274
4. Durkacz, B.W., Omidiji, O., Gray, D.A., Shall, S. (1980) Nature 283: 593-596
5. Durkacz, B.W., Irwin, J., Shall, S. (1981) Eur J Biochem 255: 10493-10501
6. Nduka, N., Skidmore, C.J., Shall, S. (1980) Eur J Biochem 105: 525- 530
7. Durrant, L.G., Boyle, J.M. (1982) Chem Biol Interactions 38: 325-338
8. Natarajan, A.T., Csukas, I., Degrassi, F., van Zeeland, A.A., Palitti, F., Tanzarella, C., da Salvia, R., Fiore, M. (1982) Chromosome Alterations and Chromatin Structure. A.T. Natarajan, G. Obe, H. Altmann (eds.) Elsevier Biomedical Press, Amsterdam, pp.47-59
9. Natarajan, A.T., van Zeeland, A.A., Zwananburg, T.S.B. (1983) ADP-ribosylation, DNA Repair, and Cancer. M. Miwa, O. Hayashi, S. Shall, M. Smulson (eds.), Japan Scientific Society Press, pp. 227-242
10. Sakamoto, H., Kawamitsu, H., Miwa, M., Terada, M., Sugimura, T. (1983) J Antibiotics 36: 296-300
11. Zwanenburg, T.S.B., Hansson, K., Darroudi, F., van Zeeland, A.A., Natarajan, A.T. (1985) Mutation Res 151: 251-262
12. Zwanenburg, T.S.B., van Buul, P.P.W. (1986) Mutation Res 175: 33-37
13. Natarajan, A.T., Mullenders, L.H.F., Zwanenburg, T.S.B. (1986) Genetic Toxicology of Environmental Chemicals, Part A, Basic Principles of Mechanisms of Action. C. Ramel, B. Lambert, J. Magnusson (eds.), Alan R. Liss Inc., New York, pp. 373-384
14. Smulson, M.E., Schein, P., Mullins, Jr. D.W., Sudhakar, S. (1977) Cancer Res 37: 3006-3012
15. Kawamitsu, H., Miwa, M., Tanaka, Y., Sakamoto, H., Terada, M., Hoshi, A., Sugimura, T. (1982) J Pharm Dyn 5: 900-904
16. Thraves, P.J., Mossman, K.L., Brennan, T., Dritschilo, A. (1986) Int J Radiat Biol 50: 961-972
17. Rümke, P.H. (1981) Cancer Chemotherapy. H.M. Pinedo (ed.), Elsevier, New York, pp.397-408
18. Martindale (1982) The Pharmaceutical Press, London 171: 204-205
19. Jacobson, E.L., Jacobson, M.K. (1976) Arch Biochem Biophys 175: 627-634

Cytotoxic Effects of 3-Aminobenzamide in CHO-K1 Cells Treated with Purine Antimetabolites

Barbara W. Durkacz, Adrian L. Harris, and Kevin Moses

Cancer Research Unit, University of Newcastle upon Tyne, Royal Victoria Infirmary, Newcastle upon Tyne NE1 4LP United Kingdom

Introduction

3-Aminobenzamide (3AB), an inhibitor of ADP-ribosyl transferase, also affects *de novo* purine biosynthesis, demonstrated by an inhibition of incorporation of radiolabel from glucose, methionine and formate into the DNA (1-3). The step(s) affected have not been identified. Mutant cell lines, deficient in salvage nucleotide synthesis, are no more sensitive to 3AB than wild-type (4). Also, there is no effect of 3AB on dNTP pool sizes (2, 5). These data indicate that the effect of 3AB on *de novo* purine biosynthesis is not normally rate-limiting. Other effects of 3AB include inhibition of cell growth and inhibition of adenosine transport (5).

It seems certain that the effect of 3AB on the cytotoxicity and DNA repair of cells treated with monofunctional alkylating agents is mediated via an inhibition of ADP-ribosyl transferase activity (7, 8). Nevertheless, the pleiotropic effects of 3AB could result in the modulation of the cytotoxicity of other drugs independent of ADP-ribosyl transferase inhibition. In this paper, we describe the cytotoxic effects of 3AB on a range of purine base and nucleoside analogues. The data lead us to hypothesize that 3AB modulates purine analogue cytotoxicity by mechanism(s) independent of DNA repair inhibition, but involving effects on *de novo* purine biosynthesis and nucleoside transport.

Purine analogues are widely used in the treatment of leukemias (6-thioguanine [6TG] and 6-mercaptopurine [6MP]), and as immunosuppressive agents (azathioprine). They are incorporated via the purine salvage pathway, and are converted to nucleotides which are incorporated into DNA and RNA (9, 10). The presence of 6TG in DNA results in unilateral chromatid damage (UCD), manifest as DNA strand breaks and "kinking" of chromosomes in the second cell cycle following incorporation (11). This is thought to be due to the inability of 6TG containing DNA to function adequately as a replication template. In addition, purine analogue metabolites inhibit steps in *de novo* purine biosynthesis, leading to depletion of purine nucleotide pools (12). Evidence suggests that the cytotoxic locus of action of 6TG is due to its presence in the DNA (10, 11). However, the cytotoxic mode of action of 6MP is in dispute, as at equitoxic doses to 6TG, very little UCD is observed, but the effect on purine nucleotide depletion is much more severe (13, 14). The role, if any, of DNA excision repair in the modulation of cytotoxicity is unknown.

Results

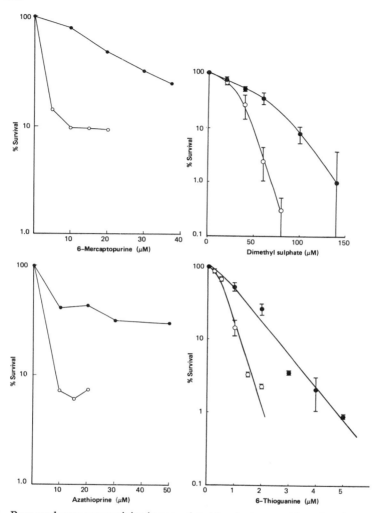

Fig. 1. Base analogue cytotoxicity is potentiated by the benzamides. Survival of CHO-K1 cells treated with base analogues in the presence (O) or absence (●) of 3mM 3-aminobenzamide (3AB). Potentiation of cytotoxicity of a monofunctional alkylating agent, dimethyl sulphate, by 3AB is shown for comparison.

Co-administration of non-toxic concentrations of 3AB (3mM) to CHO-K1 cells potentiates the cytotoxicity of 6TG, 6MP and azathioprine (Fig. 1). The Dose Enhancement Factors (DEFs) at 10% survival ranged from about 2X (6TG) to ≥10X (6Mp and azathioprine). Potentiation of the cytotoxicity of a monofunctional alkylating agent, dimethyl sulfate (DMS), by 3AB is shown for comparison (DEFX3). Similar results were obtained using 3-acetylaminobenzamide, 3-methoxybenzamide and benzamide, but no difference in survival was obtained with 3-aminobenzoic acid, which does not inhibit ADP-ribosyl transferase (data not shown).

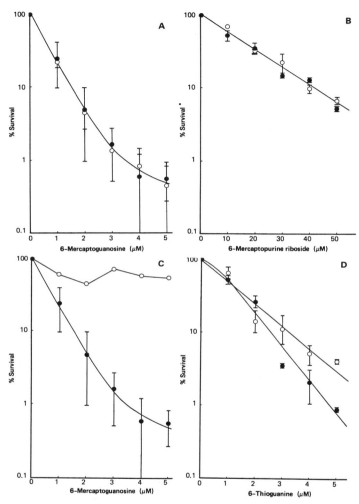

Fig. 2. Nucleoside analogue cytotoxicity is not potentiated by the benzamides. Survival of CHO-K1 cells treated with corresponding nucleoside analogues in the presence (O) or absence (●) of 3AB. (A and B). The effect of dipyridamole (20 μM) on 6-mercaptoguanosine and 6-thioguanine cytotoxicity (C and D). (O) + dipyridamole (●) control.

Similar experiments were carried out with the corresponding nucleoside analogues (Fig. 2). It can be seen that for 6-mercaptoguanosine and 6-

mercaptopurine riboside, there was no potentiation of cytotoxicity by 3AB. Since the most likely explanation for the differential effect of 3AB on base and nucleoside analogue cytotoxicity is that the benzamides specifically inhibit nucleoside, but not base, transport (see discussion), we compared the effects of 3AB with dipyridamole, a well-defined nucleoside transport inhibitor (15). Non-toxic concentrations of dipyridamole (20 μM), which block nucleoside transport in CHO-K1 cells, completely protect against the cytotoxicity of 6-mercaptoguanosine, while having little or no effect on the cytotoxicity of 6TG (Fig. 2).

We have investigated the timing of the benzamide mediated potentiation of base analogue cytotoxicity using synchronized cells treated with 2 μM 6TG (Fig. 3A). It can be seen that 3AB must be present at least from the start of S-phase in the first cell cycle of 6TG administration for potentiation of cytotoxicity to occur.

The role of ADP-ribosyl transferase activity and DNA repair in 6TG cytotoxicity was assessed. We compared the effect of equitoxic doses of 6TG and DMS on NAD levels (Fig. 3B). Whereas DMS caused a rapid and reversible lowering of NAD levels, 6TG produced no effect, even after cells had progressed through to the second cell cycle (30 hr) when 6TG induced DNA strand breaks are known to occur (11).

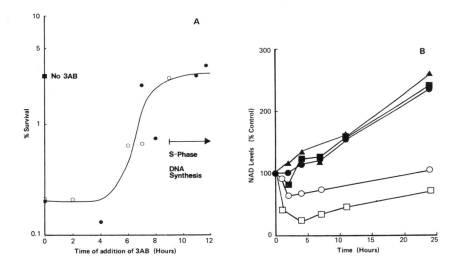

Fig. 3A. Timing of enhancement of cytotoxicity by 3AB. Cells were synchronized in G_1 by serum deprivation. Following re-addition of serum, cells were treated with 2μM TG. 3AB (3 mM) was added either at the start, or at intervals thereafter. 24 hr later, cells were plated for survivors. The time of onset of S-phase DNA synthesis was monitored by [³H]Thd incorporation. Results from 2 independent experiments (●, ○).
Fig. 3B. NAD utilization. Drugs (either DMS or TG) were added at 0 hr and cells sampled at intervals thereafter for NAD content. (●) 2.5 μM TG (■) 5.0 μM TG (▲) 10.0μM TG (○) 80 μM DMS (□) 140 μM DMS.

399

DNA strand breaks were monitored by nucleoid sedimentation (16). DMA induced DNA strand break levels were increased dramatically in the presence of 3AB (Fig. 4). Although there was a dose dependent increase in 6TG induced DNA strand breaks, no further increase was observed in the presence of 3AB. However, caution must be exercised in the interpretation of these results, since at least some 6TG induced DNA strand breaks have been shown to be protein crosslinked (18), and therefore alkaline elution studies are required to confirm these results. Note that although we can detect breaks using concentrations of DMS that yield 100% survival, doses of 6TG that reduce survival to less than 10% must be used.

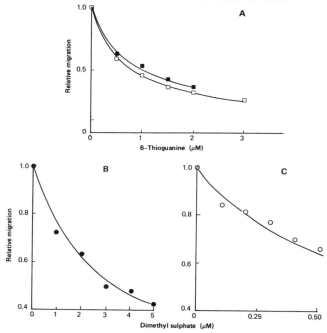

Fig. 4. DNA strand break assay. DNA strand breaks were monitored by sedimentation of nucleoids through a 15-30% neutral sucrose gradient. Migration of nucleoids is expressed relative to migration of control (undamaged) nucleoids. A: 24 h treatment with TG (□) control (■) + 3AB. B: 30 min treatment with DMS. C: 30 min treatment with DMS + 3AB.

Discussion

Is the enhancement of purine analogue cytotoxicity obtained with 3AB caused by an inhibition of ADP-ribosyl transferase activity and its consequential effect on DNA repair? The following observations suggest not: (i). 3AB is required during the first cell cycle of 6TG administration, although it has been shown that DNA strand breaks do not appear until the 2nd G_2 phase and are associated exclusively with the nascent strand (11).

400

(ii). There is no alteration in cellular NAD content or DNA strand break levels in the presence of 3AB (both associated with ADP-ribosyl transferase activity) resulting from 6TG treatment. (iii).The effect of 3AB was much greater in 6MP and azathioprine treated cells compared to 6TG (compare DEF values). This could be accounted for by the fact that 6MP has a much greater inhibitory effect on *de novo* purine biosynthesis than 6TG (14) (see below).

An alternative mechanism of 3AB action could involve its known effects on *de novo* purine biosynthesis (1, 3) and/or nucleoside transport (5). The combined inhibitory effects of 3AB and purine analogues on purine biosynthesis could result in sufficient depletion of intracellular nucleotide pools to result in enhanced cellular cytotoxicity. In addition, these effects would lead to an increased bioavailability of 5-phosphoribosyl-1-pyrophosphate (PRPP), the first enzymic product in the *de novo* pathway. Increased PRPP levels would enhance the activity of hypoxanthine phosphoribosyl transferase, leading to increased salvage of purine analogues.

An alternative mechanism may involve an inhibition of nucleoside transport. It has been shown that 3-acetylamidobenzamide inhibits adenosine transport, and we have shown that 3AB inhibits thymidine transport (unpublished results). This would explain the differential effect of 3AB on purine base and nucleoside cytotoxicity. Grem and Fischer (17) observed that nucleoside transport inhibitors enhanced the cytotoxicity of 5' fluorouracil by preventing efflux of 5'-fluorodeoxyuridine. Since purine base analogues can be interconverted to the nucleoside derivatives, a similar mechanism could apply for the potentiation of 6TG by 3AB. However in this case, we might expect to see an enhancement of 6TG cytotoxicity by dipyridamole, which is not seen.

In conclusion, 3AB enhances the cytotoxicity of a range of purine analogues by a mechanism which is probably independent of ADP-ribosyl transferase inhibition. The DEF values obtained are comparable or greater than those obtained with a monofunctional alkylating agent, DMS. Furthermore, the purine analogues are of major clinical value and these data indicate a role for the benzamides in chemotherapy.

Acknowledgments: Data shown here have been reprinted with permission from Moses *et al.*, Cancer Res. 48, 5650-5654, 1988.

References

1. Milam, K.M., Thomas, G.H., Cleaver, J.E. (1986) Expt Cell Res 165: 260-268
2. Hunting, D.J., Gowans, B.J., Henderson, J.F. (1985) Mol Pharmacol 28: 200-206
3. Milam, K.M., Cleaver, J.E. (1984) Science 223: 589
4. Schwarz, J.L., Weischelbaum, R.R. (1985) ADP-ribosylation of proteins, F.R.Althaus, H.Hilz, S.Shall (eds.), Springer-Verlag, Berlin, Heidelberg, New York, Tokyo, pp. 332-336
5. Kidwell, W.R., Noguchi, P.D., Purnell, M.R. (1985) ibid: 402-409. Also, Purnell, M.R., Kidwell, W.R. (1985) ibid: 98-105.
6. Snyder, R.D. (1984) Biochem Biophys Res Comm 124: 457-461

7. Durkacz, B.W., Omidiji, O., Gray, D., Shall, S. (1980) Nature 283: 593-596
8. James, M., Lehmann, A. (1982) Biochemistry 21: 4007-4013
9. LePage, G.A. (1960) Cancer Res 20: 403-408
10. Christie, N.T., Drake, S., Meyn, R.E. Nelson, J.A. (1984) Cancer Res 44: 3665-3671
11. Fairchild, C.R., Maybaum, J., Kennedy, K.A. (1986) Biochem Pharmacol 35: 3533-3541
12. Patterson, A.R.P., Tidd, D.M. (1975) Handbook of Experimental Pharmacology 38. A.C. Sartorelli, D.G. Johns (eds.), Springer-Verlag, Berlin, pp. 384-394
13. Maybaum, J., Mandel, G.H. (1983) Proceedings of the AACR Preclinical Pharmacol and Expt Therapeutics. Abst. 1163
14. Nelson, J.A., Carpenter, J.W., Rose, L.M., Adamson, D.J. (1975) Cancer Res 35: 2872-2878
15. Patterson, A.R.P., Lau, E.Y., Dahlig, E., Cass, C.E. (1980) Mol Pharmacol 18: 40-44
16. Durkacz, B.W., Irwin, J., Shall, S. (1981) Eur J Biochem 121: 65-69
17. Grem, J.L., Fischer, P.H. (1986) Biochem Pharmacol 35: 2651-2654
18. Covey, J.M., D'Incalci, M., Kohn, K.W. (1986) Proceedings of the AACR, Abst. 68

Hyperthermia and Poly(ADP-ribose) Metabolism

Göran Jonsson, Luc Menard[1], Elaine L. Jacobson, Patrick W. Rankin, Guy G. Poirier[1], and Myron K. Jacobson

Departments of Biochemistry and Medicine, Texas College of Osteopathic Medicine, University of North Texas, Ft. Worth, Texas 76107 USA

Introduction

Hyperthermia or heat shock, has been found to have profound effects on many cellular structures and functions (1). A striking effect is the change of RNA and protein synthesis patterns that occur following heat shock which result in suppression of synthesis of the majority of cellular proteins while a specific family of proteins, known as heat shock proteins, are induced. These proteins are assumed to have protective functions making the cell more resistant to further damage (2).

Poly(ADP-ribose) glycohydrolase is a nuclear enzyme which catalyses the degradation of poly(ADP-ribose) by splitting the ribose-ribose bonds and liberating ADP-ribose residues (3). Following DNA damage, there is a rapid synthesis and turnover of poly(ADP-ribose) in intact cells that is required for processes in the cell nucleus necessary for cellular recovery from DNA damage (4, 5). This laboratory has previously shown that poly(ADP-ribose) metabolism is altered by hyperthermia (6, 7). Following hyperthermic treatment, a large increase in the accumulation of poly(ADP-ribose) was observed. Reported here are measurements of poly(ADP-ribose) content as a function of temperature and recovery time following hyperthermia. These data have allowed the calculation of the half-life of the polymers as a function of temperature, which under these conditions, reflects the activity of poly(ADP-ribose) glycohydrolase. Further, temperature effects on glycohydrolase activity have been measured directly *in vitro* and in intact cells. The changes observed in poly(ADP-ribose) metabolism following heat shock can be accounted for by hyperthermic effects on poly(ADP-ribose) glycohydrolase, which involve thermal inactivation of the enzyme and a reversible metabolic inactivation of the enzyme.

Results

In the following studies confluent cultures of C3H10T1/2 cells were subjected to a standard hyperthermic treatment of 45°C for 30 min followed by return of cultures to 37°C. A rapid increase in the intracellular content of poly(ADP-ribose) occurred following heat shock as is shown in Fig. 1.

---

[1]Centre de Recherche de L'Hotel-Dieu Quebec, Quebec, Canada G1R2J6

Maximal accumulation of poly(ADP-ribose) occurred approximately 4 hr following hyperthermic treatment and then polymer content decreased gradually over the 24 hr recovery period. This sustained period of increased polymer levels is very different compared with the response to other types of stresses such as DNA alkylating agents (8). Even at 24 hr, polymer levels were significantly elevated. The addition of benzamide when polymer levels were near maximal accumulation resulted in a decrease to near basal levels.

The accumulation of poly(ADP-ribose) following hyperthermia must be attributable to either an increased rate of synthesis or a decreased rate of turnover of the polymers or both. To examine the effects of hyperthermia on poly(ADP-ribose) polymerase and poly(ADP-ribose) glycohydrolase activities in intact cells, we probed cells with MNNG. Treatment with 68 μM MNNG causes a rapid activation of poly(ADP-ribose) polymerase which in turn results in depletion of the cellular NAD pool (8). The rate of depletion of NAD provides an estimation of poly(ADP-ribose) polymerase activity in the cells. Poly(ADP-ribose) glycohydrolase activity in intact cells can be estimated by determining the rate of polymer turnover following addition of an inhibitor such as benzamide that effectively blocks poly(ADP-ribose) polymerase activity as is observed in Fig. 1 . Cultures were probed with MNNG following hyperthermia and compared to control samples held at 37°C and similarly probed. Hyperthermic treatment had little or no effect on the rate of NAD consumption (data not shown). These data suggest that poly(ADP-ribose) polymerase activity was unaffected by hyperthermia. Fig.

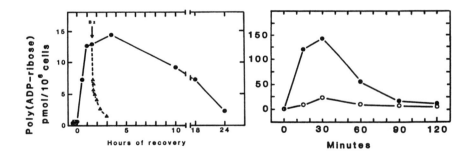

Fig. 1. (left) Poly(ADP-ribose) content as a function of recovery following hyperthermia. C3H10T1/2 cells were heated to 45°C for 30 min and then returned to 37°C (●). In parallel experiments 5 mM benzamide was added to the cultures at 1.5 hr after hyperthermia (▲).

Fig. 2. (right) Poly(ADP-ribose) content during treatment with MNNG in control and heat-shocked C3H10T1/2 cells. Cells were heated to 45°C for 30 min (●) in parallel to control samples held at 37°C (○). The cells were then rapidly returned to 37°C and at time 0 given 68 μM MNNG.

404

2 shows poly(ADP-ribose) accumulation versus time for control and hyperthermia treated cells. In both cases, a rapid increase of polymer occurred that was maximal after about 30 min. However, hyperthermic treatment resulted in a maximum polymer level approximately 10 times higher than that following MNNG alone, suggesting an effect of hyperthermia on poly(ADP-ribose) glycohydrolase activity.

The effect of hyperthermia on poly(ADP-ribose) glycohydrolase activity was examined further by administration of 5 mM benzamide at 30 min using the same experimental design described for Fig. 2. A rapid biphasic decay in polymer levels was observed with approximately 90% of the polymer turning over rapidly while the remainder was relatively stable. A semilogarithmic plot of the decay of the unstable component is shown in Fig. 3. From the slope of these plots, a half-life of 0.8 min was estimated for poly(ADP-ribose) in cells treated with MNNG alone. However, following hyperthermia the half-life increased to approximately 10 min. The turnover of polymers following hyperthermia alone also had a half-life of approximately 10 min (Fig. 1). Taken together, the data of Figs. 2 and 3 show that a major alteration of poly(ADP-ribose) metabolism by hyperthermia involves poly(ADP-ribose) glycohydrolase activity.

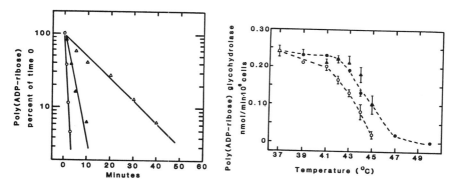

Fig. 3. (left) Turnover of poly(ADP-ribose) in control and hyperthermia treated cells. C3H10T1/2 cells were treated as in Fig. 2. After 30 min with MNNG, 5 mM benzamide was added (time 0 in the figure). Poly(ADP-ribose) was measured and calculated as a percentage of time 0. *Solid lines*, least squares fits of the experimental data; (●) 37°C control cells, (▲) cells given a hyperthermic treatment.

Fig. 4. (right) Activity of poly(ADP-ribose) glycohydrolase as a function of temperature. C3H10T1/2 cells or cell extracts were heated at each temperature for 30 min. (○) intact cells, no recovery; (▲) intact cells, 24 hr recovery at 37°C; (●) heating of cell homogenates. Data are the mean of two separate experiments; *error bars*, where the duplicates differed by a % standard deviation of more than 5%.

Activity of poly(ADP-ribose) glycohydrolase was measured directly by the release of monomers of ADP-ribose from ^{32}P-labeled poly(ADP-ribose)

405

(10). The effect of heating cell extracts was compared to the effects of hyperthermic treatment of intact cells (Table 1). Incubation of extracts from control cells at 37°C for 30 min did not result in a change of activity but incubation at 45°C for 20 and 30 min resulted in a decrease to 53 and 41% of controls, respectively. Hyperthermic treatment of intact cells at 45°C for 20 and 30 min prior to preparation of extracts resulted in greater decreases in enzyme activity than the heating of cell extracts for the corresponding times. However, when the cells were allowed to recover at 37°C before preparation of cell extracts, enzyme activity partially recovered. It was interesting that, following recovery, the activities were very similar to those of heated extracts.

Table 1. Activity of poly(ADP-ribose) glycohydrolase relative to 37°C controls following 20- and 30- min heating times at 45°C.

Treatment	Activity of poly(ADP-ribose) glycohydrolase	
	nmol/10^6 cells/min	%
Cell extracts		
37°C	0.24 ± 0.02	100
45°C, 20 min	0.13 ± 0.00	53
45°C, 30 min	0.10 ± 0.03	41
Intact cells		
45°C, 20 min		
No recovery	0.06 ± 0.01	23
24 hr recovery	0.03 ± 0.01	53
Intact cells		
45°C,30 min		
No recovery	0.02 ± 0.01	8
24 hr recovery	0.01 ± 0.02	41

Data are shown for heated cell extracts of control cells and for extracts prepared following hyperthermic treatment of intact cells. For cell extracts heated *in vitro*, the activity was measured immediately after heat treatment. For intact cells, extracts were made immediately(no recovery) and following a recovery period of 24 hr. Values are the mean of three independent determinations with standard deviations.

Fig. 3 shows activity of poly(ADP-ribose) glycohydrolyase as a function of temperature. In each case, the cell cultures or extracts were held 30 min at each temperature. Values are shown for heated extracts and for cells subjected to hyperthermia in culture and either harvested immediately for assay or allowed to recover for 24 hr prior to assay. Over the entire temperature range studied activity was more temperature resistant in heated extracts. The activities in heated extracts were very similar to those obtained from extracts of cultures made following 24 hr of recovery from hyperthermia. The enzyme activity decreased with increasing temperature in all cases in a biphasic pattern with a breakpoint around 42°C. At temperatures above 42°C the decrease in enzyme activity with increasing temperature suggested thermal inactivation of the enzyme. Utilizing the

Arrhenius equation we calculated an inactivation energy of approximately 110 kcal/mol, which is well within the range of energies known to cause denaturation of several other enzymes and proteins (10).

Discussion

Poly(ADP-ribose) glycohydrolase appears to be a ubiquitous enzyme and a variety of evidence has led to the conclusion that it is the primary degrading enzyme for poly(ADP-ribose). In this study it is shown that hyperthermia alters the rate of poly(ADP-ribose) degradation in intact cells and the activity of poly(ADP-ribose) glycohydrolase when intact cells are subjected to hyperthermia or when cell extracts are heated. From these data two mechanisms can be postulated in the decrease of poly(ADP-ribose) glycohydrolase activity following hyperthermia: (i) an irreversible partial thermal denaturation of the enzyme and (ii) a reversible metabolic modulation of enzyme activity. The data of Fig. 4 and Table 1 show that poly(ADP-ribose) glycohydrolase activity is partially thermally denatured by the hyperthermic conditions used in this study. However, thermal denaturation can account for only a portion of the decreased activity observed when intact cells are subjected to hyperthermia and assayed immediately. In contrast, when cells are allowed to recover prior to preparation of cell extracts, the failure to recover total enzyme activity can be accounted for by thermal denaturation. The recovery does not appear to be due to new synthesis of the enzyme (11). The simplest explanation for these data is that part of the decrease in activity observed in intact cells following hyperthermia and the recovery following return to 37°C is due to a reversible metabolic alteration of enzyme activity in response to hyperthermia.

Acknowledgments This work was supported in part by grants from the NIH (CA43894), The National Research Council of Canada and The Medical Research Council Stockholm, Sweden.

References

1. Magun, B.E., (1981) Radiat Res 87: 657-669
2. Asburner, M., Bonner, J.J. (1979) Cell 17:241-254
3. Miwa, M., Sugimura, T. (1971) J Biol Chem 246: 6362-6264
4. Wielckens, K., George, E., Pless, T., Hilz, H. (1982) J Biol Chem 257: 4098-4104
5. Jacobson, E.L., Antol, K.M., Juarez-Salinas, H., Jacobson, M.K. (1983) J Biol Chem 258: 103-107
6. Juarez-Salinas, H., Duran-Torres, G., Jacobson, M.K. (1984) Biochem Biophys Res Commun 122: 1381-1388
7. Jacobson, M.K., Duran-Torres, G., Juarez-Salinas, H., Jacobson, E.L. (1985) ADP-Ribosylation of Proteins, F.R. Althaus, H. Hilz, S. Shall (eds.) Springer-Verlag, Berlin, Heidelberg, pp.293-297
8. Juarez-Salinas, H., Sims, J.L., Jacobson, M.K. (1979) Nature 282: 740-741
9. Menard, L., Poirier, G. G. (1987) Biochem Cell Biol 65: 668-673
10. Dewey, W.C., Hopwood, L. E., Sapareto, S.A., Gerweck, L. E. (1977) Radiology 123:463-474
11. Jonsson G. G., Jacobson, E. L., Jacobson, M.K. (1988) Cancer Res 48: 4233-4239

Comparison of Low Dose Nicotinamide Versus Benzamide, Administered *Per Os,* as Radiosensitizers in a C3H Mammary Carcinoma

Elisabeth Kjellén [1] and Ronald W. Pero

Department of Molecular Ecogenetics, Wallenberg Laboratory, University of Lund, S-220 07 Lund, Sweden

Introduction

Nicotinamide and benzamide are inhibitors of the chromatin bound enzyme poly(ADP-ribose) polymerase, which is involved in the mechanism of DNA repair after high doses of ionizing radiation. There have been reports that both benzamide and nicotinamide radiosensitize animal tumor models at doses of 200 mg/kg or higher (2, 4, 5). In order to evaluate if any differences in tumor radiosensitization between nicotinamide and benzamide are apparent at low doses, animals with transplanted adenocarcinomas were given these drugs orally at a dose of 10 mg/kg five times a week from the day of transplantation until irradiation (in total, 50-80 mg/kg).

Results

Tumor growth measurements and tumor response. Table I shows the median time for the tumors to reach the growth end point, i.e. a volume 5 times that of the starting volume. Animals not surviving the first week after irradiation were excluded from the analysis (11/34 given 30 Gy alone, 10/38 given 30 Gy + benzamide, 0/19 given 30 Gy + nicotinamide). We found a significant time difference when comparing irradiation alone to irradiation in combination with nicotinamide, in favor of the nicotinamide treated animals. The data indicate, however, no radiosensitizing effect of benzamide, but rather a radioprotecting effect by this substance.

Table 2 shows the complete remission (CR) rates for the different treatment schedules. There was no difference in complete remission rate between tumors treated with benzamide in combination with irradiation versus irradiation alone. When comparing CR rates for tumors treated with nicotinamide combined with irradiation and benzamide in combination with irradiation there was a significant difference in favor of the nicotinamide treated tumors. The results for complete remissions comparing 30 Gy alone and 30 Gy plus nicotinamide is statistically in favor of the combined treatment which confirms our previous results (6).

---

[1]Department of Oncology, University Hospital, S-221 85 Lund, Sweden

Table 1. Median time for tumors to reach the growth endpoint, i.e. a volume five times that at the start of treatment of 30 Gy alone or when combined with either nicotinamide or benzamide.

Treatment	Number of Animals	Time (median)
A. 30 Gy	23	37.1
B. 30 Gy + nicotinamide (10 mg/kg)	19	68.9
C. 30 Gy + benzamide (10 mg/kg)	28	25.9
Statistics:	A vs. B p=0.0047 A vs. C p=0.030 B vs. C p<0.00005	

Table 2. Number of complete responses (CR) in relation to different treatments of radiation at a dose of 30 Gy alone or in combination with nicotinamide or benzamide.

Treatment	Number of Animals	CR*
A. 30 Gy	23	10
B. 30 Gy + nicotinamide (10 mg/kg)	19	15
C. 30 Gy + benzamide (10 mg/kg)	28	9
Statistics:	A vs. B p=0.029 A vs. C p=0.561 B vs. C p=0.003	

*A complete response (CR) was defined as total disappearance of this tumor for at least seven days.

Discussion

Nicotinamide and benzamide are inhibitors of the chromatin bound enzyme poly(ADP-ribose) polymerase (3, 8), and inhibitors of poly(ADP-ribose) polymerase are known radiosensitizers (1, 4). In a previous study (6) we have shown that low doses of nicotinamide given orally to mice from day of transplantation until day of radiation sensitizes a C3H mammary adenocarcinoma to radiation. Nicotinamide not only is an inhibitor of poly(ADP-ribose) polymerase but also has metabolic effects on the cell. Thus it is known that nicotinamide can increase the NAD pool in both malignant and normal cells (2, 7). Horsman et al. (4) have also indicated metabolic effects on the circulation with nicotinamide. Our results with *per os* administered benzamide indicate a difference in the radiosensitizing action of nicotinamide and benzamide. Nicotinamide can have radiosensitizing properties other than a simple direct poly(ADP-ribose)

409

polymerase inhibition since no effect was seen with benzamide in this experimental design.

References

1. Ben-Hur, E. (1984) Int J Radiat Biol 46: 659-671
2. Calcutt, G., Ting, S.M., Preece, A.W. (1970) Br J Cancer 24: 380-388
3. Hilz, H., Stone, P. (1976) Rev Physiol Biochem Pharmacol, Springer-Verlag, pp. 1-58
4. Horsman, M.R., Brown, D.M., Lemmon, M.J., Brown, J.M., Lee, W.W. (1986) Int J Radiation Oncology Biol Phys 12: 1307-1310
5. Jonsson, G.G., Kjellén, E., Pero, R.W., Cameron, R. (1985) Cancer Res 45: 3609-3614
6. Kjellén, E., Pero, R.W., Cameron, R., Ranstam, J. (1986) Acta Radiol Oncol 25: 281-284
7. Kjellén, E., Jonsson, G.G., Pero, R.W., Christensson, P.I. (1986) Int J Radiat Biol 49: 151-162
8. Preiss, J., Schlaeger, R., Hilz, H. (1971) FEBS Lett 19: 244-24

Viral Hepatitis in Mice and ADP-Ribose Metabolism

Hans Kröger, Monika Klewer, K. Noel Masihi, Werner Lange,
and Beate Rohde- Schultz

Robert Koch-Institut, D 1000 Berlin 65, West Germany

Introduction

We have shown earlier that the toxicity of endotoxin can be prevented to a large extent by antagonists of ADP-ribose metabolism (1). On the other hand there are some indications that muramyldipeptide (MDP) interferes with the this metabolism also (2). It was reported in 1980 that MDP, in combination with trehalose dimycolate, could induce resistance against influenza virus infection (3). In this chapter we are dealing with the influence of muramyldipeptide and benzamide upon the mouse hepatitis virus type 3 (MHV3).

Results

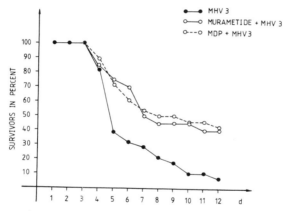

Fig. 1. Influence of muramyldipeptide and murametide upon the survival time of mice infected with MHV3 virus. Infection with MHV3 virus at time 0. Application of 1 mg (i.p.) each of MDP and murametide 1 hr, 24 hr, and 48 hr after MHV3 infection. Each group of 28 animals.

Determination of GOT and GPT were performed as described by Bergmeyer (4). The MHV3 virus belongs to the group of corona viruses. The procedure for the infection is described elsewhere (2). After the infection of mice with the virus MHV3 (2) there was a marked increase of GOT and GPT. In the animals given 1 mg of MDP or 1 mg of murametide 1 hr, 24 hr, and 48 hr after the MHV3 infection the increase of the GOT and GPT activity did not occur. Histopathological examination of the livers

411

from mice infected with MHV3 revealed extensive necrotic lesions. In contrast, MDP treated animals showed a strong suppression of the hepatic necrosis. We measured the percent survival under this treatment and with MDP. With murametide as with MDP 40% of the animals survived (Fig. 1).

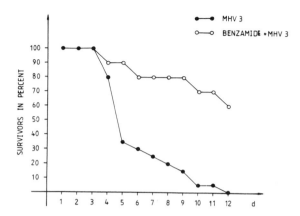

Fig. 2. Influence of benzamide upon the survival time of mice infected with MHV3 virus. Infection with MHV3 virus at time zero. Application of benzamide (i.p.): 25 mg/kg 1, 24, and 48 hr after infection. Each group of 10 animals.

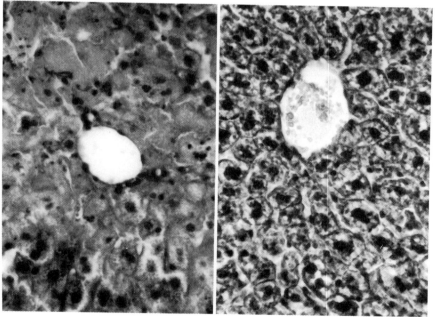

Fig. 3. Histopathological examination of the liver of mice. (Left) Injected with MHV3 (Right) Injected with MHV3 + benzamide (MDP). The same conditions as in Fig. 2 were used. Animals were killed at day 4 and liver was stained by haematoxilin eosin.

412

The acute toxicity of benzamide was tested by intraperitoneal injection. While 1.5 g/kg did not show any effect, a 2.0 g/kg injection resulted in the death of 75% of the animals. There was a marked reduction of the increase of GOT and GPT in mice infected with MHV3 virus and benzamide. Also therapeutic experiments were performed. With benzamide 60-80 % of the infected animals survived (Fig. 2). The histopathology of the liver showed that in mice infected with MHV3, necrotic lesions occured around the central artery on day 3 after infection (Fig. 3 left). This was completely inhibited by benzamide (Fig. 3 right).

Discussion

In earlier work we showed that the toxic effect of D-galactosamine on the liver can be inhibited by tryptophan and methionine (5). Furthermore, we found that D-galactosamine influences ADP-ribose metabolism (6). On the other hand it was reported that galactosamine enhances the effect of endotoxin (7). As already mentioned in the introduction, inhibitors of ADP-ribosylation reduce markedly the effect of endotoxin (1). These results stimulated us to study the effect of MDP on the ADP-ribose metabolism and we found indications for the interference of MDP with this metabolism (2). The induction of viral hepatitis by MHV3 in mice can be inhibited to a large extent by MDP and by benzamide. These are indications that the ADP-ribose metabolism plays a key role in the development of the hepatitis.

References

1. Kröger, H., Grätz, R. (1983) Falk Symposium No. 38 Mechanisms of Hepatocyte Injury and Death, Basel
2. Kröger, H., Masihi, K.N., Lange, W. (1987) Int Symp Immunomodulators and Nonspecific Host Mechanisms Against Microbial Infections, Berlin
3. Masihi, K.N., Brehmer, W., Lange, W., Ribi, E. (1980) 4th International Congress of Immunology, J.L. Prend'Homme and V.A.L. Hawken (eds.), Paris, Abstr.
4. Bergmeyer, H.U. (ed.) (1974) Methoden der Enzymatischen Analyse. Verlag Chemie, Weinheim
5. Kröger, H., Grätz, R., Musetean, C., Haase, J. (1981) Arzneimittel-Forsch 31: 989
6. Kröger, H., Grätz, R., Grahn, H. (1983) Int J Biochem 15: 1131-1136
7. Galanos, C., Freudenberg, M.A., Reutter, W. (1979) Proc Natl Acad Sci (USA) 76: 5939-5942

Abbreviations:
GOT - Glutamic-oxalacetic transaminase
GPT - Glutamic-pyruvic transaminase
MDP - Muramyldipeptide
MHV3 - Mouse hepatitis virus type 3

GTP-binding Proteins, Substrates of Pertussis Toxin-Catalyzed ADP-Ribosylation, as Mediators of Receptor-Coupled Signal Transduction

Michio Ui

Department of Physiological Chemistry, Faculty of Pharmaceutical Sciences, University of Tokyo 113, Japan

"IAP" action of pertussis toxin. Pertussis toxin is endowed with such multiple biological activities that it has been referred to as LPF (lymphocytosis-promoting factor), HSF (histamine-sensitizing factor), HA (hemagglutinin), MPA (mouse protective antigen) and IAP (islet-activating protein). The last name, IAP, was introduced by Ui and his colleagues (1) who were the first to report that a single injection of pertussis vaccine into rats resulted in extremely marked hyperinsulinemia *in vivo* when these rats were challenged with glucose or other insulin secretagogues within several days of vaccination. Also, much more insulin was released *in vitro*, from pancreatic islets excised from the vaccine-treated rats than from those of nontreated rats (2). This unique "IAP" action of pertussis vaccine was later reproduced strictly by the protein purified from the culture medium of *Bordetella pertussis* (3, 4); the protein proved to be pertussis toxin because it displayed all the biological activities of the toxin as well as LPF, HSF, MPA as well as IAP.

Interaction of pertussis toxin with mammalian cells as an A-B toxin to cause ADP-ribosylation of GTP-binding proteins in the cell membrane. IAP is a hexameric protein that is readily resolved into the largest monomer and the residual pentamer under certain conditions (5). The largest subunit is the A-protomer in the sense that it is able to become *A*ctive as an ADP-ribosyltransferase after being released from the residual B-oligomer, which *B*inds to the glycoprotein(s) on the surface of mammalian cells (6). Thus, IAP proves to possess what is called an A-B structure. The subsequent internalization of the membrane surface proteins is responsible for the gradual entrance (7) into the cells of the holotoxin bound via the B-oligomer (8). The toxin then undergoes intracellular processing to release the active A-protomer, which catalyzes ADP-ribosylation of GTP-binding proteins (henceforth referred to as G-proteins for brevity) inside the cell membrane (9, 10).

The G-protein that plays an essential role as a mediator of receptor-coupled inhibition of adenylate cyclase was the first to be identified as the specific substrate of IAP-catalyzed ADP-ribosylation (11-14). This IAP-substrate G-protein is referred to as G_i (or N_i) where the subscript, i, stands for "inhibition". Another G-protein, G_s (or N_s), involved in stimulation of

414

adenylate cyclase was then identified as the substrate of cholera toxin-catalyzed ADP-ribosylation. Once ADP-ribosylated, G_i irreversibly loses its function as a mediator of receptor-coupled inhibition of adenylate cyclase; ADP-ribosylated G_i is incapable of coupling receptors to the adenylate cyclase catalyst any longer. This is the mechanism by which pertussis toxin exerts some of its multiple biological activities including the IAP activity. The IAP activity is reasonably explained as follows (15): in pancreatic islet B-cells, there are a number of α_2-adrenergic receptors that are typically coupled via G_i to inhibit adenylate cyclase. These receptors are responsible for suppression of insulin release from the cells, since cAMP is a potent and physiological trigger of pancreatic insulin secretion (16). ADP-ribosylation of G_i by IAP hence results in reversal of receptor-coupled inhibition of the cyclase, thereby enhancing insulin secretory responses of pancreatic cells.

Role of IAP-substrate G-proteins in receptor-coupled signaling other than adenylate cyclase inhibition. As briefly described above, our finding that receptor-linked inhibition of adenylate cyclase in cells was reversed by IAP, which ADP-ribosylated a protein in the cell membrane, afforded convincing evidence for the role of this membrane protein, G_i , as an inhibitory transducer between the receptor and the adenylate cyclase catalyst. This strategy could be applied to other receptor-effector systems; blockade by IAP of a receptor-linked signaling in a other cell types lends strong support to the idea that a similar G-protein or one of the IAP-substrate G-proteins (i.e., G-proteins serving as the IAP-catalyzed ADP-ribosylation) is coupled to the receptor to initiate the signal in the cell. Taking advantage of this strategy, IAP-substrate G-proteins have been found to act as indispensable transducers in receptor-coupled activations of phospholipase C in mast cells (17-19) and neutrophils (20-22) and of phospholipase A_2 in 3T3 fibroblasts (23-25) as well as in muscarinic activation of K^+-channels in atrial cells (26).

Purification of IAP-substrate G-proteins from various sources. Gilman and his colleagues were pioneers in purification of G-proteins serving as the substrates of cholera toxin and IAP-catalyzed ADP-ribosylation (27). They purified two kinds of IAP substrates (G_i and G_o) from membrane fractions of bovine brain. Transducin, which communicates between rhodopsin and cGMP phosphodiesterase in disc membranes of rod outer segments (ROS) in vertebrate retinal cells, is a unique G-protein that serves as the substrate of both cholera toxin- and IAP-catalyzed ADP-ribosylation. Table 1 lists most of the GTP-binding proteins thus far characterized including those recently purified in our laboratory as the substrate of IAP-catalyzed ADP-ribosylation. These G-proteins are characterized by their unique trimeric structure of $\alpha\beta\gamma$ (27-30). There are a site for GTP binding and another site to be ADP-ribosylated by IAP or cholera toxin in each of the α-subunits which differ from each other among these G-proteins. The molecular weights of α-subunits were 45,000 (and 52,000), 41,000, 39,000 and 39,000 for G_s, G_i, G_o and transducin, respectively, The βγ-subunits are not resolved further, unless the proteins are denatured by boiling in SDS. Thus, G-proteins behave as if they are dimers under physiologic conditions, as will be described later. The βγ-subunits are common among the G-proteins except transducin, γ of which appears to be somewhat different from the γ of other IAP substrates. The β-subunit is a mixture of two peptides with M_r=35,000 and 36,000, while the γ-subunit is composed of at least three different peptides with M_r=5,000-8,000.

Major G-Proteins so far Purified

Table 1.

	Gs	Gi	Go	Go´	G$_{HL}$	Td
α	45K* (52K)*	41K*	39K*	40K$^{(*)}$	40K	39K* (39K)*
ADD-rib.	CT		IAP (PT)			IAP+CT
β			35K+36K*			
γ			5-8K (3 species)			8K*
Source	Liver RBC	Liver	Brain		HL-60	ROS (Cone)

* cDNA cloned

As shown in Table 1, IAP-substrate G-proteins having α-subunits of M_r=40,000 have been purified from rat brain (tentatively referred to as Go' in Table 1 (32) and from HL-60 cells that had been differentiated to neutrophils by dimethylsulfoxide [referred to as G$_{HL}$ (33)]. They differ from G-proteins previously purified, since neither of them interact with any of the antibodies currently available for the G-proteins, i.e., those raised against purified α-subunits of G i and Go and purified βγ. It remains to be determined whether these new IAP substrates with apparently identical molecular weights are really identical with each other. Nor is any decisive information available for the physiological role of Go, Go' and G$_{HL}$, although the latter one is a candidate of the G-protein acting as transducer between the chemotactic peptide receptors and phospholipase C in neutrophils. This is because activation of the phospholipase by the chemotactic peptide in neutrophils was abolished by prior treatment of the cells with IAP (20-22, 34). A similar IAP-substrate G-protein has been partially purified from sea urchin eggs (35).

cDNAs encoding α-subunits of G-proteins that serve as the substrates of toxin-catalyzed ADP-ribosylation. cDNAs (or genomic DNAs in certain cases) have been cloned for subunits of IAP or cholera toxin-substrate G-proteins by us (36) as well as in several other laboratories. Two different α-subunits (Mr=52,000 and 45,000) of Gs were found to result from alternative splicing of the same genomic DNA. In contrast, four or more different cDNAs have been cloned for the α-subunits of IAP substrates; the α-subunits of Gi , Go , Go' and GHL are likely to be proteins encoded by these different DNAs. The site in the α-peptide of the G-protein that is selectively ADP-ribosylated by IAP was identified as the cysteine residue at the fourth position from the C-termini of the α-subunits of Gi , Go and transducin (36). The modification of this cysteine residue by IAP as well as by N-ethylmaleimide impairs the capability of the G-protein to act as a transducer, or prevents the G-protein from being coupled to receptors, as will be described later.

Table 2.

Comparison between Cholera (CT) and Pertussis (IAP) Toxins

	CT	IAP
(1) Substrate of (ADP−ribosylation)	Gs	Gi, Go, Go′, GHL ···
(2) Factor required for	ARF	$\beta\gamma$
(3) Uncoupling of G from receptor	(+)	+
(4) Stimulation of GTP↔GDP (turnon)	+ (slow)	−
(5) Inhibition of GTPase (turnoff)	+	−
(6) Effect without R stimulation	cAMP↑	−

Molecular mechanism by which toxin-substrate G-proteins act as transducer in signaling. Fig. 1 illustrates the manner in which the toxin-substrate G-proteins play their roles as transducer between receptors and effectors such as adenylate cyclase, phospholipase C and cation channels. It is well known that a receptor agonist binds to its own membrane receptor with a higher affinity in the absence of GTP than in its presence. The higher affinity of binding results from association of a GDP-bound trimeric G-protein (GDP- αβγ) to the receptor as shown in the upper-left part of Fig. 1. This is an inactive state of the G-protein in the sense that the protein never

417

interacts with the effector system under this condition. This state is important, however, since it facilitates the reception by the receptor of signals brought by agonists. The binding of the agonist to this receptor-G-protein complex immediately gives rise to an exchange of the previously bound GDp with intracellular GTP on the α-subunit of the G-protein. The GTP-bound α is dissociated from $\beta\gamma$, and both are dissociated from the receptor protein as shown in the lower-left part of Fig. 1. This state is an active one because the resolved GTP-bound α and/or $\beta\gamma$ are capable of interaction with the effector system. Owing to GTPase activity of α-subunits of G-proteins, however, GTP is then hydrolyzed to GDP on the α-subunit, thereby promptly recovering the original inactive state, which is ready for reception of the next signal molecule or receptor agonist. The GTP-GDP exchange and GTPase reactions are often referred to as turn on and turn off reactions, respectively.

It remains unknown how GTP-bound α or $\beta\gamma$ of G-proteins interacts with the effector system such as phospholipase C or ion channels. In the case of adenylate cyclase, the activation involves direct interaction of the GTP-bound α-subunit of G s with the adenylate cyclase catalyst. The inhibition of adenylate cyclase, however, results from combined effects of direct and indirect interactions between some α-subunits, $\beta\gamma$-subunits and the catalytic protein of adenylate cyclase as follows. Firstly, $\beta\gamma$ liberated from G$_i$, G$_o$ and G$_o'$ may form a trimeric complex with the α-subunit of Gs, thereby decreasing the concentration of free αs, the direct activator of the cyclase catalyst (31). The inhibition by this mechanism could be expected to occur in a number of mammalian cell types, since these IAP substrates are much more abundant than Gs in these cells (37). Secondly, the α-subunit of G$_i$ competes with the α-subunit of Gs for the activation site on the cyclase catalyst, though the affinity of G$_i$ was much lower than the affinity of Gs for this site (38). No competition was observed, however, between the α-subunit of G s and α-subunits of G$_o$, G$_o'$ and G$_{HL}$. Thirdly, $\beta\gamma$ is capable of direct interaction with the adenylate cyclase catalyst in such a manner as to lower the cyclase activity (38). The interaction was observed at rather higher concentrations of $\beta\gamma$. Fourthly, $\beta\gamma$ binds to calmodulin with a high affinity (39). Calmodulin is a potent activator of the adenylate cyclase catalyst as such, but is not so after it is bound by $\beta\gamma$ of G-proteins. Thus, the inhibition of adenylate cyclase by $\beta\gamma$ of G-proteins was biphasic in the presence of calmodulin; the inhibition by lower concentrations of $\beta\gamma$ was due to prevention of calmodulin activation of the cyclase and the inhibition by the higher concentrations reflected the direct interaction with the cyclase. The relative importance of these multiple mechanisms for adenylate cyclase inhibition will be the subject of future investigations.

The α-subunits of G-proteins are ADP-ribosylated by IAP only when the subunits are tightly bound to $\beta\gamma$, i.e., in its inactive form. The ADP-ribosylated G-proteins are no longer capable of being coupled to receptors (as illustrated in the upper-right part of Fig. 1), as revealed by the lowered affinity of an agonist binding to, or the failure of the agonist to induce the GTP-GDP exchange or GTPase reactions in cell membranes treated with IAP or phospholipid vesicles into which purified receptors and ADP-ribosylated G i were reconstituted (41, 42). It should be emphasized here that the effect of ADP ribosylation by IAP is extremely selective on coupling of G-proteins to receptors; that is, the activities of purified G-proteins observable without stimulation of receptors were not impaired at all by ADP-ribosylation by IAP of their α-subunits (see the lower-right part of Fig. 1). For instance, the addition of GTPγS, a non-hydrolyzable analogue of GTP, to purified GDP-bound G-protein trimers caused their resolution into GTPγS-bound α and βγ at the same rate and the same degree regardless of whether or not the G-proteins had been ADP-ribosylated by IAP. Further, the GTPase activity of a purified G-protein is not affected by IAP-induced ADP-ribosylation.

Role of G-Proteins as Transducer

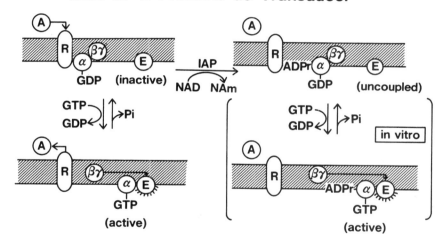

Fig. 1.

Comparison of pertussis toxin and cholera toxin as probes of G-proteins. Major differences between pertussis and cholera toxins are listed in Table 2. In contrast to IAP-induced ADP-ribosylation which uncouples the substrate G-proteins from receptors selectively (see above), the cholera toxin-catalyzed ADP-ribosylation of Gs produced multiple effects such as inhibition of GTPase or induction of spontaneous GTP-GDP exchange without receptor stimulation, although the rate of the exchange was slower than that observed upon receptor stimulation (30). ADP-ribosylated G_s was uncoupled from β–adrenergic receptors to some extent in a certain case (40) but was not in another case (13). Thus, a vast amount of cAMP accumulates spontaneously in cholera toxin-treated cells even without stimulation of any receptor in the cells, whereas there is no detectable change in the cAMP content in IAP-treated cells in the absence of receptor stimulation. This is the reason why IAP is much superior to cholera toxin as a probe for G-proteins. ADP-ribosylation by IAP substrates abolishes the involvement of these G-proteins in intracellular signaling very selectively, while ADP-ribosylation by cholera toxin of its substrate, G_s' function and tends to exert additional "nonspecific" effects on signaling owing to cAMP accumulation. Data are currently accumulating to indicate multiple involvements in cellular signaling of G-proteins serving as the specific substrates of IAP-catalyzed ADP-ribosylation.

Acknowledgements. The author thanks his colleagues for their excellent contributions to the work presented here; including Drs. F. Okajima, T. Katada, H. Itoh, T. Murayama, H. Kurose, M. Yajima, M. Tamura and T. Nakamura. A part of this work was supported by research grants from the Scientific Research Fund of the Ministry of Education, Science, and Culture, Japan and by the grant from Yamada Science Foundation.

References
1. Ui, M. (1984) Trend in Pharmacol Sci 5: 277-279
2. Katada, T., Ui, M.,(1977) Endocrinology 101: 1247-1255
3. Yajima, M., Hosoda, K., Kanbayashi, Y., Nakamura, T., Nogimori, K., Nakase, Y., Ui, M. (1978) J Biochem 83: 295-303
4. Yajima, J. Hosoda, K., Kanbayashi, Y., Nakamura, T., Takahashi, I., Ui, M. (1978) J Biochem 83: 305-312
5. Tamura, M., Nogimori, K., Murai, S., Yajima, M., Ito, K., Katada, T., Ui, M., Ishii, S. (1982) Biochemistry 21: 5516-5522
6. Katada, T., Tamura, M., Ui, M. (1983) Arch Biochem Biophys 224: 290-298
7. Katada, T., Ui, M. (1980) J Biol Chem 255: 9580-9588
8. Tamura, M., Nogimori, K., Yajima, M., Ase, K., Ui, M. (1983) J Biol Chem 258: 6756- 6761
9. Katada, T., Ui, M. (1982) Proc Natl Acad Sci USA 79: 3129-3133
10. Katada, T., Ui, M. (1982) J Biol Chem 257: 7210-7216
11. Murayama, T., Katada, T., Ui, M. (1983) Arch Biochem Biophys 221: 381-390
12. Murayama, T., Ui, M (1983) J Biol Chem 258: 3319-3326
13. Murayama, T., Ui, M. (1984) J Biol Chem 259: 761-769
14. Kurose, H., Katada, T., Amano, T., Ui, M. (1983) J Biol Chem 258: 4870-4875
15.Katada, T., Ui, M. (1981) J Biol Chem 256: 8310-8317
16. Katada, T., Ui, M. (1979) J Biol Chem 254: 469-479
17. Nakamura, T., Ui, M. (1983) Biol Chem Pharmacol 32: 3435-3441
18. Nakamura, T., Ui, M. (1984) FEBS Lett 173: 414-418
19. Nakamura, T., Ui, M. (1985) J Biol Chem 260: 3584-3593
20. Okajima, F., Ui, M. (1984) J Biol Chem 259: 13863-13871
21. Okajima, F., Katada, T., Ui, M. (1985) J Biol Chem 260: 6761-6768
22. Ohta, H., Okajima, F., Ui, M. (1985) J Biol Chem 260: 15771-15780
23. Murayama, T., Ui, M. (1985) J Biol Chem 260: 7226-7233
24. Murayama, T., Ui, M. (1987) J Biol Chem 262: in press
25. Murayama, T., Ui, M. (1987) J Biol Chem 262: in press
26. Ui, M. (1986) Phosphoinositides and Receptor Mechanisms, Vol. 7, Putney, J.W. (ed.), New York: Alan R. Liss, Inc, pp 163-195

27. Gilman, A.G. (1984) Cell 36: 577-579
28. Bokoch, G.M., Katada, T., Northup, J.K., Ui, M., Gilman, A.G. (1984) J Biol Chem 259: 3560-3567
29. Katada, T., Bokoch, G.M., Northup, J.K., Ui, M., Gilman, A.G., (1984) J Biol Chem 259: 3568-3577
30. Katada, T., Northup, J.K., Bokoch, G.M., Ui, M., Gilman, A.G. (1984) J Biol Chem 259: 3578-3585
31. Katada T., Bokoch, G.M., Smigel, M.D., Ui, M., Gilman, A.G. (1984) J Biol Chem 259: 3586-3595
32. Katada, T., Oinuma, M., Kusakabe, K., Ui, M. (1987) FEBS Lett. 213: 353-358
33. Oinuma, M., Katada, T., Ui, M. (1987) J Bio Chem 262: in press
34. Kikuchi, A., Kozawa, O., Kaibuchi, K., Katada, T., Ui, M., Takai, Y. (1986) J Biol Chem 261: 11558-11562
35. Oinuma, M., Katada, T., Yokosawa, H., Ui, M. (1986) FEBS Lett 207: 28-34
36. Itoh, H., Kozasa, T., Nagata, S., Nakamura, S., Katada, T., Ui, M., Iwai, S., Ohtsuka, E., Kawasaki, H., Suzuki, K., Kaziro, Y. (1986) Proc Natl Acad Sci USA 83: 3776-3780
37. Katada, T., Oinuma, M., Ui, M. (1986) J Biol Chem 261: 8182-8191
38. Katada, T., Oinuma, M., Ui, M. (1986) J Biol Chem 261: 5215-5221
39. Katada, T., Kusakabe, K., Oinuma, M., Ui, M. (1987) in press
40. Kurose, H., Ui, M. (1983) J Cyclic Nucleotide and Protein Phosphorylation Res 9: 305- 318
41. Haga, K., Haga, T., Ichiyama, A., Katada, T., Kurose, H., Ui, M. (1985) Nature 316: 731- 732
42. Kurose, H., Katada, T., Haga, T., Haga, K., Ichiyama, A., Ui, M. (1986): J Biol Chem 261: 6423-6428

Clostridium botulinum Exoenzyme C3 ADP-Ribosylates and Functionally Modifies p21.bot in PC12 Cells Without Affecting Neurosecretion

Eric J. Rubin, Patrice Boquet[1], Michel R. Popoff[2], and D. Michael Gill

Department of Molecular Biology and Microbiology, Tufts University School of Medicine, Boston, Massachusetts 02111 USA

Introduction

Many strains of *Clostridium botulinum* synthesize an exoenzyme (C3) which catalyzes the specific ADP-ribosylation of a 21 kilodalton substrate (1). All C3-producing strains also make either C1 or D neurotoxins. Botulinal neurotoxins block the release of neurotransmitters from nerve terminals and at the start of our work the C1 and D neurotoxins had been reported to be ADP-ribosyl transferases (2, 3). Hence we decided to determine if there was any relationship between C3-catalyzed ADP-ribosylation and neurosecretion. Here we show that when C3 is introduced into the cytosol of PC12 cells it induces a morphological differentiation but does not alter secretion of a neurotransmitter. By contrast the introduction of a neurotoxin into the cytosol of PC12 cells does block secretion but causes no shape change.

Results

Clostridium botulinum **synthesizes two ADP-ribosyl transferases.** The partially-purified culture supernatant of *C. botulinum* strain 17784 catalyzes the ADP-ribosylation of two proteins in crude PC12 cell lysates (Fig. 1, lane 1). The band at 45 kDa coincides with actin, the substrate for botulinal C2 toxin (4, 5). We call the smaller product, of apparent molecular mass 21 kDa, p21.bot. The ADP-ribose-p21.bot band is sometimes broad, perhaps representing some heterogeneity. It represents a protein as it is digested by proteinase K and trypsin, but not by RNase. Labeling of p21.bot was blocked when lysates were first incubated with unlabeled NAD before the [^{32}P]NAD, indicating that the ADP-ribosylation could be taken to completion (Fig. 1, lane 2).

p21.bot and actin are unrelated structurally and functionally. Only a part of the available actin had been silently ADP-ribosylated in Fig. 1, lane 2. Other differences between the labeling conditions for actin and p21.bot

---

[1]Unité des Antigènes Bactériens, Institute Pasteur, Paris Cedex 15, France
[2]Unité des Anaérobies, Institue Pasteur, Paris Cedex 15, France

soon became evident. Agents which interact with actin and reduce its effectiveness as a substrate had little effect on the labeling of p21.bot (Table 1). Such agents included myosin (6), DNase I (7), and phalloidin (5) a fungal metabolite that stabilizes F actin: actin is only a substrate for C2 in the globular, G form. While C2, the enzyme which ADP-ribosylates actin, is rapidly destroyed by boiling, C3, which catalyzes the modification of p21.bot is stable. Finally we and others (1, 8) purified C3 to homogeneity. Pure C3 does not modify actin.

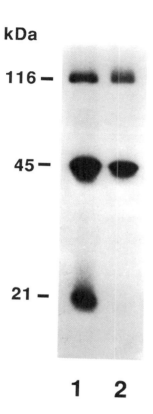

Fig. 1. ADP-ribosylation of PC12 cell lysates by *Clostridium botulinum* culture supernatant. Portions of lysates were [^{32}P]ADP-ribosylated without (lane 1) or with (lane 2) a preceding incubation with unlabeled NAD and toxin. Both samples contained 20 μM unlabeled NAD and 100 μg/ml supernatant protein. 5 μM [^{32}P]NAD was added to the mixture in sample 1 immediately and to sample 2 after first preincubating for 30 min at 37°C. Both samples were incubated with [^{32}P]NAD for 30 min at 37°C. Culture supernatant of *C. botulinum* strain 17784 was partially purified by precipitation with ammonium sulfate and acid (5). PC12 cells (12) were pelleted by centrifugation and were lysed by freezing and thawing. Labeling was performed using ca. 50 μg of cellular protein and 5μM [^{32}P]NAD (250,000 c.p.m.) Samples were incubated for 30 min at 37°C and were then precipitated with 10% cold trichloroacetic acid. Precipitates were dissolved, neutralized and analyzed by SDS-PAGE and autoradiography. For quantitation, bands were cut out of dried gels and counted. The uppermost radioactive band represents poly(ADP-ribose) polymerase which is labeled independently of the toxin.

It remained possible that p21.bot and actin might be structurally related. For example, p21.bot might represent a proteolytic fragment of actin. However, the proteolytic fragmentation patterns of the two labeled proteins are quite dissimilar (Fig. 2). Additionally actin is much more protease sensitive than p21.bot (Fig. 2). Impure enzyme preparations (concentrated culture supernatant) often generate an ADP-ribosylated 36 kDa protein. Partial proteolytic mapping suggests that this is a fragment of actin (Fig. 2).

423

Properties of p21.bot. We have used pure C3 to further characterize the p21.bot ADP-ribosylation free of interference from actin labeling. Upon complete digestion with snake venom phosphodiesterase, [^{32}P]ADP-ribosylated p21.bot releases [^{32}P]AMP but not [^{32}P]phosphoribosyl AMP, indicating that C3 catalyzes a mono-ADP-ribosylation.

Table 1. ADP-ribosylation in the presence of actin-binding compounds

Addition	Relative incorporation	
	p21^{bot}	Actin
None	100	100
Phalloidin, 1 μM	95	0
DNase I, 0.1 mg/ml	118	56
Myosin, 2.5 mg/ml	72	8

PC12 lysates were preincubated alone or with phalloidin, deoxyribonuclease I, or myosin for 10 min at 25°C before [^{32}P]ADP-ribosylation as described in Fig. 1. Values are expressed relative to the control that had no addition.

Since the ADP-ribosylation of p21.bot goes to completion, we can quantitate the substrate in cell lysates. It constitutes about 0.01% of the cellular protein. PC12 cells have both particulate and soluble forms: the cells were broken by freeze-thaw and fractionated by differential centrifugation. About 90% of the p21.bot was in the cytosol (100,000 x g supernatant), while the other 10% was largely in the membrane fraction (100,000 x g pellet).

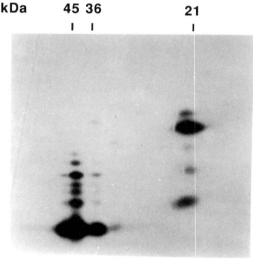

Fig. 2. Partial proteolysis of *C. botulinum* enzyme substrates with Staphylococcus V8 protease. A portion of PC12 lysate was labeled as described in Fig. 1 and fractionated on a 10% SDS-PAGE tube gel. The tube was placed on top of a 15% SDS-PAGE slab gel and overlaid with 100 μg/ml Staphylococcus V8 protease (14) before fractionation in the second dimension.

Introduction of C3 into the cytosol. The purified form of C3 does not efficiently enter cells. If it is part of a toxin it may lack a B chain required for binding and entering cells. To determine the physiological effects of p21.bot modification we had, therefore, to develop a method of introducing C3 into the cytosol of cells in bulk. We screened possible procedures with the use of fragment A, the enzymically active portion of diphtheria toxin which was kindly provided by John Collier. Fragment A cannot enter cells and so is not toxic on its own but a single cytosolic molecule is lethal (9).

Electroporation (a high voltage electric shock which causes a transient increase in permeability) is a technique used to transfect cells with DNA (10). We shocked PC12 cells in the presence of different concentrations of fragment A and measured protein synthesis 24 hr later. Under optimal conditions (of shock voltage, cell concentration and buffer composition), protein synthesis was reduced by only 66% at the highest fragment A concentration tested (Fig. 3) while from 30-50% of cells were killed by the procedure even in the absence of Fragment A. There was also great day to day and sample to sample variation.

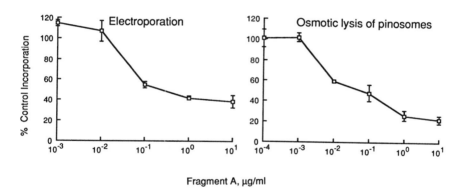

Fragment A, µg/ml

Fig. 3. Introduction of diphtheria toxin fragment A into PC12 cells by electroporation and osmotic lysis. For electroporation (10), approximately 10^6 cells were suspended in phosphate-buffered saline and were plated a 24-well plate. For osmotic lysis, cells plated in polylysine-coated plates, were incubated for 30 min in uptake medium (Dulbecco's modified Eagle's medium (DME), 0.5 M sucrose, 10% polyethylene glycol, 0.1 mg/ml ovalbumin) containing fragment A as shown, washed briefly in lysis medium (60% DME), and incubated with a fresh portion of lysis medium before being returned to growth medium. Twenty four hrs later, the rate of protein synthesis was measured by incubating the cells with [^{14}C]leucine (1 µCi/ml) in leucine-free medium for 1 hr, precipitating in TCA, filtering and counting. Values are the means of triplicate determinations ± SEM.

The technique of osmotic lysis of pinosomes (11) was more successful. Cells were allowed to endocytose in hypertonic medium containing fragment A and were then exposed to a hypotonic medium that swelled endosomes and, in some cases, released their contents into the cytosol. Twenty four hr later, [^{14}C]leucine incorporation had been reduced by about 80% (Fig. 3).

425

This method seemed fairly benign; few cells were killed without fragment A. As this method also proved to be reproducible, we used it for our further studies.

Effects of intracellular C3.   To confirm that active C3 was introduced into cells by the method, and to quantify the effect, cells that had been osmotically shocked after endocytosing C3 were later lysed in the presence of C3 and [^{32}P]NAD. This measured the amount of p21.bot remaining to be labeled and, hence, the amount modified *in situ*.   Six hrs after the introduction of C3 the available p21.bot was considerably decreased (Table 2).   As expected, the available actin was unaffected.

Table 2. Intracellular C3 reduces p21.bot available for reaction *in vitro*.

C3, ng/ml	p21.bot	Actin
	[^{32}P]ADP-ribose incorporated, fmol	
0	133	188
60	32	188
200	23	195

Pure C3 was introduced into PC12 cells by osmotic shock. After 6 hr, cells were removed, centrifuged and lysed. The lysates were incubated with [^{32}P]NAD and with a mixture of C2 and C3 as in Fig. 1, plus 0.5% NP-40. Values represent means of duplicate samples. A reduced incorporation *in vitro* imples that p21.bot had been modified intracellularly.

Purified C3 was not cytotoxic when introduced into PC12 cells. Cells treated with C3 remained viable and continued to divide for at least three days. However, C3 did induce morphological changes. As shown in Fig. 4, cells produced short processes reminiscent of the earliest stages of morphological differentiation induced by, for example, nerve growth factor (12). Visible changes began within 4 hrs. Two days later, the cells had reverted to their normal morphology.

C3 does not affect secretion.   The Clostridial neurotoxins, botulinum and tetanus toxin, both inhibit secretion. When we introduced tetanus toxin into PC12 cells, secretion of norepinephrine was inhibited. There was reason to believe that C3 might be a fragment of C1 or D neurotoxins (see Introduction) or act in a similar way. Accordingly we asked if intracellular C3 inhibited secretion but we found that it did not. Cells which had taken up C3 and in which p21.bot had been substantially ADP-ribosylated still secreted norepinephrine normally (Fig. 5).

We later showed that C3 and neurotoxins are distinct entities. Purified botulinum neurotoxins and tetanus toxin have no ADP-ribosyltransferase activity. Furthermore, C3 and C1 are antigenically unrelated (1). It seems likely that the previously observed enzyme activity of the neurotoxins (2,3) had been due to contaminating C3 and that ADP-ribosylation of p21.bot does not cause neurotoxicity.

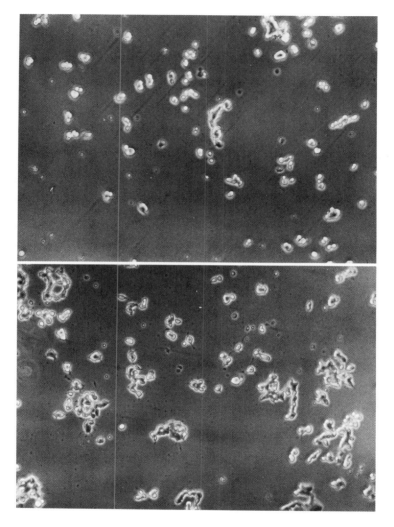

Fig. 4. Partial neuronal morphology induced by C3. Purified C3 (4 μg/ml) was introduced into cells by osmotic lysis as described in Fig. 3. Phase-contrast micrographs (x160) were taken after 24 hr. Upper panel: shock control. Lower panel: shocked in the presence of C3.

427

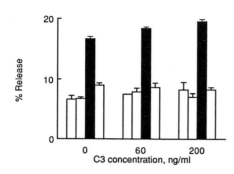

Fig. 5. C3 does not block secretion of norepinephrine. Purified C3 was introduced into cells by osmotic lysis as in Fig. 3. Secretion was assayed after 6 hr as described by Figliomeni and Grasso (15). The cells were allowed to take up [³H]norepinephrine (1 µCi/ml) for one hr, then washed in buffer containing 5 mM K⁺. The chart represents the [³H]norepinephrine released during successive five min incubation in buffer containing 5 mM K⁺, 5 mM K⁺, 80 mM K⁺ (solid bars), and 5 mM K⁺. The residue was also solubilized with SDS and counted. The released counts are expressed as % of the radioactivity present at the particular stage. Means of three ± SEM.

Discussion

C3 is an ADP-ribosyltransferase produced by *C. botulinum* which is distinct from C2 toxin. Its substrate, p21.bot, is structurally and functionally unrelated to actin. To study its physiological effect we have introduced C3 into cell cytosol by osmotic lysis of pinosomes and have shown that this technique is effective. Intracellular C3 does not alter the secretory ability of cells, and thus is distinct in mechanism from the botulinal neurotoxins. On the other hand, C3 induces shape changes in PC12 cells which may represent the beginning of a morphological differentiation. Previously we reported that NIH 3T3 cells also change shape upon introduction of C3 into their cytosol (1). NIH 3T3 cells start with an extended morphology and retract with C3, leaving dendritic processes behind. The end result is similar for PC12 cells but since these start out round they must actively form the processes. The injection of *ras* proteins causes PC12 cells to acquire neurone-like morphologies (13), and 3T3 cells to retract with dendritic processes. Thus in both cases the ADP-ribosylation of p21.bot has an effect similar to that of *ras* oncogene protein.

References

1. Rubin, E.J., Gill, D.M., Boquet, P., Popoff, M.R. (1988) Mol Cell Biol 8: 418-426
2. Ohashi, Y., Narumiya, S. (1987) J Biol Chem 262: 1430-1433
3. Ohashi, Y., Kamiya, T., Fujiwara, M., Narumiya, S. (1987) Biochem Biophys Res Comm 142: 1032-1038
4. Aktories, K., Barmann, M., Ohishi, I., Tsuyama, S., Jakobs, K.H., Habermann, E. (1986) Nature 322: 390-392
5. Ohishi, I., Tsuyama, S. (1986) Biochem Biophys Res Comm 136: 802-806
6. Weeds, A. (1982) Nature 296: 811-816
7. Lazarides, E., Linberg, U. (1974) Proc Natl Acad Sci 71: 4742-4486
8. Aktories, K., Weller, U., Chatwal, G.S. (1987) FEBS Letts 212: 109-113

9. Yamaizumi, M., Mekada, E., Uchida, T., Okada, Y. (1978) Cell 15: 245-250
10. Potter, H., Weir, L., Leder, P. (1984) Proc Natl Acad Sci 81: 7161-7165
11. Okada, C.Y., Rechsteiner, M. (1982) Cell 29: 33-41
12. Greene, L.A., Tischler, A.S. (1976) Proc Natl Acad Sci 73: 2424-2428
13. Bar-Sagi, D., Feramisco, J.R. (1985) Cell 42: 841-848
14. Cleveland, D.W., Fisher, S.D., Kirshner, M.W., Laemlli, U.K. (1977) J Biol Chem 252: 1102-1106
15. Figliomeni, B., Grasso, A. (1985) Biochem Biophys Res Comm 128: 249-256

ADP-Ribosylation of Human c-Ha-*ras* Protein by Hen Liver ADP-Ribosyl Transferase

Hisae Kawamitsu, Masanao Miwa, Yoshinori Tanigawa[1],
Makoto Shimoyama[1], Shigeru Noguchi[2], Susumu Nishimura[2],
and Takashi Sugimura

Virology Division, National Cancer Center Research Institute, Tsukiji, Chuo-ku,
Tokyo 104, Japan

Introduction

A portion of the amino acid sequence of human c-Ha-*ras* proto-oncogene product p21 is highly homologous with the corresponding region of a family of guanine binding membrane proteins, G-proteins, that are involved in signal transduction (1). Therefore, an analogous function is suggested for *ras* proteins and G-proteins. Although G-proteins are ADP-ribosylated by cholera toxin (2) or pertussis toxin (3), there has been no data of ADP-ribosylation of c-Ha-*ras* product by bacterial toxins. On the other hand, both Tsai *et al.* (4) and our group (5) found ADP-ribosylation of *Escherichia coli* synthesized c-Ha-*ras* protein by eukaryotic ADP-ribosyl transferases. Here we identify the amino acid residue which is ADP-ribosylated by a purified hen liver enzyme.

Results and Discussion

ADP-Ribosylation of human c-Ha-*ras* protein by hen liver ADP-ribosyl transferase. Human c-Ha-*ras* protein, normal or mutated from glycine to valine at the 12th position, was produced in *Escherichia coli* and purified (6). These proteins were incubated with ADP-ribosyl transferase purified from hen liver nuclei (7) in the presence of [adenylate-^{32}P]NAD. Incorporation of the radioactivity derived from [^{32}P]NAD was clearly shown at the positions corresponding to the protein bands of normal and mutated c-Ha-*ras* proteins (Fig. 1A and 1B, lanes 3 and 4). However no significant radioactive band was formed in the absence of c-Ha-*ras* proteins (Fig. 1B, lane 5). The incorporation of the radioactivity into c-Ha-*ras* protein was inhibited by more than 50% in the presence of 40 mM arginine methylester or 40 mM nicotinamide.

---

[1]Department of Biochemistry, Shimane Medical University, Enya, Izumo 693, Japan
[2]Biology Division, National Cancer Center Research Institute, Tsukiji, Chuo-ku, Tokyo 104, Japan

Isolation of ADP-ribosylated peptide. After normal human c-Ha-*ras* protein was ADP-ribosylated by incubation with [^{32}P]NAD and purified hen liver enzyme, it was separated from unreacted [^{32}P]NAD and the hen liver enzyme with high performance liquid chromatography (HPLC) using a TSK G 2000 SW (Toyo Soda) column. Then the ADP-ribosylated c-Ha-*ras* protein was reduced with 2-mercaptoethanol and alkylated with 4-vinylpyridine and then separated by HPLC on a reverse phase column (µBONDAPAK™C18, Waters) in a gradient of 0-60% acetonitrile in 0.1% trifluoroacetic acid. Then the ADP-ribosylated c-Ha-*ras* protein was treated with cyanogen bromide in the presence of formic acid. The radioactive peptide peak was separated by HPLC with a µBONDAPACK™C18 column. Then the peptide with radioactivity was digested with trypsin. The tryptic peptides were fractionated by HPLC with a µBONDAPACK™C18 column with a gradient of 0-40% acetonitrile in 0.1% trifluoroacetic acid. The single radioactive peptide peak was analyzed for amino acid sequence using an Applied Biosystems gas-phase sequencer, Model 470A.

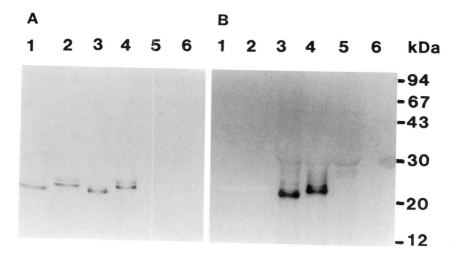

Fig. 1. Enzymatic ADP-ribosylation of c-Ha-*ras* oncogene products, p21. Electrophoretogram of purified p21 incubated with [^{32}P]NAD in the presence or absence of the hen liver enzyme. A, protein staining with Coomassie brilliant blue. B, autoradiogram. Lanes: 1, Normal c-Ha-*ras* p21 without enzyme; 2, mutated p21 without enzyme; 3, normal p21 with enzyme; 4, mutated p21 with enzyme; 5, enzyme without p21; 6, [^{32}P]NAD with neither p21 nor enzyme.

Amino acid sequence of ADP-ribosylated peptide. The radioactive tryptic peptide peak yielded a dodecapeptide with a sequence of Thr-Val-Glu-Ser-X-Gln-Ala-Gln-Asp-Leu-Ala-Arg, which corresponded to the Thr(124)-Arg(135) as deduced from the human c-Ha-*ras* protein as shown

431

in Fig. 2. Therefore, the presence of radioactivity in this peptide and the apparent absence of the unmodified Arg in the 128th position strongly indicate that Arg(128) is the site of ADP-ribosylation. The resistance to tryptic digestion at Arg(128)-Gln(129) bond might be due to ADP-ribosylation to an Arg(128) residue. The Arg residue at 128 position was suggested to be close to the GTP binding site (8) but our preliminary data suggest that there is no significant change of GTP binding after ADP-ribosylation. Further studies on ADP-ribosylation of c-Ha-*ras* protein will help clarify the function of ADP-ribosylation on signal transduction.

Met Thr Glu Tyr Lys Leu Val Val Val Gly Ala Gly Gly Val Gly   15

Lys Ser Ala Leu Thr Ile Gln Leu Ile Gln Asn His Phe Val Asp   30

Glu Tyr Asp Pro Thr Ile Glu Asp Ser Tyr Arg Lys Gln Val Val   45

Ile Asp Gly Glu Thr Cys Leu Leu Asp Ile Leu Asp Thr Ala Gly   60

Gln Glu Glu Tyr Ser Ala Met Arg Asp Gln Tyr Met Arg Thr Gly   75

Glu Gly Phe Leu Cys Val Phe Ala Ile Asn Asn Thr Lys Ser Phe   90

Glu Asp Ile His Gln Tyr Arg Glu Gln Ile Lys Arg Val Lys Asp   105

Ser Asp Asp Val Pro Met Val Leu Val Gly Asn Lys Cys Asp Leu 120

Ala Ala Arg <u>Thr Val Glu Ser Arg Gln Ala Gln Asp Leu Ala Arg</u> 135

Ser Tyr Gly Ile Pro Tyr Ile Glu Thr Ser Ala Lys Thr Arg Gln 150

Gly Val Glu Asp Ala Phe Tyr Thr Leu Val Arg Glu Ile Arg Gln 165

His Lys Leu Arg Lys Leu Asn Pro Pro Asp Glu Ser Gly Pro Gly 180

Cys Met Ser Cys Lys Cys Val Leu Ser 189

Fig. 2. Site of ADP-ribosylated arginine residue in normal human c-Ha-*ras* protein. The underlined sequence corresponds to the isolated ADP-ribosylated tryptic peptide.

References

1. Hurley, J.B., Simon, M.I., Teplow, D.B., Robishaw, J.D., Gilman, A.G. (1984) Science 226: 860-862
2. Dop, C.V., Tsubokawa, M., Bourne, H.R., Ramachandran, J. (1983) J Biol Chem 259: 696-698
3. West, R.E., Moss, J., Vaughan, M., Liu, T., Liu, T-Y. (1985) J Biol Chem 260: 14428-14430
4. Tsai, S-C., Adamik, R., Moss, J., Vaughan, M., Mann, V., Kung, H. (1985) Proc Natl Acad Sci USA 82: 8310-8314
5. Kawamitsu, H., Miwa, M., Tanigawa, Y., Shimoyama, M., Noguchi, S., Nishimura, S., Ohtsuka, E., Sugimura, T. (1986) Proc Japan Acad 62: Ser B, 102-104
6. Miura, K., Inoue, Y., Nakamori, H., Iwai, S., Ohtsuka, E., Ikehara, M., Noguchi, S., Nishimura, S. (1986) Jpn J Cancer Res (Gann) 77: 45-51
7. Tanigawa, Y., Tsuchiya, M., Imai, Y., Shimoyama, M., (1984) J Biol Chem 259: 2022-2029
8. McCormick, F., Clark, B.F., LaCour, T.F.M., Kjeldgaard, M., Norskov-Lauritsen, L., Nyborg, J. (1985) Science 230: 78-82

Mono(ADP-Ribosyl)ation in Rat Liver Mitochondria

Balz Frei[1] and Christoph Richter

Laboratorium für Biochemie, Eidgenössische Technische Hochschule, ETH-Zentrum, CH-8092 Zürich, Switzerland

Introduction

In rat liver most of the cellular mono(ADP-ribosylated) proteins are associated with the mitochondrial fraction (1). Two mono(ADP-ribosyl)ating systems have been described in mitochondria, one in the soluble (matrix) fraction (2, 3), the other in submitochondrial particles (SMP, inverted inner membrane vesicles) (3, 4). The ADP-ribosylated matrix protein has a molecular mass of 100 kDa and appears to consist of two major subunits of equal mass. In SMP of both rat liver (4) and beef heart (3), there is one major acceptor protein for mono(ADP-ribose), which migrates with an apparent molecular mass of 30 kDa in sodium dodecyl sulfate-polyacrylamide gel electrophoresis (SDS-PAGE). Mono(ADP-ribosylation) of the acceptor protein of beef heart SMP was suggested to occur non-enzymically (3). In rat liver SMP, ADP-ribosylation of the 30 kDa protein most probably occurs at an arginine residue, and is readily reversible in the presence of ATP (4). The characteristics of this ADP-ribosylation reaction, i.e. protein specificity and sensitivity to ATP, together with the observation that intramitochondrial hydrolysis of $NAD(P)^+$ is accompanied by release of Ca^{2+} from mitochondria suggests a functional link between mitochondrial protein ADP-ribosylation and Ca^{2+} release (5, 6).

Using a newly developed fluorescent technique (7, 8) we have investigated protein ADP-ribosylation in intact mitochondria. Our findings indicate the existence of at least three classes of ADP-ribosylated proteins in rat liver mitochondria and give evidence for a transient increase of protein-bound mono(ADP-ribose) during the $NAD(P)^+$- linked Ca^{2+} release.

Results

Incubation of SMP with [adenine-2,8-^{3}H]NAD^+ yields $179 + 47$ (n=20) pmol of ADP-ribose/mg of protein, as judged from acid-precipitable radioactivity. To get information about the stability of the ADP-ribose adduct, the modified, acid-precipitated SMP were immediately washed and dried. The resulting powder was completely dissolved in a buffer containing 6 M guanidine. After separation from residual noncovalently bound radioactivity (7) the proteins were subjected to various treatments. The

[1]Present address: Department of Biochemistry, University of California, Berkeley, CA

ADP-ribose-protein conjugate proved stable for at least 12 hr at pH 4.0 and 4°C, as well as in the presence of 1 M ammonium chloride at neutral pH and 37°C (the latter condition henceforth will be referred to as "in the absence of hydroxylamine"). The linkage is moderately labile to 1 M and 3 M neutral hydroxylamine at 37°C, with a half-life in 3 M hydroxylamine of 6.0 hr. In 4 M hydroxylamine the half-life of the linkage is 4.0 hr. At basic pH, radioactivity is released very fast from the protein, with a half-life at pH 9.3 of 50 min, and of 20 min in 1 M sodium hydroxide. This is in agreement with and extends our previous findings (4). The release of protein-bound radioactivity follows single first-order kinetics under all conditions. This indicates that only one class of mono(ADP-ribosylated) proteins has been formed by incubation of SMP with NAD^+.

Several lines of evidence suggest that ADP-ribosylation in rat liver SMP can be due to nonenzymatic covalent protein modification with free ADP-ribose previously formed by enzymatic hydrolysis of NAD^+. A similar reaction sequence has been proposed (3) for the mono(ADP-ribosylation) of a 30 kDa acceptor protein in beef heart SMP. The covalent modification with free ADP-ribose has the same specificity with respect to the acceptor protein as with NAD^+. It is not a Schiff base adduct. Although possibly nonenzymatic, the specific mono(ADP-ribosylation) of the 30 kDa protein may fullfill a physiological function in intact mitochondria.

Table 1. Protein-bound mono(ADP-ribose) in rat liver mitochondria[a]

Treatment	ADP-ribose released (pmol/mg of protein)		
1 M NH_4Cl, pH 7.0, 37°C, 12 hr	6.0	±	2.1
1 M NH_2OH, pH 7.0, 37°C, 12 hr	23.8	±	5.7
3 M NH_2OH, pH 7.0, 37°C, 12 hr	54.3	±	6.9
1 M NaOH, 37°C, 2 hr	205.6	±	17.7

[a]The acid insoluble material from rat liver mitochondria was dissolved and subjected to G-25 (superfine) column centrifugation. The samples (containing 0.8-2.0 mg of protein) were incubated under the conditions given in the Table and assessed for protein-bound mono(ADP-ribose) fluorimetrically as described by Jacobson *et al.* and Payne *et al.* Each value represents the average ± standard deviation of three mitochondrial preparations. For each mitochondrial preparation and incubation condition at least three samples were examined.

When freshly isolated rat liver miotochondria were assessed for endogenous protein-bound mono(ADP-ribose) using the method of Jacobson and co-workers (7, 8), 23.8 and 6.0 pmol of ADP-ribose/mg of protein were detected following incubation of the mitochondrial proteins for 12 hr in the presence and absence, respectively, of 1 M hydroxylamine (Table 1). Incubation with 3 M neutral hydroxylamine during 12 hr liberated 54.3 pmol of ADP-ribose/mg of mitochondrial protein. Treatment of the mitochondrial proteins with 1 M sodium hydroxide for 2 hr released approximately 200 pmol of ADP-ribose equivalents/mg of protein (Table 1). The portion of covalently bound ADP-ribose residues released in the absence of hydroxylamine most probably is linked to the mitochondrial acceptor protein(s) via a carboxylate ester linkage similar to that between ADP-ribose and histones formed in chromatin of rat liver by poly(ADP-ribose)synthetase (9, 10). Since this linkage is extremely labile both in the absence and the presence of hydroxylamine (9, 10), the portion of ADP-ribose residues released from the mitochondrial proteins in the absence of hydroxylamine are presumably also released in its presence. Therefore, the "hydroxylamine-requiring" ADP-ribose residues in rat liver mitochondria amount to about (23.8 minus 6.0) = 17.8 pmol/mg of protein when 1 M neutral hydroxylamine is used for chemical release, compared to about (54.3 minus 6.0) = 48.3 pmol/mg of protein when 3 M hydroxylmine is used.

Oxidation of pyridine nucleotides by a variety of compounds, e.g. t-butylhydroperoxide, menadione, or alloxan, and subsequent hydrolysis of oxidized pyridine nucleotides leads to release of Ca^{2+} from rat liver mitochondria (11-13). Since NAD^+ hydrolysis is accompanied by protein ADP-ribosylation in the isolated inner mitochondrial membrane, we proposed regulation of the physiological Ca^{2+} release pathway by mono(ADP-ribosylation) (5, 6). To examine this hypothesis, we challenged Ca^{2+}-loaded mitochondria with t-butylhydroperoxide and assessed the level of mitochondrial protein-bound mono(ADP-ribose) before and during Ca^{2+} release. For the release of the ADP-ribose residues from mitochondrial proteins, 1 M sodium hydroxide at 37°C for 2 hr was used. A moderate increase (about 30 pmol/mg of protein) of protein-bound mono(ADP-ribose) was observed during the t-butylhydroperoxide-induced release of Ca^{2+}. The level of protein-bound ADP-ribose remains enhanced after the addition of t-butylhydroperoxide until Ca^{2+} release is completed. It reaches two peak values which coincide with the two phases of significant release of Ca^{2+}. Omission of t-butylhydroperoxide results in retention of Ca^{2+} and in an unchanged level of protein-bound ADP-ribose. These findings are fully consistent with and strongly support the suggested participation of ADP-ribosylation in the $NAD(P)^+$-stimulated release of Ca^{2+} from rat liver mitochondria.

Acknowledgements. This work was supported by Grant 3.503-0.83 from the Schweizerischer Nationalfonds. B. F. was also supported by an EMBO short term fellowship.

References

1. Adamietz, P., Wielckens, K., Bredehorst, R., Lengyel, H., Hilz, H. (1981) Biochem Biophys Res Commun 101: 96-103
2. Kun, E., Zimber, P.H., Chang, A.C.Y., Puschendorf, B., Grunicke, H. (1975) Proc Natl Acad Sci USA 72: 1436-1440
3. Hilz, H., Koch, R., Fanick, W., Klapproth, K., Adamietz, P. (1984) Proc Natl Acad Sci USA 81: 3929-3933

4. Richter, C., Winterhalter, K.H., Baumhüter, S., Lötsher, H.R., Moser, B. (1983) Proc Natl Acad Sci USA 80: 3188-3192
5. Hofstetter, W., Mühlebach, T., Lötscher, H.R., Winterhalter, K.H., Richter, C. (1981) Eur J Biochem 117: 361-367
6. Richter, C., Schlegel, J., Frei, B. (1985) Adp-ribosylation of Protein, F.R. Althaus, H. Hilz, S. Shall, (eds.), Springer Verlag, Berlin, Heidelberg, pp. 530-535
7. Jacobson, M.K., Payne, D.M., Alvarez-Gonzalez, R., Juarez-Salinas, H., Sims, J.L., Jacobson, E.L. (1984) Methods Enzymol 106: 483-494
8. Payne, D.M., Jacobson, E.L., Moss, J., Jacobson, M.K. (1985) Biochemistry 24: 7540-7549
9. Ogata, N., Ueda, K., Hayaishi, O. (1980) J Biol Chem 255: 7610-7615
10. Ogata, N., Ueda, K., Kagamiyama, H., Hayaishi, O. (1980) J Biol Chem 255: 7616-7620
11. Frei, B., Winterhalter, K.H., Richter, C. (1985) J Biol Chem 260: 7394-7401
12. Frei, B., Winterhalter, K.H., Richter, C. (1985) Eur J Biochem 149: 633-639
13. Frei, B., Winterhalter, K.H., Richter, C. (1986) Biochemistry 25: 4438-4443

Abbreviations:

kDa	kilodaltons
PAGE	polyacrylamide gel electrophoresis
SDS	sodium dodecyl sulfate
SMP	submitochondrial particles

Botulinum Neurotoxins ADP-Ribosylate A Novel GTP-Binding Protein (Gb)

Shuh Narumiya, Yasuhiro Ohashi[1], and Motohatsu Fujiwara

Department of Pharmacology, Faculty of Medicine, Kyoto University, Sakyo-ku, Kyoto 606, Japan

Introduction

Botulinum neurotoxin is an exotoxin produced by *Clostridium botulinum* which, when ingested, causes an acute and fatal poisoning characterized by progressive descending muscle paralysis (see review 1 and 2). There are several types of botulinum neurotoxins termed A, B, C1, D, E, F and G, which are distinct antigenically but similar in molecular size and structure. Pharmacologically all types of the toxin act similarly on various nerve endings, especially those of cholinergic nerves, and block the release of neurotransmitters. Though the exact mechanism of this action has not been clarified yet, Knight *et al.* (3), using permeabilized adrenal chromaffin cells in culture, have shown that the toxin acts on some target(s) downstream from calcium entry to the cells. Since virtually nothing is known about molecular events in the exocytotic process after the rise in free calcium ion concentration in cells, elucidation of a botulinum toxin target(s) would contribute to clarifying the mechanism of this basic cell function. A hypothesis that the botulinum neurotoxin is an enzyme has been proposed on the basis of its extreme potency and long duration of action (1,4). Recently we have found ADP-ribosyl transferase activity in types C1 and D botulinum neurotoxin (5,6). Here we review these findings and present more detailed analyses on the target protein(s) and catalytic nature of this reaction. Our results are discussed in terms of the present knowledge on the mechanism of the exocytotic process.

Results

Type C1 and D botulinum neurotoxins as ADP-ribosyl transferases. As shown by Knight *et al.* (3), type D botulinum neurotoxin was able to inhibit exocytosis in cultured chromaffin cells. Fig. 1 represents the results of our experiments showing the time course of this inhibition. When cultured bovine adrenal chromaffin cells were incubated with type D botulinum neurotoxin, inhibition of acetylcholine-evoked catecholamine release appeared. This inhibition, however, did not occur instantaneously but appeared and increased with days of incubation, suggesting involvement

---

[1]On leave from the Department of Biochemistry, Nara Medical University

of a slow catalytic process in this intoxication. In order to investigate the underlying mechanism of this inhibition, we incubated crude membranes of bovine adrenal gland with type D botulinum neurotoxin in the presence of [^{32}P]NAD. A M$_r$ 21,000 membrane protein was specifically ADP-ribosylated under these conditions (Fig. 2). The amount of ADP-ribosylation increased with dose of the toxin used in the assay and was abolished by prior boiling. The reaction was also dependent on concentrations of NAD and membranes in the mixture.

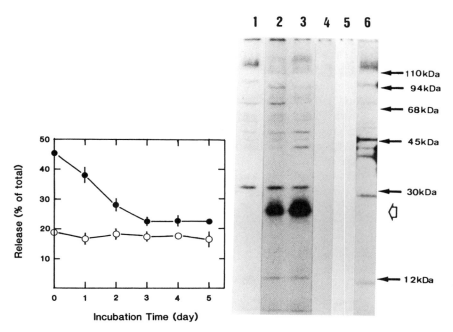

Fig. 1. (left) Inhibition of acetylcholine-induced catecholamine release by botulinum toxin in cultured adrenal chromaffin cells. Bovine adrenal chromaffin cells were isolated and cultured at a density of 5 x 10^5/ml. After two days of culture, the cells were washed and suspended in a fresh medium containing 20 μg/ml of type D botulinum neurotoxin. After incubation for indicated days, the cells were washed and stimulated 500 with μM acetylcholine. Catecholamines in the media and cells were extracted separately and quantified electrochemically. Catecholamine release is expressed as % of the total amount. ●, acetylcholine-evoked release; ○, basal release.

Fig. 2. (right) ADP-ribosylation of a specific protein in bovine adrenal membranes by type D botulinum toxin. Membranes of bovine adrenal gland (100 μg) were incubated with 20 μg of type D botulinum neurotoxin and [^{32}P]NAD at 30°C for 45 min, acid-precipitated, and subjected to SDS-polyacrylamide gel electrophoresis. Lane 1, control without the toxin; lane 2, with the toxin; lane 3, with the toxin plus 10 mM dithiothreitol; lane 4, with the toxin plus 25 mM agmatine; lane 5, with the toxin plus 25 mM L-arginine methyl ester; lane 6, with 20 μg of cholera toxin. Reproduced from ref 5.

438

When the concentration of NAD was changed from 10 to 160 μM or membrane concentration from 100 μg to 2 mg/200 μl, a Michaelis-Menten type of saturation kinetics was observed in both experiments (Fig. 3). The K_m for NAD and membranes were about 18 μM and 1 mg/ml, respectively. These results clearly demonstrated that the type D botulinum toxin worked as an enzyme and ADP-ribosylated a specific protein in the membranes. We have observed similar ADP-ribosyl transferase activity in type C1 botulinum neurotoxin and found that this type of the toxin ADP-ribosylated the same M_r 21,000 protein as type D toxin (6).

Since the preparation of the toxin used was a progenitor toxin (M-toxin) which was composed of 7S toxic and 7S nontoxic components (2), we examined whether this enzyme activity was associated with the toxic component itself. We separated the toxic and nontoxic components by chromatography on DEAE-Sephadex, and examined the enzyme activity in the two fractions. As reported previously (5), the enzyme activity was solely found in the toxic component and not in the nontoxic component. Since the 7S toxic component of the botulinum neurotoxin was composed of two polypeptide chains bound by an intramolecular disulfide bond (Fig. 4), we next examined whether the activity was associated with the light or the heavy chain. The results are shown in Fig. 5. As shown in the figure, the light chain exhibited activity about three fold higher specific activity than the original 7S toxin, whereas no enzyme activity was found in the heavy chain. Since the molecular weight of the light chain was almost three times smaller than that of the native toxin, these results suggest that the light chain contained all the enzyme activity found in the native toxin.

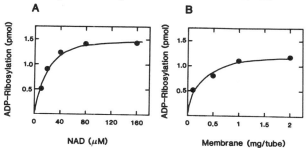

Fig. 3. Saturation of ADP-ribosyl transferase activity with NAD and membrane. Adrenal membrane (100 μg) was incubated with various concentrations of NAD (A) or various amounts of the membrane was incubated with 10 μm NAD (B) in a total volume of 200 μl at 30°C for 30 min. After the reaction, [^{32}P]labelled products were analyzed and quantified as described (6).

Characterization of botulinum toxin substrate. After we published two initial reports on ADP-ribosylation by botulinum toxin (5, 6), we have become aware that the substrate protein for botulinum toxin was present not only in the membrane fraction but also in the cytosol. Table 1 illustrates the subcellular distribution of the substrate protein in the homogenate of bovine

439

adrenal gland. This protein was recovered mainly in the 10,000 x g pellet and 105,000 x g supernatant; these two fractions comprised about 30 and 35% of the total amount, respectively. The ADP-ribosylated proteins in the cytosol fraction were not distinguishable from those in the membrane fraction on SDS-polyacrylamide gel electrophoresis.

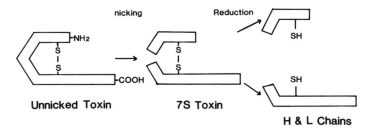

Fig. 4. Organization and fragmentation of botulinum neurotoxin. Botulinum neurotoxin is a protein with a molecular weight of 150,000 to 160,000. It is synthesized as a single chain polypeptide in the *Clostridium* bacteria. It is nicked by endogenous protease to yield a dichain molecule linked by a disulfide bond. This nicking is usually associated with activation of its' toxicity. The nicked molecule is separated into heavy and light chains by reduction of the disulfide bond.

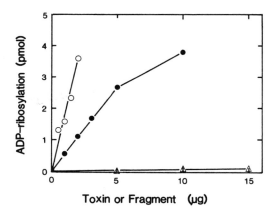

Fig. 5. Localization of ADP-ribosyl transferase activity in the light chain of type C1 botulinum toxin. Various amounts of type C1 botulinum toxin (●) and its light (○), and heavy chains (△) were separately incubated with bovine brain membranes and [^{32}P]NAD. Products were acid-precipitated and analyzed.

We have also noticed that, when we used fresh homogenate, ADP-ribosylation occured not only on the 21 kDa protein we had reported but also on a 23 kDa protein. Fig. 6 shows the time course of the labelling of the two proteins in the cytosol fraction. The labelling of the 23 kDa protein occurred first, and, with time, the labelling of the 21 kDa protein became more prominent. These results may suggest that the 23 kDa protein was the original form of the substrate and the 21 kDa was its proteolytic fragment, although we do not exclude the possibility that both proteins serve as native substrates for the reaction. Labelling of the 23 and 21 kDa doublet was also observed in freshly prepared membranes. The pI values of these proteins were determined by isoelectric focusing on polyacrylamide gels. They were between pH 5.5 to 6.0 in the ADP-ribosylated forms. These substrate proteins are present in various tissues and cells. They include bovine and mouse brain, mouse pancreas, rabbit blood platelet, cultured bovine adrenal chromaffin cells and GOTO mouse neuroblastoma cells.

Table 1. Subcellular distribution of botulinum toxin substrate.

Fraction	Protein	ADP-ribosylated 21 & 23 kDa Proteins	
	(mg)	(pmol)	(pmol/mg)
Homogenate	2,379	11,470	4.84
1,000 x g Pellet	325	1,380	4.25
10,000 x g Pellet	423	3,300	7.80
105,000 x g Pellet	139	940	6.76
105,000 x g Supernatant	661	3,650	5.52

[a] Bovine adrenal glands, 35 g, were cut into small pieces in 175 ml of cold 0.25 M sucrose-3.3 mM $CaCl_2$, and homogenized by a Potter-Elvehjem homogenizer. The homogenate was filtered through 4 layers of cheese cloth, and centrifuged successively at 1,000 x g for 15 min., at 12,000 x g for 20 min., and at 105,000 x g for 60 min. One hundred µg protein of each fraction was used in ADP-ribosylation by type D botulinum toxin. After autoradiography, a band containing the ADP-ribosylated protein(s) was cut from the gel and the radioactivity was measured in toluene scintillator.

Incubation Time (h)

1 2 3 4

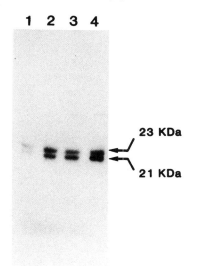

23 KDa ←
21 KDa ←

Fig. 6. Time course of ADP-ribosylation of 23 and 21 kDa cytosolic protein by type D botulinum toxin. Bovine adrenal cytosol (5 mg protein) was incubated with 50 µg of type D botulinum toxin and 50 µM [^{32}P]NAD in a total volume of 1 ml at 30°C. At 1, 2, 3 and 4 h, 50 µl of aliquots were taken and subjected to 15% SDS-polyacrylamide gel electrophoresis.

In order to characterize the nature of this substrate protein, we examined the effects of various nucleotides on the ADP-ribosylation by type D toxin. As shown in Fig. 7, GTP stimulated the reaction in a concentration

441

dependent manner; stimulation was observed at 10 nM GTP and reached a maximum at 100 μM. At this concentration the incorporated radioactivity was 6 to 7 fold higher than the control. GDP showed a stimulation similar to GTP, but its potency was ten times weaker than GTP. GMP, on the other hand, showed no stimulation. ADP and ATP showed only slight stimulation above 100 μM. AMP was without effect.

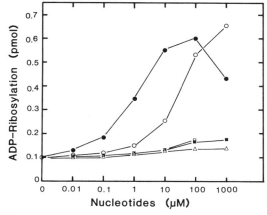

Fig. 7. Stimulation of botulinum toxin-catalyzed ADP-ribosylation by guanine nucleotides. Charcoal-treated mouse brain membrane was incubated with 10 μg of type D botulinum neurotoxin in the presence of various nucleotides and 10 mM dithiothreitol at 30°C for 45 min. ●, GTP; ○, GDP; △, GMP; ■, ATP; □, ADP. Reproduced from Biochem Biophys Res Commun (6).

Discussion

The present study has demonstrated that types C1 and D botulinum neurotoxin have ADP-ribosyl transferase activity and that activity was localized in the light chain of the toxin molecule. Simpson has proposed a three step model for botulinum toxin intoxication (4). This model consists of the binding step in which the toxin binds to a receptor on the plasma membrane of a target cell, the internalization step in which it enters into the cytoplasm and the poisoning step in which it damages the release function of the cell, probably by a catalytic process. If this is true, the toxin should consist of at least three functional domains which are responsible for each of the three steps. Experiments suggesting such a possibility have been reported. DasGupta and collaborators (7), using a phrenic nerve-hemi-diaphragm preparation, recently showed that neither L- chain nor H-chain caused paralysis if they were added separately, but that L- chain was active when the preparation was pretreated with H-chain. Our finding that the light chain of the toxin has ADP-ribosyl transferase activity suggests that the light chain contains a catalytic domain of the molecule and is responsible for the intracellular poisoning step. For the heavy chain, on the other hand, Kozaki *et al.* (8) reported that botulinum neurotoxin action was blocked by ganglioside G_{T1b}, a component of nerve cell membrane and that this ganglioside is bound to the carboxyl portion of the heavy chain (8, 9). Since gangliosides are known as receptors for many bacterial toxins such as cholera toxin (10), these results suggest that G_{T1b} serves as a receptor for the botulinum toxin and its heavy chain contains the binding domain of the molecule. Thus, it is likely that the botulinum toxin belongs to the A-B toxin group as proposed by Neville and Hudson for diphteria and other toxins (11).

Our studies have also demonstrated that the botulinum toxin has a specific target protein(s) inside cells and that it ADP-ribosylates this protein(s). Fig. 8 illustrates the possible function of this protein in the exocytotic process. The experiments by Knight *et al.* (3) imply that this protein lies between calcium and exocytosis and conveys the calcium signal to exocytotic machinery. Thus, it is likely that this protein plays a crucial role in stimulus-exocytosis coupling. Since this is considered to be a novel protein with novel function, we propose a name of "Gb" for this protein. At present the biochemical function of this protein is not clear. Recently, several groups demonstrated that guanine nucleotides directly stimulate the exocytosis process in various permeabilized cells (12-14). It is in this context interesting that ADP-ribosylation of this protein is stimulated by both GTP and GDP. These results may imply that Gb itself is a GTP-binding protein or a protein coupled to some GTP-binding protein(s).

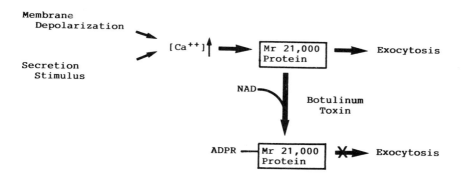

Fig. 8. Proposed molecular mechanism of botulinum neurotoxin action.

The present study has raised the possibility of multiple forms of Gb. It is present not only in the membranes but also in the cytosol. In addition, it is present as a doublet of 23 and 21 kDa in the two fractions. Since the protein in the cytosol is indistinguishable from that in the membrane, it is likely that they are the same protein. However, whether this distribution represents a functional interaction of the molecule with the membrane or is an artifact produced during homogenization and fractionation remains unanswered. It also remains to be elucidated whether both 23 and 21 kDa proteins are native targets of botulinum toxins or whether the 21 kDa protein is the result of *in vitro* proteolysis. These issues will be clarified by purification and sequencing of these proteins in near future.

443

Acknowledgements. The authors thank Drs. Y. Kamata and S. Kozaki and Prof. G. Sakaguchi of Osaka Prefectural University for their generous supply of botulinum toxin and fragments. We also thank Miss H. Nishio for her technical assistance. This work was supported in part by a grant in aid from the Ministry of Education, Science and Culture of Japan (no. 62570105) and grants from Mochida Memorial Foundation for Medical and Pharmaceutical Research, Uehara Memorial Foundation and Japanese Foundation on Metabolism and Diseases. Data appearing in this manuscript have been reprinted with permission from refs. 5 and 6.

References

1. Simpson, L.L. (1981) Pharmacol Rev 33:155-188
2. Sakaguchi, G. (1983) Pharmac Ther 19: 165-194
3. Knight, D.E., Tonge, D.A., Baker, P.F. (1985) Nature 317: 719-721
4. Simpson, L.L. (1986) Ann Rev Pharmacol Toxicol 26: 427-453
5. Ohashi, Y., Narumiya, S., (1987) J Biol Chem 262: 1430-1433
6. Ohashi, Y., Kamiya, T., Fujiwara, M., Narumiya, S. (1987) Biochem Biophys Res Commun 142: 1032-1038
7. Bandyopadhyay, S., Clark, A. W., DasGupta, B. R., Sathyamoorthy, V. (1987) J Biol Chem 262: 2660-2663
8. Kozaki, S., Sakaguchi, G., Nishimura, M., Iwamori, M., Nagai, Y. (1984) FEMS Microbiol Lett 21: 219-223
9. Kamata, Y., Kozaki, S., Sakaguchi, G., Iwamori, M., Nagai, Y. (1986) Biochem Biophys Res Commun 140: 1015-1019
10. Moss, J., Vaughan, M. (1979) Ann Rev Biochem 48: 581-600
11. Neville, D.M., Hudson, T.H. (1986) Ann Rev Biochem 55: 195-224
12. Haslam, R.J., Davidson, M.M.L. (1984) FEBS Lett 174: 90-95
13. Knight, D.E., Baker, P.F. (1985) FEBS Lett 189: 345-349
14. Barrowman, M.M., Cockcroft, S., Gomperts, B.D. (1986) Nature 319: 504-507

444

C3: A Novel Botulinum ADP-Ribosyltransferase Modifies a Putative 21 kDa G-Protein

Klaus Aktories, Monika Laux, and Sigrid Rösener

Rudolf-Bucheim-Institut für Pharmakologie der Justus-Liebig-Universität, Gießen D-6300, West Germany

Introduction

So far eight different botulinum toxins (A, B, C1, C2, D, E, F, G) have been described which are produced by various strains of *Clostridium botulinum* (1). Whereas seven of the botulinum toxins are neurotoxins and block the release at the cholinergic synapses, botulinum C2 toxin is not neurotoxic and acts on various non-neuronal tissues (1-3). It has been shown that component I of the binary botulinum C2 toxin possesses ADP-ribosyltransferase activity (4) on the eukaryotic substrate non-muscle actin (5). Here we describe another ADP-ribosyltransferase which is produced by certain strains of *Clostridium botulinum* type C. In order to distinguish the novel ADP-ribosyltransferase from botulinum neurotoxin C1 and botulinum C2 toxin we termed this enzyme C3.

Results

Characterization of ADP-ribosylation by C3. The novel ADP-ribosyltransferase was purified to apparent homogeneity from the supernatant of *Clostridum botulinum* type C (strain 4/12) as described (6). The purified enzyme migrated as a single band in sodium dodecyl sulfate (SDS)-polyacrylamide gel electrophoresis corresponding to a molecular mass of approximately 25 kDa. As shown in Fig. 1, in the presence of [^{32}P]NAD the enzyme catalyzed the labelling of an approximately 21 kDa protein in membranes of human platelets. Similarly, in various other cell types such as S49 lymphoma cells, neuroblastoma x glioma hybrid cells, sperm, fibroblasts and leukocytes, a 21 kDa protein was labelled. Whereas the labelling was slightly increased with ADP-ribose (100 µM), the addition of unlabelled NAD or pretreatment of [^{32}P]NAD with NADase almost completely blocked the labelling. Nicotinamide apparently reversed the ADP-ribosylation by C3. As shown in Fig. 2 the addition of nicotinamide released the bound radioactivity from the ADP-ribosylated substrate in a dose dependent manner. Incubation of platelet membranes with 100-300 ng/ml C3 for 15 min at 37°C resulted in a maximal ADP-ribosylation. The finding that the amount of label increased with time without any change in the apparent molecular mass of the target protein strongly suggests that the 21 kDa substrate is mono-ADP-ribosylated. Pretreatment of platelet membranes with trypsin (100 μg/ml) or heating the membranes for 5 min at 95°C completely prevented the ADP-ribosylation by C3, indicating the proteinaceous nature of the 21 kDa substrate. In contrast the ADP-ribosyltransferase C3 was largely stable against trypsin treatment and heating.

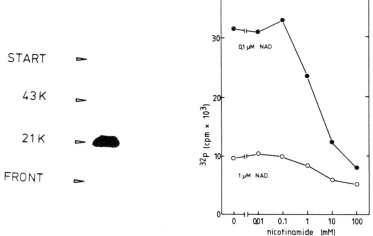

START

43K

21K

FRONT

Fig. 1. (left) ADP-ribosylation of human platelet membrane proteins by C3. Crude human platelet membranes were incubated with 1.4 μg/ml C3 in the presence of 0.5 μM [^{32}P]NAD (about 0.2 μCi) for 30 min at 37°C as described (6). The autoradiogram of the SDS polyacrylamide gel analysis of the labelled proteins is shown.

Fig. 2. (right) Influence of nicotinamide on C3-catalyzed ADP-ribosylation. Human platelet membranes were ADP-ribosylated with 1 μg/ml C3 in the presence of 0.1 and 1 μM [^{32}P]NAD (about 0.5 μCi), respectively, for 30 min as described (6). Thereafter, nicotinamide was added to give the indicated final concentrations and the incubation was continued for further 30 min. The amount of bound [^{32}P]ADP-ribose was determined by stopping the reaction with 400 μl SDS (2%, w/v), BSA (9 mg/ml) and precipitation of the proteins with 500 μl trichloroacetic acid (30%, w/v). Proteins were collected onto nitrocellulose filters. The filters were washed with 20 ml of 6% trichloroacetic acid and placed in scintillation fluid for determination of retained radioactivity (6).

Influences of guanine nucleotides on C3-catalyzed ADP-ribosylation. It is a well-known phenomenon that the ADP-ribosylation of GTP-binding proteins (G-proteins) by cholera- and pertussis toxin is regulated by guanine nucleotides (7-10). Therefore we studied the influence of the stable GTP-analog GTPγS on the ADP-ribosylation by C3. As shown in Fig. 3 GTPγS decreased the C3-induced ADP-ribosylation of the 21 kDa substrate in platelet membranes. A maximal effect was found at about 10 μM GTPγS (Fig. 4). GTP (100 μM) *per se* had almost no effect on the ADP-ribosylation. However the addition of GTP shifted the concentration-inhibition curve of GTPγS to the right. Also Gpp(NH)p and Gpp(CH$_2$)p decreased the ADP-ribosylation by C3. However, GTPγS was much more potent and efficient. At least three different molecular levels are potential sites for the action of guanine nucleotides on the ADP-ribosylation by C3. First, guanine nucleotides may effect the ADP-ribosyltransferase C3 itself, second the 21 kDa substrate may be the target and finally NAD metabolism may be affected by guanine nucleotides, ultimately decreasing the ADP-ribosylation of the 21 kDa substrate. As shown in Fig. 5, under control conditions the ADP-ribosylation was about 70% inhibited by GTPγS. In GTPγS-pretreated platelet membranes ADP-ribosylation of the 21 kDa substrate was largely reduced without addition of GTPγS to the ADP-ribosylation assay. Further addition of GTPγS caused only a marginal inhibitory effect. In order to exclude that the NAD metabolism is affected by guanine nucleotides we compared the effects of GTPγS on the ADP-ribosylation of the 21 kDa substrate by C3 with the influence of GTP analogues on the ADP-ribosylation of actin by botulinum C2 toxin. Using this protocol we found that GTPγS inhibited the ADP-ribosylation by C3 but not by botulinum C2 toxin.

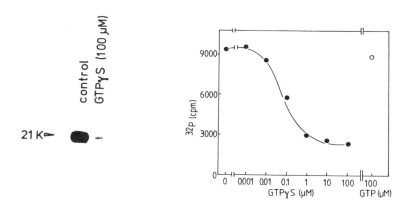

Fig. 3. (left) Inhibition of C3-catalyzed ADP-ribosylation of the 21 kDa platelet membrane protein by GTPγS. Crude human platelet membranes were ADP-ribosylated in the presence of 1 μg/ml C3 and of 1 μM [^{32}P]NAD without and with 100 μM GTPγS as in (6). The autoradiogram of the SDS polyacrylamide gel analysis of the labelled 21 kDa protein is shown.

Fig. 4. (right) Influence of GTPγS and GTP on the C3-catalyzed ADP-ribosylation of the 21 kDa protein in human platelet membranes. Human platelet membranes were preincubated with the indicated concentrations of GTPγS (●) or with GTP (100 μM; O) for 5 min. Thereafter, the ADP-ribosylation was initiated by the addition of C3 (9 μg/ml) and [^{32}P]NAD (0.1 μM) and the reaction continued for 5 min. The radioactivity of the labelled proteins were determined as described (6).

Discussion

Certain strains of *Clostridium botulinum* type C produce a novel ADP-ribosyltransferase C3, which ADP-ribosylates a 21 kDa protein in membranes of various cell types. Thus C3 is clearly distinct from component I of botulinum C2 toxin, which was shown to ADP-ribosylate actin (5, 11).

Cholera toxin and pertussis toxin ADP-ribosylate GTP binding proteins involved in transmembrane signal transduction, which are regulated by guanine nucleotides (7, 8). Furthermore, guanine nucleotides drastically alter the ability of the G-proteins to serve as substrates of these toxins (7-10). Also ADP-ribosylation by C3 was affected by guanine nucleotides. The finding that ADP-ribosylation by C3 was reduced even after pretreatment of platelet membranes with GTPγS suggest that the 21 kDa substrate and not C3 itself is the target of the stable GTP analogue. It is fascinating to speculate that the substrate of C3 is also a G-protein. However, purification of the 21 kDa substrate and/or functional studies with C3 have to be carried out, to clarify the hypothesis.

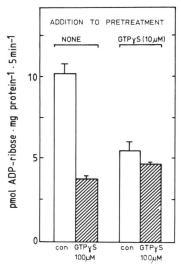

Fig. 5. Influence of pretreatment of human platelet membranes with GTPγS on the ADP-ribosylation by C3. Platelet membranes were pretreated without and with GTPγS (10 μM) for 5 min at 37°C. Thereafter, treatment was stopped by the addition of 1 ml ice-cold triethanolamine-HCl (50 mM, pH 7.5). The pelleted membranes were washed several times, resuspended and ADP-ribosylated by C3 without and with GTPγS (100 μM) as in (6). Data are given as means ± SEM (n=3).

Recently, it has been reported that botulinum neurotoxins C1 and D also possess ADP-ribosyltransferase activity (12, 13). Furthermore, 21 kDa proteins were also found to be ADP-ribosylated by these neurotoxins in a guanine nucleotide dependent manner (13). We have compared the ADP-ribosyltransferase activities of C3 to that of a highly purified C1 preparation (14). However we observed only a marginal ADP-ribosylation of 21 kDa proteins in platelet membranes by C1. C3 was about 1000 times more potent than C1 in labelling the 21 kDa substrate. Thus, further studies are necessary to examine a possible relation between C3 and C1. Antibodies against C3 and/or C1 will help to clarify whether C3 is an active species of C1 or a mere contaminant of the neurotoxin C1 preparation used.

References

1. Habermann, E., Dreyer, F. (1986) Curr Top Microbiol Immunol 129: 93-179
2. Simpson, L.L. (1982) J Pharmacol Exp Ther 223: 695-701
3. Ohishi, I. (1983) Infect Immun 40: 691-695
4. Simpson, L.L. (1984) J Pharmacol Exp Ther 230: 665-669
5. Aktories, K., Bärmann, M., Ohishi, I., Tsuyama, S., Jakobs, K.H., Habermann, E. (1986) Nature 322: 390-392
6. Aktories, K., Weller, U., Chhatwal, G.S. (1987) FEBS Lett 212: 109-113
7. Ui, M. (1984) Trends Pharmacol Sci 5: 277-279
8. Gill, M. (1982) ADP-Ribosylation Reactions, O. Hayaishi, K.Ueda (eds.), Academic Press, New York, pp. 592-621
9. Kahn, R.A., Gilman, A.G. (1984) J Biol Chem 259: 6235-6240
10. Aktories, K., Hungerer, K.D., Robbel, L., Jakobs, K.H. (1984) IUPHAR 9th International Congress of Pharmacology, London 1984, Abstr Vol 1655P
11. Aktories, K., Ankenbauer, T., Schering, B., Jakobs, K.H. (1986) Eur J Biochem 161: 155-162
12. Ohashi, Y., Narumiya, S. (1987) J Biol Chem 262: 1430-1433
13. Ohashi, Y., Kamiya, T., Fujiwara, M., Narumiya, S. (1987) Biochem Biophys Res Commun 142: 1032-1038
14. Aktories, K., Frevert, J. (1987) Biochem J 247: 363-368

Stimulation of Choleragen Enzymatic Activities by GTP and a Membrane Protein from Bovine Brain

Su-Chen Tsai, Masatoshi Noda, Ronald Adamik, Joel Moss, and Martha Vaughan

Laboratory of Cellular Metabolism, National Heart, Lung, and Blood Institute, National Institutes of Health, Bethesda, Maryland 20892 USA

Introduction

Activation of adenylate cyclase by choleragen results from the toxin-catalyzed ADP-ribosylation of a regulatory component of the cyclase system, $G_s\alpha$, a guanine nucleotide-binding protein involved in stimulation of the cyclase catalytic unit (1). The ADP-ribosylation reaction and cyclase activation are enhanced by soluble and membrane components (2-9). One membrane protein, known as ADP-ribosylation factor or ARF, was extensively purified by Kahn and Gilman and shown to be a GTP-binding protein (8, 9).

To test the possibility that the function of ARF was to activate directly the ADP-ribosyltransferase activity of toxin, advantage was taken of the fact that choleragen, in addition to catalyzing the ADP-ribosylation of $G_s\alpha$ (10) also possesses a number of other enzymatic activities (1). The toxin ADP-ribosylates arginine and other low molecular weight guanidino compounds (11); it is believed that the ADP-ribose moiety is attached to the guanidino group. Proteins unrelated to the cyclase system, are also modified presumably because they contain suitably accessible arginine residues (12). In addition, the toxin auto-ADP-ribosylates the A_1 protein (13), with at least three mono(ADP-ribosyl)ation sites having been observed. The toxin also utilizes water as an acceptor, catalyzing the hydrolysis of NAD to ADP-ribose and nicotinamide (14).

Results and Discussion

ARF was purified from bovine brain membranes by a modification of the procedure of Kahn and Gilman (9). In the final purification step on Ultrogel AcA 54, a protein peak was observed consistent with the approximate molecular weight of ARF (19,000). The protein size was also similar to that of ARF by sodium dodecyl sulfate-polyacrylamide gel electrophoresis (SDS-PAGE). When the ARF-activity across the Ultrogel AcA 54 protein peak in a reaction containing [^{32}P]NAD, Gs, DMPC, and choleragen was analyzed by SDS-PAGE, it was observed that not only was the ADP-ribosylation of $G_s\alpha$ increased, the auto-ADP-ribosylation of the choleragen A_1 protein was also enhanced by fractions containing ARF (Fig. 1). The effect of ARF on auto-ADP-ribosylation was dependent on GTP, and its nonhydrolyzable analogues, GTPγ S [guanosine-5'-0-(3-thiotriphosphate)] and Gpp(NH)p (guanylyl-β,γ-imidodiphosphate); GDP and GDPβS [guanosine-5'-0-(2-thiodiphosphate)] were inactive, as was App(NH)p (adenylyl-β,γ-imidodiphosphate) (15). These data were consistent with the participation of a guanine nucleotide-binding protein such as ARF in the stimulation of the auto-ADP-ribosylation reaction.

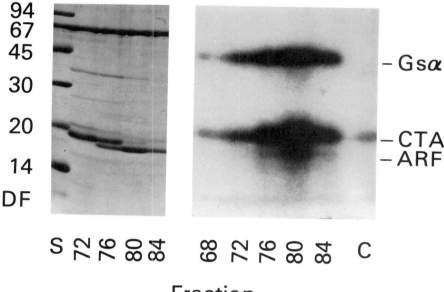

Fraction

Fig. 1. A preparation of ARF, purified by a modification of the procedure of Kahn and Gilman (9), was chromatographed on Ultrogel AcA 54. Samples of fractions across a peak of protein corresponding in apparent size to ARF were taken for analysis by SDS-PAGE (left) and assay of their ability to stimulate choleragen-catalyzed ADP-ribosylation of Gsα. For the latter, samples were incubated with choleragen, Gsα (0.1 μg), 10 μM [^{32}P]NAD, 250 μM GTP and 1 mM dimyristoyl phosphatidylcholine (15). Proteins were separated by SDS-PAGE. An autoradiogram of the gel is on the right. Fractions that enhanced ADP-ribosylation of Gsα also increased auto-ADP-ribosylation of choleragen A1. Both of these activities were correlated with the presence of a single protein of approximately 19 kDa. Data are from reference 15.

To test the ability of ARF to enhance the ADP-ribosylation of simple guanidino compounds, its activity was examined in the NAD:agmatine ADP-ribosyltransferase assay (Table 1). ARF increased the toxin-catalyzed ADP-ribosylation of agmatine approximately 3-fold. As noted with the auto-ADP-ribosylation reaction, activation of ADP-ribosylagmatine formation was dependent on GTP or its non-hydrolyzable analogues. GDP, GDPβS, ATP and App(NH)p were consistently less effective. ARF also increased the toxin-catalyzed ADP-ribosylation of proteins unrelated to the adenylate cyclase system as well as the hydrolysis of NAD to nicotinamide and ADP-ribose (data not shown). These data are consistent with the conclusion that the primary effect of ARF is on the catalytic activity of the toxin. Both $G_s\alpha$-dependent as well as $G_s\alpha$-independent activities are thus stimulated.

Table 1. Effect of nucleotides on NAD:agmatine ADP-ribosyltransferase activity of choleragen A subunit in the presence of ARF

Additions (100 μM)	Transferase Activity (nmol/μg/60 min)
None	1.09
GTP	3.16
Gpp(NH)p	2.88
GTPγS	2.67
GDP	1.28
GDPβS	1.16
App(NH)p	1.20
ATP	1.07

Assays contained 1 μg of choleragen A subunit and 1 μg of ARF. Data are from reference 15.

To determine the possible mechanism for stimulation of toxin by ARF, NAD:agmatine ADP-ribosyltransferase activity was examined at different substrate concentrations. The principal effect of ARF was to decrease the K_m's for both NAD and agmatine, with little change in the V_{max}. Thus, at subsaturating substrate concentrations, as might be expected to occur *in vivo*, ARF would enhance the rate of ADP-ribosylation.

These studies are consistent with a role for the membrane ADP-ribosylation factor in enhancing toxin-catalyzed reactions by direct interaction with toxin rather than through binding to one of the substrates such as $G_s\alpha$. These conclusions are supported by the findings that $G_s\alpha$-independent as well as $G_s\alpha$-dependent reactions were enhanced by ARF. ARF, previously shown to be a guanine nucleotide-binding protein, required GTP or its analogues for activity; GDP, ATP and their analogues were inactive. Based on these studies, it would appear that activation of adenylate cyclase by choleragen involves a guanine nucleotide-binding protein cascade where the first guanine nucleotide-binding protein, ARF, stimulates the enzymatic activity of the A_1 protein of toxin, thereby promoting ADP-ribosylation of a second guanine nucleotide-binding protein, $G_s\alpha$, which then directly activates the catalytic unit. It is interesting to speculate as to the physiological role of ARF in animal tissues. Conceivably, it may be involved in the GTP-dependent regulation of endogenous ADP-ribosyl-transferases, or similar enzymes.

References

1. Moss, J., Vaughan, M. (1988) Adv Enzymology 61: 303-379
2. Enomoto, K., Gill, D.M. (1980) J Biol Chem 255: 1252-1258
3. Le Vine III, H., Cuatrecasas, P. (1981) Biochim Biophys Acta 672: 248-261
4. Pinkett, M.O., Anderson, W.B. (1982) Biochim Biophys Acta 714: 337-343
5. Schleifer, L.S., Kahn, R., Hanski, E., Northup, J.K., Sternweis, P.C., Gilman, A.G. (1982) J Biol Chem 257: 20-23
6. Gill, D.M., Meren, R. (1983) J Biol Chem 258: 11908-11914
7. Kahn, R.A., Gilman, A.G. (1984) J Biol Chem 259: 6235-6240
8. Kahn, R.A., Gilman, A.G. (1984) J Biol Chem 259: 6228-6234
9. Kahn, R.A., Gilman, A.G. (1986) J Biol Chem 261: 7906-7911
10. Northup, J.K., Sternweis, P.C., Smigel, M.D., Schleifer, L.S., Ross, E.M., Gilman, A.G. (1980) Proc Natl Acad Sci USA 77: 6516-6520
11. Moss, J., Vaughan, M. (1977) J Biol Chem 252: 2455-2457
12. Moss, J., Vaughan, M. (1978) Proc Natl Acad Sci USA 75: 3621-3624
13. Moss, J., Stanley, S.J., Watkins, P.A., Vaughan, M. (1980) J Biol Chem 255: 7835-7837
14. Moss, J., Manganiello, V.C., Vaughan, M. (1976) Proc Natl Acad Sci USA 73: 4424-4427
15. Tsai, S.C., Noda, M., Adamik, R., Moss, J., Vaughan, M. (1987) Proc Natl Acad Sci USA 84: 5139-5142

Activation of the NAD Glycohydrolase, NAD:Agmatine and NAD:Gsα ADP-Ribosyltransferase and Auto-ADP-Ribosylation Activities of Choleragen by Guanyl Nucleotide and Soluble Proteins Purified from Bovine Brain

Masatoshi Noda, Su-Chen Tsai, Ronald Adamik, Patrick P. Chang, Barbara C. Kunz, Joel Moss, and Martha Vaughan

Laboratory of Cellular Metabolism, National Heart, Lung, and Blood Institute, National Institutes of Health, Bethesda, Maryland 20892 USA

Introduction

Choleragen (cholera toxin) exerts its effects on animal cells by activating adenylate cyclase, thereby increasing intracellular cAMP content (1). The A1 protein of choleragen, released from the holotoxin by reduction of a single disulfide bond linking the A1 and A2 proteins, catalyzes the mono-ADP-ribosylation of Gsα, a regulatory component of the adenylate cyclase system that is responsible for the GTP-dependent activation of the cyclase catalytic unit. ADP-ribosylation of Gsα apparently increases its sensitivity to GTP and its dissociation from the inhibitory Gβγ complex (1).

The ADP-ribosylation reaction is stimulated by GTP, phospholipids, and various cellular factors, both membrane and soluble (2-9). Kahn and Gilman (7, 9) isolated a membrane protein termed ADP-ribosylation factor (ARF) that promoted the toxin-catalyzed ADP-ribosylation of Gsα. The protein bound GTP in a reaction which was enhanced by NaCl and dimyristoyl phosphatidylcholine (9). It was proposed that ARF bound directly to Gsα (7) and that the ARF•Gsα complex served as the actual substrate in the toxin-catalyzed reaction (7). Tsai et al. (10) demonstrated that ARF from bovine brain membranes activated the toxin directly rather than interacting with the substrate Gsα. Other factors that enhanced the ability of choleragen to ADP-ribosylate Gsα were identified in a soluble fraction from bovine brain. Two proteins that accounted for most of the choleragen activation by the soluble fraction were resolved by ion exchange chromatography and separately purified. Each exhibited one major band by sodium dodecyl sulfate-polyacrylamide gel electrophoresis.

Results and Discussion

The effects of these two soluble ADP-ribosylation factors (termed sARF I and sARF II), on the Gsα-dependent and Gsα-independent choleragen-catalyzed reactions were determined, as previously done by Tsai et al. (10)

for the membrane ADP-ribosylation factor (mARF). The purified sARF I enhanced ADP-ribosylation of $G_{s\alpha}$ as well as $G\beta$ (Fig. 1). In the presence of sARF I, dimyristoyl phosphatidycholine enhanced the toxin-catalyzed ADP-ribosylation of $G_{s\alpha}$; it inhibited, however, the auto-ADP-ribosylation reaction (Fig. 1). The activity co-chromatographed with an approximately 19 kDa protein as determined by sodium dodecyl sulfate-polyacrylamide gel electrophoresis of fractions obtained by gel permeation chromatography (Fig. 1).

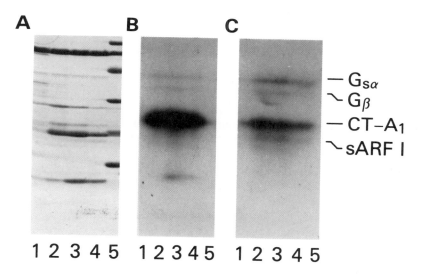

Fig. 1. Ultrogel AcA 54 chromatography of sARF I; effect of dimyristoyl phosphatidylcholine (DMPC) on ADP-ribosylation of $G_{s\alpha}$ and choleragen A_1 protein. Purified sARF I was chromatographed on a column (1.2 x 104 cm) of Ultrogel AcA 54. Fractions (1 ml) were collected and samples were (A) analyzed by sodium dodecylsulfate-polyacrylamide gel electrophoresis (SDS-PAGE) and assayed for ARF activity in a reaction mixture containing G_s (0.4 μg), choleragen (25 μg), and 100 μM GTP without (B) or with (C) 1 mM DMPC. (A) SDS-PAGE of 250 μl of fraction plus bovine serum albumin, 10 μg. Lanes 1-4, fractions 68, 72, 76 and 80; Lane 5, standard proteins, phosphorylase b, bovine serum albumin, ovalbumin, carbonic anhydrase, soybean trypsin inhibitor, α-lactalbumin. (B) and (C) Autoradiograms of ADP-ribosylated proteins. Lanes 1-4, fractions 68, 72, 76 and 80; Lane 5, column buffer.

Like mARF, sARF I and sARF II increased the auto-ADP-ribosylation of choleragen A_1; in both instances, GTP or its nonhydrolyzable analogues GTPγS [guanosine-5'-0-(3-thiotriphosphate)] and Gpp(NH)p (guanylyl-imidodiphosphate) were required for stimulation: GDP, GDPβS [guanosine-5'-0-(2-thiodiphosphate)], ATP, and App(NH)p (adenylyl-imidodiphosphate) were ineffective.

Table 1. Effect of nucleotides on NAD:agmatine ADP-ribosyltransferase activity of choleragen A subunit in the presence of sARF I or sARF II

| Nucleotide Added | [Carbonyl-^{14}C]Nicotinamide Released | |
	sARF I	sARF II
(30 µM)	(nmol•µg A subunit $^{-1}$•h^{-1})	
None	2.10	1.22
GTP	5.48	2.12
GTPγS	4.70	1.91
Gpp(NH)p	4.70	1.73
ATP	2.15	1.26
App(NH)p	2.00	1.11
GDP	2.00	1.16
GDPβS	1.89	1.11

Assays containing choleragen A (1 µg), sARF I (2.8 µg) or sARF II (2.4 µg) with nucleotide as indicated were performed as described in reference 10.

Effects of sARF I and II on the NAD:agmatine ADP-ribosyltransferase activity of toxin were also similar to those of mARF. Both stimulated ADP-ribosylation of agmatine approximately 3-fold with maximal activity observed at a molar ratio of approximately 3:1 (ARF:choleragen A subunit). Activation by both sARF I and II was dependent on GTP or its analogues, as noted for stimulation of the ADP-ribosylation of Gsα and auto-ADP-ribosylation (Table 1). The products of the ADP-ribosylation of agmatine were resolved by SAX - high performance liquid chromatography. sARF enhanced the choleragen-catalyzed formation of the α-anomer of ADP-ribosylagmatine; the product was identified (a) by its retention time on HPLC relative to a standard synthesized in the presence of a highly purified turkey erythrocyte NAD:agmatine ADP-ribosyltransferase (11) and (b) by its selective degradation by a stereospecific erythrocyte ADP-ribosylarginine hydrolase (12). In the absence of agmatine, choleragen catalyzes the hydrolysis of NAD to ADP-ribose and nicotinamide; sARF stimulated this reaction as well. Thus sARF I and sARF II, like mARF, increased activity in each of the assays employed, consistent with the conclusion that these proteins interact directly with the toxin and alter its catalytic properties.

These studies demonstrate the existence of both membrane and soluble proteins having similar activities and capable of activating the catalytic unit of choleragen. Since this activation is dependent on GTP and since the substrate for the toxin-catalyzed reaction is also a GTP-binding protein, it appears that a guanine nucleotide-binding protein cascade is involved in toxin-catalyzed activation of adenylate cyclase (Fig. 2). In the first step, the A1 protein of toxin is activated by ARF; it is not clear whether the membrane or soluble species is involved preferentially in the cell. The ARF-choleragen complex then catalyzes the ADP-ribosylation of Gsα, enhancing the sensitivity of Gsα to GTP and its ability to activate the cyclase catalytic unit.

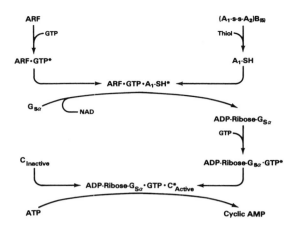

Fig. 2. A guanine nucleotide-binding protein cascade participates in the activation of adenylate cyclase by choleragen.

References

1. Moss, J., Vaughan, M. (1988) Adv Enzymology 61: 303-379
2. Enomoto, K., Gill, D.M. (1980) J Biol Chem 255: 1252-1258
3. Nakaya, S., Moss, J., Vaughan, M. (1980) Biochemistry 19: 4871-4874
4. Le Vine III, H., Cuatrecasas, P. (1981) Biochim Biophys Acta 672: 248-261
5. Schleifer, L.S., Kahn, R.A., Hanski, E., Northup, J.K., Sternweis, P.C., Gilman, A.G. (1982) J Biol Chem 257: 20-23
6. Gill, D.M., Meren, R. (1983) J Biol Chem 258: 11908-11914
7. Kahn, R.A., Gilman, A.G. (1984) J Biol Chem 259: 6228-6234
8. Kahn, R.A., Gilman, A.G. (1984) J Biol Chem 259: 6235-6240
9. Kahn, R.A., Gilman, A.G. (1986) J Biol Chem 261: 7906-7911
10. Tsai, S.-C., Noda, M., Adamik, R., Moss, J., Vaughan, M. (1987) Proc Natl Acad Sci USA 84: 5139-5142
11. Moss, J., Stanley, S.J., Watkins, P.A. (1980) J Biol Chem 255: 5838-5840
12. Moss, J., Oppenheimer, N.J., West Jr., R.E., Stanley, S.J. (1986) Biochemistry 25: 5408-5414

Active Sites and Homology of Diphtheria Toxin and *Pseudomonas aeruginosa* Exotoxin A

R. John Collier

Department of Microbiology and Molecular Genetics, Harvard Medical School and The Shipley Institute of Medicine, Boston, Massachusetts 02115 USA

Introduction

A major goal of our current work is to understand the actions of bacterial toxins with ADP-ribosyltransferase activity at the level of 3-dimensional protein structure. Recently an x-ray crystallographic map of Exotoxin A (ETA) of *Pseudomanas aeruginosa* at 3.0 Å resolution has been obtained by Davis McKay (University of Colorado, Boulder), in collaboration with our group (1). In addition, a map of diptheria toxin dimer is expected to emerge soon, through collaborative studies (2) with the group of David Eisenberg (University of California, Los Angeles) and/or through independent work (3) of R. Sarma and coworkers (State University of New York, Stony Brook). In this report we summarize research leading to an emerging understanding of active site structure and mechanism of catalysis in these two toxins.

Diphtheria toxin (DT) and ETA have remarkably similar enzymatic properties (4). Both toxins block protein synthesis in mammalian cells by catalyzing ADP-ribosylation of the diphthamide group, a post-translationally modified histidine, on Elongation Factor-2 (EF-2). Despite this similarity in substrate specificity, there are major differences in structure between the two toxins. Thus, for example, in DT the catalytic domain is amino-terminal, whereas in ETA it is carboxyl-terminal; immunological crossreactivity is vanishingly low; and sequence homology is undetectable at the nucleic acid level and so weak at the protein level that it was originally overlooked. As shown below, however, significant amino acid sequence homology does in fact exist between the catalytic domains of the DT and ETA. This, together with our recent results on active site photolabelling, argues strongly for the notion that the two toxins share common ancestry, but are distantly related.

Several years ago we set out to identify residues in the active site of diphtheria toxin. We knew that the enzymically active component of DT, fragment A, contained a single NAD binding site with K_d of 8 μM (5). Fragment A (93 residues) is generated from intact DT (535 residues) by mild tryptic digestion and reduction. As an initial approach to identifying active site residues, we decided to test the possibility that unmodified NAD might be used as a photoaffinity label. Each of three preparations of NAD, containing radiolabel in the nicotinamide moiety, the adenine moiety, or the adenylate phosphate, was mixed with fragment A, and the mixtures were irradiated under a low pressure Hg lamp (wavelength maximum, 254 nm) and assayed for incorporation of label into the protein. Incorporation from the nicotinamide-labelled NAD was at least 5-fold higher than that from the other preparations and approached stoichiometric equivalence under optimal conditions (6). Both DT, fragment A and activated ETA were efficiently photolabelled, whereas several control proteins, including the NAD-linked dehydrogenases tested were not.

By protein sequencing methods we found that the radiolabel from nicotinamide-labelled NAD was covalently attached at a single site within fragment A, position 148, corresponding to glutamic acid in the native protein (6). Next the structure of the photoproduct generated at this position was determined (7). A tripeptide prepared by thermolytic digestion of photolabelled fragment A was analyzed by a combination of biochemical, nuclear magnetic resonance, and mass spectrometric methods. The results showed that the product at position 148 was an α-amino-γ-(6-nicotinamidyl) butyric acid residue (7). This structure is generated by decarboxylation of the Glu-148 side chain, formation of a new C-C bond between C-6 of the nicotinamide ring and the gamma methylene carbon of Glu-148, and rupture of the bond linking nicotinamide to ADP-ribose.

The structure of the photoproduct is consistent with the notion that Glu-148 could be involved in catalysis. If the gamma methylene group of Glu-148 were in close proximity to C-6 of the nicotinamide ring of NAD in the enzyme-substrate complex, then the carboxyl group of Glu-148 would be within a short radius of C-6, and, depending on the geometry of the complex, might be in direct contact with member atoms of the bond disrupted in the ADP-ribosylation reaction. Also, the fact that the photoproduct was inactive in ADP-ribosylation was consistent with a role of Glu-148 in catalysis.

To test the role of Glu-148 we mutated this residue to aspartic acid (E148D) by site-directed mutagenesis of a cloned DNA fragment of the toxin gene. The Asp-148 mutant form of fragment A, expressed in *E. coli*, was found to have less than 1% the NAD:EF-2 ADP-ribosyltransferase activity of wild-type fragment A (8). Despite this drastic reduction in enzymic activity, Asp-148 fragment A binds NAD with slightly higher affinity than the wild type A and quenches tryptophan fluorescence to the same degree (ref. 8 K. Reich and R.J. Collier, unpublished results). This provides strong evidence that the mutation did not cause a major disruption in protein folding.

We have not been able to test the effect of the E148D mutation on toxicity because cloning of the whole DT gene is prohibited, except under the highest level containment, and has not yet been performed. However, we obtained permission to reconstruct the whole DT gene in *E. coli*, provided that the enzymic function had been disabled by a 3-base mutation (essentially non-reversible) at position 148. We first mutated Glu-148 to Ser by the mutation GAA → GAT in gene fragment F2, encoding fragment A plus part of B, and showed that the Ser-148 form of fragment A had less than 1% the ADP-ribosylation activity of the wild-type (9). Next the mutant fragment F2 was recombined with another gene fragment encoding the complimentary C-terminal region of DT, to form Ser-148 DT. We expressed this mutant toxin in *E. coli* and showed that it is secreted to the periplasmic compartment. Purified Ser-148 DT has about 1,500-fold lower toxicity than wild-type DT in a 6-hr tissue culture assay, and Ser-148 fragment A shows about the same level of reduction in NAD:EF-2 ADP-ribosyl transferase activity. The Ser-148 fragment A binds NAD with only slightly lower affinity ($K_d \sim 12$ μM) than the wild type, however, implying that the mutation did not cause major changes in protein conformation (K. Reich and R.J. Collier, unpublished results).

These studies implicating Glu-148 as a crucial active site residue in DT provided the impetus to test ETA for capacity to be photolabelled by the same procedure. Photolabelling of activated ETA with nicotinamide-labelled NAD was about as efficient as that with DT, and we set out to locate the label in this toxin (10). Initial experiments involved ETA that had been activated by incubation with urea plus dithiothreitol. However when the activated full-length toxin was subjected to the same photolabelling conditions used with DT fragment A, extensive protein-protein crosslinking was observed, and it was difficult to purify photolabelled peptides after proteolytic digestion of the preparation. Ultimately this problem was solved by developing a procedure for isolating an enzymically active fragment of activated ETA, roughly analogous to DT fragment A. This was done by treating activated intact ETA with thermolysin in the presence of saturating levels of NAD. The resulting fragment could be photolabelled without significant protein-protein crosslinking, and the efficiency of photolabelling was similar to that found with DT fragment A.

Sequencing of various proteolytic peptides isolated from the photolabelled thermolysin fragment showed that all the label was located at position 553, which, like position 148 of DT, corresponded to a glutamic acid residue in the unlabelled protein (10). The photoproduct at position 553 was chromatographically the same as that in photolabelled fragment A. Lastly, we found that substitution of aspartic acid for glutamic acid at position 553 by site-directed mutagenesis resulted in a drastic loss of NAD:EF-2 ADP-ribosyltransferase activity, even greater than that seen in the DT fragment A (C. Douglas and R. J. Collier, unpublished results).

In contrast to DT, the intact ETA gene may be cloned and expressed under BL-1 + EK-1 containment conditions. This has enabled us to construct the Glu $\rightarrow$ Asp mutant at position 553 and measure the toxicity of the mutant protein (C. Douglas and R. J. Collier, unpublished results). Toxicity of the mutant, as measured by inhibition of protein synthesis in tissue culture, is at least 4 orders of magnitude lower than that of the wild-type toxin. The diminution of toxicity appears to be somewhat greater than that of ADP-ribosylation activity, conceivably because the mutation may also affect ability of the catalytic domain to penetrate to the cytoplasm or its stability within the cytoplasm.

Availibility of the crystallographic structure of ETA has permitted localization of Glu-553 to a prominent cleft in Domain III of the toxin. Other crystallographic evidence has shown that this cleft corresponds to the NAD binding site (11), although the conformation of NAD within this site is still unknown. Tyr-470 is very close (5-6 Å) to the Glu-553 side chain, and it may be that the nicotinamide ring of NAD binds between these two residues, forming stacking interactions with the tyrosine ring and ionic interactions with the cationic nicotinamide ring. There are also other aromatic groups nearby (Tyr-481, Phe-469, Trp-466), one or more of which may contribute to binding of the adenine ring of NAD.

How do Glu-148 of DT and Glu-553 of ETA contribute to catalysis? The ADP-ribosyl group of NAD becomes attached via alpha linkage to N-1 of the imidazole ring of diphthamide. Assuming the gamma-carboxyl group of the active-site glutamic acid is ionized and is in close proximity to the N-glycosidic linkage of NAD ruptured during ADP-ribosylation, the carboxylate can be conceived to promote ADP-ribosylation by any of several mechanisms. By analogy with lysozyme, the carboxylate might stabilize a developing oxocarbonium ion on the ribose. Alternatively it might stabilize some other transition state intermediate in the transfer of the positive charge from the nicotinamide ring to the imidazole ring of diphthamide. Finally it might act as a general base either to abstract a proton from the attacking diphthamide imidazole (if it is initially protonated) or to act as a sink for the ionizable proton generated on the imidazole ring during ADP-ribosylation. Further studies to discriminate among these mechanisms are under way.

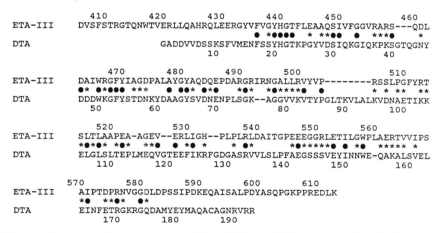

Fig. 1. Homology between Domain III of ETA and DT fragment A. Solid circles represent identical residues; asterisks represent conservative replacements.

461

Given the parallels in the photolabelling reactions of DT and ETA, together with the virtual identity of their enzymic properties, it would be surprising indeed if these two toxins were not evolutionarily related proteins. Although an initial search for significant amino acid sequence homology was negative (12), we have detected significant homology between the catalytic domains of DT and ETA (10 and S.E. Carrol and R.J. Collier, unpublished results). As shown in Fig. 1, there is homology in the immediate vicinity of the active site glutamic acids, and even stronger homology elsewhere in the catalytic domains. All of the residues mentioned above as elements of the NAD site are conserved, and other conserved sequences are concentrated in and around the active-site cleft. We predict that some of the exposed, conserved residues are involved in EF-2 binding.

References

1. Allured, V.S., Collier, R.J., Carrol, S.F., McKay, D.B. (1986) Proc Natl Acad Sci USA 83: 1320-1324
2. Collier, R.J., Westbrook, E.M., McKay, D.B., Eisenberg, D. (1982) J Biol Chem 257: 5283-5285
3. McKeever, B., Sarma, R. (1982) J Biol Cem 257: 6923-6925
4. Chung, D.W., Collier, R.J. (1977) Infect Immun 16: 832-841
5. Kandel, J., Collier, R.J., Chung, D.W. (1974) J Biol Chem 249: 2088-2097
6. Carroll, S.F., Collier, R.J. (1984) Proc Natl Acad Sci USA 81: 3307-3311
7. Carroll, S.F., McCloskey, J.A., Crain, P.F., Oppenheimer, N.J., Manschner, T.M., Collier, R.J. (1985) Proc Natl Acad Sci USA 82: 7237-7241
8. Tween, R.K., Barbieri, J.T., Collier, R.J. (1985) J Biol Chem 260: 10392-10394
9. Barbieri, J.T., Collier, R.J. (1987) Infect Immun 55: 1647-1651
10. Carroll, S.F., Collier, R.J. (1987) J Biol Chem 262: 8707-8711
11. Allured, V.S., Brandhuber, B.J., McKay D.B. (1987) B. Bonavida, R.J. Collier (eds.) Cell-Mediated Cytotoxicity. Alan R. Liss, New York
12. Gray, G.L., Smith, D.H., Baldridge, J.S., Harkins, R.N., Vasil, M.L., Chen, E.Y., Heyneker, H.L. (1984) Proc Natl Acad Sci USA 81: 2645-2649

The Cloning of the cDNA and the Gene for Human Poly(ADP-Ribose) Polymerase: Status on the Biological Function(s) Using Recombinant Probes

Mark Smulson, Hussein Alkhatib, Kishor Bhatia, Defeng Chen, Barry Cherney, Vincente Notario, Caroline Tahourdin, Anatoly Dritschilo, Preston Hensley, Ted Breitman[1], Gary Stein[2], Yves Pommier[2], O. Wesley McBride[2], Michael Bustin[2], and Chandrakant Giri[3]

Department of Biochemistry, Georgetown University Schools of Medicine and Dentistry, Washington, D.C. 20007

Introduction

In March, 1987, our laboratory was the initial group to report on the isolation and expression of a full-length cDNA encoding poly(ADP-ribose) polymerase (1, 2). The focus of this chapter will be to review some of the evidence leading up to the cloning of this cDNA as well as to provide a progress report on the various approaches presently underway (Spring, 1987) using the cDNA for various projects directed at elucidation of the function of poly(ADP-ribose) polymerase in cells.

Results and Discussion

cDNAs encoding poly(ADP-ribose) polymerase from a human hepatoma lambda gtll cDNA library were isolated by immunological screening (1, 2). One insert of 1.3 kb consistently hybridized on RNA gel blots to mRNA species of 3.6-3.7 kb, which was consistent with the size of RNA necessary to code for the polymerase protein (116 kDa). This cDNA was subsequently used in both *in vitro* hybrid selection and hybrid-arrested translation studies. A mRNA species from HeLa cells of 3.6-3.7 kb was selected that was translated into a 116-kDa protein, which was selectively immunoprecipitated with anti-poly(ADP-ribose) polymerase.

To confirm that the 1.3 kb insert from lambda gtll encoded poly(ADP-ribose) polymerase, the insert was used to screen a 3- to 4-kb subset of a transformed human fibroblast cDNA library in the Okayama-Berg vector. One of the vectors, clone pcD-12 yielded a *Bam*H1 fragment within the additive size 3.6-7 kb, which corresponded with the size of mRNAs noted earlier for poly(ADP-ribose) polymerase. This vector [pcD-poly(ADPR)p]

---

[1]Division of Virology, Food and Drug Administration, Bethesda, MD 21205; and Department of Biochemistry
[2]Molecular Biology, University of Florida College of Medicine, Gainesville, FL 32610
[3]National Institutes of Health, Bethesda, MD 21205

was tested in a transient transfection of Cos cells. The cDNA we isolated contained the complete coding sequence as indicated by the following criteria: (i) A 3-fold increase in *in vitro* activity was noted in extracts from transfected cells compared to mock or pSV2-CAT transfected cells; (ii) a 6-fold increase in polymerase activity in pcD-poly(ADPR)p transfected cell extracts compared to controls was observed by activity gel analysis of electrophoretically separated proteins at 116 kDa (Fig. 1). (iii) a 10-15 fold increase in newly synthesized polymerase was detected by immunoprecipitation of labeled transfected cell extracts (Fig. 1-C).

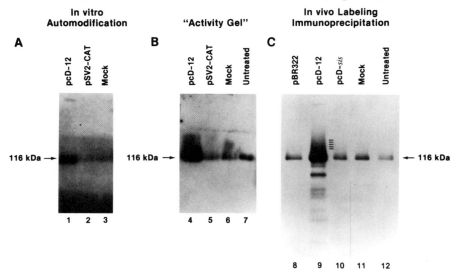

Fig. 1. Transfection of COS cells with pcD-12 produces enhanced expression of polymerase activity and immunoprecipitable polymerase protein. 1 x 10^6 COS cells in duplicate flasks were treated in the presence or absence of plasmid DNA. A. Sonicated samples were assayed for activity for 10 sec with 2.8 µCi [^{32}P]NAD by electrophoresis on 7.5% NaDodSO$_4$/polyacrylamide gels. B. The sonicated samples were separated by electrophoresis in 7.5% NaDodSO$_4$/polyacrylamide (containing 100 µg/ml sonicated salmon sperm DNA). The gels were subsequently renatured and the gel assayed directly with [^{32}P]NAD for *in situ* poly(ADP-ribose) polymerase activity. The washed gels were exposed to X-ray film at -70°C. Lanes 4-7 show cells transfected with pcD-12, pSV2-CAT, cells mock-transfected or untreated, respectively. C. The COS cells (2 x 10^7/flask) were treated as above for 45 hr after which they were starved for methionine at 37°C in methionine-free medium. [^{35}S]Methionine (250 µCi/flask) was added and the cells were incubated at 37°C for 3 hr, after which they were lysed; the proteins immunoprecipitated with anti-polymerase, analyzed by NaDodSO$_4$/PAGE and autoradiography. (Taken from Ref. 2).

To our knowledge this paper (2) and earlier abstracts (1) described the first cloning and expression of a cDNA encoding poly(ADP-ribose) polymerase. In a more recent paper (3) we have reported the total amino acid sequence for human poly(ADP-ribose) polymerase.

DNA sequence and derived amino acid sequence of human poly(ADP-ribose) polymerase cDNA. The Okayama-Berg derived cDNA insert (3681 bp) encoding poly(ADP-ribose) polymerase was digested with the appropriate restriction enzymes and subcloned in phage M13 for sequencing. The nucleotide sequence of the cDNA is shown in Fig. 2. A single long open reading frame was observed extending 3042 bp from the first ATG codon to a TAA termination codon at position 3202-3204. This ATG is flanked by sequences that fulfill the Kozak criteria for initiation codons. The protein deduced from this sequence contained 1014 amino acids, and is close to the molecular weight (Fig. 2, dashes) estimates of the purified human protein (112-116 kDa). The 3042-nucleotide open reading frame is flanked by untranslated sequences of 159 bp 5' and 459 bp 3'. The 5' untranslated sequences have a 74% GC content. A potential polyadenylation signal (AATAA) is present 16 bp upstream from the poly(A) sequence.

Comparison of derived amino acid sequence with poly(ADP-ribose) polymerase peptide fragments. Human poly(ADP-ribose) polymerase was purified to homogeneity (4), digested with endoproteinase Lys C, and the resulting peptides were isolated by reverse phase HPLC. Limiting quantities (i.e. 50 ng) of 4 different oligopeptides were obtained. The sequence deduced from the long open reading frame of the polymerase cDNA revealed a match for 3 peptides (Fig. 2, dashes).

Secondary structure predictions for human poly(ADP-ribose) polymerase. A computer-derived secondary structure prediction for the polymerase based on a Chou-Fasman analysis predicts a very low β-sheet (< 9%), a high content of random coil (38%), a large number of β-turns and approximately 25% helical content. The structure shows several regions of helix-turn-helix (i.e. centered at residues 82, 97, 232, 258 and 275) which is characteristic of a number of DNA binding proteins (Fig. 3). The predicted structure does not show a B-A-B structure, which is characteristic of NAD binding domains.

A hydropathic plot for poly(ADP-ribose) polymerase (Fig. 4) based on the procedure of Hopp and Woods which has been used to predict antigenic determinants, predicts a protein with quite low hydrophobic character.

Poly(ADP-ribose) polymerase homology to other proteins. Sequence similarity comparison of the polymerase with the National Biomedical Center's protein data bases revealed no extensive identities with other proteins (> 4200 proteins searched). Although no extensive similarities were observed, the polymerase exhibits some short but interesting identities. The most statistically significant identity is with the catalytic site of the ricin A chain, a cytotoxic plant protein which inactivates the 60S ribosome. This may represent the active site of the enzyme, based upon the catalytic

465

```
                                              -150        -140        -130
                                      AATCTATCAGGAACGGCCGGTGGCCCGGTCGGCGTGTTC
          -40        -30        -20        -10
      CTGGCTCCTGGCTCCGGCTTCCGGAGCTTTGGCGCAGCTAGGGGAGG
                                              40

 -120       -110       -100        -90        -80        -70        -60        -50
GGTGCGGCTTCTGGCCGCTGCGGCTGGGTGAGCCGCCGCAAGCGTGTT
                                                              10
MetAlaGluSerSerAspLysLeuTyrArgValGluTyrArgIleSerGlyArgIleArgIleSerProLysAspSerLeuArgMetAlaIleMetAlaIleMetValGln
ATGGCGGAGTCTTCGGATAAGCTCTATCGAGTCGAGTACCGCAAGAGCGGGCGCCCTTCTGCAAGAAATGCAGCGAGAGCATCCCCAAGGACTCGCTCCGGATGGCCATCATGGTGCAG
      GLU——THR                          50
            #1

SerProMetPheAspGlyTyrHisTrpTyrHisPheSerCysPheTrpLysValGlyHisSerIleArgHisProAspValGlyValAspGlyHisProSerGluLeuArgTrpAsp
TCGCCCATGTTTGATGGATACCACTGGTATCACTTCTCCTGCTTCTGGAAGGTGGGCCACTCCATCCGGCACCCTGACGTGGGTGTGGATGGGCACCCTGAGCGAGCTTCGGTGGGAT
                     90                              110                            120
                                     60                              70

AspGlnLysValLysThrAlaGluAlaGlyLysValThrArgGlyLysValAlaGluGlyLysValAspGlnLysLysValLysThrAlaGlyLysValAlaLysSer
GACCAGAAGTCAAGACAGCACAGAGCCGGGAAGGTGACGAGAGGCAAGGCCGAAGGTGGAAGGCCAGAAGGGTGCAGCAAGGCCAGGAGTTAGCAAGGCCCAGTATGCCAAGTCC
                            130                            150                            160
                  #2

AsnArgSerThrCysLysGlyCysMetGluLysIleLeuLysGlyGlnValArgLeuSerLysLysMetValAspAspProGlnLeuGlyMetIleGluAspArgTrpTyrHisPro
AACAGAAGTACCTGCAAGGGGTGTATGGAGAAGATAGAAAAAGGCCAGGTGCGCTGTCCAAGAAGATGGTGGACGACCCAGAGCCAGCATGATTGACCCTGTACCATCCA
                            170                            190                            200
                                            180

GlyCysPheValLysAsnArgGluGlyLeuGlyPheArgProGluTyrSerAlaSerGlnLeuLysGlyPheSerLeuLeuAlaThrGluAspArgTrpGlnLeuSerGlnLeuPro
GGCTGCTTTGTCAAGAACAGGGAGGGTCTGGGTTTCCGGCCCGAGTACAGTGCCAGCGAGTCAGCTCAAGGGCTTCAGCCTCCTTGCTCACGAGGATAAAGAGCCCTGAAGAGCAGCTCCCA
                            210                            230                            240
            #3                              220

GlyValLysSerGluGluLysGlyLysGlyArgLysGlyAspGlyValAlaLysAlaGluGluAlaAlaLysLysLysSerLysLeuGluLysLysAlaAlaLeuLysAlaAla
GGAGTCAAGAGTGAAGAGAAAGAGACGGTGGATGGAGTGGAGCAAGGTGAGAAAGAACAAGGATAGTAAGCTTGAAAAAGCCTAAAGGCT
                            250                            270                            280
                                            260

GlnAsnAspSerLeuIleTrpAsnIleLeuLysLysGluThrAspSerLeuLysLeuAspPheAsnLeuIlePheAsnLysValGlnLeuValProSerGlyLysAlaLeuLeu
CAGAACGATAGTCTGATCTGGAACATCCTGAAGAAGGAGACTGATAGCCTCAAGCTGGACTTCACTCGTGCTGTCCGGGAGTCCGGGAGTCCCTGCGGGGATTCTGG
                            290                            310                            320
                                            300

AspArgValAlaAlaAspGlyMetValPheGlyValAlaLeuGluProCysLeuCysSerGlyGlnLeuValAspPheSerSerAspAlaTyrTyrCysThrGlyAspValAlaTrpPheLysLys
GACCAGAGCTGATGGCATGGTTCGGGTAGCACTCGGGAATGCTCGAGGAATGGAGTCTGGGTCAGCTGGTGGACTTCAGCTCTGACGCCTACTAGGGACCTCTGCTGACCAAG
                            330                            350                            360
                                            340

CysMetValLysThrGlnHisThrProAsnArgLysGluTrpValThrProValThrProLysProLysPheArgGluArgIleGluSerTyrLeuLysValLeuLysGlnAsnAspArgIlePheProGluProGlu
TGTATGGTCAAGACACAGCAACACCCGGAAGAATTCCGAGAATTCTTACCTCAAGAATTTAAGGTTAAAGACAAGCATATTCCCCAGAA
                            370                            390                            400
                                            380

ThrArgAlaSerValAlaAlaAlaThrProProSerThrAlaSerAlaAlaProAlaAlaAlaValAlaAsnSerSerAlaAspAlaSerProLeuSerAsnMetLysSerArgIleLeuThrTrpLeuGlyTyrLys
ACCAGCGCTCCTGGCGGCCACCCTGCGGCCCTCCACACCTCGGCCTGTGAACTCCTGCTCGTGGAACTCCTGCTGCCAGATAAGCCATATTCCAACATGAAGATCCTGACTCGCGGGAAG
                            410                            430                            440
                                            420

LeuSerArgAsnLysAspGluValLysLysGluGlyLeuValLeuLeuLeuGlyLeuLysLeuGlyLysGluLysLeuLysGlyGluLeuThrGlyThrArgAlaAlaAsnLysSerAlaSerThrLysSerLeuSerIleSerThrLeuCysLysGluCysLeuValLeuValGluLysLysGluGluLysSerLysLysGluIleSerLeuGluThrLeuLysLeuValGlyLysLysMetMetAsn
CTGTCCCGGAACAAGGATGAAGTGAAGAAGGGCCTGGTCTTCGGGGGAGTTGACGGCGGCCAACAAGCGCTTCCTGTGATCAGCAGCCAACAAGGAGGTGGAAAAGATGAAT
                            450                            470                            480
                                            460

LysLysMetGluGluValLysLysGluGluAlaAsnIleArgValAlaValSerGluGluGlyAspPheLeuGluLeuAsnAspValSerGluAlaSerThrLysSerLeuSerGlySerLysLysGlyLysGlyValValLeuLeuSerGlyLysGlyLeuGlyLysAlaSerLeuGlyGluGlyLysLeuSerGlyGluLysLysLysGlyProGlyGlnValLeuGlyLysLysHisIleLeuSerPro
AAGAAGATGGAGGAAGTAAAGGAAGCCAACATCCGAGTTGTGTCAGAGGAGGGCGACTTCCTCGAGCTGAATGATGTCTTTAGCGAAGCTTCAGGAGTTGTCTTAGCGACCATCTTGTCCCCT
                            490                            510                            520
                                            500

TrpProAlaGluGluLysLysAlaGluGluProValGluValAlaValAlaLeuAlaProArgAlaArgGlyLysGlyGlyArgAsnLysLeuValSerGlyGluLeuLysGluSerLysGluLysSerLysGluLysSerLysGluLysSerLysGluLysSerLys
TGGGGGGCAGAGGGTGAAGGCGCAGAGCCAGTGGCCCCAGAAGTGGAAGGCTGCCTCCCAAGAGTGCTCGGCGTGGGCCTGCGCGCCAGGGGCCAGGCAGGAAGGGTATCAACAAATCTAAA
                            530                            550                            560
                                            540

LysArgMetLysLeuThrLeuLysGluThrLeuLysGlyValLysGlyAlaLysHisIleSerAlaHisValLeuGlyLysProAspSerProSerGlyLeuGluHisSerAlaHisValLeuGlyLysProAspSerProSerGlyLeuLysHisAla
AAGAGAATGAAATTAACTCTTAAAGGAGGACGAAGCTGTGGACTGAAGGTGGCCAGCTGGATCCAGTTGGACTGAAGGTGGCCAGCTGGATCCAGTTGGACTGAAGGTGGCCAGCTGGATC

                                      466
```

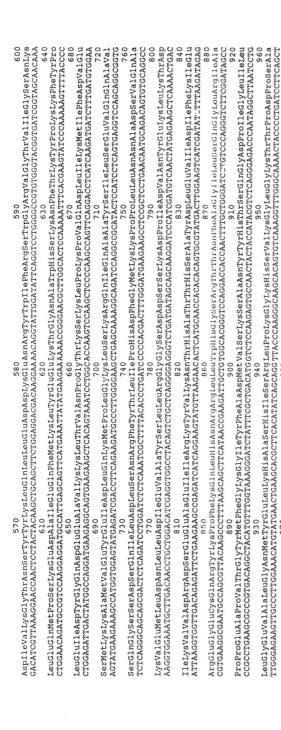

Fig. 2. Nucleotide sequence of the Okayayma-Berg pcD-poly(ADPR)p cDNA insert and the deduced amino acid sequence of the 113, 135-kDa protein. The protein contains sequences coding for three poly(ADP-ribose) polymerase peptides (underlined and sequentially numbered), in the 3' untranslated region; a putative mRNA processing signal (AATAAA) is underlined. In the 5' region two nucleotides that correspond to the Kozak criteria for initiation are underlined. (Taken from Ref. 3).

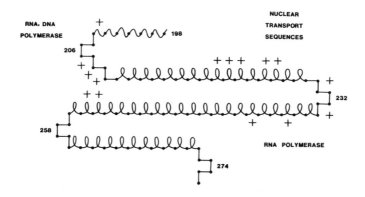

Fig. 3. Schematic diagram of a computer-derived secondary structure for poly(ADP-ribose) polymerase illustrating two helix structures together with regions which slow identity to the indicated proteins.

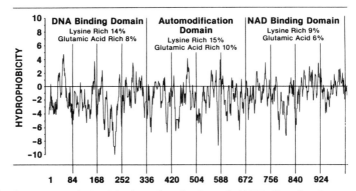

Fig. 4. A computer-derived hydropathy plot for poly(ADP-ribose) polymerase. Seven-residue averages have been plotted by using a program based on the procedure of Hopp and Woods. All values above the horizontal line in the hydropathy plot indicated hydrophobic character, value below the line indicated hydrophilic character of the peptide. The three domains of the polymerase are illustrated along with the atypical amino acid compositions. (Taken from Ref. 3).

reaction (cleavage of a glycosidic bond) of ricin. Other observed identities include two regions spanning amino acid residues 47-120 and 160-260 which exhibit a slight identity with a number of polymerases including the various subunits of *E. coli* RNA polymerase and phase T7 RNA polymerase as well as probable DNA polymerase from duck (residues 47-64 with 47% identity) and, DNA polymerase from Hepatitis B virus (residues 189-220 with 26% identity). Also a number of oncogenes such as N-Myc show some identity. Although of low significance (Z values 1.7-4.0) these proteins do not show some identities with other portions of the polymerase and may reflect a distant relationship.

468

Another interesting feature of this protein is the presence of a putative nucleotide binding fold similar to many DNA repair enzymes. A comparison of ATP-requiring enzymes has led to the suggestion of a consensus nucleotide binding fold. Similar sites are observed in the NAD binding domain of the polymerase (A site, residues 888-901 and B site, residues 940-957). The polymerase also contains regions within the DNA-binding domain which are suggestive of a Zn^{2+} binding finger. These sequences are of the form $CysX_2CysX_{25}HisX_2His$ (residues 21-53), $HisX_2CysX_5HisX_3His$ (residues 53-66) and $CysX_2CysX_{30}HisX_2Cys$ (residues 125-162). Since the NH_2- terminal portion of the polymerase specifically binds Zn^{2+}, Zn^{2+} binding fingers may be involved in polymerase-DNA recognition events.

We have used the 3.7 kb polymerase cDNA as a probe in Northern analysis to investigate the possible induction of mRNA for the enzyme during various cellular events.

Endogenous DNA strand breaks during DNA replication and induction of poly(ADP-ribose) polymerase mRNA. A number of normal biological processes commence via DNA breaks; accordingly, it was of importance to test whether in these cases induction of poly(ADP-ribose) polymerase transcripts occurred. Based upon experiments suggesting that cells accumulate DNA strand breaks when progressing from exponential toward stationary growth, we examined the levels of polymerase transcripts in logarithmically growing HeLa cells. Hybridization was noted solely to a 3.7 kb mRNA species which in earlier studies was shown to encode the polymerase (2). mRNA levels were highest at early log phase cell growth and decreased slowly as cells approached late exponential phase. At stationary densities, negligible polymerase mRNA was apparent in the cells, although significant quantities of potentially catalytically active enzyme are known to be present at this stage. In our experience, the polymerase mRNA levels are relatively low in cultured cells and seem relatively unstable compared to other transcripts examined. This property of the mRNA is in contrast to the high abundance of enzyme protein in the nuclei of eukaryotic cells and may be related to a low turnover of the enzyme protein in cells (N. Berger, personal communication).

Serum stimulation of WI-38 cells. mRNA levels in the fibroblast cell line WI-38 induced to enter the cell cycle by serum addition to starved cultures were investigated. The amount of RNA blotted to the filter was monitored by hybridization to a control cDNA, which hybridizes to a 0.9 kb transcript, and has been shown to be expressed equally during all phases of the cell cycle. Two major points emerge from the investigation of polymerase transcripts in this system. Even quiescent cells appear to contain basal levels of polymerase mRNA. Secondly, consistent with our earlier data with synchronized HeLa cells (2), higher levels of polymerase transcript accumulated in cells during the DNA replicative stage of the cell cycle (10).

Cell cycle analysis of polymerase transcription. Earlier data from our laboratory showed that in synchronized HeLa cells polymerase mRNA levels are low during the first 4 hr of S phase, however, a significant increase occurs at 5 hr and then again at 7 hr of S phase, which is preceeded by the peak of newly synthesized histone mRNA. These mRNA levels are consistent with high levels of enzymatic activity for poly(ADP-ribose) polymerase and for levels of poly(ADP-ribose) polymer *per se* occurring during the same time frames of S phase.

The relative transcript levels and transcription rates of the polymerase gene during the cell cycle in HeLa cells were determined using cells synchronized by a double thymidine block. Total RNA as well as nuclei from cells were isolated at hourly intervals during the progression of S phase and examined either by nuclear run-off transcription or Northern analysis (10).

We used cDNA to establish acceptors for poly(ADP-ribosyl)ation. Histones H3 and H4 were used as hybridization controls since the abundance of transcription of these histones during the S phase is well established (6). Transcripts for histones H3 and H4 were low at the beginning of S phase, became progressively abundant and reached a plateau about mid S phase. The histone transcripts then decreased during the end of S phase and disappeared by the G_2/S boundary. In contrast the polymerase transcripts peaked at mid to late S phase. Exactly the same fluctuations in polymerase transcripts were noted in an earlier study (2).

Measurements of transcription rates in isolated nuclei during the cell cycle. To determine whether the increased levels of polymerase mRNA noted above during S phase were due to increased rates of transcription, we performed nuclear run-off transcription assays. Nuclei were prepared hourly from 0 to 11 hr after the induction of the synchronized S phase. RNA was transcribed *in vitro* with [^{32}P]UTP, purified and hybridized to filters containing 2 μg of various non-radioactive cDNAs. In agreement with earlier data, the amount of control histone H4 gene (i.e. multi-gene family) run-off transcript was very high relative to other probes and its transcription peaked at 5 hr of S phase, consistent with the high levels for this transcript noted in Northern analysis above. Among the various transcripts tested, histone mRNA and poly(ADP-ribose) polymerase mRNA transcription rates were highest in S phase; however, the peak of polymerase mRNA transcription was at mid S and late S while histone H4 transcription peaked at early S phase. The transcription rate peak levels corresponded in time with peak levels in the Northern analysis. We have consistently observed, by both techniques, 3 distinct peaks of polymerase mRNA levels (i.e. approximately mid and late S phase, and at 10 hr after the onset of S phase).

Thus, the kinetics of polymerase transcription rate suggest that poly(ADP-ribosyl)ation may play a significant biological role during selected stages of DNA replication, as suggested by earlier work. For further controls, transcription of the non-histone proteins, HMG 14 and 17 was assessed during S phase. In general, the kinetics of run-off and mRNA levels for HMG 17 more closely followed the pattern of histone transcription than of poly(ADP-ribose) polymerase. In contrast to the HMG 17 regulation, it appears that the polymerase mRNA is rather unstable. The signal intensities for HMG 17 and polymerase are almost equivalent in run-off transcriptions. Secondly, in Northern analysis, the signals obtained for polymerase are not as high as those for HMG 17. Finally, polymerase transcription (by run-off) occurs throughout the periods studied (10), yet the steady state mRNA levels were almost negligible at certain time periods (i.e. hr 0-4, and 9, 11).

Replication-dependent polymerase induction. Nuclear transcription of histone genes is terminated when DNA replication is inhibited by *in vivo* incubation of nuclei of synchronized S phase cells with various DNA replication inhibitors (7). To test whether poly(ADP-ribose) polymerase transcripts are also coupled to DNA replication, HeLa cells were synchronized in S phase and at the G_1/S boundary (i.e. zero time). The cells were incubated in the presence or absence of hydroxyurea and the cultures were allowed to incubate for either 2 or 5 hr. The effect of the inhibitor of DNA synthesis on histone H4 transcription during S phase was striking. Neglible transcription of this gene was evident by 5 hrs (Fig. 5 A, B). In contrast, transcription of polymerase occurred to the same extent either in the presence or absence of DNA replication.

Enhanced rate of DNA strand break repair by hyper-expression of poly(ADP-ribose) polymerase cDNA. In the data in Fig. 1 above, we established that a significantly increased potential for poly(ADP-ribosyl)ation would exist around 2 days after transfection with pcD-12. We decided to use this time window to directly test whether poly(ADP-ribosyl)ation is involved in DNA repair by measuring the resealing of X-ray induced DNA strand breaks (8). Cos cells were prelabeled with [^{14}C]thymidine for two days, after which cells were either mock transfected, transfected with pcC-12 or carried through without any transfection protocol. After 44 hr, the cells were irradiated with 2000 rads on ice and immediately allowed to repair at 37°C. Cells were maintained on ice until assayed by alkaline elution. Unrepaired SSB (single strand breaks) in experimental DNA were compared to the control gamma-ray induced SSB and expressed as the radiation dose that would produce Rad-equivalent breaks. The alkaline elution method used was the short assay to detect SSB up to 3000 SSB rad-equivalents.

Cells were either mock transfected, transfected with pcD-12, or pcD-19 transfected (deleted in part of the first exon of polymerase). The alkaline elution curves are shown in Fig. 6 and a % DNA break frequency repair was calculated (Table 1). In this experiment the break frequency repair due to the hyper-expression of poly(ADP-ribose) polymerase was striking with approximately 96% calculated SSB rejoining by 15 min compared to 43%

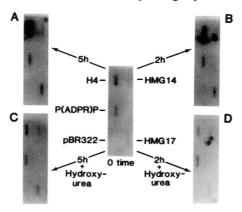

Fig. 5. Replication-dependence of histone, nonhistones and poly(ADP-ribose) polymerase mRNA transcription rates. HeLa cells were synchronized and S phase cells (0-Time) were incubated in the presence or absence of 0.1 mM hydroxyurea for 2 hr (B, D) or 5 hr (A, C). Nuclei were prepared and assayed for nuclear run-off transcription. (Taken from Ref. 10).

rejoining for mock transfected cells and 55% repair in the case of the 5' deleted plasmid. These differences reflect normal experimental variations in transfection efficiencies as well as in endogenous DNA breaks in cells receiving no radiation. To confirm that the increased repair of X-ray induced DNA strand breaks was due to increased poly(ADP-ribosyl)ation of Cos cells, the effects of the poly(ADP-ribose) inhibitor, 3-aminobenzamide (3-AB) was studied (Fig. 8, Table 1). DNA SSB rejoining was only moderately affected due to either 2.5 or 5 mM of 3-AB in both control transfections with about a 20% reduction in DNA repair rate at 15 min in cells transfected with pcD-19. This partial effect is consistent with earlier data which indicates that under similar conditions poly(ADP-ribosyl)ation inhibitors slow, but do not prevent X-ray induced DNA strand break resealing (9). Consistent with these data, the current approach utilizing a cloned expression vector suggests that the initial rate of DNA strand rejoining is accelerated due to the poly(ADP-ribosyl)ation modification. The results presented above indicate the potential of utilizing the human poly(ADP-ribose) polymerase clone to study DNA repair mechanisms.

Chromosome mapping and polymorphisms. Using the 3.7 kb cDNA insert as a probe for Southern analysis of EcoRI digests, a complex pattern of 2.3, 5.3, 6.8, 7.0, 8.0, and 25 kb hybridizing bands were detected in

472

human DNA while 2.0 and 25 kb or 1.2, 2.3, 2.9, 10, and 11 kb cross-hybridizing fragments were found in Chinese hamster or mouse DNAs, respectively. Analysis of 93 man-mouse and human-hamster *Eco*RI digested somatic cell hybrid DNAs with the 3.7 kb cDNA probe demonstrated that the 2.3, 7.0, 8.0, and 25 kb sequence all segregate onco-dantly with chromosomal DNA while the 5.3 kb fragment is located on chromosome 13 and the 6.8 kb band is present on chromosome 14. Examination of hybrids containing random breaks or specific translocations permitted assignment of the sequences on chromosome 1 to the long arm or short arm proximal to the *N-ras* proto-oncogene while the sequence on chromosome 14 is located above 14q32 (3). A representative Southern blot of a restriction digest is shown in Fig. 7.

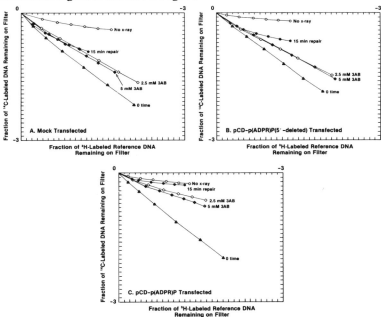

Fig. 6. Alkaline elution of X-ray induced SSB following repair for 15 min in mock (A), pcD-19 (B), or pcD-12 (C) transfected Cos cells. Each of the cells were allowed to repair in the absence (●) or presence of 2.5 mM (◇) or 5 mM 3-AB (◆). The alkaline elution curves for cells receiving no damage (○) and cells not subjected to repair after damage (▲) are also depicted. Cos cells, labelled with [^{14}C]thymidine for 48 hr were transfected with the appropriate plasmid or mock transfected. Following transfection 42 hr cells were detached and suspended in cold medium. Each sample, except controls, was irradiated with 2000 rads of X-ray. For assessing DNA breaks, cells were lysed in SDS in the presence of proteinase K. Elution curves, normalized with respect to internal standards, were plotted on a double log scale to represent retention of ^{14}C-labelled DNA versus retention of reference DNA (^{3}H-labelled). (Taken from Ref. 8).

After removing the probe, the same blots were sequentially hybridized with an 0.9 kb 5' cDNA fragment (*Pst*I-*Hind*III), a 1.95 kb 5' cDNA (*Pst*I-*Pst*I) sequence, a 1.6 kb 3' cDNA (*Pst*I-*Hind*III) fragment, and a

1.3 kb internal, predominantly 3' cDNA fragment (about 2.0 kb - 3.3 kb cDNA map). The 5.3 sequence on chromosome 13 was detected with each of these probes whereas the 6.8 kb band on chromosome 14 was not observed with either of the 5' cDNA probes. The analysis also allowed the cluster of EcoRI fragments on chromosome 1q to be ordered as follows: 5' flank - 25 kb - 7.0 kb - 8.0 kb - 2.3 kb - 3' flank. The cDNA contains no EcoRI site as the junction of the 7.0 kb and 8.0 kb fragments and this site must occur within an intron. Any fragments which are entirely intronic would not be detected with the cDNA probe, of course. A summary of the percent discordancy is shown in Table 2.

Table 1. 5' deletion of polymerase cDNA or 3-aminobenzamide abolishes increased DNA repair.

Transfection Conditions	DNA Break Frequency in 15 min		
	0	2.5	5.0
	3-Aminobenzamide (mM)		
Mock	43*	32	38
pcD-19	55	29	30
pcD-12	96	73	64

*43% represents a rad equivalent of 1136.

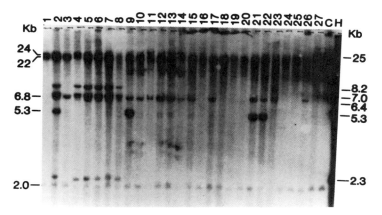

Fig. 7. Southern hybridization of representative EcoRI-digested human-hamster somatic cell hybrid DNAs with a full length 3.7 kb human polymerase cDNA probe. A different hybrid cell DNA is present in lanes 1-27. Parental Chinese hamster (lane C) and human placental (lane H) DNAs are also shown. The size of hybridizing human (2.3, 5.3, 6.4, 6.8, 7.0, 8.2, and 25 kb) and hamster (2.0, 22, and 24 kb) fragments are depicted. The 5.3 kb human band was detected in lanes 2, 9, 21, and 22, whereas the 6.8 kb band was observed in lanes 2-15, 17, 21-23, and 26; the remaining human bands were found in lanes 2, 4-8, and 23. The independent segregation of hybridizing human sequences indicates that they are present at three different loci. (Taken from Ref. 3).

We interpret these results to suggest that a large (> 18-42 kb) functional gene is located on chromosome 1q and the sequences on chromosomes 13

474

and 14 most likely represent processed pseudogenes. A tentative map for the structure of the active poly(ADP-ribose) polymerase gene is given in Fig. 8. The localization of sequences to chromosomes 1q, 13 and 14 was confirmed by Southern analysis of *Hind*III digests of the somatic cell hybrid DNAs (3).

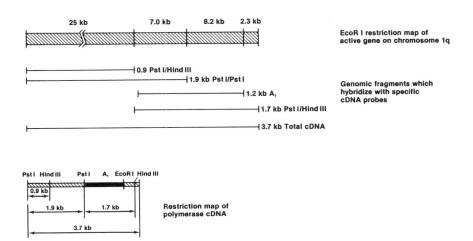

Fig. 8. Genomic organization of poly(ADP-ribose) polymerase. The alignment of similar *Eco*RI sites present in the genomic and cDNA map is illustrated together with cDNA probes used to order the *Eco*RI fragments. The approximate location in the genomic map for the *Pst*I cDNA site is also shown. (Separate scales have been used to construct the genomic and cDNA maps.) b, Base. (Taken from Ref. 3).

Several simple two allele polymorphisms were detected with the poly(ADP-ribose) polymerase cDNA probes including 2.6 kb and 2.9 kb *Hind*III alleles on chromosome 13 (0.12:0.88 frequencies) using the 3.7 kb probe. Allelic 5.6 kb and 7.9 kb *Pst*I bands (0.10:0.90 frequencies) are probably located on chromosome 13 since there is strong linkage disequilibrium with the *Hind*III alleles. Allelic 0.9 kb and 1.2 kb *Pst*I bands (0.72:0.28 frequency) are probably located on chromosome 1 and this polymorphism is detected only with the 0.9 kb 5' cDNA probe; an overlapping invariant 0.9 kb band with the 3.7 kb probe obscures the polymorphism. Allelic 1.4 kb and 11.7 kb *Sst*I bands (0.55:0.45) are best detected with the 0.9 kb 5' cDNA probe and probably are located on chromosome 1. Allelic 4.2 kb and 4.4 kb *Taq*I bands (0.85:0.15 frequency) have not been chromosomally localized and the polymorphic site lies < 200 bp 3' to the *Pst*I site at 1.9 kb and the sequence extends 4.2 kb 3' to this site. Four different polymorphisms were detected with *Msp*I but none have been chromosomally localized.

475

Table 2. Segregation of poly(ADP-ribose) polymerase gene in human-rodent hybrids.

Human Chromosome	% Discordancy		
	Gene	P1	P2
1	0	29	45
2	13	33	45
3	29	46	60
4	32	54	50
5	20	45	40
6	24	43	24
7	40	21	45
8	36	47	39
9	23	34	45
10	27	38	61
11	23	40	50
12	27	34	34
13	32	4	51
14	35	35	7
15	38	27	32
16	36	38	31
17	40	34	46
18	43	28	44
19	19	35	44
20	23	43	43
21	51	42	48
22	24	44	50
X	52	56	45

The poly(ADP-ribose) polymerase gene (column 2) was detected as 2.3, 7.0, 8.0, and 25 kb hybridizing bands (27 positive hybrids) in EcoRI digests of human-rodent somatic cell hybrid DNAs or as a 5.3 kb sequence (column 3) or 6.8 kb band (column 4). Detection of each sequence, or group of sequences, is correlated with the presence or absence of each human chromosome in the somatic cell hybrids. Discordancy indicates the presence of hybridizing sequences in the absence of the chromosome or absence of the hybridizing bands despite the presence of the chromosome; the sum of these numbers divided by total hybrids examined (X100) represents percent discordancy. The human-hamster hybrids consisted of 26 primary clones and 15 subclones and the human-mouse hybrids contained 13 primary hybrids and 42 subclones. The 5.3 kb and 6.8 kb human poly(ADP-ribose) polymerase sequences were detected in 35 and 54 hybrid cell DNAs, respectively.

Acknowledgements. We gratefully acknowledge Drs. Elizabeth Slattery and Robert Roeder for their generous gifts of poly(ADP-ribose) polymerase and antibody. We thank Dr. Hiroto Okayama for useful discussions and for sharing his library with us. This research was supported by grants CA13195 and CA25344 from the National Cancer Institute and grant OSR-86-0024 from the U.S. Air Force, Office of Scientific Research.

References

1. Alkhatib, H., Smulson, M. (1986) Fed Proc 45: 1589
2. Alkhatib, H.M., Chen, D., Cherney, B., Bhatia, K., Notario, V., Giri, C., Stein, G., Slattery, E., Roeder, R.G., Smulson, M.E. (1987) Proc Natl Acad Sci USA 84: 1224-1228
3. Alkhatib, H.M., Chen, D.F., Cherney, B., Bhatia, K., Notario, V., Giri, C., Stein, G., Slattery, E., Roeder, R.C., Smulson, M.E. (1987) Proc Natl Acad Sci USA 84: 1224-1228

4. Slattery, E., Dignam, J.D., Matsui, T., Roeder, R.G. (1983) J Biol Chem 258: 5955-5959
5. Rittling, S.R., Brooks, K.M., Cristofalo, V.J., Baserga, R. (1986) Proc Natl Acad Sci USA 83: 3316-3320
6. Stein, G.S., Borun, T.W. (1972) J Cell Biol 52: 292-307
7. Plumb, M.J., Stein, G. (1983) Nucleic Acids Res 1: 2391-2410
8. Bhatia, K., Giri, C., Pommier, Y., Cherney, B., Dritschilo, A., Alkhatib, H., Smulson, M. (1987), Submitted
9. Zwelling, L.A., Kerrigan, D., Pommier, Y. (1982) Biochem Biophys Res Commun 104: 897-902
10. Bhatia, K., Busten, M., Breitman, T., Cherney, B., Dritschilo, A., Forance, A., Stein, G., Smulson, M., Submitted

Adenosine Diphosphate Ribosylation of Elongation Factor-2 as a Therapeutic Target: Genetic Construction and Selective Action of a Diphtheria Toxin-Related Interleukin-2 Fusion Protein

Diane P. Williams[1], Patricia Bacha[2], Vicki Kelley[3], Terry B. Strom[4], and John R. Murphy

Evans Department of Clinical Research and Department of Medicine, The University Hospital, Boston, Massachusetts 02118 USA

Introduction

The mature form of diphtheria toxin (535 amino acid protein) is selectively cleaved by trypsin into two polypeptides. The N-terminal 192 amino acid polypeptide, fragment A, has been shown to carry the catalytic center for ADP-ribosylation of eukaryotic elongation factor 2 (EF-2). Fragment B of diphtheria toxin, a 343 amino acid polypeptide, contains the lipid associating regions that facilitate the membrane translocation of fragment A into the cytosol of sensitive cells, as well as the diphtheria toxin receptor binding domain (1). Murphy *et al.* (2) have recently shown that the C-terminal 50 amino acids of the toxin molecule are required for the formation of the receptor binding domain. The process by which diphtheria toxin intoxicates sensitive eukaryotic cells involves at least the following steps: (i) the binding of diphtheria toxin to the cell surface receptor (3), (ii) internalization of toxin by receptor mediated endocytosis (4), and upon acidification of the endosome, (iii) a partial unfolding of fragment B, thereby exposing the lipid associating regions of toxin (5, 6), and facilitating (iv) the membrane translocation of fragment A to the cytosol. Yamiazumi *et al.* (7) have elegantly demonstrated that the introduction of a single molecule of fragment A into the cytosol is sufficient to be lethal for that cell.

Bacha *et al.* (8) demonstrated that a conjugate toxin composed of the non-toxic diphtheria toxin-related protein CRM45 which was disulfide cross-linked to the polypeptide thyrotropin releasing hormone (TRH) was selectively toxic for TRH receptor bearing eukaryotic cells *in vitro*. Moreover, the administration of the CRM45-TRH conjugate to rats resulted

---

[1]Department of Microbiology, Boston University School of Medicine, Boston, MA 02118

[2]Seragen, Inc., Hopkinton, MA 01748

[3]Department of Medicine, Brigham and Women's Hospital, Boston, MA 02115

[4]Charles A. Dana Research Institute, Harvard Thorndike Laboratory of the Beth Israel Hospital, Harvard Medical School, Boston, MA 02215

in organ-specific binding (9). These studies, as well as many other investigations with immunotoxins and monoclonal antibodies disulfide cross-linked to fragments of microbial or plant toxins (10-12), suggested that a molecular genetic approach to chimeric toxin development might be feasible. Since the first step in the intoxication process is the binding of diphtheria toxin to its receptor on the cell surface, we reasoned that receptor binding domain substitution would result in the formation of recombinant toxins with unique target cell specificity. In order to design these recombinant fusion proteins we have used protein engineering and recombinant DNA methodologies in order to replace the diphtheria toxin receptor binding domain with either α-melanocyte stimulating hormone (2), or the T-cell specific growth factor interleukin-2 (IL-2) (13). In these instances it was well known that the respective ligand bound to a specific receptor on the surface of target cells, and that bound ligand was internalized by receptor mediated endocytosis in a manner analogous to diphtheria toxin itself. Once cloned on plasmid vectors in *E. coli* K12, the diphtheria toxin-related polypeptide ligand fusion genes have been found to direct the expression of their respective chimeric toxin in recombinant strains of *E. coli* (19, 31).

In the present report, we describe the genetic construction, expression, and cell receptor targeted toxicity of a fusion protein composed of the first 485 amino acids of diphtheria toxin linked to amino acids 2 through 133 of IL-2. The fusion protein, IL-2-toxin, is shown to be selectively toxic toward high affinity IL-2 receptor bearing cells *in vitro,* and to block an activated T-cell mediated delayed type hypersensitivity (DTH) reaction *in vivo*.

Results and Discussion

The expression of IL-2 and the appearance of the IL-2 receptor on the T-cell surface mark critical events in the induction of an immune response (14). Upon stimulation of the resting T-cell with either antigen or mitogen in the presence of interleukin-1 (IL-1), IL-2 receptors are expressed *de novo* (15, 16). The IL-2 receptor has been shown to fall into "high" and "low" affinity classes (17, 18), and recent studies have demonstrated that the high-affinity receptor is composed of at least a 55,000 (p55) and 70,000 (p70) dalton glycoprotein complex (19, 20). Individually, each of the two subunits has been shown to bind IL-2 with "low" to "moderate" affinity; however, only the high-affinity receptor appears to mediate internalization of the growth factor by receptor mediated endocytosis (21, 22). Since the high-affinity IL-2 receptor appears to be limited in distribution to activated T-cells, recently activated B-cells and macrophages, a potent cytotoxic agent directed toward this receptor would be anticipated to be a potent immunosuppressive agent. As such, IL-2-toxin may have clinical application for a variety of autoimmune disorders, acute allograft rejection, and for treatment of T-cell dyscracias which are characterized by the presence of high-affinity IL-2 cell surface receptors.

A chimeric diphtheria toxin-related IL-2 fusion gene was assembled from plasmid pABC508 and a synthetic gene encoding IL-2 and a translational signal. Plasmid pABC508 encodes the diphtheria *tox* promoter, the *tox* signal sequence, and Glyl through Ala485 of the mature form of diphtheria toxin (23). The region of the *tox* gene encoding Ala485 is also defined by a unique *Sph*l restriction endonuclease site. Williams *et al.* (24) have designed, synthesized, and cloned a synthetic gene encoding human IL-2 using *E. coli* preferred codon usage. The synthetic IL-2 gene is carried on plasmid pDW15, and its 5'-end is defined by a *Sph*l restriction site that is positioned to retain IL-2 translational reading frame following ligation to the *Sph*l site of the truncated diphtheria *tox* gene. As shown in Fig. 1, the chimeric diphtheria toxin-related IL-2 fusion gene was constructed by the molecular cloning of a 428 base pair (bp) *Sph*l-*Sal*l fragment of pABC508 into the *Sph*l-*Sal*l sites of pDW15. The resulting recombinant plasmid, pABI508, encodes the IL-2-toxin fusion protein.

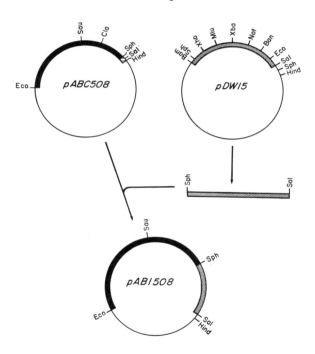

Fig. 1. Plasmid constructs used in the assembly of a diphtheria toxin-related/interleukin-2 fusion gene. Plasmid pABC508 carries the genetic information encoding the diphtheria *tox* promoter, *tox* signal sequence, and the *tox* structural gene from Glyl to Ala485. Plasmid pDW15 carries the synthetic human IL-2 encoding gene. Vectorial cloning of the 428 bp *Sph*l-*Sal*l fragment of pDW15 into *Sph*l-*Sal*l digested pABC508 results in the formation of the IL-2-toxin encoding pABI508 (from ref. 13).

480

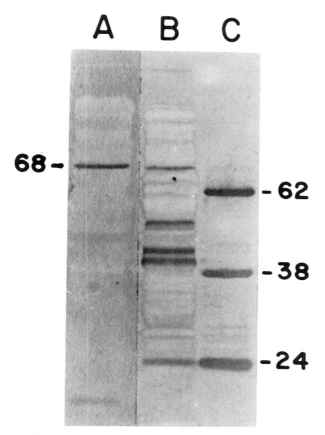

Fig. 2. Immunoblot analysis of crude periplasmic extracts of *E. coli* (pABI508) following SDS-polyacrylamide gel electrophoresis and transfer to nitrocellulose paper. Lane A, *E. coli*(pABI508) extracts probed with monoclonal antibody to recombinant IL-2; Lane B, *E. coli*(pABI508) extracts probed with polyclonal anti-diphtheria toxin serum; Lane C, partially "nicked" diphtheria toxin probed with anti-toxin. Apparent molecular weights x 10^{-3} are indicated.

Cloned diphtheria *tox* gene fragments that contain the *tox* promoter and *tox* signal sequence have been shown to be constitutively expressed in recombinant strains of *E. coli* (25-27). In addition, *tox* gene products that contain the 25 amino acid signal sequence have been shown to be exported to the periplasmic compartment by the SecA apparatus in *E. coli* K12 (26). *E. coli* (pABI508) was grown in 10 liter volumes of Luria broth at 30°C to an absorbance of 1.8 - 2.2. Periplasmic extracts were prepared as previously described (25), and analyzed by sodium dodecyl sulfate (SDS) polyacrylamide gel electrophoresis and immunoblotting. As can be seen in Fig. 2 (lane A), a single protein with an apparent molecular weight of 68,000 was detected in immunoblots probed with monoclonal antibodies against IL-

2. In addition, immunoblots probed with anti-diphtheria toxin sera also reveal an M_r 68,000 (lane B). Several additional protein bands with an M_r that range between 56,000 to 24,000 are also visualized on immunoblots probed with anti-diphtheria toxin sera. Since immunoblots probed with anti-IL-2 reveal only a single protein, we conclude that the multiple anti-toxin reactive proteins are the result of proteolytic degradation of the M_r 68,000 full length fusion protein by an *E. coli* endoprotease(s) (28). Furthermore, the presence of only a single IL-2 immunoreactive species strongly suggests that there is a proteolytic cleavage site close to, or at the toxin/IL-2 fusion junction, and that the additional cleavage sites are at specific amino acid residues within the fragment B portion of the fusion protein. Based upon the nucleic acid sequence analysis the molecular weight of the mature form of IL-2-toxin was anticipated to be 68,086. The appearance of a 68,000 dalton protein that retains immunologic determinants intrinsic to both IL-2 and diphtheria toxin is in excellent agreement with the expected molecular weight and properties of the IL-2-toxin fusion protein.

Table 1. Eukaryotic cell lines used in this study.

Cell Line	Species	Tissue Type	High affinity IL-2 receptor	Sensitivity to IL-2-toxin
C91/PL	human	cord T-cell HTLV-I transformed	+	+
C10/MJ	human	cord T-cell HTLV-I transformed	+	+
MT-2	human	cord T-cell HTLV-I transformed	+	+
Hut 102/6TG	human	T-cell (ATL)	+	+
CTLL-2	murine	T-cell (cytotoxic)	+	+
CEM-EM3	human	T-cell (ALL)	-	-
CCRF HSB-2	human	T-cell (ALL)	-	-
K562	human	granulocyte (CML)	-	-

ATL, adult T-cell leukemia; ALL, acute lymphoblastic leukemia; CML, chronic myelogenous leukemia.

Williams *et al.* (13) have described the partial purification of IL-2-toxin from periplasmic extracts of *E. coli* (pABI508) by immunoaffinity chromatography using immobilized monoclonal anti-IL-2. Following partial purification, IL-2-toxin was tested for biologic activity against a variety of eukaryotic cell lines. As seen in Table 1, there is a direct correlation between the sensitivity of a given cell line to IL-2-toxin and the presence of high-affinity IL-2 receptors. In order to further characterize the action of IL-2-toxin, Bacha *et al.* (29) have used a series of cell receptor specific agents as potential competitive inhibitors. Table 2 shows that both excess free recombinant IL-2 (rIL-2), as well as a monoclonal antibody that binds to the p55 subunit of the high-affinity IL-2 receptor were able to block the IL-2-toxin mediated inhibition of protein synthesis in C91/PL cells. In contrast, transferrin which binds to the transferrin receptor, or the monoclonal antibody 4F2, which is directed to an early activation antigen on the T-cell surface, fail to block the action of IL-2-toxin. These results clearly suggest that the IL-2-toxin mediated inhibition of protein synthesis is mediated through the IL-2 receptor.

It is widely known that following binding to its receptor on the cell surface, diphtheria toxin is internalized by receptor mediated endocytosis from clathrin coated pits (4). In addition, the acidification of the eukaryotic vesicle by a cellular proton pump is an essential step in the diphtherial intoxication process. Lysosomatrophic agents that block the acidification process have also been shown to block the action of diphtheria toxin. In similar experiments, Bacha *et al.* (29) have shown that the addition of chloroquine to C91/PL cells in culture completely blocks the action of IL-2-toxin (Table 2). These results clearly demonstrate that IL-2-toxin, like diphtheria toxin, must pass through an acidic vesicle in order to deliver its fragment A to the cytosol of target cells.

Since we have shown that IL-2-toxin action is effected through the IL-2 receptor on HTLV-I infected transformed T-cell lines, one could argue that the IL-2-toxin mediated inhibition of protein synthesis was due to steric hindrance of the receptor rather than ADP-ribosylation of target cell EF-2. In order to examine the mechanism of inhibition of protein synthesis, we have exposed both the IL-2 dependent murine CTLL-2 cell line, as well as IL-2 independent human C91/PL cells, to both diphtheria toxin and IL-2-toxin. Following a 24 hr incubation, cells were lysed and the level of EF-2 available for ADP-ribosylation was determined by the addition of purified diphtheria toxin fragment A and [^{32}P]NAD, and measurement of the transfer of [^{32}P]ADP to elongation factor 2, as described by Moynihan and Pappenheimer (30). As seen in Table 3, CTLL-2 T-cells, which are resistant to diphtheria toxin, have reduced levels of elongation factor 2 available for ADP-ribosylation only after exposure to IL-2-toxin. In contrast, human

Table 2. Incorporation of [14C]leucine by human C91/PL IL-2R+ T-cells following a 24 hr exposure to IL-2-toxin in the presence or absence of various additions (adapted from ref. 29).

IL-2-toxin concentration	Additions	% Control Incorporation
-	-	100
4×10^{-9} M	-	14
"	IL-2 (10^{-7})	111
"	33B.3 (10^{-7})	72
"	Transferrin (10^{-7})	17
"	4F2 (10^{-7})	15
-	Chloroquine (6×10^{-6})	85
1×10^{-9} M	-	38
"	Chloroquine (6×10^{-6})	90

C91/PL T-cells, which are sensitive to both diphtheria toxin and IL-2-toxin, have reduced levels of elongation factor 2 following exposure to either toxin. These experiments strongly argue that the cytotoxic action of IL-2-toxin is due to the ADP-ribosylation of target cell elongation factor 2, rather than simply blocking the IL-2 receptor.

Since the *de novo* expression of the IL-2 receptor is a critical event in the initiation of an immune response, several investigators have examined the effect of targeting this receptor with monoclonal antibodies as a means of immunosuppression. Indeed, it has been recently shown that targeting the low affinity p55 subunit of the IL-2 receptor with the rat anti-mouse IgM monoclonal antibody M7/20 effectively suppresses the immune response in a delayed type hypersensitivity (DTH) model (31), prolongs allograft survival in both mouse and rat models of cardiac transplantation (32, 33), and blocks diabetic insulitis in the NOD mouse and lupus nephritis in the NZB/w mouse (31). In these instances it is doubtful that immunosuppression is induced by blocking the IL-2 receptor, since successful therapy required both complement activation and a monoclonal antibody that recognizes an IL-2 binding domain of the IL-2 receptor (34). Since the affinity of IL-2 for the high-affinity IL-2 receptor is 100 - 1000 fold higher than the affinity of monoclonal anti-IL-2 receptor antibodies for the receptor, we anticipated that target cell killing by the ADP-ribosylation of elongation factor 2 would make IL-2-toxin an even more potent immunosuppressive agent. Moreover, since target cell killing by IL-2-toxin does not require the action of complement, we reasoned that IL-2-toxin should be a more versatile therapeutic agent.

Table 3. Murine CTLL-2 and human C91/PL T-cell elongation factor 2 available for ADP-ribosylation following 24 hr exposure to either diphtheria toxin or IL-2-toxin.

Cell Line	Toxin and Concentration	% control level of EF-2 available for ADP-ribosylation
CTLL-2	Diphtheria (10^{-7} M)	98
"	IL-2-toxin (10^{-8} M)	8
C91/PL	Diphtheria (10^{-6} M)	< 5
"	IL-2-toxin (10^{-8} M)	< 5

We have examined the immunosuppressive effect of IL-2-toxin in a murine DTH model. BALB/c mice were immunized by subcutaneous injection of the hapten trinitrobenzenesulfonic acid (TNBS). Seven days following immunization, mice were challenged with a 25 µl injection of TNBS into the right footpad. After 24 hr, bilateral footpad thickness was measured with a micrometer in order to assess the degree of the immune response. DTH units were defined as differences of 0.01 mm in the thickness between the injected and the non-injected footpad of each animal. Phosphate buffered saline (25 µl) was injected into the footpads of nonimmunized mice to determine background values. In the case of those animals that were treated, IL-2-toxin was administered daily by intravenous injection from the day of priming to the day of challenge.

Table 4. IL-2-toxin suppresses delayed type hypersensitivity in BALB/c mice.

IL-2-toxin ng/day/mouse	DTH units
-	45 ± 3.5
500	1 ± 2.1 *
50	2 ± 2.1 *
5	15 ± 4.1 *

* $p < 0.001$; values = mean $\pm$ SEM; 5 mice/group.

As shown in Table 4, treatment of TNBS immunized BALB/c mice with IL-2-toxin was found to induce a marked immunosuppression. Indeed, administration of the 50 - 500 ng IL-2-toxin per day for 6 days was found to virtually block the immune response. Even administration of IL-2-toxin at a concentration of 5 ng per day was found to induce a marked immunosuppression in the treated animals. Moreover, flow cytometric analysis of T-cell subsets in the draining lymph nodes of control and IL-2-toxin treated animals revealed that the only difference between these two groups was the elimination of the activated high-affinity IL-2 receptor bearing subset of cells from the T-cell population in the treated group (data not shown). These observations strongly suggest that IL-2-toxin induced immunosuppression results from the selective elimination of those T-cell clones that were activated by TNBS immunization.

In summary, we have designed and genetically constructed a novel chimeric toxin that is composed of portions of diphtheria toxin linked to IL-2 through a peptide bond. IL-2-toxin has been found to be a potent cytotoxin against high-affinity IL-2 receptor bearing T-cells *in vitro*. We have shown that the action of IL-2-toxin is mediated through the IL-2 receptor, and further that inhibition of protein synthesis results from the ADP-ribosylation of target cell elongation factor 2. Studies of IL-2-toxin action in a delayed type hypersensitivity model in the mouse have shown that this chimeric toxin is directed to high-affinity IL-2 receptor bearing cells *in vivo*.

Acknowledgements. The work was supported in part by Public Health Service grants AI-21628 (J.R.M.) and AI-22882 (T.B.S.) from the National Institute of Allergy and Infectious Diseases, and CA-41746 (J.R.M.) from the National Cancer Institute, and a grant from Seragen, Inc., Hopkinton, MA.

References

1. Pappenheimer, A.M., Jr. (1977) Ann Rev Biochem 46: 69-94
2. Murphy, J.R., Bishai, W., Borowski, M., Miyanohara, A., Boyd, J., Nagle, S. (1986) Proc Natl Acad Sci USA 83: 8258-8262

3. Middlebrook, J.L., Dorland, R.B., Leppla, S.H. (1978) J Biol Chem 253: 7325-7330
4. Moya, M., Dautry-Varsat, A., Goud, B., Louvard, D., Boquet, P. (1985) J Cell Biol 101: 548-559
5. Sandvig, K., Tonnessen, T.I., Sand, S., Olsnes, S. (1986) J Biol Chem 261: 11639-11644
6. Eisenberg, D., Schwartz, E., Komaromy, M., Wall, R. (1984) J Mol Biol 179: 125-142
7. Yamaizumi, M., Mekada, M., Uchida, T., Okada, Y. (1978) Cell 15: 245-250
8. Bacha, P., Murphy, J.R., Reichlin, S. (1983) J Biol Chem 258: 1565-1570
9. Bacha, P., Murphy, J.R., Reichlin, S. (1983) Endocrinol 113: 1072-1076
10. Vitetta, E.S., Krolick, K.A., Miyama-Inaba, M., Cushley, W., Uhr, J.W. (1983) Science 219: 644-650
11. Neville, D.M., Jr. (1986) Crit Rev Ther Drug Carrier Sys 2: 329-344
12. Pastan, I., Willingham, M.C., Fitzgerald, D.J. (1986) Cell 46: 641-648
13. Williams, D.P., Parker, K., Bishai, W., Borowski, M., Genbauffe, F., Strom, T.B., Murphy, J.R. Submitted
14. Smith, K.A. (1984) Ann Rev Immunol 2: 319-333
15. Cotner, T., Williams, J.M., Christensen, L., Reddish, T., Shapiro, H.J., Strom, T.B., Strominger, J. (1983) J Exp Med 157: 461-472
16. Williams, J.M., Loertscher, R., Cotner, T., Reddish, M., Shapiro, H.J., Carpenter, C.B., Strominger, J., Strom, T.B. (1984) J Immunol 132: 2330-2337
17. Robb, R.J., Greene, W.C., Rusk, C.M. (1984) J Exp Med 160: 1126-1146
18. Lowenthal, J.W., Zubler, R.H., Nabholz, M., Mac Donald, H.R. (1985) Nature 325: 669-672
19. Robb, R.J., Rusk, C.M., Yodoi, J., Greene, W.C. (1987) Proc Natl Acad Sci USA 84: 2002-2006
20. Kuo, L-M., Rusk, C.M., Robb, R.J. (1986) J Immunol 137: 1544-1551
21. Fugii, M., Sugimura, K., Sano, K., Naki, M., Sagita, K., Hinuma, Y. (1986) J Exp Med 163: 550-562
22. Weissman, A.M., Harford, J.B., Svetlik, P.B., Leonard, W.L., Depper, J.M., Waldmann, T.A., Greene, W.C. (1986) Proc Natl Acad Sci USA 83: 1463-1466
23. Bishai, W.R., Miyanohara, A., Murphy, J.R. (1987) J Bacteriol 169: 1554-1563
24. Williams, D.P., Regier, D.A., Akiyoshi, D., Genbauffe, F., Murphy, J.R. In preparation
25. Leong, D., Coleman, K.D., Murphy, J.R. (1983) J Biol Chem 258: 15016-15020
26. Leong, D., Coleman, K.D., Murphy, J.R. (1983) Science 220: 515-517
27. Tweten, R.K., Collier, R.J. (1983) J Bacteriol 156: 680-685
28. Swamy, K.H.S., Goldberg, A.L. (1982) J Bacteriol 149: 1027-1033
29. Bacha, P., Waters, C., Williams, J.M., Murphy, J.R., Strom, T.B. J Exp Med Submitted
30. Moynihan, M., Pappenheimer, A.M., Jr. (1981) Infect Immun 32: 575-582
31. Kelley, V.E., Gaulton, G.N., Strom, T.B. (1987) J Immunol 138: 2771-2775
32. Kirkman, R.L., Barrett, L.V., Gaulton, G.N., Kelley, V.E., Ythier, A., Strom, T.B. (1985) J Exp Med 162: 358-362
33. Kupiec-Weglinski, J.W., Diamantstein, T., Tilney, N.L., Strom, T.B. (1986) Proc Natl Acad Sci USA 83: 2624-2627
34. Kelley, V.E., Gaulton, G.N., Strom, T.B. (1987) J Immunol 138: 2771-2778

Molecular Structure of Poly(ADP-Ribose) Synthetase

Yutaka Shizuta, Tomohiro Kurosaki, Hiroshi Ushiro, Shigetaka Suzuki, Yasuhiro Mitsuuchi, Michiko Matsuda, Katsumi Toda, Yuichi Yokoyama, Yasutake Yamamoto, and Kenichi Ito

Departments of Medical Chemistry and Internal Medicine, Kochi Medical School, Kochi 781-51, Japan

Introduction

During the past several years, it has become clear that poly(ADP-ribose) synthetase has two unique features (1, 2). One is that the enzyme requires DNA for catalytic activity and another is that the enzyme is subjected to automodification during the reaction. These two unique features will provide us with a key for clarifying the physiological function of this enzyme *in vivo*. In this article, we will present our recent data on molecular cloning of human poly(ADP-ribose) synthetase and will discuss the physiological functions of this enzyme on the basis of its structural characteristics.

Results

Poly(ADP-ribose) synthetase can be cleaved by limited proteolysis into three domains, the first for binding of DNA, the second for automodification and the third for binding of the substrate, NAD (3-5). As the initial step for cDNA cloning, we first determined the partial amino acid sequence of the NAD binding domain since the N terminus of the DNA binding domain as well as that of the native enzyme is blocked. We then prepared 4 sets of oligonucleotide probes and isolated a cDNA clone by screening a human fibroblast cDNA library. Since the enzyme was originally purified from human placenta, we simultaneously screened the human placental cDNA libraries and isolated 6 different clones.

Fig. 1 shows a total of 3,800 nucleotides excluding the poly(dA) tract of the cDNA for human poly(ADP-ribose) synthetase. Above the nucleotide sequence, the deduced amino acid sequence is also presented. Six triangles show the sites cleaved by papain and by α-chymotrypsin. The major site cleaved by papain is shown by the second triangle and the major site split by α-chymotrypsin is shown by the fifth triangle. Therefore, the region starting from the initial methionine to the residue at 373 corresponds to the DNA binding domain, the region from the residue at 374 through the residue at 525 consists of the automodification domain and the residual sequence from the residue at 526 through the C terminal tryptophan corresponds to the NAD binding domain.

487

```
5'----------------TGCGGCTGGGTGAGCGACGCACGCGAGGCGGCGAGGCGGCGAGCGTGTTTCTTAGGTCGTGGCGTCGGGCTTCCGGAGCTTTGGCGGCAGCTAGGGGAGG       -1

                  10            20            30            40            50
MetAlaGluSerAspLysTyrArgValTyrArgValAlaLysSerGluSerIleProLysSerLysCysSerGluSerIleMetMetValIleMetValIleSerProMetProPheAspGlyLysValProHis
ATGGCGGAGTCTTCGAGTACGCCAAGAGCGAGCGGCTCTTGCAAGAAATGCAGGAGGACCTGCGCTCCGGATGGCCATCATGGTGACGTCGCCCATGTTTGATGGAAAAGTCCCACAC    150

                  60            70            80            90            100
TrpTyrHisPheSerCysPheTrpLysValGlyHisHisSerIleArgHisProAspValGlyValAspGlyPheGluLeuArgTrpAspAspProAspGlnGlnLysGlnLysThrGlyLysGlyLysValGlnAsp
TGGTACCACTTCTCCTGCTTCTGGAAGGTGGGCCACTCCATCCGGCACCCTGACGTTGAGGTGGCGGTTCTGGATGACCAGCAGAAGCAGAAGACAGGCAAGGGCAAGGTGCAGGAT    300

                  110           120           130           140           150
GlyIleGlySerLysAlaGluLeuLysThrLeuTyrAlaLeuTyrAlaLysAspPheAlaAlaAlaGlyGlnSerThrCysLysGlyMetGlyLeuLysIleGluGlyLysGlnValArgLeuAsnArgSerMetAsnValArgLeuArgLeuAsnValAspAspProGln
GGAATTGGTAGCAAGGCAGAGAAGACTCTGGGTGACTTTGCAGCAGCTGCCAAGTCAACAGAGTACGTGCAAGGGGTGTATGGAGAAGATAGAAAAGGGCAAGTGTTCGATCTTGATGGAGAAGAAGAGCGGCCACCAG    450

                  160           170           180           190           200
LeuGlyMetIleAspArgTrpTyrHisProGlyCysPheValLysSerAsnArgGluValLeuGlnGlyLysGlyPheGluTyrSerAlaSerGlnLeuLysGlyPheSerLeuLeuAlaThrGluAspLysGluAlaLeuLysLysGlnLeuPro
CTAGGCATGATTGACCGCTGGTACCATCCAGGCTGCTTTGTCAAGAGCAACAGGGAGGTGGTTTCCGGCGAGTACAGTGCGAGTCAGCTCAAGGGCTTCAGCCTCCTTGCTACAGAAGATAAAGAAGCCCTGAAGAAGCAGCTCCCA    600

                  210           220           230           240           250
GlyValLysSerGluGlyLysArgLysAspGlyValAlaAspLysAlaLysAlaLysSerGlyLysAspLysAlaLysArgAlaIleLysGlyGluLysLysGlnGluGlnGlnLysGlnLysThrGlyLysGluTrpAsnIleLeuLysAspAsp
GGAGTCAAGAGTGAAGGAAAGAGAAAAGGCGATAAGGTTGAGATGGAGTGGTGAAGTGCGGAAGAACAAGGATAGTAGCTGTAAAGAAAAAGCCTTGAAAAAGCCTTAAGGATAGTTACAGAACATCAGGAC    750

                  260           270           280           290           300
GluLeuLysLysValCysSerThrAsnAspSerLeuLeuGlyLeuLeuIlePheAsnLysGlnGlnAlaProSerGlyLysGlnIleAlaLysProSerGlyProSerThrValLysProSerGlyGluPheArgGlyValAlaLeuLysProCysGluCysSerGly
GAGCTAAAGAAAGTGTTCAACATAATGACCGAGGAGTCACTATCTTCAACAAGCAACCGTGCCTTCTGGGGGATCGGCGATCTTGGTGCCTTCCCTTCCCTGCGAGGAATGCTCGGGT    900

                  310           320           330           340           350
GlnLeuValPheLysSerArgAlaThrTyrTyrCysThrGlyAspValThrArgTyrTyrProPheProSerAlaAlaLeuThrAlaAlaProGluProSerThrValLysMetValThrArgTrpValThrGlnGlnProAsnArgLysGlyIleLysGluSerLeuLysLysVal
CAGCTGGTCTTCAAGAGCCGATGCTATTACTGCACTGGGGACGTCACTCGCTACTATCCATTTCCCTCAAGTGTATGGTCAAGACACAGAAGGAGTGGGTAACCCCAAGGAATTCCGAGAAATCTTCTTACCTCAAGAAATTGAAGTT    1050

                  360           370           380           390           400
LysLysGlnAsnAspArgLysIleProProArgIleAlaProProProSerGluThrArgSerAlaSerValAlaAlaLeuAlaSerValAlaAlaAsnSerSerAlaAlaSerProLeuSerSerAsnMetLysIleLeuGluThrLeuLys
AAAAGCAGGACCGTATATTCCCCCAGAAACACGCCTCCGTGCGGCCACCTCGCGGTGATTGAGAAAACCCGGTCGGCAGGAAGTGAGCCATGATGGAAGATGAAGATAAAGGAGAATCTCACTCTCGGGAG    1200

                  410           420           430           440           450
LeuSerArgAspAsnLysAsnAspGluValLysSerAlaMetIleGluGlyLysLysLeuGlyLysLysGlyTyrThrGlyAlaAlaAsnLysAlaAlaSerLeuGluLeuProSerLysAsnLysSerThrLysLysGluGlyValLysMetAsnLysLysGluAlaAsn
CTGTCCCGAACAAGGATGATGAAGTCAAGAGCGATGAATCATCGGGGGACGGAGAAGTTGAGAACTCCGGGGGCGGAGCAAGGCTTCCTGTGCATCAGCACCAAAGGAGGTGGAAGATGAATAAGGAAGGAAGGCAGAGGCAGAGTCAGG    1350

                  460           470           480           490           500
IleArgValValSerGluAlaSerPheLeuGluGlyIleAsnAspValSerArgAlaSerThrLysLysSerLeuGlnGluLeuProPheLeuGluLeuAlaHisIleLeuGluSerProTrpProGlyAlaGlyValLysAlaGluProArgGlyLysLeuSerGluGly
ATCCGAGTTGTGTCTGAGGACTTCCTCCGACCTCTCCGCCTCCACCAGAGCCTTCGAGTTGTTCTTAGCGCACATCTTGTCCCCTGGGGGCAGGTCGTTGAAGTTGTGGCCCCAAGGAGGAAGTCAGG    1500

                  510           520           530           540           550
AlaAlaLeuSerLysLysSerLysGlnIleValValLysGlyLeuValAspAlaIleValAlaValAlaAlaValAlaAspProAspSerGlyLeuGluHisSerSerAlaHisValLeuGluThrLeuArgMetLysArgMetGluProAsnLysGlyLysGlyLysGluTrpProIlePheGlyAspLysGly
GCTGCCCTCTCCAAAAAAGCAAGGCCAGGTCAGGAGGAAGGTATCAACAAATGAAAAGAGAATGAAATTAACTCTTAAAGGAGGAGGAGCAGCTGTGGATCCTGATTCTCGGACTGGAAAGCCATGTCCTGGACAGAAAAGGTGGG    1650

                  560           570           580           590           600
LysValPheSerAlaThrLeuGlyLeuValAspIleValLysGlyLeuTyrAsnSerTyrTyrTyrLysGluGlnLeuGluGluGlnTyrSerTrpIlePheGlnArgTyrTrpIleValLeuAsnArgTyrTrpGlyArgValGlyThrValIleGlyTyrGlySerAsnLys
AAGGTCTTCAGTGCAACCTTGGGCTTGGTGGACATCGTTAAAGGACACATCTTACAAGCTACGAGTACAGTTGGATATTCAGGTCCTGGGGGCGTGTGGGACAGTCATCGGTACGGTACGGTGATCGGTAGCAACAAA    1800
```

488

Fig. 1. Nucleotide sequence of cloned cDNA encoding human poly(ADP-ribose) synthetase. Nucleotide residues are numbered in the 5' to 3' direction, beginning with the first residue of the ATG triplet encoding the initiative methionine and the nucleotides on the 5' side of residue 1 are indicated by negative numbers; the number of the nucleotide residue at the right end of each line is given. The deduced amino acid sequence of poly(ADP-ribose) synthetase is shown above the nucleotide sequence and the amino acid residues are numbered beginning with the initiative methionine. Open triangles denote the sites cleaved by papain, and closed triangles indicate the sites cleaved by α-chymotrypsin.

Table 1 is a summary of the amino acid composition of poly(ADP-ribose) synthetase. It reveals that the enzyme consists of 1,014 amino acid residues with a calculated molecular weight of 113,153. The total amino acid composition calculated from the deduced amino acid sequence coincided well with that determined experimentally.

Table 1. Amino acid composition of poly(ADP-ribose) synthetase.

Domain Residue	DNA Binding Residues	Auto- modification Residues	NAD Binding Residues	Native Enzyme Calculated Residues	(%)	Determined (%)
Lys	54	24	49	127	12.52	13.34
His	5	1	13	19	1.87	2.06
Arg	15	4	14	33	3.25	3.43
Asp	25	4	33	62	6.11 (9.66)	9.94
Asn	7	7	22	36	3.55	
Thr	15	5	22	42	4.14	4.48
Ser	30	17	38	85	8.38	8.46
Glu	30	15	30	75	7.40 (10.75)	10.96
Gln	14	3	17	34	3.35	
Pro	18	5	21	44	4.34	4.69
Gly	25	9	36	70	6.90	7.43
Ala	24	17	26	67	6.61	6.94
Val	25	12	31	68	6.71	6.22
Met	9	5	11	25	2.47	2.14
Ile	12	6	30	48	4.73	4.61
Leu	25	14	51	90	8.88	9.09
Tyr	9	0	23	32	3.16	3.22
Phe	13	2	15	30	2.96	2.99
Cys	11	1	2	14	1.38	N.D.
Trp	7	1	5	13	1.28	N.D.
Total	373	152	489	1,014		
Mr	41,968	16,304	54,881	113,153		116,000
±	+14	+9	0	+23		pI = 10

In regard to the composition of each domain, the DNA binding domain consists of 373 residues with a calculated molecular weight of 41,968, and this domain is quite basic. The automodification domain consists of 152 residues with a molecular weight of 16,304. This domain is also basic. The NAD binding domain contains 489 residues and its molecular weight is 54,881. In contrast to the other two domains, the net charge of the NAD binding domain is neutral.

Among these three domains, the DNA binding domain is most characteristic not only because it is quite basic but because it contains a unique sequence element as shown in Fig. 2. Namely, a cluster of lysine residues is involved in the region between sequence residues 220 and 227, and this element resembles the so-called essential peptide sequences for the nuclear location of SV40 and polyoma large T antigens. It was further noted that the DNA binding domain involves a homologous repeat in the sequence as shown in Fig. 3. Furthermore, of particular interest is the fact that the region including the homologous repeat is quite similar to the sequences of several oncogene-products such as c-*fos* and v-*fos* (Fig. 3 and Table 2).

PARS	Ala - Lys - Lys - Lys - Ser- Lys - Lys - Glu	220-227
T-Antigen (SV40)	Pro - Lys - Lys - Lys - Arg - Lys - Val - Glu	126-133
T-Antigen (PMV)	Pro - Lys - Lys - Ala - Arg - Glu - Asp - Pro	280-287

Fig 2. Homologous elements in poly(ADP-ribose) synthetase (PARS) and T antigens for nuclear location.

```
              ↓  ↓                                      ↓  ↓
        YRVEYAKSERASCKKCSESIPKDSLRMAI-MVQSPMFD-GKVPHWYHFSCFWKV-GH (  9- 62)
 PARS   ::::::  :  :: : :  : :  :     :      ::        :  :::  :: :
        FAAEYAKSNRSTCKGCMEKIEKGQVRLSKKMVDPEKPQLGMIDRWYHPGCFVKNREE (113-169)


 PARS   EKPQLGMIDRWYHPGCFVKNREELGFRPEYS-AS-QLKG-FSLLAT-EDKEALKKQL   (147-199)
        ::    :  :  : : :   ::: : :: : :       :: :  ::  :
 fos    EKEKLEFILAAHRPAC--KIPDDLGFPEEMSVASLDLTGGLPEVATPESEEAFTLPL   (189-243)
```

Fig. 3. Homologous repeat and homology of poly(ADP-ribose) synthetase (PARS) with c-*fos*. One letter amino acid notation is used. Arrows show putative metal fingers for zinc as poly(ADP-ribose) synthetase is shown to be a zinc metallo enzyme (6).

Table 2. Comparison of homologous regions of oncogene products and poly(ADP-ribose) synthetase

Oncogene	Species	Homologous Region[a]	Homology Score[b]	R*/R[c]	Percent Match
fos	Human	172-267 (129-224)	-67	29/100	29
	Mouse	172-265 (129-222)	-63	25/97	26
	FBJ murine osteosarcoma virus	189-265 (147-222)	-60	22/79	28
myc	Human	341-393 (204-254)	-59	12/53	23
		368-440 (188-261)	-54	15/74	20
	Mouse	316-372 (196-254)	-64	14/59	24
	Chicken	297-348 (204-253)	-55	12/52	23
N-*myc*	Human	39-93 (437-494)	-53	12/58	21
	Mouse	348-426 (202-285)	-52	19/85	22
myb	Drosophila melanogaster	16-109 (794-893)	-61	22/101	22
		172-327 (400-552)	-56	35/160	22
ras	Drosophila melanogaster	65-176 (166-274)	-51	26/114	23
	Dictyostelium discoideum	97-168 (512-582)	-63	16/74	22
	Harvey murine sarcoma virus	6-165 (559-718)	-54	34/167	20

[a]Residue numbers of homologous sequence of oncogene product are presented in comparison with those of poly(ADP-ribose) synthetase shown in parenthesis.
[b]See Ref. 7.
[c]R* denotes numbers of matched amino acid residues and R represents those of total amino acid residues including gaps in the homologous region.

Discussion

At present, it is still difficult to definitely clarify the exact physiological significance of poly(ADP-ribose) synthetase *in vivo*. It is now clear, however, that the functions of this enzyme as determined *in vitro* can be divided into 3 categories as shown in Fig. 4. The first function is the reaction itself. Namely, the synthesis of poly(ADP-ribose) occurs with concomitant consumption of NAD. Whether or not poly(ADP-ribose) itself has any physiological function is still unknown. Nevertheless, the consumption of the substrate NAD will influence cell viability.

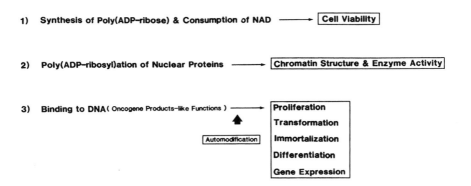

Fig. 4. Major functions of poly(ADP-ribose) synthetase.

The second function, poly(ADP-ribosyl)ation of nuclear proteins, has been most extensively studied by many investigators. However, poly(ADP-ribosyl)ation of nuclear proteins, in general, will lead to death of cells, because chromatin structure is highly modified and catalytic functions of many important nuclear enzymes are lost due to steric hindrance associated with poly(ADP-ribosyl)ation.

Therefore, these two functions do not appear to be natural or physiological. We rather consider that the third function, namely its ability to bind to DNA, is the most important. By binding to DNA, it could control proliferation, transformation, immortalization, differentiation, and, in general, gene expression of living eukaryotic cells similar to oncogene products such as c-*fos* and v-*fos*. In other words, poly(ADP-ribose) synthetase is a bifunctional protein; one function lies in its catalytic activity and another function lies in its ability to bind to DNA. Under the natural or physiological conditions, the binding ability DNA would play a major role in living eukaryotic cells. In such a case, automodification might be the most

492

important feedback mechanism because this enzyme loses or reduces its ability to bind to DNA after automodification.

Our detailed original data presented at the 8th International Symposium on ADP-ribosylation held in Texas appeared in Ref. 8. Subsequently, however, the cDNA sequence has been corrected at three positions as follows:

1. The coding sequence corresponding to the deduced amino acid residue No. 237 (C<u>G</u>C : Arg) should read (<u>G</u>CC:Ala).

2. In the coding sequence corresponding to the deduced amino acid residues No. 366 and 367 (GCC'CAC : Ala-His), three G's are deleted. The correct sequence is GC<u>G</u>•<u>G</u>CC•AC<u>G</u>(Ala-Ala-Thr). Accordingly, plus 1 is required for numbering the deduced amino acid residues after No. 368 (eg. 368→369, 369→370, ..., 1013→1014). Also, plus 3 is required for numbering the nucleotide sequence after nucleotide No. 1102 (eg. 1200→1203, etc.). Thus, this correction leads to the conclusion that the open reading frame for this enzyme encodes a protein of 1,014 amino acid residues with a calculated molecular weight of 113,153.

3. The coding sequence corresponding to leucine at the corrected amino acid position No. 419 (TT<u>A</u>) should read TT<u>G</u> (leucine).

Acknowledgements. This article is dedicated to Dr. Osamu Hayaishi (President of Osaka Medical College) under whose guidance this work has been developed.

References

1. Shizuta, Y., Ito, S., Nakata, K., Hayaishi, O. (1980) Methods Enzymol 66: 159-165
2. Shizuta, Y., Kameshita, I., Ushiro, H., Matsuda, M., Suzuki, S., Mitsuuchi, Y., Yokoyama, Y., Kurosaki, T. (1985) Adv Enz Regul 25: 377-384
3. Kameshita, I., Matsuda, Z., Taniguchi, T., Shizuta, Y. (1984) J Biol Chem 259: 4770-4776
4. Kameshita, I., Matsuda, M., Nishikimi, M., Ushiro, H., Shizuta, Y. (1986) J Biol Chem 261: 3863-3868
5. Ushiro, H., Yokoyama, Y., Shizuta, Y. (1987) J Biol Chem 262: 2352-2357
6. Zahradka, P., Ebisuzaki, K. (1984) Eur J Biochem 142: 503-509
7. Kanehisa, M. (1986) IDEAS User Manual, Program Resources, Inc.
8. Kurosaki, M., Ushiro, H., Mitsuuchi, Y., Suzuki, S., Matsuda, M., Matsuda, Y., Katunuma, N., Kangawa, K., Matsuo, H., Hirose, T., Inayama, S., Shizuta, Y. (1987) J Biol Chem 262: 15990-15997

The Biological Function(s) of Poly(ADP-ribose) Polymerase as Observed by Transient Expression of the cDNA and Transcriptional Regulation of the Gene

Kishor Bhatia, Yves Pommier[1], Al Fornace[1], Ted Breitman, Barry Cherney, Chandrakant Giri[2], and Mark Smulson

Department of Biochemistry, Georgetown University, Washington, D.C. 20007 USA

Introduction

The poly(ADP-ribosyl)ation reaction was discovered more then twenty years ago in the laboratories of Sugimura, Mandel and Hayaishi (1-3). Since then much of the data has implicated a close relationship between this nuclear reaction (4, 5) and important biological events in the nucleus including DNA repair, cell cycle regulation, cell differentiation, oncogenesis, and chromatin modification (6-11). The proposed biological roles for the poly(ADP-ribose) polymerase have largely been based on the use of inhibitors (12). However the basis of these inhibitors in delineating the functional aspects of the ribosylation reactions have recently been subject to criticism (13). With a view to bridge the gap between the proposed roles for the polymerase and the factual molecular basis of its function our laboratory initiated cloning of the polymerase cDNA. We reported a partial cDNA clone in 1986 (14) and subsequently were the first to report the isolation of a full length cDNA for the polymerase (15). Details of the cloning, sequencing and other studies initiated in our laboratory have been described (Smulson *et al.* and Thraves *et al.*) elsewhere in this monograph (16, 17).

In this chapter we present new approaches utilizing molecular probes and expression vectors to study the relationship between poly(ADP-ribose) polymerase activity, DNA breaks, and various putative biological functions of the enzyme.

Results and Discussion

A large number of proteins involved in DNA repair, recombination and replication have been shown to be induced when required (18-20). The SOS response in *E. coli*, which coordinates the expression of over twenty genes, is probably the best understood model system demonstrating the functional orientation of transcriptionally induced genes (21). This response to

[1]National Institutes of Health, Bethesda, MD 21205
[2]United States Food and Drug Administration, Bethesda, MD 21205

induction has been successfully used to isolate unknown genes, the products of which may be involved in DNA repair, regulation of the cell cycle or response to other forms of stress. Hence we decided to study the pattern of transcription of the polymerase under various cellular conditions which would allow a precise determination of its functions. The availability of a full length cDNA for the polymerase gave us the opportunity to over-express the enzyme in a feasible cell system and follow the response of this "over-expressing cell" with respect to relevant parameters such as DNA repair.

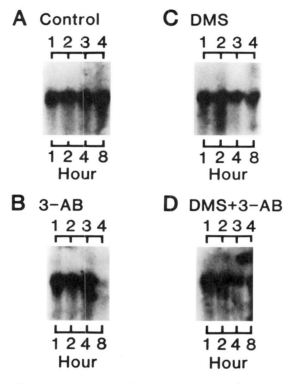

Fig. 1. HeLa cells at a concentration of approximately 5 x 10^5 cells/ml were treated with 200 μM DMS (panel C) or 5 mM 3-aminobenzamide (panel B) or a combination of the two (panel D). Cells were isolated at various times indicated (1, 2, 4 and 8 hr) by centrifugation and processed for isolation of total cellular RNA. These RNA samples were electrophoresed on an agarose formaldehyde gel, blotted on a nylon membrane and hybridized to a 3.7 kb *Xho* cDNA insert from pcd-(ADPR)P. Signal intensities of the 3.7 kb band mRNA in treated cells were compared to control untreated cells (panel A).

Response of steady state polymerase mRNA levels to exogenously induced DNA strand breaks. Elsewhere in this volume Smulson *et al.* discussed experiments on S phase related expression of poly(ADP-ribose) polymerase (16). Induction of the polymerase transcripts which occur at mid

S and S/G2 phases of the cell cycle may be related to the discontinuous nature of DNA replication.

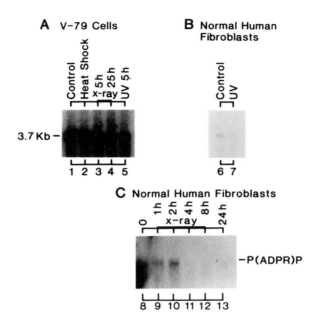

Fig. 2. CHO V-79 cells (panel A) or human fibroblast cells (panels B and C) were treated with various DNA damaging protocols as described earlier. RNA isolated from these was subjected to Northern Analysis using the *Xho* fragment from poly(ADP-ribose) polymerase that was radiolabeled as described. To confirm that equal amounts of RNA are being analyzed, RNA was hybridized also to an actin probe (data not shown).

Time dependent changes in polymerase transcripts in HeLa cells treated with different concentrations of DMS, an alkylating agent known to activate the catalytic activity of the enzyme, was investigated (Fig. 1). Cells were exposed to 200 μM DMS, isolated at 1, 2, 4, 8 hr, and RNA was prepared and Northern hybridization analysis performed as described (22). The results of these experiments are shown in Fig. 1, and indicate that no detectable change in the levels of polymerase transcripts occurred with respect to those in untreated controls. 3-Aminobenzamide is a potent inhibitor of poly(ADP-ribose) polymerase. It has been shown to cause persistent DNA strand breaks in the presence of DNA damaging agents, as well as to induce increased cellular transformation and sister chromatid exchanges (23). As shown by the data in Fig. 1B essentially no changes in

496

steady state levels for poly(ADP-ribose) polymerase transcripts were noted by this treatment. Thus even when cells were unable to synthesize *de novo* polymer, this did not cause an induction of the mRNA for the polymerase.

Several other types of DNA damaging agents were tested to confirm that the transcription of this enzyme is not regulated in the cells by exogeneously caused DNA strand breaks. Contrary to this observation it is of significance that Schermold and Wiestler have recently shown that the induction of rat liver O^6-alkylguanine-DNA alkyl transferase is a result of X-irradiation (24). Thus in the case of this higher eukaryotic DNA repair enzyme, an induction by strand breaking events is operative. In the experiment in Fig. 2A, V-79 CHO cells were treated under a variety of conditions known to induce DNA strand breaks; RNA was isolated and Northern analysis performed. Lane 1 shows the transcript levels from untreated cells, lane 2 shows the mRNA levels from cells heat shocked, while lanes 3 and 4 measure transcript levels in cells exposed to 500 rad X-ray for 5 and 25 hr respectively. Normal human fibroblasts were independently tested for induction of poly(ADP-ribose) polymerase transcripts during a period of 1-24 hr after X-irradiation (Fig. 2C). Except for a slight change in the levels of the mRNA in cells treated with X-irradiation, no significant elevation in transcript levels was observed in either V-79 cells or normal human fibroblasts. The effect of UV radiation was tested in both V-79 cells (A) and normal human fibroblasts (B). Again, under quite different conditions and utilizing different cell types, no significant induction of transcripts for this DNA repair associated enzyme was detected in our experiments. In normal human fibroblasts however we observed a decrease in mRNA levels following UV and X-irradiation (lane 7, panel B and lanes 11-13, panel C).

Response of polymerase transcript levels to differentiation of HL60 cells. HL60, a human promyelocytic leukemia cell line, is capable of terminally differentiating *in vitro* to functionally mature granulocytes (25). Since DNA strand breaking events may be involved in this differentiation process (26), NADase activity is possibly also associated with the differentiation process in HL60 cells, we investigated the mRNA levels for poly(ADP-ribose) polymerase in HL60 cells induced for differentiation by retinoic acid and DMSO (Fig. 3).

HL60 cells were induced to differentiate by either 10^{-7} retinoic acid or 180 mM DMSO. RNA was isolated from the cells at various times during the differentiation process. Differentiation was confirmed morphologically as described earlier (27). Fig. 3A shows the levels of the polymerase following retinoic acid induction. Immediately before differentiation the levels of the mRNA are relatively low. At around 40 min after induction the mRNA levels increased and remained high for up to 12-24 hr following which there was a steady decline. Essentially the same type of pattern was observed with the induction by DMSO, with the notable exception that the mRNA levels rose much later in the differentiation process (Fig. 3B).

497

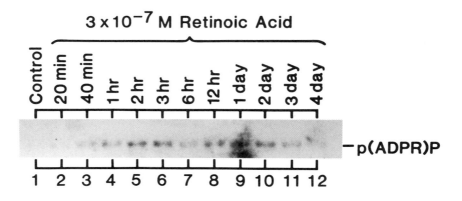

Fig. 3. HL60 cells were induced to differentiate by addition of 3 x 10⁻⁷ retinoic acid or 180 mM DMSO. Approximately 5×10^7 cells were isolated at various time points following addition of retinoic acid or DMSO and centrifuged, washed and the pellet was stored at 4°C until all samples were ready for RNA processing . Total cellular RNA was isolated from the cells by the guanidium isothiocyanate method. RNA was resolved on an agarose formaldehyde gel and mRNA for poly(ADP-ribose) polymerase was visualized by hybridization and autoradiography as described earlier. Hybridization with actin was also performed to confirm that equal amounts of RNA were being compared (data not shown).

The use of an expression vector to study the involvement of the polymerase in repair of single strand breaks. To assess whether significant expression of the cloned polymerase occurred at a defined period when subsequent DNA repair measurements were made, a transient transfection of Cos cells using DEAE-dextran was performed utilizing pcD-12 and pcD-19 (a partial cDNA for poly(ADP-ribose) polymerase deleted by approximately 400 base pairs at the 5' region of the cDNA) (Fig. 4, ref. 15). Additional controls included duplicate plates of mock-transfected cells, and cells transfected with a plasmid containing chloramphencol acetyl transferase cDNA (pSV2-CAT). After 38 hr, the cells were rinsed, scraped, sonicated, and assayed for poly(ADP-ribose) polymerase activity utilizing [³²P]NAD in an *in vitro* assay. The data (Table 1) indicated that nearly a 3-fold increase of specific activity for poly(ADP-ribose) polymerase above endogenous levels would be present in cells during the subsequent DNA repair studies outlined below. The control expression plasmid (pcD-19) did not show an increase in endogenous enzymatic activity and hence was a good control for subsequent radiation and strand break repair experiments.

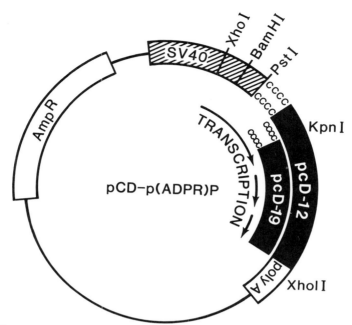

Fig. 4. The structure and component parts of the two vectors encoding complete (pcD-12) and partial (pcD-19) sequences for poly(ADP-ribose) polymerase used in the current experiments. These vectors have been characterized and described recently (15). Compared to pcD-12, pcD-19 is missing the entire 159 bp 5' untranslated region, as well as the initial 151 bp of the coding sequence including the start ATG signal. SV-40 refers to the position of the late SV-40 poly promoter and poly A refers to SV-40 poly(adenylation) signal of the vector.

Table 1. Specific activity of poly(ADP-ribose) polymerase of transfected and control Cos cells.

Cell Treatment	Poly(ADP-ribose) polymerase activity pmol(ADP-ribose)/min/mg protein
Mock Transfection	1530 ± 300[1]
pSV2-CAT[2]	1590 ± 200
pcD-19	1660 ± 64
pcD-12	4140 ± 350

Duplicate T-175 flasks of Cos cells were incubated in the presence or absence of plasmid DNA (25 μg per flask), with DEAE dextran for 3 hr; cells were harvested at 48 hr, washed, detached and sonicated in 0.25 M sucrose buffer containing 50 mM Tris pH 8.0, 2 mM MgCl$_2$, 1 mM DTT and 0.1 mM PMSF. Initial velocity (45 sec) assays with [^{32}P]NAD were performed as previously described (20).

[1]The S.D. is shown for triplicate assays on two flasks/conditions.
[2]Chloramphenicol acetyl transferase activity showed greater than 93% conversion of substrate per mg cell extract.

Having established that a significantly increased potential for poly(ADP-ribosyl)ation would exist around 2 days after transfection with pcD-12, we decided to use this time window to measure the resealing of X-ray induced DNA strand breaks.

Table 2. Hyper-expression of poly(ADP-ribose) polymerase increases the rate of X-ray induced DNA break resealing in Cos cells.

Cell Treatment	Time (min) to reseal at 37°	DNA Break Frequency (Rad-Equivalents)
None	0	2000
	15	766
Mock-Transfected	0	2000
	15	723
pcD-12	0	2000
	15	387

Cos cells (1×10^6) in 75 cm^2 flasks were labelled with [^{14}C] thymidine (0.01 μCi/ml) for 48 hr. Following this they were either mock transfected or transfected with pcD-12 DNA (25 μg/flask) for 4 hr at 37°C in presence of DEAE dextran. One flask was carried through without transfection. Following transfection, 40 hr cells were irradiated with 2000 rads and DNA repair followed by alkaline elution as described.

The resealing of DNA SSB in normal eukaryotic cells following X-irradiation is a rapid process, where a measurement of rate of repair, rather than extent, is required. In Table 2, differences in the amount of SSB resealed during a 15 min repair period in cells hyper-expressing poly(ADP-ribose) polymerase were compared to control treatments. Residual DNA breaks for both controls were 766 and 723, respectively. In contrast, unsealed DNA breaks at the same time point for cells transfected with the plasmid encoding poly(ADP-ribose) polymerase was 387, suggesting that the DNA repair rate was nearly twice that of the control cells. Since the efficiency of DNA uptake during transient transfection may vary considerably, this may represent a low estimate of increased repair following over-expression of the polymerase. After 15 min no further repair was evident in cells transfected with the cDNA while both controls showed continued repair. However, the extent of repair at 30 min in the three experimental conditions was about the same (ie. 352 rad-equivalents) suggesting that the over-expression of polymerase stimulated the rate of repair rather than the capacity of re-ligation. Further details of on this system are provided by Smulson et al. (16) in this volume.

Conclusion

A major clue to the role of the polymerase centers around DNA strand breaks. These can be caused by both external agents (e.g., physical or chemical insults to the DNA) or indigenously by naturally occurring events (e.g., DNA replication and cell differentiation). We examined cells undergoing these processes. Our results indicate endogenous break related regulation of the polymerase expression during cell growth (15) and differentiation events. No induction of the message is evident in any process causing externally induced cell damage. This does not preclude the involvement of the polymerase in DNA repair since the catalytic activity of the enzyme is induced and an over-expression of the enzyme is shown to accelerate DNA repair.

References

1. Sugimura, T. (1973) Prog Nucleic Acid Res Mol Biology 13: 95-116
2. Chambon, P., Weill, J.D., Doly, J., Strosser, M.T., Mandel, P. (1966) Biochem Biophys Res Commun 25: 638-643
3. Hayaishi, O., Ueda, K. (1977) Ann Rev Biochem 46: 95-116
4. Aubin, R.J., Dam, V.T., Miclette, J., Brousseau, Y., Poirier, G.G. (1982) Can J Biochem 60: 295-305
5. Butt, T.R., Brothers, J.F., Giri, C.P., Smulson, M. (1978) Nuc Acids Res 5: 2775-2778
6. Durkacz, B.W., Omidiji, O., Gray, D.A., Shall, S. (1980) Nature 283: 593-596
7. Kun, E., Kirsten, E., Milo, G.E., Jurian, P., Kunari, H.L. (1983) Proc Natl Acad Sci USA 81: 243-247
8. Kasid, U.N., Stefanik, D.F., Lubet, A.R., Dritschilo, A., Smulson, M. (1986) Carcinogenesis 7: 327-330
9. Smulson, M., Butt, T., Nolan, N., Jump, D., Decoste, B. (1980) Novel ADP-Ribosylations of Regulatory Enzymes and Proteins, M. Smulson, T. Sugimura (eds.), Elsevier, North Holland, pp. 59-71
10. Oikawa, A., Tohda, M., Kanai, M., Miwa, M., Sugimura, T. (1980) Biochem Biophys Res Commun 97: 1311-1316
11. Rastl, E., Swetly, P. (1978) J Biol Chem 253: 4333-4340
12. Purnell, M.R., Whish, W.J.D. (1980) Biochem J. 185: 775-777
13. Cleaver, J.E., Milam, D.M., Morgan, W.F. (1985) Radiat Res 101: 16-28
14. Alkhatib, H., Smulson, M. (1986) Fed Proc 45: 1589
15. Alkhatib, H., Chen, D., Cherney, B., Bhatia, K., Notario, V., Giri, C., Stein, G., Slattery, E., Roeder, R., Smulson, M. (1987) Proc Natl Acad Sci USA 84: 1224-1228
16. Smulson, M., et al. (1987) This volume
17. Traves, P., et al. (1987) This volume
18. McClanahan, T., McEntee, D. (1984) Mol Cell Biol 5: 75-84
19. Ruby, S.W., Szostak, J.W. (1985) Mol Cell Biol 5: 75-84
20. Fogliano, M., Schendel, P.F. (1981) Nature 289: 196-198
21. Little, J.W., Mount, D.W. (1982) Cell 29: 11
22. Church, G., Gilbert, W. (1984) Proc Natl Acad Sci USA 81: 1991-1995
23. Natarajan, A.T., van Zeeland, A.A., Zwanenburg, T.S.B. (1983) ADP-Ribosylation, DNA Repair and Cancer, M.Miwa, O. Hayaishi, S. Shall, M. Smulson, T. Sugimura (eds.), Japan Sci Soc Press, pp. 227-242
24. Schmerold, I., Wiestler, O.D. (1986) Can Res 46: 245-249
25. Collins, S.J., Gallo, R.C., Gallagher, R.E. (1977) Nature 270: 347-349
26. Farzaneh, F., Meldrum, R., Shall, S. (1987) Nuc Acids Res 15: 3493-3502
27. Brietman, T.R., Selonick, S.E., Collins, S.J. (1980) Proc Natl Acad Sci USA 77: 2936-2940

Rapid Isolation of ADP-Ribosyl Transferase by Specific Precipitation and Partial Sequencing of its DNA-Binding Domain

Kalman G. Buki, Eva Kirsten, and Ernest Kun

Department of Pharmacology and the Cardiovascular Research Institute, The University of California-San Francisco, San Francisco, California 94143 USA

Introduction

ADP-ribosyl transferase is a specific DNA-associating nuclear protein, whose enzymatic activity depends on DNA-binding, probably to certain DNA sequences (1, 2). The study of molecular and biochemical mechanisms involving ADP-ribosyl transferase depends on experimentally efficient isolation of the enzyme. We present here a method that yields 8-9 mg of 95% homogeneous enzyme per kg starting material (calf thymus) within 3 days. In order to determine the polypeptide structure of ADP-ribosyl transferase, (with special reference to catalytically active and binding domains), we prepared stable polypeptides from ADP-ribosyl transferase by digestion with plasmin. We also report here on the partial sequencing of a polypeptide (36 kDa) that participates in the DNA-binding function of ADP-ribosyl transferase.

Results and Discussion

The starting material is a 40-70% saturated $(NH_4)_2SO_4$ cut (3) which upon dilution to an ionic strength of 9-10 m Siemens is subjected to absorption and precipitation by dihydroxy Reactive Red-120 (4). The enzyme and other DNA associating proteins are extracted from the red precipitate with a buffer containing 2.0 M KCl, which leaves the hexa-K^+ salt of the dye as a precipitate and extracts proteins.

Table 1. Purification of ADP-ribosyl transferase.

	Total protein mg	Total units nmol ADPR/min	units per mg protein	average purification factor
1. Crude Extract	18,500 - 24,500	13,200 - 23,500	0.7 - 1.0	
2. $(NH_4)_2SO_4$ precipitation	2,800 - 3,500	12,200 - 23,000	4.3 - 6.6	~ 6
3. Eluate from Red Dye	300 - 650	6,200 - 16,000	20 - 27	~ 27
4. Product of chromatography	2 - 3	1,300 - 3,600	750 - 1200	~ 1070

Average yield 10-15%; purification 1000-1400 fold. Results are compiled from four complete preparative experiments. Reprinted with permission from ref. 4.

502

Table 2. Partial sequence of the 36 kDa basic polypeptide.

KSKKEKDKEIKLEKALKAQNDLIWNVKDELKKA(C)STNDLKE(LL)IFNKQEVP

Further purification is achieved by passing the preceding extract through an in-line series of chromatography and dialysis steps. This consists of an hydroxy-apatite column, a UV-monitor, hollow-fiber dialysis (Spectra-Por, MW cut-off 9000), a 10 ml-bed of DEAE cellulose and finally a column of benzamide-Sepharose 4B (containing a 1, 6-hexenediamine linker), that specifically binds ADP-ribosyl transferase without loss of activity. The retained enzyme is eluted with a buffer containing 1 mM 3-methoxybenzamide and concentrated on a Centricon-30 device. The yield and purification is shown in Table 1. Gel electrophoresis, immunotransblot and ELISA-test analysis of the enzyme (120 kDa) indicated at least 95% homogeneity.

ADP-ribosyl transferase was digested with plasmin (ADP-ribosyl transferase/plasmin ratio = 60) and distinct polypeptides of 29 kDa, 36 kDa, 56 kDa, and 42 kDa were separated by gel electrophoresis and HPLC. The first two polypeptides were basic and bound to DNA-cellulose, whereas the 42 and 56 kDa peptides had affinity to the benzamide affinity column. The

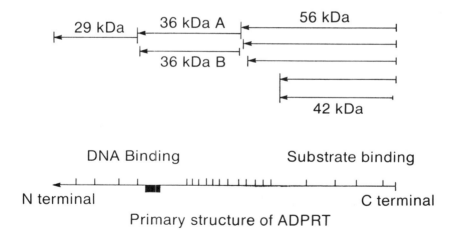

Fig. 1. Suggested structure of ADP-ribosyl transferase. The upper scheme illustrates the proposed polypeptide structure whereas the lower line indicates DNA- and NAD+-binding sites. The vertical lines on the lower line represent sites of ADP-ribosylation and the solid black bar the sequenced region.

503

29 kDa polypeptide represents the blocked amino terminal fragment of the ADP-ribosyl transferase molecule. The 36 kDa fragment was a duplex, differing only in 1 lysine residue, and could be sequenced to 51 residues. The sequence is shown in Table 2, and the proposed reconstructed ADP-ribosyl transferase molecule is shown in Fig. 1. Various DNA probes were synthesized based on the amino acid sequence (Table 2) and are being utilized for the identification of the ADP-ribosyl transferase gene in genomic libraries.

References

1. Hakam, A., McLick, J., Buki, K., Kun, E. (1987) FEBS Lett 212: 73-78
2. Gaal, J.C., Pearson, C.K. (1985) Biochem J 230: 1-18
3. Yoshihara, K., Hashida, T., Tanaka, Y., Ohgushi, H., Yoshihara, H., Kamiya, T. (1978) J Biol Chem 253: 6459-6466
4. Buki, K.G., Kirsten, E., Kun, E. (1987) Anal Biochem 167: 160-166

Sequence Analysis of the DNA Associated with Poly(ADP-Ribose) Polymerase Molecules

Marie-Elisabeth Ittel, Jean-Marc Jeltsch[1], and Claude P. Niedergang

Centre de Neurochemie du CNRS, 67084 Strasbourg Cédex, France

Introduction

Since the first description of an eukaryotic ADP-ribosylation system it has been shown that DNA is required for the synthesis of the polymer (1, 2). During the past several years, the enzyme has been purified to homogeneity from various sources (3, 4). It has been demonstrated that the purified enzyme absolutely requires double stranded DNA to express its activity *in vitro* (5-8), nevertheless covalently closed circular double-stranded DNA does not have any effect on the enzyme activity (9, 10). It has been shown that poly(ADP-ribose) polymerase is activated by and binds to, single or double stranded breaks on DNA (10).

In previous studies we described the purification to homogeneity of calf thymus poly(ADP-ribose) polymerase (7, 11, 12). This pure enzyme preparation still contains some DNA called sDNA which, separated from the enzyme molecules by hydroxypatite chromatography, has been shown to activate the enzyme more efficiently than total calf thymus DNA when added to the enzyme (7). In a morphological investigation of the poly(ADP-ribose) polymerase-sDNA complex we showed that the DNA wrapped the enzyme molecules (13). Moreover, hydrolysis of this complex by micrococcal nuclease (14) yielded two DNA populations having a mean size of 140 bp and 70 bp called the "140 bp DNA" and the "70 bp DNA" respectively. These results suggest an association of these DNAs with the enzyme molecules. Since nothing is known about the positioning of the poly(ADP-ribose) polymerase molecules relative to specific sequences in tissue, we isolated the "140 bp DNA" associated to the enzyme molecules, cloned it in bacteriophage M13mp8 and analyzed the nucleotide chains obtained from different clones.

Results

Isolation and characterization of the "140 bp DNA" associated with polymerase molecules. As described previously (14) the sDNA-poly(ADP-ribose) polymerase complex has been digested by micrococcal nuclease for

[1]Laboratoire de Génétique Moléculaire des Eucaryotes du CNRS et U-184 de l'INSERM, 67085 Strasbourg Cédex, France

505

30 min and the phenol extracted DNA submitted to polyacrylamide gel electrophoresis. The "140 bp DNA" protected from the micrococcal nuclease digestion was cut out of the gels and electro-eluted according to Maniatis *et al.* (15). Electrophoresis of the isolated "140 bp DNA" on a denaturing urea polyacrylamide gel (Fig. 1) shows DNA fragments centered around 130-138 bases suggesting no appreciable number of single chain scissions.

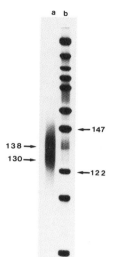

Fig. 1. Denaturing urea polyacrylamide gel. a: ^{32}P 5' end-labeled "140 bp DNA" with T4 polynucleotide kinase as indicated in Ref. 15. b: ^{32}P 5' end- labeled marker DNA fragments generated after digestion of pBR 322 DNA with Mspl. The lengths of some of the marker DNA fragments are indicated in bases.

Cloning of the "140 bp DNA" in M13mp8 vector. DNA from the "140 bp DNA" was subjected to S1 nuclease treatment followed by a dephosphorylation using bacterial alkaline phosphatase. The DNA was then treated with T4 polynucleotide kinase and the ends repaired with T4 DNA polymerase. After each enzymatic step, the solutions were extracted with phenol and chloroform and the DNA recovered by ethanol precipitation. The blunt-ended DNA was cloned by blunt-end ligation in the *Hinc*II site of M13mp8 vector. All these steps were achieved according to standard procedures (15). M13 clones (26 total) were obtained.

Analysis of the cloned DNA by dot hybridization. The single stranded DNA of each M13 clone was spotted onto a nylon membrane (Hyband-N, Amersham). Hybridizations were done according to the supplier of the nylon membrane using nick translated ^{32}P-labeled probes. The probes were either the "140 bp DNA" (Fig. 2A), the "70 bp DNA" (Fig. 2B) or the total calf thymus DNA (Fig. 2C). It is noteworthy that 12 out of the 26 clones reacted strongly with the three probes. This result suggests that some of the DNA sequences associated to poly(ADP-ribose) polymerase molecules are highly repeated in the genome. Furthermore dot hybridization results with the "70 bp DNA" used as probe (Fig. 2B) strongly support the previous suggestion (14) that the "70 bp DNA" derives from the "140 bp DNA". The 14 clones which reacted only very weakly with total calf thymus DNA probe, (giving a barely visible signal on Fig. 2C) may correspond to sequences of very low frequency in the genome.

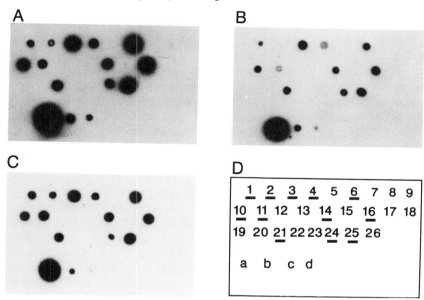

Fig. 2. Dot hybridization of cloned sequences from the "140 bp DNA". The single stranded DNA of 26 clones (~50 ng) were spotted as indicated in panel D. Controls, alkali denatured prior to spotting were the following: a) M13mp8 vector DNA (~50 ng) b) calf thymus total DNA (~100 ng) c) "140 bp DNA" (~50 ng) d) "70 bp DNA" (~20 ng). The probes were: "140 bp DNA" (panel A), "70 bp DNA" (panel B) and calf thymus total DNA (panel C). The 12 strong positive clones are underlined in panel D.

Sequence analysis. The sequences of the 12 clones which are highly repeated in the genome were determined from the M13 single stranded recombinant template using the dideoxynucleotide sequencing technique (16). Fig. 3 shows that the inserts are in the range of 169 bp (clone 1) and 100 bp (clone 24). No striking homology could be found between these sequences. Nevertheless the sequence of clone 6 and clone 16 was built by imperfect repeats of analogous motifs. In clone 6 the 5'-GATCATATGA-3' building block is repeated nearly 11 times. In clone 16 the 5'-GATCAGTGTA-3' building block is repeated about 6 times. Both motifs carry the MboI recognition sequence: 5' GATC-3' which is also present in clones 2, 3 (three times) 10 and 21.

507

Discussion

The double stranded DNA sequence of approximately "140 bp" length, associated with purified enzyme molecules, has been cloned and sequenced into bacteriophage M13. The isolated "140 bp DNA" has been obtained from highly purified calf thymus nuclear poly(ADP-ribose) polymerase preparations still containing some DNA, called sDNA. These preparations contain no more than 1% contamination by other proteins (7, 18). Taking into account the amount of DNA ("140 bp DNA" and "70 bp DNA") obtained after micrococcal nuclease digestion and the fact that naked sDNA is completely hydrolyzed under the same conditions, (14) there is no doubt about the association of these two DNA populations to poly(ADP-ribose) polymerase molecules. It could be argued that repositioning occurs during purification steps. Even so, some of the polymerase molecules could be expected to reassociate with preferred sequences if they exist at all. We found two clones (6 and 16) with analogous repeated motifs which possess the MboI recognition sequence present also in four other clones. It is noteworthy that the coenzymic octamer duplex described recently by Tseng and coworkers (17) possesses this same recognition sequence. Hybridization experiments of the different clones with the "140 bp DNA", "70 bp DNA" and total calf thymus DNA used as probes gave rise to about 50% clones which reacted with the three probes. These data suggest that some of the DNA sequences associated with poly(ADP-ribose) polymerase molecules are highly repeated in the genome. The sequence analysis of the positive clones showed no striking homology. The same conclusion could be reached by the sequence analysis of the clones which reacted only very weakly with calf thymus total DNA probe (data not shown). These results suggest that the enzyme molecules are associated to non-homologous nucleotide chains. Nevertheless single strand break(s) (9, 10) and/or the ternary structure of DNA or methylation of some bases can be implicated in the binding of poly(ADP-ribose) polymerase molecules to the DNA.

Several studies have been published on the stimulatory effect of the DNA on poly(ADP-ribose) polymerase. They have shown that the ability of DNA to stimulate the activity of poly(ADP-ribose) polymerase is dependent on the number and/or nature of strand breaks and ends in the DNA (9, 10, 19). Berger and Petzold (20) suggested that base sequence and/or composition may also be important in determining the ability of a DNA fragment to activate poly(ADP-ribose)polymerase. Recently Tseng *et al.* (17) described a sequence preference of the polymerase for an octanucleotide consensus sequence from the dexamethasone binding domain. We have observed that incubation with varying amounts of different cloned DNA ("140 bp DNA" cloned in vector pUC18 or excised by *Eco*RI + *Pst*I) activates the purified DNA-free poly(ADP-ribose) polymerase to a very low extent. A slight activation is observed when the vector (with or without the insert) is cut to produce dephosphorylated blunt ends (data not shown). The nucleotide

CL. 1
```
        10        20        30        40        50        60        70        80        90        100
CAGACTTTATTTTTGGGGGCTTCAAAATCACTGCAGATTGTGACTGCAGCCATGAAATTAAAAAACGCTTACTCCTTGGAAGGACAGTTATGACCAACCT
GTCTGAAATAAAAACCCCCGAAGTTTTAGTGACGTCTAACACTGACGTCGGTACTTTAATTTTTTGCGAATGAGGAACCTTCCTGTCAATACTGGTTGGA
        110       120       130       140       150       160
AGATAGCATATTGAAAAGCAGAGAGCTGTGCCACACCTGGTCCTGGATGCAGACTGCTACTCCTCACAG
TCTATCGTATAACTTTTCGTCTCTCGACACGGTGTGGACCAGGACCTACGTCTGACGATGAGGAGTGTC
```

CL. 2
```
        10        20        30        40        50        60        70        80        90        100
ATTCTTTTCATAAGCAACTAAAAGTCAGAGGAACCAGAGATCAAATTGCCAAAATCCCCTGGATCATGGAAAAAGCAGGAGAGTTCCAGAAAAACATTTA
TAAGAAAAGTATTCGTTGATTTTCAGTCTCCTTGGTCTCTAGTTTAACGGTTTTAGGGGACCTAGTACCTTTTTCGTCCTCTCAAGGTCTTTTTGTAAAT
        110       120
CTGCTTTATTTACTATACCAAGCTTTTG
GACGAAATAAATGATATGGTTCGAAAAC
```

CL. 3
```
        10        20        30        40        50        60        70        80        90        100
CTGGCAATTGTCAAGCAGATGAGCAGACAGTAATCACGCAGCTCAGCAGGCCCTGATCACGTGACTGAGCATGCACTGTTCACTTGACTGATCATGTACA
GACCGTTAACAGTTCGTCTACTGTCTGTCATTAGTGCGTCGAGTCGTCCGGGACTAGTGCACTGACTCGTACGTGACAAGTGAACTGACTAGTACATGT
        110       120
GATCACGTGAATAATCATGAAGTGATCAC
CTAGTGCACTTATTAGTACTTCACTAGTG
```

CL. 4
```
        10        20        30        40        50        60        70        80        90        100
GTGTGTGTGTGTGTGCCGGATGTGTGTGTGTGTGTGCACGTGCGAGCGCACATGCGTGTGCTACAGCCGCTACAGTTGTGTCGACTCTTTGCGACCCATG
CACACACACACACGCCTACACACACACACGTGCACGCTCGCGTGTACGCACACGATGTCGGCGATGTCAACACAGGCTGAGAAACGCTGGGTAC
        110       120       130
GACTGTAGCCTGAGAGGCTCCTCTGTCCGT
CTGACATCGGACTCTCCGAGGAGACAGGCA
```

CL. 6
```
        10        20        30        40        50        60        70        80        90        100
ACTGATCACGTCTATATCATGCACTGATCACATGACTTCCCATGCACTGATCACGTGGATGATCATACACTAATCATGTGACAGATCATGCACTGATCAC
TGACTAGTGCAGATATAGTACGTGACTAGTGTACTGAAGGGTACGTGACTAGTGCACCTACTAGTATGTGATTAGTACACTGTCTAGTACGTGACTAGTG
        110       120       130
GTGGCTGATCATGCACTGATACCGTGACTGAG
CACCGACTAGTACGTGACTATGGCACTGACTC
```

CL. 10
```
        10        20        30        40        50        60        70        80        90        100
GTCGAGATAGAGAGCAGAGGCGGAGTTCTTTGCTTCAATAGAGATGAATGCTGTCTCCCGGGTGCGTCTGGATGCAACCCGAGATCCTGTCGCCCTGGAG
CAGCTCTATCTCTCGTCTCCCGCCTCAAGAAACGAAGTTATCTCTACTTACGACAGAGGGCCCACGCAGACCTACGTTGGGCTCTAGGACAGCGGGACCTC
        110
AGACATTGCTTCTGAC
TCTGTAACGAAGACTG
```

CL. 11
```
        10        20        30        40        50        60        70        80        90        100
GCATTGATACTGGATGCTTGGGGCTAGTGCATCTGGGATGACCCAGAGGATGGTATGGGGAGGGAGGAGGGGAGGAGGGTTCAGAATGGGGAACACATGTA
CGTAACTATGACCTACGAACCCCGATCACGTAGACCCTACTGGGTCTCCTACCATACCCCTCCCTCCTCCCTCCTCCCAAGTCTTACCCCTTGTGTACAT
        110       120       130
TACCTGTGGTGGATTTCATTTTTGATATTTGGC
ATGGACACCACCTAAAGTAAAAACTATAAACCG
```

CL. 14
```
        10        20        30        40        50        60        70        80        90        100
CGGAGAGGTGTCCGGGCACTGGGGTTCTTATCAAGAGGGGATCGGGAAATCGGGGTGCTTCGCAAACGTGGTAGCACCCACGAGGCCACGTCTGGAATGT
GCCTCTCCACAGGCCCGTGACCCCAAGAATAGTTCTCCCCTAGCCCTTTAGCCCCACGAAGCGTTTGCACCATCGTGGGTGCTCCGGTGCAGACCTTACA
        110
CGTCGTGAGACCGGC
GCAGCACTCTGGCCG
```

CL. 16
```
        10        20        30        40        50        60        70        80        90        100
ATATCCATGTGATCAGTGCACTGATCAATCACGTGATCAGTGTATGATCAGCCACGTGAACAGTGCACTGCCACAGTCACGTGATCCGGGCCTGCTGAGC
TATAGGTACACTAGTCACGTGACTAGTTAGTGCACTAGTCACATACTAGTCGGTGCACTTGTCACGTGACGGTGTCAGTGCACTAGGCCCGGACGACTCG
        110       120
TGCGTGACTTAGTGTCTGCTCACTCTG
ACGCACTGAATCACAGACGAGTGAGAC
```

CL. 21
```
        10        20        30        40        50        60        70        80        90        100
GTGAGACGCCTCATCCTGCAGGTGCGACCGGAGGTCGGGAACCCCTTCCAGACAAAGCAGGGGAGTCGACCCTCCTGTCCAGATCAGGAGGGGAGAAAGGG
CACTCTGCCGGAGTAGGACTCCACGCTGGCCTCCACGCTTGGGGAAGGTCTGTTTCGTCCCCTCAGCTGGGAGGACAGGTCTAGTCCTCCCCTCTTTCCC
        110       120       140
CTCAGAGGAGGGGTGCCGGAAAACCTCAGTGTTCCTCTCGAGGG
GAGTCTCCTCCCCCACGGCCTTTTGGAGTCACAAGGAGAGCTCCC
```

CL. 24
```
        10        20        30        40        50        60        70        80        90        100
GTATCAGGCAGATGAGCGGTCTGGTGTCGCGCGGCTCAGCTGGCGAGTATCAGGCAGATGAGCAGTAAGTGTGGCGTGGCTCAGCTTGCGAGTATCAGG
CATAGTCCGTCTACTCGCCAGACCACAGCGCGCCGAGTCGACCGCTCATAGTCCGTCTACTCGTCATTCCACACCGCACCGAGTCGAACGCTCATAGTCC
```

CL. 25
```
        10        20        30        40        50        60        70        80        90        100
TGCTGCTGCTGCTGCTGCTAAGTCGCTTCAGTTCGTGTCCGGCTCTGTGCGACCCATAGACGGCAGCCCAGCAGGCTCCGCGTGCCCTGGATTTCTCCGGC
ACGACGACGACGACGACGATTCAGCGAAGTCAAGCACAGGCGAGACACGCTGGGTATCTGCCGTCGGGTCGTCCGAGGCGCACGGGACCTAAAGAGGCCG
        110       120
AAGTGCCTACGTAATATACCCAGAGCT
TTCACGGATGCATTATATGGGTCTCGA
```

Fig. 3. Sequences of the 12 selected clones.

sequences associated with poly(ADP-ribose) polymerase molecules are not responsible *per se* for the activation of the enzyme molecules. We have previously shown (14) that the total DNA obtained after micrococcal nuclease digestion of the sDNA-poly(ADP-ribose) polymerase complex activates more efficiently the DNA-free poly(ADP-ribose) polymerase compared to the total purified sDNA. This is probably due to the fact that micrococcal nuclease generates shorter DNA fragments with 3' phosphorylated termini which are known to be enzyme activators (20). In conclusion poly(ADP-ribose) polymerase molecules seem to be associated with heterologous sequences. Some of them are highly repeated in the genome and some are built by imperfect repeats of analogous motifs. In view of the previous conclusions that poly(ADP-ribose) synthesis may be implicated in DNA repair, cell differentiation and proliferation (4, 21), it is conceivable that DNA rearrangements could cause the enzyme to shift to damaged regions, where it is bound and activated.

References

1. Chambon, P., Weill, J.D., Mandel, P. (1963) Biochem Biophys Res Commun 11: 39-43
2. Chambon, P., Weill, J.D., Doly, J., Strosser, M.T., Mandel, P. (1966) Biochem Biophys Res Commun 25: 638-643
3. Mandel, P., Okazaki, H., Niedergang, C. (1982) Prog Nucleic Acid Res Mol Biol 27: 1-51
4. Ueda, K., Hayaishi, O. (1985) Ann Rev Biochem 54: 73-100
5. Okayama, H., Edson, C.M., Fukushima, M., Ueda, K., Hayaishi, O. (1977) J Biol Chem 252: 7000-7005
6. Yoshihara, K., Hahida, T., Tanaka, Y., Ohgushi, H., Yoshihara, H., Kamiya, T. (1978) J Biol Chem 253: 6459-6466
7. Niedergang, C., Okazaki, H., Mandel, P. (1979) Eur J Biochem 102: 43-57
8. Shizuta, Y., Ito, S., Nakata, K., Hayaishi, O. (1980) Methods Enzymol 66: 159-165
9. Benjamin, R.C., Gill, D.M. (1980) J Biol Chem 255: 10502-10508
10. Ohgushi, H., Yoshihara, K., Kamiya, T. (1980) J Biol Chem 255: 6205-6211
11. Mandel, P., Okazaki, H., Niedergang, C. (1977) FEBS Lett 84: 331-336
12. Okazaki, H., Niedergang, C., Mandel, P. (1977) C R Acad Sc Paris 285 Série D 1545-1548
13. de Murcia, G., Jongstra-Bilen, J., Ittel, M.E., Mandel, P., Delain, E. (1983) Embo J 2: 543-548
14. Ittel, M.E., Jongstra-Bilen, J., Niedergang, C., Mandel, P., Delain, E. (1985) ADP-Ribosylation of Proteins F.R. Althaus, H. Hilz, S. Shall (eds.). Springer, Berlin, Heidelberg, pp. 60-68
15. Maniatis, T., Fritsch, E.F., Sambrook, J. (1982) Molecular Cloning, a Laboratory Manual, Cold Spring Harbor Laboratory, New York
16. Messing, J. (1983) Methods Enzymol 101 (part C): 20-78
17. Tseng, A., Lee, W.M.F., Kirsten, E., Hakam, A., McLick, J., Buki, K., Kun, E. (1987) Proc Natl Acad Sci USA 84: 1107-1111
18. Jongstra-Bilen, J., Ittel, M.E., Niedergang, C., Vosberg, H.P., Mandel, P. (1983) Eur J Biochem 136: 391-396
19. Petzold, S.J., Booth, B.A., Leimbach, G.A., Berger, N.A. (1981) Biochemistry 20: 7075-7081
20. Berger, N.A., Petzold, S.J. (1985) Biochemistry 24: 4352-4355
21. Gaal, J.C., Pearson, C.K. (1986) TIBS 11: 171-175

Molecular Cloning of cDNA for Human Placental Poly(ADP-Ribose) Polymerase and Decreased Expression of its Gene during Retinoic Acid-Induced Granulocytic Differentiation of HL-60 Cells

Hisanori Suzuki[1], Kazuhiko Uchida, Hiroshi Shima, Takako Sato, Takashi Okamoto, Teruyuki Kimura[2], Takashi Sugimura, and Masanao Miwa

Virology Division, National Cancer Center Research Institute, Tsukiji, Chuo-ku, Tokyo 104, Japan

Introduction

Poly(ADP-ribose) is suggested to be involved in various physiological phenomena, such as DNA repair, sister chromatid exchanges, differentiation, proliferation and transformation of eukaryotic cells (1). To find a means of clarifying the biological function of poly(ADP-ribose), we purified poly(ADP-ribose) polymerase from human placenta. We also isolated the cDNA for poly(ADP-ribose) polymerase and determined the nucleotide sequence. Using the cDNA probe we found a decrease in the amount of transcript for the poly(ADP-ribose) polymerase gene during the retinoic acid-induced granulocytic differentiation of HL-60 cells.

Results and Discussion

Purification of poly(ADP-ribose) polymerase from human placenta and amino acid sequences of α-chymotryptic peptides. The purification of the poly(ADP-ribose) polymerase from two placentas (500-600 g) was carried out in two days yielding 100-200 μg purified enzyme with 10% recovery; a 3,500-fold purification was achieved using the procedure of Burtscher *et al.* (2) with a slight modification (3). The purified enzyme revealed a single band at the position corresponding to a molecular weight of 116 kDa (Fig. 1). The amino acid composition of the human placental poly(ADP-ribose) polymerase was very similar to that reported by Ushiro *et al.* (4).

By partial digestion of the purified poly(ADP-ribose) polymerase with α-chymotrypsin, four peptides with molecular weights of 63, 53, 40 and 16 kDa were observed. No NH2-terminal amino acid was detected from either

---

[1]Biophysics Division, National Cancer Center Research Institute, Tsukiji, Cho-ku, Tokyo 104, Japan

[2]Present address: Institute of Biochemistry, Policlinic of Borgo Roma, University of Verona, 37134 Verona, Italy

511

the intact poly(ADP-ribose) polymerase or from the 63 kDa peptide, strongly suggesting that the NH2-terminal amino acid of poly(ADP-ribose) polymerase is blocked. The NH2-terminal amino acid sequences of the 40 and the 16 kDa peptides were determined (Fig. 2). The NH2-terminal amino acid sequence of the 53 kDa peptide was the same as that of the 16 kDa peptide, indicating that the 16 kDa peptide was the NH2-terminal portion of the 53 kDa peptide. Therefore, the 40 kDa peptide was considered to be a COOH-terminal peptide of the 53 kDa peptide.

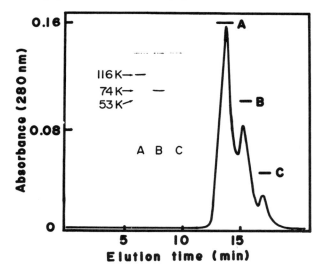

Fig. 1. HPLC pattern of poly(ADP-ribose) polymerase. Inset: polyacrylamide gel electrophoretic patterns of each peak.

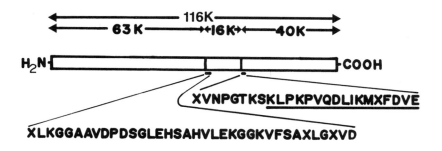

Fig. 2. Partial amino acid sequence of the α-chymotryptic digestion products of poly(ADP-ribose) polymerase. A 51-mer oligodeoxyribonucleotide probe was synthesized corresponding to the underlined amino acid sequence.

512

Isolation of cDNA clones and nucleotide sequencing. A 51-mer oligodeoxyribonucleotide was synthesized according to the partial amino acid sequence (Fig. 2) and was used to screen λgtll cDNA library of human placenta. Three cDNA clones, one containing an insert of 1.8 Kb (λPAP803), and the others containing inserts of 2.1 Kb (λPAP802 and λPAP222), were obtained (Fig. 3). Restriction endonuclease analysis showed that three *Bam*HI sites and a single *Pst*I site were located at identical positions among all three cDNA clones. The amino acid sequence predicted from the nucleotide sequence of the insert of λPAP803 completely matched the determined sequences of the NH2-terminal 36 amino acids of the 16 kDa and the NH2-terminal 25 amino acids of the 40 kDa chymotryptic peptides. We therefore concluded that the inserted sequence of λPAP803 is of a human poly(ADP-ribose) polymerase cDNA and that it covered the COOH-terminal half of the enzyme. Northern blot analysis of mRNA from HL-60 cells using λPAP803 clone as a probe revealed a single band of 3.6 Kb (Fig. 4). Moreover, the level of expression of the poly(ADP-ribose) polymerase gene during granulocytic differentiation of HL-60 cells was reduced to about 30% of the initial level after treatment with 1 μM retinoic acid for 3 days when 45% of cells showed granulocytic morphology.

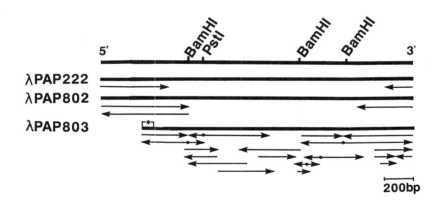

Fig. 3. Three cDNA clones with restriction enzyme sites and DNA sequencing strategies. Asterisk signifies the 36 bp which is unique to the cDNA insert in λPAP803 (3). Reprinted with permission from ref. 3.

Alkhatib *et al.* (5) also reported the cloning of cDNA for human poly(ADP-ribose) polymerase using an antibody to poly(ADP-ribose) polymerase.

513

Molecular biological studies of the poly(ADP-ribose) polymerase gene using the isolated cDNA clone should be useful in clarifying the regulatory roles of poly(ADP-ribose) polymerase in cell differentiation, DNA repair and cell transformation.

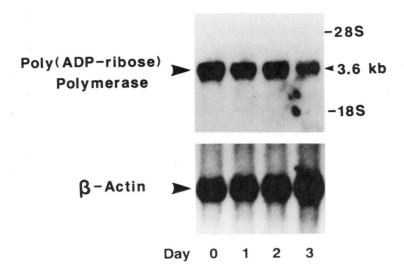

Fig. 4. The expression of the poly(ADP-ribose) polymerase gene during granulocytic differentiation of HL-60 cells in gel electrophoretic patterns. The expression of β-actin gene is shown as a positive control.

Note: After submission of this manuscript we and other investigators published the nucleotide sequence of full-length cDNAs for human fibroblast and placental poly(ADP-ribose) polymerases (Biochem Biophys Res Commun 148: 617-622, 1987; J Biol Chem 262: 15990-15997, 1987; Proc Natl Acad Sci USA 84: 8370-8374, 1987).

References

1. Ueda, K., Hayaishi, O. (1985) Ann Rev Biochem 54: 73-100
2. Burtscher, H.J., Auer, B., Klocker, H., Schweiger, M., Hirsch-Kauffman, M. (1986) Anal Biochem 152: 285-290
3. Suzuki, H., Uchida, K., Shima, H., Sato, T., Okamoto, T., Kimura, T., Miwa, M. (1987) Biochem Biophys Res Commun 146: 403-409; 148: 1549-1550
4. Ushiro, H., Yokoyama, Y., Shizuta, Y. (1987) J Biol Chem 262: 2352-2357
5. Alkhatib, H.M., Chen, D., Cherney, B., Bhatia, K., Notario, V., Giri, C., Stein, G., Slattery, E., Roeder, R.G. (1987) Proc Natl Acad Sci USA 84: 1224-1228

Isolation of a cDNA Clone Encoding Bovine Poly(ADP-Ribose) Synthetase by Immunological Screening of a cDNA Expression Library and its Application as a Probe

Taketoshi Taniguchi, Kiyoshi Yamauchi[1], Tadashi Yamamoto[1],
Kumao Toyoshima[1], Nobuhiro Harada[2], Hideaki Tanaka[2],
Seiichi Takahashi[3], Hiroshi Yamamoto[3], and Shigeyoshi Fujimoto[3]

Medical Research Laboratory, Kochi Medical School, Nankoku, Kochi 781-51, Japan

Introduction

Poly(ADP-ribosyl)ation has been proposed to regulate a variety of cellular functions, such as DNA repair, chromatin condensation, cell differentiation, and gene expression. Involvement of poly(ADP-ribosyl)ation in cell differentiation has been reported in a variety of cells, but a consensus with respect to these results has not yet been obtained. In particular, the change in the enzyme activity during erythroleukemia cell differentiation has been contradictory (1-3). Enzyme activity has been reported to increase 10- to 20-fold due to the fragmentation of DNA in cells (4). The physiological enzyme activity was determined by mildly permeabilizing cells, and the total potential enzyme activity was assayed by treating cells with detergent and DNase I. In muscle cell differentiation (5) the total potential enzyme activity did not change but the physiological enzyme activity increased, probably because DNA strand breaks formed during the differentiation. In contrast, both enzyme activities decreased during murine teratocarcinoma cell differentiation (6) and rooster spermatogenesis (7). Enzyme activity does not provide a definitive method for changes in the level of enzyme. Thus, we attempted to isolate a cDNA encoding the enzyme in order to evaluate the critical change in the level of mRNA for the enzyme during the cellular processes.

Results and Discussion

Isolation and identification of the cDNA as the enzyme structural gene. A 2.7-kb cDNA clone (ARS-1) coding for bovine poly(ADP-

[1]The Institute of Medical Science, The University of Tokyo, Minato-ku, Tokyo 108, Japan
[2]National Chemical Laboratory for Industry, Tsukuba Research Center, Yatabe, Ibaraki 305, Japan
[3]Department of Immunology, Kochi Medical School, Nankoku, Kochi 781-51, Japan

ribose) synthetase was isolated from a λ gtll expression library by immunological screening with antibodies specific for the enzyme. To identify the cDNA as the enzyme structural gene, we first determined the size of mRNA for the enzyme and examined whether the cDNA hybridized with the mRNA of the enzyme. Bovine thymus poly(A)+ RNAs were size-fractionated by sucrose density gradient centrifugation and *in vitro* translated proteins of the RNA in each fraction were immunoprecipitated with antiserum to the enzyme. The immunoprecipitated translation products were separated on a SDS-polyacrylamide gel and fluorography of the gel was performed (Fig. 1B). The largest size of the RNA which produced the 120 kDa native enzyme, as shown by the bar in Fig. 1, panel A, was estimated as 3.8 kb. The poly(A)+ RNAs were also size-fractionated on a 1.2% agarose gel, transferred to a nitrocellulose filter, and then hybridized with the ^{32}P-labelled insert DNA isolated from the clone (ARS-1). Fig. 1C shows that the insert DNA hybridized with approximately 4 kb-RNA.

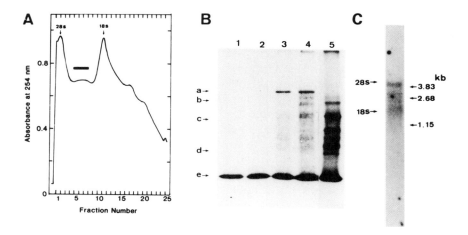

Fig. 1. Estimation of the size of mRNA for bovine poly(ADP-ribose) synthetase by *in vitro* translation of size fractionated poly(A)+ RNA and Northern blot analysis. A. Size-fractionation of poly(A)+ RNA and location of mRNA for the enzyme. Bovine thymus poly(A)+ RNA was size-fractionated by neutral sucrose density gradient centrifugation and RNA in each fraction was translated *in vitro*. The translated products were immunoprecipitated, and separated on a 7.5% SDS-polyacrylamide gel. B. Typical fluorogram of the gel. Lanes 1, 2, 3, 4, and 5, correspond to fractions #1, 3, 5, 7, and 9, respectively. Molecular weight markers a, β-galactosidase (116K), b, phosphorylase a(95K), c, bovine serum albumin (68K), d, ovalbumin (43K), e, lysozyme (14.3K). C. Northern blot analysis of bovine thymus poly(A)+ RNA. RNA (2 μg per lane) was separated on a 1.2% agarose/formaldehyde gel, transferred to a nitrocellulose filter, and hybridized with the ^{32}P-labelled 2.7 kb insert cDNA prepared from the clone ARS-1. Size markers used were 28 S(4.9 kb), 18 S(2.0 kb)rRNA, 3.8 kb, 2.7 kb, and 1.1 kb denatured DNA. Reprinted from Taniguchi *et al.*, Eur J Biochem 171: 571-575, 1988.

This result indicates that the insert DNA from the clone (ARS-1) recognizes the mRNA of the enzyme. Secondly, we determined the partial DNA sequence of the insert DNA by Sanger's method and compared the predicted amino acid sequence with the partial amino acid sequence data obtained from the 41 kDa α-chymotryptic peptide and CNBr-cleaved peptides of the enzyme as listed in Table 1. Portions of the predicted amino acid sequence exactly matched the sequence of the 26 amino acids from the N-terminal of the 41 kDa peptide and several amino acids of CNBr-cleaved peptides, as shown in Fig. 2. Thus, we identified the insert cDNA of the isolated clone (ARS-1) as a part of the structural gene of bovine poly(ADP-ribose) synthetase. The 2.7 kb-cDNA encodes the C-terminal catalytic domain, the automodification domain and a part of the DNA binding domain on the basis of the position of the N-terminal amino acid of the 41 kDa α-chymotryptic peptide of the enzyme (8). The mRNA for the enzyme is approximately 4 kb and the coding region is deduced as 3.3 kb from the molecular weight. Thus, the isolated cDNA is still missing the coding region of at least 600 bases. Alkhatib *et al.* (9) reported that mRNA for human poly(ADP-ribose) synthetase was 3.6-3.7 kb and a cDNA containing the complete coding sequence for the enzyme was isolated.

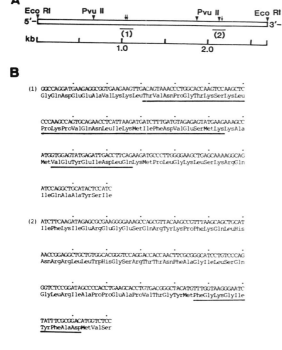

Fig. 2. A. Schematic representation of the cDNA structure. Restriction sites with *Eco*RI and *Pvu*II are indicated. B. Partial nucleotide sequence and the predicted amino acid sequence, as shown by (1) and (2) in panel A, of clone (ARS-1) are depicted. The positions of peptides that match predicted amino acids are indicated by arrows in panel A and by solid lines in panel B.

517

RNA blot analysis. The Ia antigens, class II major histocompatibility antigens, are key molecules in sensitizing thymus-derived lymphocytes in the context of nominal antigens. Expression of Ia antigens on antigen

Table 1. Amino acid sequence of α-chymotryptic, CNBr-cleaved peptides isolated from bovine poly(ADP-ribose) synthetase.

Peptide	Sequence
41 kDa α-chymotryptic	Thr-Val-Asn-Pro-Gly-Thr-Lys-Ser-Lys-Leu-Pro-Lsy-Pro-Val-Gln-Asn-Leu-Ile-Lys-Met-Ile-Phe-Asp-Val- X - X -Met-Lys
CNBr-13	Val-Glu-Tyr-Glu-Ile-Asp-Leu-Gln
CNBr-14	Phe-Gly-Lys-Gly-Ile-Tyr-Phe-Ala-Asp

presenting cells, such as macrophages, has been known to be regulated by a lymphokine. The Ia antigen has been shown to be induced in murine macrophage tumor P388D1 cells by the addition of interferon-γ to the culture medium (10). In order to elucidate the role of poly(ADP-ribose) synthetase

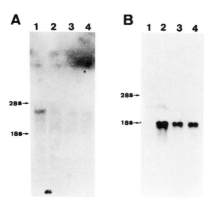

Fig. 3. RNA blot analyses with the cDNA for poly(ADP-ribose) synthetase and the cDNA for Ia antigen. Total cellular RNA (20 μg per lane), isolated from mouse macrophage tumor P388D1 cells at the indicated culture period in the presence of 10 unit/ml interferon-γ, was separated on a 0.8% agarose/formaldehyde gel and then blotted on to a nitrocellulose filter. The RNA on the filter was analyzed by the ^{32}P-labelled cDNA of poly(ADP-ribose) synthetase in panel A and the I-Aβ cDNA (11) in panel B, as a probe, respectively. Size markers used were 28 S(4.9 kb) and 18 S(2.0 kb) rRNA. Lanes 1, 2, 3, and 4 represent the RNA isolated from untreated cells, cells cultured in the presence of interferon-γ for 1, 2, and 5 day(s), respectively. Reprinted from Taniguchi *et al.*, Eur J Biochem 171: 571-575, 1988.

in the cellular activation or differentiation process, we used the cDNA as a probe to examine the level of mRNA for the enzyme during the interferon-γ-induced activiation process of Ia antigen in murine macrophage tumor cells. The cells were harvested at days 1, 2, and 5 after culturing with interferon-γ and mRNA levels for poly(ADP-ribose) synthetase and Ia antigen were determined. The total RNA isolated from the cells was separated on a 0.8% agarose gel, transferred to a nitrocellulose filter and hybridized separately with the ^{32}P-labelled cDNA of poly(ADP-ribose) synthetase or the ^{32}P-labelled I-Aβ cDNA, which codes for the β subunit of I-A molecule of Ia antigens (11), as probes. Fig. 3 shows the autoradiograms of the Northern analyses. The mRNA for the Ia antigen was maximally induced 24 hr after the incubation with interferon-γ, while the mRNA for poly(ADP-ribose) synthetase was diminished nearly completely within 24 hr. These results suggest that interferon-γ causes the depression of poly(ADP-ribose) synthetase gene expression.

References

1. Rastle, E., Swetly, P. (1978) J Biol Chem 253: 4333-4340
2. Terada, M., Fujiki, H., Marks, P.A., Sugimura, T. (1979) Proc Natl Acad Sci USA 76: 6411-6414
3. Pulito, V.L., Miller, D.L., Sassa, S., Yamane, T. (1983) J Biol Chem 258: 14756-14758
4. Benjamin, R.C., Gill, D.M. (1980) J Biol Chem 255: 10502-10508
5. Farzaneh, F., Zalin, R., Brill, D., Shall, S. (1982) Nature 300: 362-366
6. Ohashi, Y., Ueda, K., Hayaishi, O., Ikai, K., Niwa, O. (1984) Proc Natl Acad Sci USA 81: 7132-7136
7. Corominas, M., Mezquita, C. (1985) J Biol Chem 260: 16269-16273
8. Kameshita, I., Matsuda, Z., Taniguchi, T., Shizuta, Y. (1984) J Biol Chem 259: 4770-4776
9. Alkhatib, H.M., Chen, D., Cherney, B., Bhatia, K., Notario, V., Giri, C., Stein, G., Slattery, E., Roeder, R.G., Smulson, M.E. (1987) Proc Natl Acad Sci USA 84: 1224-1228
10. Steeg, P.S., Moore, R.N., Johnson, H.M., Oppenheim, J.J. (1982) J Exp Med 156: 1780-1793
11. Steinmetz, M., Minard, K., Horvath, S., McNicholas, J., Selinger, J., Wake, C., Long, E., Mach, B., Hood, L. (1982) Nature 300: 35-42

Radiation Sensitivity of Human Cells and the Poly(ADP-Ribose) Polymerase Gene

Peter J. Thraves, Kishor Bhatia, Mark E. Smulson, and Anatoly Dritschilo

Departments of Radiation Medicine and Biochemistry, Vincent T. Lombardi Cancer Research Center, Georgetown University Medical Center, Washington, D.C.   20007 USA

Introduction

Until recently much of radiation biology has centered about cellular responses to radiation using the clonogenic assay to study radiation sensitivity and radiation repair processes.   The availability of molecular techniques, allowing direct genomic analysis of cellular function, permits a much more detailed view of events taking place in the DNA as a result of radiation exposure.   The development of a complementary cDNA to the poly(ADP-ribose) polymerase gene (3) has provided a unique opportunity to study the genomic events associated with poly(ADP-ribose) metabolism.

Recent data suggests that poly(ADP-ribose) metabolism may be involved in the expression of radiation sensitivity and in the repair of radiation damage (4-11).   Therefore, ADP-ribosylation is a likely candidate for one of the biochemical mechanisms involved in the repair of radiation damage, and in the intrinsic radiation sensitivity (resistance) of cells.

Results

Cell survival studies from our laboratories have shown that the poly(ADP-ribose) polymerase inhibitor 3-aminobenzamide (3AB) increased the radiation sensitivity of normal human fibroblasts (Fig. 1) (1).   In addition, 3AB was also found to inhibit the repair of potentially lethal damage (PLD) (Fig. 1).   When these inhibitor studies were extended to human tumor cell lines of different clinical radiation curability, a differential effect was observed.   Ewing's sarcoma cells demonstrated a radiation sensitization,   whereas lung adenocarcinoma cells were found to show no radiation sensitization (Fig. 1A) (2).   Studies with these cell lines were also extended to post-radiation repair processes, observing PLD repair inhibition in Ewing's sarcoma, but not in lung adenocarcinoma (Fig. 1B).

Following the inhibitor studies, we have determined the capacity for poly(ADP-ribosyl)ation in these human tumor cell lines (Table 1).   Using whole cell sonicates, we found the polymerase activity to be 4-5 fold higher in the Ewing's sarcoma and cervical carcinoma than in the other cell lines tested.   Further, these two cell lines also had lower NAD levels (Table 1).

Table 1. Specific activity of poly(ADP-ribose) polymerase in sonicated cell extracts of human tumors

Cells	ADP-ribose pmol/min/mg protein	NAD pmol/mg protein
Human fibroblasts (NHF)	180 ± 10	3948 ± 107
Cervical carcinoma (HeLaS₃)	998 ± 307	2600 ± 223
Ewing's sarcoma (A4573)	1127 ± 74	1317 ± 124
Laryngeal squamous carcinoma (SQ-20B)	253 ± 81	2163 ± 169
Lung adenocarcinoma (A549)	212 ± 66	7449 ± 280

Human tumor cells were harvested by trypsinization, washed and sonicated in 50 mM Tris-HCl pH 8.0, 0.25 M sucrose, 2 mM $MgCl_2$, 1 mM DTT and 0.1 mM PMSF. Initial velocity (45 sec) assay with [^{32}P]NAD (final NAD concentration 100 μM), were performed as described by Cherney *et al.* (14). Data points are means ± standard deviation of three separate determinations. NAD determinations as in reference 15.

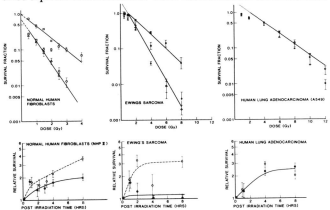

Fig 1A. Normal human fibroblasts, Ewing's sarcoma (A4573) and lung adenocarcinoma (A549) were preincubated with either 8 mM 3AB (open circles), 4 mM BZ (open triangles), or with no inhibitor (open squares) and irradiated with 0-1.2 Gy.

Fig. 1B. The effect of 3AB on the repair kinetics of potentially lethal damage in normal human fibroblasts, Ewing's sarcoma, and human lung adenocarcinoma. No inhibitor (open circles); 8 mM 3AB (closed circles). Data points are means ±1 standard deviation. Both normal fibroblast and human tumor cell cultures were grown and maintained in Eagle's MEM supplemented with 10% (V/V) fetal bovine serum, streptomycin and penicillin, at 37°C in a humidified chamber of 95% air and 5% CO_2. Stock cultures in exponential growth were detached by trypsinization and appropriate cell numbers plated in 25 cm² flasks and incubated to permit cell attachment. Inhibitors of poly(ADP-ribose) synthetase (3AB or BZ) were added to the appropriate cultures 2 hr prior to irradiation in air at room temperature. Following irradiation, the cells were exposed to the inhibitors for 24 hr, washed, and replenished with fresh medium. Refeeding was performed every 3 days and after 10-14 days, cells were fixed and the colonies stained with Giemsa. Only colonies of 50 cells or more were scored as survivors. All experiments were performed in triplicate and the survival data expressed as the mean ±S.E. of three separate survival experiments.

These observations of polymerase activity and NAD levels led us to address the following question. Are these differences in the poly(ADP-ribose) biochemical systems of these cells at the genomic or transcriptional level? Restriction enzyme analysis, Southern transfer and hybridization of these genomic DNA's with a labeled cDNA for poly(ADP-ribose) polymerase yielded, upon autoradiography, 6 expected bands in the DNA's tested (Fig. 2B). With the exception of the normal human fibroblasts, there appears to be no difference in both the intensity or number of DNA bands obtained. The absence of a 24 Kb band in the fibroblast DNA may have been due to a smaller amount of loaded DNA (Fig. 2A). Slot-blot analysis of these same genomic DNA's probed with either the poly(ADP-ribose) gene or a cDNA for actin yielded no quantitative differences in the signal for the polymerase gene between these cell types (data not shown).

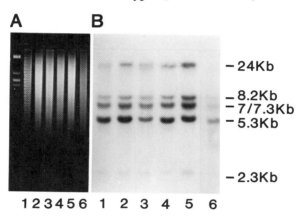

Fig. 2. Restriction enzyme analysis and Southern-blot hybridization of human tumor DNA. Human tumor DNA (20 μg) was isolated (16) and digested with *Eco*RI restriction enzyme, electrophoresed on a 0.8% agarose gel (A shows the ethidium-bromide stain) and after Southern transfer, probed with ^{32}P-labeled pcD-p(ADP-ribose) polymerase cDNA, (B). Lanes 1, Human placental DNA; 2, Laryngeal squamous cell carcinoma (SQ-20B); 3, Lung adenocarcinoma (A549); 4, Cervical carcinoma (HeLaS3); Ewing's sarcoma (A4573); 6, normal human fibroblasts (NHF). Marker-lambda-DNA.

Northern analysis of the RNA species with the cDNA for the poly(ADP-ribose) polymerase gene, however, revealed that there were more mRNA species for the polymerase in the tumor cells than in the normal human fibroblasts (Fig. 3B). The signal for the Ewing's sarcoma (Lane 3), and possibly the laryngeal tumor (Lane 5), were higher than for the normal fibroblasts (Lane 1). Further, the signal for the polymerase-mRNA species was higher in log-phase than in plateau-phase cells. The signal for plateau phase normal fibroblasts was almost nonexistent (Lane 2), while there were significant polymerase-mRNA species present in both the Ewing's sarcoma and laryngeal squamous carcinoma when at confluence in monolayer (Lanes 4 and 6).

Discussion

Our studies evaluating the role of poly(ADP-ribosylation) in the repair of radiation injury have shown that inhibitors of poly(ADP-ribose) synthetase, 3AB, and BZ, differentially sensitized various human tumor cell lines to ionizing radiation (1, 2). The sensitization observed correlated directly with the radiosensitivity *in vitro* of the human tumor cell lines tested. Our initial hypothesis to explain this observation was that the degree of sensitization by these compounds may reflect the ability of these human tumors to synthesize poly(ADP-ribose) in response to ionizing radiation. Therefore, the less radiosensitive lines could have a higher polymerase activity and may require higher inhibitor concentrations to obtain radiosensitization. In such cell lines, the synthesis of poly(ADP-ribose) would not be limiting, therefore, only a complete inhibition of polymer synthesis could potentiate cell killing.

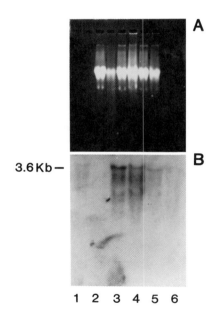

A

B

3.6 Kb —

1 2 3 4 5 6

Fig. 3. Levels of poly(ADP-ribose) polymerase mRNA in human tumor cell lines during exponential growth and at confluence. Total cellular RNA samples were isolated (17) and 20 µg of RNA from each tumor type was electrophoresed on a 1% agarose - 2.2 M, formaldehyde gel (18), blotted and subsequently probed with [^{32}P]-labeled pcD-p(ADP-ribose) polymerase cDNA. Lanes 1, 3, and 5 RNA from exponentially dividing normal human fibroblasts (NHF), Ewing's sarcoma (A4573) and laryngeal squamous cell sarcoma (SQ-20B), respectively. Lanes 2, 4, and 6, RNA from confluent cultures of normal human fibroblasts, Ewing's sarcoma, and laryngeal squamous cell sarcoma, respectively.

However, our studies on the intrinsic polymerase activity in these cell lines do not support this view. Ewing's sarcoma cells have a higher polymerase activity than do normal fibroblast, yet both cell lines are sensitized by inhibitors of the polymerase enzyme. Further, tumor cell lines

SQ-20B and A549 which have proven to be resistant to the enhancement of radiation-induced cell killing by 3AB, demonstrate polymerase activity similar to that of normal fibroblasts. The analysis of the DNA indicates that there is no amplification or rearrangement of the poly(ADP-ribose) polymerase gene in any of the cell lines irrespective of either their intrinsic radiosensitivity or response to their inhibitors of polymerase following irradiation. The analysis of the RNA species demonstrates a higher titer of polymerase mRNA in the Ewing's sarcoma that that found in either the normal fibroblasts or the laryngeal carcinoma (SQ-20B). This observation suggests that there are higher levels of polymerase enzyme in these cells and consequently a higher polymerase activity.

Our previous observations using inhibitors of poly(ADP-ribose) polymerase require further studies to elucidate the mechanism by which 3AB is working. It has been demonstrated that 3AB has a number of other mechanisms of action, e.g., a reduction in *de novo* DNA synthesis and glucose oxidation (12, 13). Further studies using additional molecular probes will be needed to address the biochemic basis of radiation sensitivity.

Acknowledgements. This work was supported by Grant No. PDT-279, American Cancer Society, Inc. The authors thank Ms. Sandra Hawkins for her assistance in the preparation of this manuscript.

References

1. Thraves, P., Mossman, K.L., Brennan, T., Dritschilo, A. (1985) Radiat Res 104: 119-127
2. Thraves, P., Mossman, K.L., Brennan, T., Dritschilo, A. (1986) Int J Radiat Biol 50(6): 961-972
3. Alkhatib, H.M., Chen, D., Cherney, B., Bhatia, K., Notario, V., Slattery, E., Roeder, R.G., Giri, C., Stein, G., Smulson, M.E. (1987) Proc Natl Acad Sci USA 84: 1224-1228
4. Ben-Hur, E. (1984) Int J Radiat Biol 46: 659-671
5. Ben-Hur, E., Elkind, E.E. (1984) Int J Radiat Biol 45: 515-523
6. Ben-Hur, E., Utsumi, H., Elkind, M.M. (1984) Br J Cancer 49(Suppl VI): 39-42
7. Brown, D.M., Evans, J.W., Brown, J.M. (1984) Br J Cancer 49(Suppl VI): 27-31
8. Huet, J., Laval, F. (1985) Int J Radiat Biol 47(6): 655-662
9. Szumiel, I., Wlodek, D., Johnson, J.L., Sundell-Bergman, S. (1984) Br J Cancer 49(Suppl VI): 33-38
10. Ueno, A.M., Tanaka, O., Matsudaria, H. (1984) Radiat Res 98: 574-582
11. Kumar, A., Keifer, J., Schneider, E.L., Crompton, N.E.A. (1985) Int J Radiat Biol 47(1): 103-112
12. Cleaver, J.E. (1984) Mutat Res 131: 123-172
13. Milam, K.M., Cleaver, J.E. (1984) Science 223: 589-591
14. Cherney, B.W., Midura, R.J., Caplan, A.I. (1985) Dev Biol 112: 115-125
15. Bernofsky, C., Swan, M. (1973) Anal Biochem 53: 452-458
16. Blin, N., Stafford, D.W. (1976) Nucleic Acids Res 3: 2303
17. Glisin, V., Crkvenjakov, R., Byrs, C. (1974) Biochemistry 13: 2633
18. Lehrach, H.D., Diamond, D., Wozney, J.M., Boedtker, H. (1977) Biochemistry 16: 4743

Abbreviations:
PLD - Potentially lethal damage
3AB - 3-Aminobenzamide
BZ - Benzamide

Author Index

DATE DUE

APR 1 7 1993			

DEMCO NO. 38-298